Looking into the Earth

Geophysics, the application of physics to the study of the Earth from its surface to its centre, is an essential part of modern earth science. This book covers the principles and applications of geophysics on all scales, ranging from deep Earth structure in relation to plate tectonics; to the search for oil, water, and minerals; to detailed studies of the near-surface.

Looking into the Earth is an introduction to geophysics suitable for those who do not necessarily intend to become professional geophysicists. These include geologists, and other earth scientists such as civil engineers, environmental scientists, and field archaeologists. Unlike other books that deal with either "global" or "exploration" geophysics, this book comprehensively covers both branches of geophysics. The book is organised into two parts: Part I describes the geophysical methods; Part II illustrates their use in a number of extended case histories. The authors recognise that many students taking introductory courses in geophysics are not necessarily fluent in mathematics or physics, so mathematical and physical principles are introduced at an elementary level. Throughout, the emphasis is on what geological (or archaeological or civil engineering) information geophysical methods can yield. Student problems and exercises are included at the end of many chapters.

Looking into the Earth is aimed primarily at introductory and intermediate university (and college) students taking courses in geology, earth science, environmental science, and engineering. It will also form an excellent introductory textbook in geophysics departments, and will help practising geologists, archaeologists, and engineers understand what geophysics can offer their work.

Alan E. Mussett has taught geophysics in the Departments of Physics and Earth Sciences at the University of Liverpool for 30 years. He is a former Secretary of the British Geophysical Association, and coauthor (with G. C. Brown) of the first and second editions of the undergraduate textbook *The Inaccessible Earth* (Allen and Unwin, London, 1981 and 1993). Alan's research interests are mainly in palaeomagnetism and radiometric dating, and he has published many scientific papers, notably on East Africa, Iceland, and the British Isles.

M. Aftab Khan has taught geophysics at the University of Leicester for 37 years. He has been editor and managing editor of the *Geophysical Journal International* since 1985. He has held many academic posts, including vice president of the Royal Astronomical Society, Dean of the Faculty of Science at the University of Leicester, Chairman of the British Geophysical Association, coleader of UNESCO IGCP 400 on Continental Rifts, and coordinator of the Kenya Rift International Seismic Project (KRISP). Aftab is the author of *Global Geology* (Wykeham Publications, London, 1976). His research interests are mainly in seismology, palaeomagnetism, exploration geophysics, and geophysical studies of the continental crust. He has published over 100 scientific papers, notably on East Africa, Cyprus, and the British Isles.

LOOKING INTO THE EARTH

An introduction to geological geophysics

Alan E. Mussett

Department of Earth Sciences,
Liverpool University, UK

M. Aftab Khan

Department of Geology,
Leicester University, UK

Illustrations by
Sue Button

CAMBRIDGE UNIVERSITY PRESS
Cambridge, New York, Melbourne, Madrid, Cape Town, Singapore, São Paulo

Cambridge University Press
32 Avenue of the Americas, New York, NY 10013-2473, USA

www.cambridge.org
Information on this title: www.cambridge.org/9780521780858

First published 2000
Reprinted 2007

Printed in the United States of America

A catalog record for this publication is available from the British Library.

Library of Congress Cataloging in Publication data

Mussett, A. E. (Alan E.)
Looking into the earth : an introduction to geological geophysics / Alan E.
Mussett, M. Aftab Khan ; [illustrations by Sue Button].
    p.   cm.
Includes bibliographical references and index.
ISBN 0-521-78085-3 (hc.) – ISBN 0-521-78574-X (pbk.)
1. Geophysics.   I. Khan, M. Aftab.   II. Title.
QE501.M87   2000
550 – dc22   00-020382

ISBN   978-0-521-78085-8 hardback
ISBN   978-0-521-78574-7 paperback

Contents

Preface: Turning a Magician into an Expert

Geophysics is essential to understanding the solid Earth, particularly on a global scale. Modern ideas of the structure and evolution of continents and oceans, or of the formation of mountain chains on land and below the oceans, for instance, are based extensively on discoveries made using geophysics. But geophysics can contribute to geological knowledge on all scales, from the global, through the medium-scale such as regional mapping or the search for oil and minerals, down to the small-scale, such as civil engineering, archaeology, and groundwater pollution, as well as detailed geological mapping.

Geophysics differs from other methods for studying the Earth because it can 'look into the Earth', for its measurements are mostly made remotely from the target, usually at the surface. It is able to do this because it measures differences in the physical properties of the subsurface rocks or structures, which are revealed by their effects at the surface, such as the magnetic field of some rocks. But it describes the subsurface in physical terms – density, electrical resistivity, magnetism, and so on, not in terms of compositions, minerals, grain-sizes, and so on, which are familiar to the geologist. Because geologists are often unfamiliar with physics (and the associated mathematics), there is a tendency either to ignore geophysics or to accept what a geophysicist says without understanding the qualifications.

This last is simply to treat the geophysicist as some sort of a magician. All subjects have their assumptions, often unstated but known to practitioners. Anyone who has used a geological map knows that just because some area has the same colour is no guarantee that the same rock will be found everywhere within it, or that faults occur exactly where marked and nowhere else. Similarly, there are reservations and limitations on what geophysics tells you, which need to be understood.

The intention of this book is not to turn you into a geophysicist (you may be glad to know), but simply to provide a basic grasp of the subject, so that, for instance, if a seismic section is described as 'unmigrated' you know whether it matters (it well might!). Or you may need to know whether a geophysical survey could help solve some geological problem of yours, so when you call on the services of a geophysicist (and worse things may happen!) you need to be able to explain your problem and understand his or her advice. You would not buy a car without specifying the number of seats, engine size, whether a sedan or sports model, and so on that you would like, and the price you are prepared to pay; and you would listen to claims about its fuel consumption, insurance group, and the choice of radios. Similarly, you need an understanding of geophysics.

The purpose of this book is to explain how geophysics helps us understand the solid Earth beneath our feet, and how – combined with the traditional methods of geology – greatly extends what can be learned about the Earth.

Acknowledgements

We wish to thank the following individuals for their help, which variously included providing advice, reprints or figures, reading and commenting on draft chapters, carrying out computing and literature searching, and agreeing to be photographed with geophysical instruments. Unfortunately, they are too numerous for us to acknowledge their help in detail, but without it this book would have been much inferior, if possible at all. If there is anyone we have forgotten, we apologise and thank them.

C. Adam, M. Atherton, R. D. Barker, H. Basford, P. D. Bauman, C. Birt, P. Brabham, P. Brenchley, A. Brock, J. Cassidy, P. N. Chroston, M. Cheadle, P. Dagley, E. A. Davis, J. H. Davies, Mr. P. Denton, J. Dickens, P. Fenning, D. Flinn, C. Gaffney, K. Games, J. Gowlett, D. H. Griffiths, E. A. Hailwood, K. Harrhy, V. Hatzichristodulu, R. J. Heitzmann, G. Hey, A. R. Hildebrand, I. Hill, K. M. Hiscock, I. P. Holman, Mr. M. Hudson, A. D. Khan, R. F. King, A. Latham, M. Lee, C. Locke, M. Lovell, P. K. H. Maguire, J. Marshall, M. Meju, J. A. Miller, I. Patterson, J. Peder, G. Potts, J. Reynolds, T. Rolfe, V. Sakkas, D. S. Sharma, P. Styles, R. Swift, E. G. Thomas, N. Thomas, R. Thompson, D. Whitcombe.

We also wish to thank the following institutions for, variously, use of libraries, drawing office and computing facilities, and providing unpublished figures: Atlantic Richfield Corporation; British Geological Survey; British Petroleum; Gardline Surveys; Leicester University, Department of Geology; Liverpool University, Department of Earth Sciences; Myanma Oil & Gas Enterprise.

Looking into the Earth

Chapter 1

Introducing Geophysics and This Book

1.1 What is geophysics?

As the word suggests, geophysics is the application of the methods of physics to the study of the Earth. But which methods, and how are they applied?

To the extent that rocks and their structures are formed by physical, chemical, and biological processes – for instance, rocks are deformed or fractured by physical forces, the compositions of volcanic rocks are determined largely by chemical processes, while oil, coal, and many limestones derive from living organisms – you might think that geophysics is all that part of geology that is not chemical or biological. But that is not how the term 'geophysics' is normally used. Before we attempt a definition, a few examples will give its flavour.

Some iron ores and some other rocks are sufficiently magnetic to deflect a compass, occasionally so much that it makes a compass an unreliable guide to north. Though this can be a nuisance for navigators, it allows us to detect the presence of the ores; however, to be able to predict where in the subsurface the magnetic rocks are located requires an understanding of how the magnetism of rocks affects a compass at the surface, and also why a compass usually points north. By replacing a compass with a much more sensitive instrument, the method can be extended to a much wider range of rocks, ones that are less magnetic than iron ores.

The first important study of the Earth's magnetism was made in the time of Queen Elizabeth I by her physician William Gilbert, with the purpose of improving global navigation. He realised that the Earth itself must be a magnet, and he investigated how it affects a compass. His work partly inspired Sir Isaac Newton to his theory of gravitation: a realisation that the force that causes object to fall (including apples upon heads) also holds the planets in their orbits about the Sun, and the Moon about the Earth. Newton deduced how this force must depend upon the distance apart of such bodies. At the time, no one could build an apparatus sensitive enough to test his theory by measuring the force between two objects different distances apart; however, Newton realised that if he were correct the Earth should pull itself nearly into a sphere, but with a slight bulge around the equator, spun out by its rotation. But some scholars thought that the Earth was lemon-shaped, longest between the poles, not across the equator. Expeditions were sent to different latitudes to measure the shape of the Earth, and their results confirmed the presence of the equatorial bulge. Newton's discovery also allows the geophysicist – using extremely sensitive instruments – to detect intrusions and cavities at depth, by the small force they exert at the surface, and even to 'weigh' the Earth itself.

These examples show that there are three aspects to geophysics. There is 'pure' geophysics (such as studying how the Earth's magnetic field is produced), there is geophysics as an aid to geology (such as finding ores), and there is the Earth as a giant laboratory (as when its shape was shown to support Newton's theory of gravitation). The geophysics of this book is about the second aspect; however, this usually requires some understanding of 'pure' geophysics, as well as of physical principles; for example, finding magnetic ores requires an understanding of both magnetism in general and the Earth's magnetism in particular.

The great advantage of geophysics to the geologist is that it can be used to make observations about the subsurface using measurements taken (usually) at the surface, as in the magnetic ore example above. In fact, geophysics is the only branch of the earth sciences that can truly 'look' into the Earth's interior, that is, remotely detect the presence of buried bodies and structures. In contrast, geology can only *infer* them. For example, are the exposed dipping strata shown in Figure 1.1 the limbs of a simple syncline as shown in Figure 1.1a, or are there at depth unsuspected structures such as faults, intrusions, or salt domes (Fig. 1.1b to d), all of which can be detected by appropriate geophysical measurements at the surface. Therefore, geophysics is able to add the third dimension, depth, in a way that traditional geology often cannot.

primarily concerned with the second, the geological applications of geophysics, but with 'pure' geophysics too, because the second cannot be understood without it.

2. Geophysics is the only branch of the earth sciences that can truly 'look' into the solid Earth.

3. Geophysics 'sees' the Earth in terms of its physical properties, which complement the usual types of geological information.

4. Geophysics mostly describes the Earth only as it is now (radiometric dating and palaeomagnetism are exceptions) but can help reveal current dynamic processes.

5. Geophysics does not replace geology but complements it.

6. The layout of the book is as described in Section 1.3. It has two parts: In Part I chapters that share many physical principles are grouped into subparts, while Part II presents case studies with introductory sections.

Further reading

The rest of this book.

PART I

GEOPHYSICAL METHODS

Data Acquisition and Processing

The first stage of most of the geophysical methods described in this book is making measurements in the field. Then usually follow further stages of working with the data, before they are ready for geological deductions to be drawn. As these stages are common to many geophysical methods they are described in this subpart.

*Chapter 2 – **Data acquisition and reduction** – describes the necessary basic stages, from taking measurements to converting the data into a relevant form. It also includes graphical ways of displaying the results more clearly. Chapter 3 – **Data processing** – describes special mathematical ways for separating wanted from unwanted parts of the results.*

Chapter 2

Data Acquisition and Reduction: Carrying Out a Geophysical Survey

This chapter describes the general principles of taking measurements – data acquisition – and the subsequent stages of correction and calculation – data reduction – often needed to make deductions about the subsurface.

A number of important geophysical terms are introduced, some of which are used in everyday language but have specialised meanings in geophysics.

2.1 Data acquisition: Taking measurements

Most geophysical measurements are made at the Earth's surface, either to save the expense and time of drilling, or because it is not feasible to go deep enough. Therefore the first actual step, after planning, to carry out a geophysical survey, is **data acquisition,** a set of measurements made with a geophysical instrument. Often the instrumental readings are taken along a line or **traverse** (Fig. 2.1a). Usually, readings are not taken continuously along the traverse but are taken at intervals – usually regular – and each place where a reading is taken is called a **station.** When the readings are plotted – often after calculations – they form a **profile** (Fig. 2.1b).

(a)

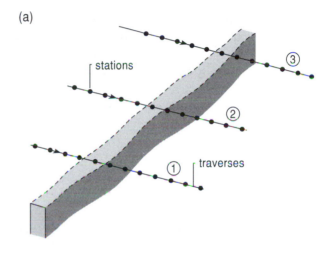

(b)

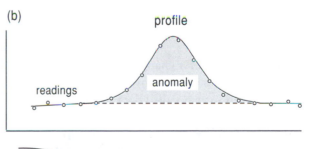

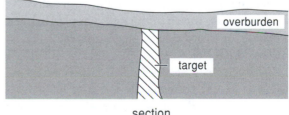

section

Figure 2.1 Traverses, stations, and profiles.

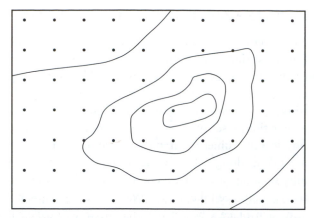

Figure 2.2 Grid of stations and contoured results.

If the **causative body** or **target** – the subsurface feature to be detected by the survey – is elongated, such as a mineral vein or a fault, the profile is best taken across it, perpendicular to the strike, so far as this can be estimated. Often, several parallel profiles are taken to see how far the body continues, or whether it changes its dimensions; if the target is not elongated, and especially if it is irregular, traverses may be close to one another, the stations then forming an array or **grid**, and the results are often contoured (Fig. 2.2).

2.2 Data reduction

Often the 'raw readings' provided by the instrument are not directly useful. For instance, in a gravity survey to detect the presence of a dense ore body by the extra pull of gravity above it, allowances, or corrections, have to be made for any undulations of the surface, for the pull of gravity also varies with height above sea level (Fig. 2.3); in a magnetic survey allowance is made for change of the Earth's magnetic field during the survey. Converting the readings into a more useful form is called **data reduction.**

The presence of a target is often revealed by an **anomaly.** In everyday language an 'anomaly' is something out of the ordinary, but in geophysics it is very common, for an anomaly is simply that part of a profile, or contour map, that is above or below the surrounding average (e.g., Figs. 2.1b and 2.3b).

Not all types of geophysical targets reveal themselves as spatial anomalies. In a seismic refraction survey (Chapter 6), the measured travel-times are plotted on a graph and from this the depths to interfaces are calculated. This too is a form of data reduction.

(a) before reduction

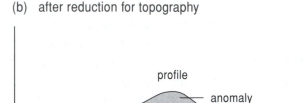

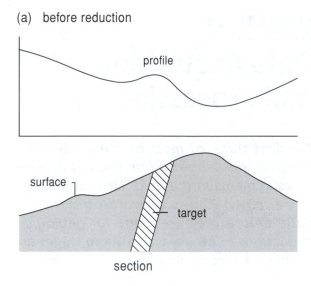

(b) after reduction for topography

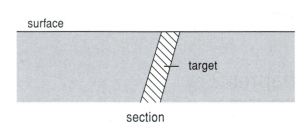

Figure 2.3 Gravity survey before and after reduction.

2.3 Signal and noise

Even after the data has been reduced, the profile may not reveal the presence of the target as clearly as one would like, because of noise. **Noise** is not sound but simply unwanted variations or fluctuations in the quantities being measured; it contrasts with the wanted part, the **signal.** In seismology, where small ground movements have to be detected, noise can be vibrations due to passing traffic or anything else that shakes the ground (Fig. 2.4a). This is not unlike the background noise at a party through which you have to try to hear your partner's conversation. Noise can also be spatial: In a magnetic survey noise may be

(a) in time

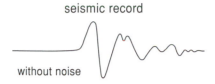

seismic record

without noise

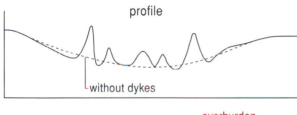

with noise

(b) in space

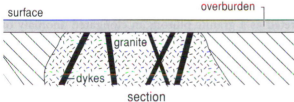

Figure 2.4 Noise (a) in time and (b) in space.

tion seismology (Section 7.3). A more general method to make the wanted signal clearer is to use **signal processing**. However, as this often uses fairly sophisticated mathematics, it has been described separately in the following chapter. Improving the sound of old sound recordings, to remove hisses and scratches, is done by signal processing.

2.4 Modelling

Because a geophysical anomaly is two steps remote from a geological description (being measured at the surface, and of a physical rather than a geological quantity), there are usually two further stages. The first is to model the reduced data in physical terms. In common language, a **model** is often a small version of the real thing, such as a model boat, but in geophysical usage it is a body or structure (described by such physical properties as depth, size, density, etc.) that could account for the data measured. Figure 2.5 shows a section that could account for the observed negative (gravity) anomaly: Values calculated from the model are compared with the actual measured values to test how well it accounts for the observations.

The model is almost always simpler than the reality. This is for several reasons. Firstly, the signal observed is often 'blurred' compared with the causative body; for example, the anomaly of Figure 2.1b is wider than the target and tails off without sharp margins, which makes it hard to decide the

due to wire fences, or buried bits of defunct cars, which obscure the signal (i.e., the anomaly due to the buried body that has to be detected). What is noise – and what signal – may depend on the purpose of the survey; when searching for a weak magnetic anomaly of a granite intrusion the strong anomalies of basaltic dykes would be noise, whereas to a person looking for dykes they are the signal (Fig. 2.4b). Here, 'noise' is like the definition of a weed, as a plant unwanted in its place.

One common method to improve the **signal-to-noise ratio** is to repeat readings and take their average: The signal parts of each reading add, whereas the noise, usually being random, tends to cancel. This is called **stacking**, and it can also be done with profiles, which is particularly important in reflec-

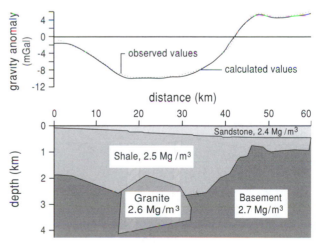

Figure 2.5 A simple density model to account for the variation in the pull of gravity.

unwanted noise, though which is which will partly depend upon what geological information is being sought.

4. Modelling is finding a physical body or structure that would approximately account for the observed data, and models usually are simpler than reality. Modelling may be forward or inverse.

5. Geological interpretation translates physical models into geologically feasible bodies or structures, taking account of all available information.

6. Gridded data may be displayed in several ways: Contouring, often with shading or colour, can be done; features can be further enhanced by using false illumination; or isometric projection may be used. Fence projection is useful when profiles are widely spaced.

7. Important terms: survey, station, traverse, profile, grid; data acquisition, data reduction; target, causative body, anomaly; signal, noise, signal-to-noise ratio, stacking; resolution; models, inversion problem, forward and inverse modelling, interpretation; contouring, isometric projection, false illumination, stacked profiles, fence projection.

Further reading

Milsom (1996) covers some of the preceding points in his first chapter; otherwise they are distributed throughout standard textbooks such as Kearey and Brooks (1991), Robinson and Coruh (1988), and Telford et al. (1990).

Problems

1. What is the difference between a positive and a negative anomaly?

2. A profile across a known subsurface body fails to show an anomaly. This might be because of which of the following?
   (i) Small signal-to-noise ratio.
   (ii) Data has not been reduced.
   (iii) Station spacing was too large.
   (iv) Instrument was not sensitive enough.

   (v) The body does not differ from its surrounding in the physical property being measured.

3. The purpose of stacking is to:
   (i) Display the result more clearly.
   (ii) Improve the signal-to-noise ratio.
   (iii) Help reduce the data.

4. To deduce the possible shape of the body producing a very elongated anomaly, the observed anomaly found by surveying along a single traverse is compared with one calculated for a body assumed to be uniform and horizontal, and to extend a long way to either side of the traverse. This is an example of:
   (i) Forward modelling in 2D.
   (ii) Inverse modelling in 2D.
   (iii) Forward modelling in 2½D.
   (iv) Inverse modelling in 2½D.
   (v) Forward modelling in 3D.
   (vi) Inverse modelling in 3D.

5. Ore veins in an area strike roughly E–W. The area is surveyed on a grid and, after reduction, any anomalies are to be made more obvious by use of false illumination. This should be from which of the following directions?
   (i) N. (ii) NE. (iii) E. (iv) SE. (v) S. (vi) SW. (vii) W. (viii) NW.

6. In a survey negative but not positive anomalies could be significant. How could you display the results to pick up only the negative anomalies?

7. An area has been surveyed along a few traverses, some of which intersect. The most appropriate way to display the results would be by:
   (i) Contouring.
   (ii) Stacked profiles.
   (iii) Fence projection.
   (iv) Isometric projection.

8. An anomaly has been found for a survey over a grid. You wish to know not just where the anomaly is but also if it has any particularly large values and where these are. You could achieve this by which of the following?
   (i) Contouring.
   (ii) Contouring plus colour.
   (iii) Use of false illumination.
   (iv) Isometric projection.

Chapter 3

Data Processing: Getting More Information from the Data

Even after the results of a survey have been reduced and displayed, as described in the previous chapter, the features of interest may not be obvious. If so, there may be further stages of processing that will enhance the features.

The methods described, which are mathematical, are applicable to the results of most geophysical techniques (and widely outside geophysics). This presentation emphasises the underlying concepts, which are referred to in later parts of the book.

3.1 Fourier analysis

3.1.1 Wavelength

To fix our ideas, we invent a simple example. Suppose a granite was intruded below some area in the past and later was exposed by erosion. Today, the uneven surface of the granite and surrounding country rock is buried beneath overburden, which – to keep it simple – has a level surface (Fig. 3.1). To detect the presence and extent of the granite, we exploit the granite having a lower density than the country rock, by measuring the variation in gravity (g) along a traverse (the method will be explained in Chapter 8, but detailed understanding is not needed here).

The granite alone causes the value of gravity to be lower in the middle of the profile, a negative anomaly (Fig. 3.1b); however, the varying thickness of overburden produces a varying value of gravity, least where it is thickest because its density is lower than that of the rocks beneath (Fig. 3.1c). Together (Fig. 3.1d), they produce a profile in which the anomaly of the granite is not obvious; however, it is responsible for the dip in the middle part of the profile because it is much wider than the anomalies due to the variations in overburden thickness, which con-

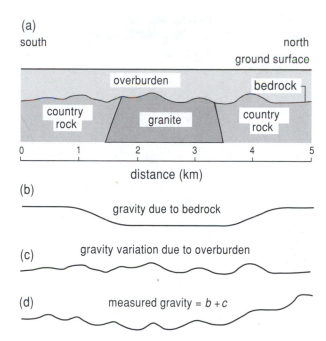

Figure 3.1 Components of a gravity anomaly.

tribute fairly narrow peaks and troughs. The essence of **Fourier analysis** is to sort features by their widths, from which we can then select the ones we want.

These widths are measured as wavelengths. Waves are common in nature: One obvious example is water waves; the undulations that move along a rope shaken at one end are another. In geophysics, we shall meet both seismic and electromagnetic waves. For the present, we ignore that they move; what is important here is that they have the same general shape, shown in Figure 3.2. The repeat distance – conveniently measured between successive crests – is the **wavelength**, λ (pronounced lamda), and the maximum deviation from the undisturbed position is the **amplitude**, a (the trough-to-crest height is therefore twice the amplitude). This curve is called a **sinusoid**, because it is described by the mathematical sine function (Box 3.1).

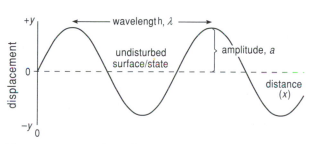

Figure 3.2 Wavelength and amplitude.

BOX 3.1 Fourier harmonic analysis equations

A sinusoid has the mathematical form described by a sine curve:

$$y = a \sin\left(\frac{2\pi x}{\lambda}\right) \qquad \text{Eq. 1}$$

where x is the distance along a profile. We can use the equation to calculate the value of the displacement, y, for any distance, x.

When x is zero, the displacement y is zero. As x increases, y increases, up to a maximum of a at $x = \lambda/4$; then it decreases, becoming zero at $x = \lambda/2$, and continues to decrease to a minimum of $-a$ at $x = 3\lambda/4$, after which it increases again, reaching zero at $x = \lambda$. This cycle repeats, as shown in Figure 3.2. Whenever the quantity in brackets equals π, 2π, 3π, . . . , y passes through zero. (A cosine curve has the same shape but is displaced a quarter wavelength along the x axis, compared to a sine curve.)

A harmonic series consists of several sinusoids, each such that an exact number of half-wavelengths equals the length, L, of the signal (Fig. 3.3), that is, $\lambda/2 = L$, $\lambda = L$, $3\lambda/2 = L$, $2\lambda = L$, . . .

$$\begin{aligned}
y = a_0 &+ a_1 \sin\left(2\pi x \frac{1}{2L}\right) \\
&+ a_2 \sin\left(2\pi x \frac{2}{2L}\right) \\
&+ a_3 \sin\left(2\pi x \frac{3}{2L}\right) + \dots
\end{aligned} \qquad \text{Eq. 2}$$

Therefore their wavelengths are $2L$, L, ⅔L, $L/2$, . . . In a Fourier series, these add together to match the signal, which is done by giving the amplitude of each harmonic a particular value. Therefore the series is found by replacing λ of Eq. 1 by $2L$, L, . . . in turn and giving different values to a: a_1, a_2, . . . ; a_0 is an extra term used to adjust the whole curve up or down; y is the addition of the series and gives the value of the signal at any value of x along the signal.

The same ideas apply to time-varying signals as well as spatial ones. The equations can be converted by replacing each spatial quantity by the corresponding temporal one:

Spatial quantities	Temporal quantities
L, length of signal	T, duration of signal
x, distance along signal	t, time since signal began
λ, wavelength	τ, period of harmonic = 1/frequency, f

τ (tau, to rhyme with cow) is the time it takes the oscillation at a place to complete one cycle (e.g., the time for a cork on the sea to bob from crest to trough to the next crest). However, it is usually more convenient to use the frequency, f, so that Eq. 1 becomes, for time-varying signals,

$$y = a \sin\left(2\pi f t\right) \qquad \text{Eq. 3}$$

The values of the amplitudes are found from the signal using these expressions:

$$a_0 = \frac{1}{L}\int_0^L y\, dx \quad \text{or} \quad \frac{1}{T}\int_0^T y\, dt$$
$$= \text{average value of signal} \qquad \text{Eq. 4a}$$

$$a_n = \frac{2}{L}\int_0^L y \sin\left(\frac{n\pi x}{L}\right) dx$$
$$\text{or} \quad \frac{2}{T}\int_0^T y \sin\left(\frac{n\pi t}{T}\right) dt \qquad \text{Eq. 4b}$$

3.1.2 Harmonic analysis

Few profiles look like Figure 3.2, so why bother with waves? The essence of Fourier analysis is that a wiggly line, such as the gravity profile of Figure 3.1, can be reproduced by adding together a series of waves. Only certain wavelengths are used, such that 1, 2, 3, . . . half-wavelengths exactly fit the length of the profile; these are called **harmonics.** Their amplitudes are adjusted so that, added together, they match the required shape.

Figure 3.3a shows that an approximate sawtooth shape can be built up using the first 10 harmonics, because at different distances the harmonics have different values, which may add or partially cancel. The approximation improves as more harmonics are added. Figure 3.3b shows another profile or waveform produced using different proportions of the same harmonics. Remarkably, any wiggly line can be matched, though the finer details require higher harmonics, which have shorter wavelengths.

(a)

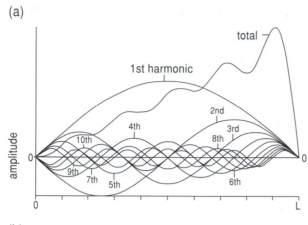

(b)

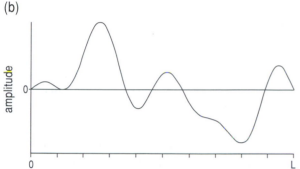

Figure 3.3 Addition of waves.

How the values of the amplitudes of the harmonics depend on the curve to be reproduced is explained in Box 3.1.

Signals in time. Some signals vary with *time* rather than *space,* such as the variation with time of the height of the sea surface at some place (Fig. 3.4), rather than the shape of the surface at some instant: The height fluctuates due to the passage of waves and also rises and falls over a period of about 13 hr due to the tides. This can be analysed in exactly the same way as the spatial gravity profile except that time replaces distance (Box 3.1). However, instead of describing features in terms of wavelength we now use **frequency,** *f*, which is measured in Hertz, Hz, the number of times a complete cycle repeats in 1 sec.

In signal processing, time-varying signals are often referred to as if they were in space, or vice versa, so a low-frequency signal (one that varies slowly with time) may be described as having a long wavelength, or a wide granite as producing a low-frequency anomaly.

(a)

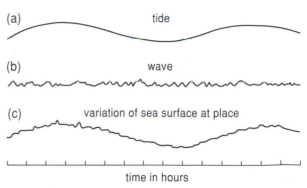

Figure 3.4 Variation in sea level at a point, due to waves and tides.

3.1.3 Fourier analysis of a profile

We now apply this idea of wavelength analysis to the gravity profiles of Figure 3.1. The profile due to the granite alone consists mostly of long wavelengths (Fig. 3.5a) – the main one being about 6 km long – while the overburden profile clearly has mostly shorter wavelengths. Therefore we can make the anomaly of the granite clearer by analysing the measured profile, Figure 3.1d, into harmonics, then rejecting wavelengths shorter than the fifth harmonic and finally recombining the remaining harmonics into a single wiggly curve (Fig. 3.5a). How well the recombined harmonics match the observed profile can be judged by showing their differences as a residual anomaly (Fig. 3.5b); this also indicates whether the unmatched part consists mainly of unwanted short wavelengths.

(a) granite anomaly and matching harmonics

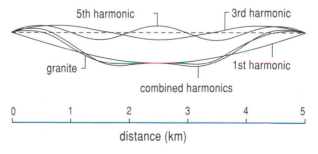

(b) residual (unmatched) anomaly

Figure 3.5 Analysis of anomaly.

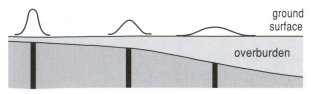

Figure 3.6 An anomaly broadens as the body is buried more deeply.

Though this enhances the wanted anomaly compared to unwanted ones, there are some points to note. Each component curve due to a different geological body usually contains a range of wavelengths, so separation of wanted from unwanted anomalies, or signals, is only partial; for instance, the overburden and granite have some wavelengths in common. Secondly, the wavelengths of the harmonics deduced depend upon the length of profile being analysed, for they have to fit its length exactly, and as the length of the profile is rather arbitrary so are the harmonics. These two points emphasise that Fourier harmonic analysis (to give it its full title) is a mathematical procedure and does not analyse the measured signal into its geophysical or geological components as such.

Thirdly, rejecting the shorter wavelengths is often regarded as removing the anomalies of bodies near the surface to leave deeper ones, but this is not necessarily so. It is true that for many geophysical techniques a body near the surface gives a narrower (as well as stronger) anomaly than when buried deeper (Fig. 3.6; also Figs. 8.9 and 8.10). But whereas a narrow anomaly cannot be due to a deep body, the converse is not true, for a broad anomaly can be due to either a narrow body at depth or to a broad body near the surface (Fig. 3.7; also Fig. 8.17a).

The unwanted anomalies of the overburden of Figure 3.1 are an example of noise (Section 2.3). As noise has a shorter wavelength than the signal, removing shorter wavelengths often improves the signal-to-noise ratio, so making the anomaly more obvious.

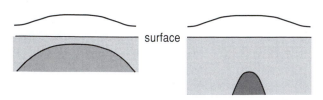

Figure 3.7 Similar anomalies from shallow and deep bodies.

3.1.4 Fourier analysis in 2D: Gridded data

The examples so far have been in one dimension, such as a profile, but readings may be taken on a grid and contoured, as described in Section 2.6, and we may wish to remove the short-wavelength anomalies in much the same way.

Such surveys are analysed by sets of waves at right angles to one another, parallel to the sides of the rectangle. Waves with a single wavelength, λ, extending over an area with their crests and troughs parallel, look rather like a ploughed field or piece of corrugated iron (Fig. 3.8a and b); two such sets at right angles, when added, resemble an egg tray (Fig. 3.8c). The readings of an areal survey are analysed into harmonic series in each of the two directions, and as with analysis of profiles, various ranges of wavelengths can be eliminated and the remaining wavelengths recombined. (An example using the related process of filtering is shown in Figure 3.13.) However, there are other possibilities, such as eliminating *all* wavelengths in one of the two directions: This will emphasise features elongated in the other direction.

Fourier analysis is a large subject, just touched upon here. It can, for instance, be extended to readings taken over the whole surface of the Earth (surface spherical harmonic analysis, mentioned in Section 11.1.4), and even throughout the Earth (spherical harmonic analysis), but these will not be considered here.

3.1.5 Why a harmonic series?

What is special about harmonics? Why not use sinusoidal waves that don't fit an exact number of half-wavelengths into the length of the profile or some shape of curve other than a sinusoid? There are two reasons. The first is that some natural signals are made up of such a series. For instance, oscillations of musical instruments, such as a vibrating string of a guitar, can be analysed into harmonics that correspond to an exact number of half-wavelengths fitting into the length of the string. It is the different proportions of the harmonics that make a guitar sound different from another type of instrument playing the same note, and a trained musician can hear the individual harmonics. In fact, it is from music that the term 'harmonic' derives.

The second and more important reason is a mathematical one, that the amount of each har-

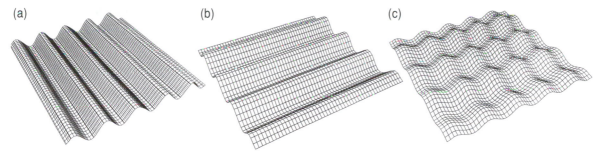

Figure 3.8 Combination of waves at right angles.

monic can be calculated independently. Thus, for example, to calculate the amount of the third harmonic, it is not necessary to know the amounts of the first, second, and fourth harmonics, and so on. So it is necessary only to calculate the amplitudes of relevant harmonics and not every one down to the extremely short wavelengths that may be present.

3.2. Digital filtering

3.2.1 Simple filters

An alternative way of rejecting unwanted wavelengths or frequencies is by **filtering**. In everyday language, a filter can be used to separate large from small particles; for example, if a mixture of rice and beans were put into a filter (or sieve) with suitable mesh size, the beans would be retained while the rice grains would be let through. In electronics, filters are used to remove unwanted frequencies, such as sharp 'spikes' and pulses from the electricity supply, or to change the proportions of bass and treble when recording music. Though neither of these filters works in the same way as the digital filters we shall be considering, they embody the same idea of separating wanted from unwanted things by size. Digital filters do this mathematically and have much in common with Fourier analysis.

Digital filters are usually applied to values taken at regularly spaced **sampling intervals,** either along a line or on a grid. (If readings are continuous, values at regular intervals are just read off; if irregularly spaced, regular ones are found by interpolation.)

One of the simplest digital filters takes the average of three successive readings along a profile and records the result at their midpoint (Fig. 3.9; 3-pt filter). This is repeated at each sampling position in turn, so that each reading is 'used' three times. This simple 3-point average is sometimes called a running average; or we talk of a 'moving window': Imagine a card with a hole just wide enough to see three readings, and average only the points that you can 'see through the window' as the card is moved point by point along the profile.

The average value of y, $\bar{y}$, is:

$$\bar{y}_n = \frac{1}{3}\left(y_{n-1} + y_n + y_{n+1}\right) \qquad \text{Eq. 3.1}$$

The subscript n denotes the point where we shall record the result (i.e., the midpoint of the three readings), $n-1$ is the previous point, and $n+1$ the following point. Then the process is repeated with the window centred on the next point (i.e., with n increased by 1).

The short wavelength 'jaggedness' on the left part of the unfiltered line is 'smoothed' much more than the hump in the middle, while the straight line at the right is not affected at all. In terms of Fourier analysis, we would say that shorter-wavelength amplitudes (or higher frequencies, for a time-varying signal) are the most reduced. A slightly more complex filter of this type could have a window of five points. Figure 3.9 shows that it smooths short wavelengths more effectively than the 3-point filter.

Other filters could have yet more points, but another way to change their effectiveness is to use a weighted average rather than a simple average (i.e., the values of the points are multiplied by different amounts, or coefficients). For example, a weighted 7-point filter could be

$$\bar{y}_n = (-0.115y_{n-3} + 0y_{n-2} + 0.344y_{n-1} + 0.541y_n$$
$$+ 0.344y_{n+1} + 0y_{n+2} - 0.115y_{n+3} \qquad \text{Eq. 3.2}$$

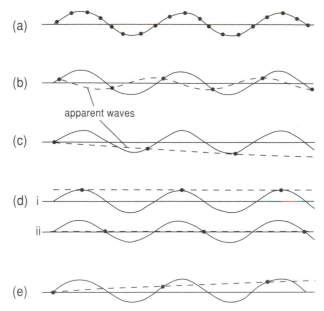

Figure 3.11 Effect of sampling intervals.

aliasing produces spurious periods when a harmonic has a period less than twice the sampling interval. Since period is the reciprocal of the frequency, $1/f$, this is equivalent to saying that spurious frequencies are produced when the number of samplings each second is less than twice the frequency of a harmonic. The critical frequency, half the sampling frequency, is called the **Nyquist frequency.** To avoid the spurious frequencies, any higher frequencies have to be removed before sampling, usually by using a nondigital electronic filter.

In geophysics, aliasing is more commonly a potential problem with time-varying signals – such as seismic recording of ground motions – than with spatial anomalies.

3.2.3 Designing a simple filter

Filters can be designed to remove, to a considerable extent, whichever range of wavelengths we wish, by our choice of the sampling interval, the number of points in the window, and the values of the coefficients of the points. As we saw earlier, the sampling interval is important because wavelengths much less than the sampling interval are automatically largely rejected (though aliasing can occur). The sampling interval is therefore chosen

to be somewhat shorter than the shortest wavelength we wish to retain; a quarter of this length is a suitable sampling distance. Provided the correct sampling interval has been chosen, there is little advantage in having a filter with more than seven points. Then the coefficients are chosen (by mathematical calculations beyond the scope of this book) to give the maximum discrimination between wanted and unwanted wavelengths, or frequencies, though this cannot be done as well as by Fourier analysis.

3.2.4 Filtering in 2D: Gridded data

In Section 3.1.4 we explained that Fourier analysis could be applied to gridded or 2D data; similarly, filters can be used in 2D, with points sampled all round the point in question. The larger dots of Figure 3.12 show the sampling positions for one particular filter, while the numbers are the weightings, which are the same for all points on a circle. Figure 3.13 shows the effect of applying a low-pass filter that progressively reduces wavelengths in the range 16 to 10 km and entirely removes those shorter than 10 km. It reveals a large, near-circular anomaly. (The anomaly is actually centred somewhat to the north, because there is also a steady increase from north to south – a 'regional anomaly' in the terminology of gravity and magnetic surveying; see Section 8.6.2 – which also needs to be removed.)

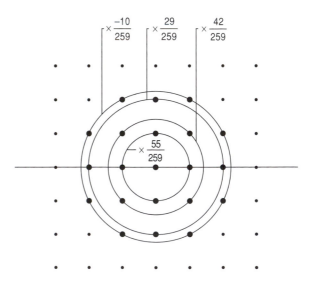

Figure 3.12 Filter window and coefficients for gridded data.

(a) unfiltered

(b) after low-pass filtering

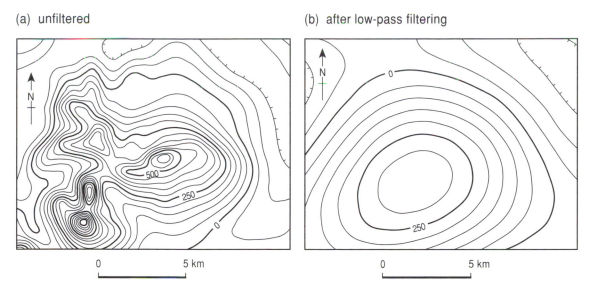

Figure 3.13 Gridded data before and after low-pass filtering.

3.2.5 Using filters to enhance various types of features

The filters described so far can enhance the signal simply by reducing unwanted wavelengths, such as the short wavelengths due to noise, as is done with Fourier analysis. However, with 2D data there are other possibilities, as with Fourier analysis. One is to use directional filters to separate elongated features by their direction. For instance, this technique could be used to emphasise anomalies due to ore veins in a region where their likely direction is known. An example of directional filtering is given in Section 27.4.2.

Another type of filter can be used to emphasise edges of an anomaly, and so help outline the positions of the causative body, by selecting where short wavelengths are concentrated. This is in contrast to enhancing large anomalies by filtering out the short wavelengths, as described in Sections 3.1.3 and 3.2.4, which tends to deemphasise the edges.

An alternative way to pick out edges is by finding where values are changing most rapidly. In Figure 3.1 the value of the anomaly changes most rapidly near the edges of the granite, so picking out where this occurs may outline a body; similarly, the steepest slopes or gradients occur around the edge of a broad hill. An example using gradient to pick out edges is shown in Figure 25.5a, in Section 25.3.2.

3.3 Summing up: Fourier analysis and filtering

A filter achieves much the same result as a Fourier analysis, so which should be used? Fourier analysis requires that the whole signal or profile be analysed, whereas a filter needs only a few successive readings at a time, those in the 'window'. Therefore, a Fourier analysis cannot be made before the signal is complete, and it requires more computing as it uses more data at a time. However, Fourier analysis gives the more complete separation of frequencies or wavelengths, and the performance of a filter is specified in the terms of Fourier analysis, rather than vice versa, so Fourier analysis underpins filters. Fourier analysis can analyse a continuous curve, whereas a filter operates on discrete data, but as much geophysical data is discontinuous this is of little consequence, provided the sampling interval is chosen suitably and precautions are taken against aliasing. In practice, filters are more often used than Fourier analysis.

Fourier harmonic analysis and the various forms of filtering are used to emphasise wanted features in 1D or 2D, but none of the methods are guaranteed to enhance the desired feature, because the causative body often generates a signal with a range of wavelengths that may overlap those due to 'noise'. Signal processing is no substitute for careful design of a survey to optimise the quality of data that can be

Seismology

Seismology is primarily concerned with determining the structure of the Earth – on all scales – and only subsidiarily with earthquakes. It uses the ability of seismic waves – vibrations of rocks – to propagate through the Earth. They do not generally travel in straight lines but are deflected, by refraction or reflection, by the layers they encounter, before they return to the surface of the Earth, and this allows the internal structure to be determined. Seismology is particularly useful for determining the positions of roughly horizontal interfaces between layers, which makes it the most useful single geophysical technique.

This subpart is divided into four chapters: **Global seismology and seismic waves** *first explains how seismology works and then uses it to explore the Earth's structure on the largest scale.* **Refraction seismology** *exploits the special case of seismic waves propagating along an interface after refraction, while* **Reflection seismology** *depends on waves reflected back up from interfaces.* **Earthquakes and seismotectonics** *– which comes second – is concerned with earthquakes and what they reveal about the tectonic processes that produce them.*

Chapter 4

Global Seismology and Seismic Waves

Global seismology reveals that, on the largest scale, the Earth is concentrically layered, with the primary divisions of crust, mantle, and core, plus other concentric features. It also shows that the core is divided into a liquid outer core and a solid inner core.

As seismology depends upon seismic waves, which travel in the Earth, the chapter begins by introducing these, together with their generation, propagation, and detection.

4.1 Waves, pulses, and rays

Waves. An example of a **wave** is a water wave, but so also are waves produced by shaking a rope attached at its further end, or pushing in and out a long spring (Fig. 4.1). If the end is moved rhythmically, a series of disturbances travels along the rope or spring; Figure 4.1 shows them before the disturbances have reached the further end.

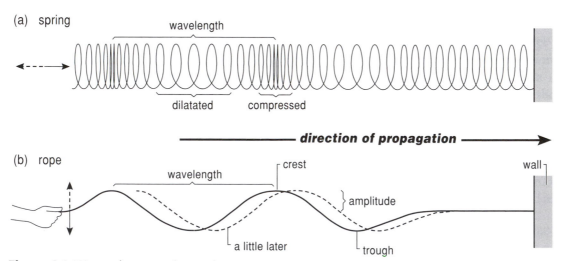

Figure 4.1 Waves along a spring and a rope.

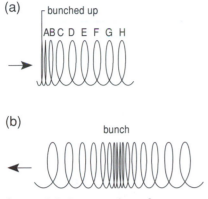

(a)

Figure 4.2 Propagation of waves.

Consider the spring in Figure 4.2. When it is held stretched but stationary, the turns will be equally spaced because the tension of the spring upon each turn is the same from either side. Pushing the end in quickly a short distance causes the first couple of turns, *A* and *B*, to bunch up (Fig. 4.2a), but then the next turn, *C*, being no longer pulled so hard from the left, moves to the right and bunches up against *D*, and so on, propagating a compression to the right. Then, if the end is pulled out, stretching the spring, a dilatation propagates along the spring, following the compression. So a rhythmic pushing in and pulling out of the end of the spring produces a regular series of compressions and dilatations that move along, forming waves.

Though the waves travel along, the spring does not, each turn only oscillating about its stationary position. Similarly, water is not moved along by water waves, which is why you cannot propel a ball across a pond by generating waves behind it by throwing in stones.

There are some terms you need to learn, most of them illustrated in Figure 4.1 (some were introduced in Section 3.1.1). **Wavelength, λ,** is the repeat length, conveniently measured between successive crests or compressions. The **amplitude, *a*,** is the maximum displacement from the stationary position. The waves travel along at some speed called, in seismology, the **seismic velocity, *v*** (a velocity should specify the direction of propagation as well as its speed, but this is often neglected in seismology). The number of crests or compressions that pass any fixed point on the rope, spring, and so on, in one second is the **frequency, *f*,** measured in Hz (Hertz,

oscillations, or cycles, per second). In one second, *f* wave crests, each λ metres apart, will pass a point; and by the time the last one has passed, the first one will have travelled a distance of *f* times λ metres. As velocity is the distance travelled in one second,

$$v = f \times \lambda$$
$$\text{velocity} = \text{frequency} \times \text{wavelength} \qquad \text{Eq. 4.1}$$

Pulses. You can see waves travelling along springs, ropes, and water, so it so easy to measure how fast they are moving. In seismology, we can learn a lot just by timing how long it takes seismic waves to travel different distances through the Earth, but, of course, we can't see them moving inside the Earth. We therefore need a way of 'marking' waves so that their progress can be observed. The simplest way is to generate just a few waves and time how long it is before the ground some distance away begins to move. A very short series of waves is called a **pulse.** Pulses can have various shapes, but a very simple one is just one compression (or crest) followed by one dilatation (or trough), as shown in Figure 4.3.

Seismic pulses are easily generated: Any quick, sharp disturbance of the ground does it, from a hammer blow to the onset of an earthquake. One common way is to fire a charge of explosive buried at the bottom of a hole (Fig. 4.4). The rapid expansion produces a compression that travels spherically outwards; then the ground tends to spring back into the cavity produced, causing a dilatation. The compression has a spherical surface called a **wave front,** and it expands away from the source. (Waves due to explosions are often described as shock waves, but shock waves exist only very close to the explosion, where material has been forced bodily outwards, moving faster than the natural speed of propagation of ground oscillations. They slow down within a metre or less and become normal seismic waves.)

Typical waves studied in seismology have velocities measured in km/sec and frequencies of some tens of Hertz. For example, a wave with velocity of 2 km/sec and frequency of 10 Hz has a wavelength, found using Eq. 4.1, of 0.2 km or 200 m. So seismic wavelengths can be quite long compared to thicknesses of strata or other types of layers. This has important consequences, explained in Section 7.8.2.

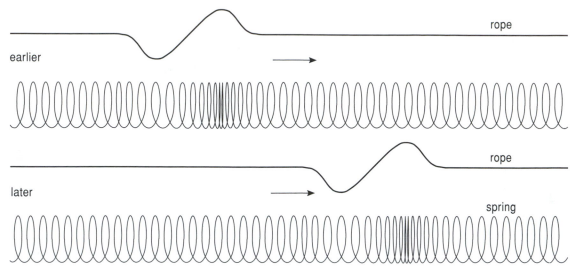

Figure 4.3 A pulse.

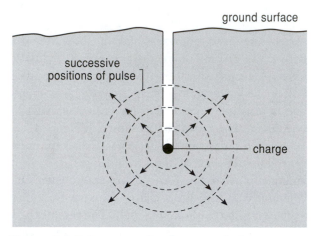

Figure 4.4 Pulse generation by an explosion.

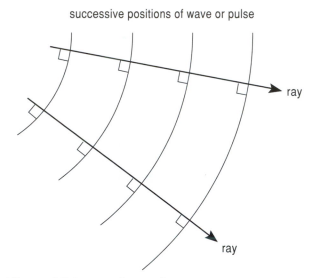

Figure 4.5 Rays and wave fronts.

Rays. In the solid Earth, waves travel outwards from their source in all directions (Fig. 4.4). If we are interested only in what happens in one direction we need consider only part of the wave front. The path of a tiny portion of the wave front, or pulse, forms a **ray** (Fig. 4.5). *Rays are always perpendicular to wave fronts, and vice versa.* As rays are simpler to consider than waves, most seismology theory will be explained using them.

4.2 Detecting seismic waves: Seismometers and geophones

When a seismic wave passes a point – perhaps after travelling deep in the Earth – it causes the ground to oscillate. It is easy to measure the movement of

water waves if we are on the stationary shore, but how can we measure ground motion when we ourselves are moving up and down, or from side to side?

Consider a plumb bob hanging from a frame resting firmly on the ground (Fig. 4.6a). If the ground suddenly moves – to the left, say – the frame moves with it, but the bob tends to remain still. Thus the scale moves left past the bob, giving a positive deflection on the scale shown. This design of instrument detects horizontal ground motion; vertical ground motion could be measured using a bob on a spring (Fig. 4.6b).

(a) horizontal motion

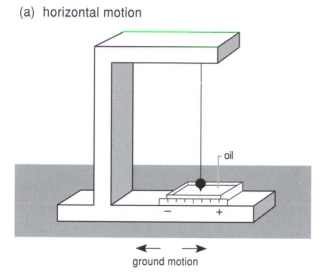

(b) vertical motion

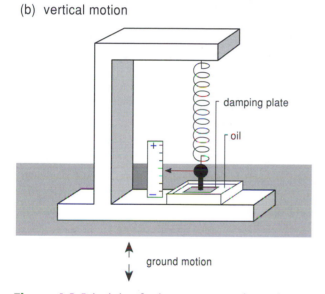

Figure 4.6 Principle of seismometers and geophones.

Of course, shortly after the frame has moved the bob will begin to move too, and thereafter it will tend to go on oscillating, obscuring the record of any subsequent ground motion. To reduce this effect the instrument is 'damped', which could be done by having part of the bob immersed in oil, causing the oscillations to die away once the ground stops moving.

Actual instruments have to be compact, damped in some way, and very sensitive, but depend on the above principles. Moving-coil and moving-magnet instruments are used. In Figure 4.7, the moving mass is a magnet suspended inside a coil of wire by a compact spring; relative movement of magnet and

coil generates a small electric current (as described in Section 14.1.1), which is amplified electronically; this is the principle used in some record pickups. The amplified signal is sometimes recorded on paper on a moving drum for immediate inspection, but records are usually stored electronically, on magnetic tape, often employing digital recording (described in Section 7.7.1) because it is easier to process. Digital recording is also used, for instance, in compact discs.

Instruments are classed as **seismometers** or **geophones**, but function similarly. Compared to geophones, seismometers are much more sensitive – but are less robust and compact and have to be set up carefully – so they are used to measure weak signals, as is often the case with global seismology, because sources are distant. A modern seismometer can detect the ground motion caused by a person walking a kilometre away, if there is little other disturbance. A seismometer record – a **seismogram** – is shown later, in Figure 4.17. Geophones are used in small-scale seismic surveys where many instruments have to be set out quickly but the highest sensitivity is not needed. We call both seismometers and geophones seismic **receivers.**

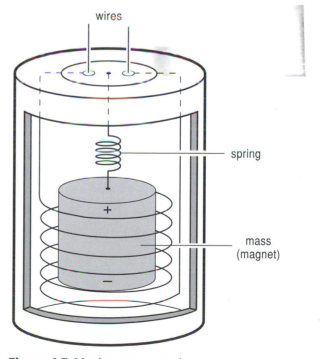

Figure 4.7 Moving-magnet seismometer.

To obtain full information about the ground motion, three seismometers are needed (sometimes combined into a single instrument): one to measure the vertical part or component of ground motion, the other two to measure horizontal motions at right angles, often N–S and E–W (Fig. 5.22 shows a three-component seismogram). Often, single-component instruments are used for economy.

4.3 The Earth is concentrically layered

4.3.1 Spherical symmetry of the Earth's interior

The Earth is nearly a sphere, as pictures taken from space show. Of course, it is not quite a sphere because of mountains and other topographic features, but the height of even Mount Everest is only a thousandth of the Earth's radius.

A much larger departure from sphericity is the equatorial bulge, caused by the centrifugal force of the Earth's rotation spinning out its material (see Fig. 9.17). The difference in radii is only about a third of one percent. The bulge has a smooth shape, unlike the jaggedness of mountains, and so can be allowed for in calculations; we shall assume this has been done and so shall seldom mention the bulge again in this chapter.

The first question to ask about the interior of the Earth is whether it also is spherically symmetrical: Is it layered like an onion, or are there large lateral differences as found for rocks near the Earth's surface? We can test this by timing how long it takes a pulse of seismic energy to travel between pairs of points *separated by the same distance*. In Figure 4.8 the ray paths are shown dashed because we don't yet know where they are; they serve just to connect the pairs of points, A_1B_1, A_2B_2, ... Be clear that we are not at this point testing whether the interior is seismologically uniform but only if it has the same velocity everywhere at a given depth.

In global seismology distances are usually given, not as kilometres around the surface, but as the angle subtended at the centre, **the epicentral angle, Δ.** Measurements show that the times to travel all paths with the same epicentral angle are nearly the same, and this is true for both large and small angles (Fig. 4.8a and b), so the Earth is indeed spherically symmetrical, like an onion. (There are very small differences, which will be considered in Section 4.6.) This is convenient because now we need not bother where the actual locations of sources and seismometers are, only how far apart they are, so we can combine results from all over the world to answer the next question.

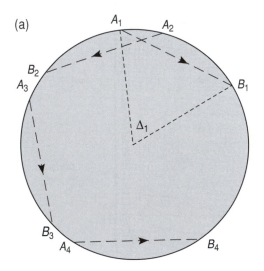 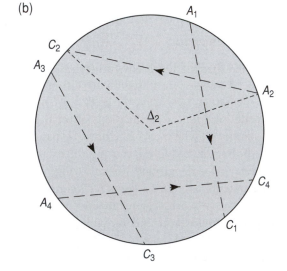

Figure 4.8 Same-length paths through the Earth.

(a)

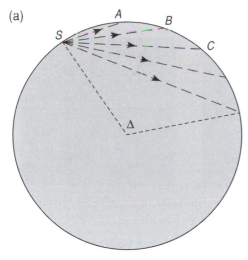

(b)

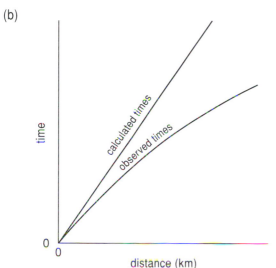

Figure 4.9 Travel-times for a uniform and the actual Earth.

4.3.2 Concentric layering

Having established that the Earth is symmetrical, we next ask, Is it perhaps uniform, like an onion with only one layer? If so, the seismic velocity would be the same at all depths, rays would be straight lines (Fig. 4.9a), and the time that rays take would be simply proportional to the distance they travel. This is easily tested by comparing times calculated for a uniform Earth with actual observed times.

The calculated times assume that all the Earth has a seismic velocity equal to the average of surface rocks, which we can measure. Figure 4.9b shows that as the source-to-receiver distance (straight through the Earth) is increased, the time observed is progressively *less*

than for a uniform Earth. So the Earth is not seismically uniform, and as longer rays have travelled deeper into the Earth the seismic velocity is faster at depth.

How much faster? This is not so easy to answer because when the seismic velocity is not uniform, ray paths are not straight. So we need to know how wave propagation and rays are affected when they encounter a change of velocity, the topic of the next section.

4.4 Finding the path of a ray through the Earth

4.4.1 Refraction: Snell's law

When wave fronts cross obliquely into a rock with a higher seismic velocity, they speed up, which causes them to slew around and change direction (Fig. 4.10a), just as a line of people, arms linked, would slew if each person went faster once they had crossed a line. This bending is called **refraction**.

How much is the change of direction? First we note that a long way from the source a small part of a wave front will be nearly a flat plane, so two successive wave fronts – or successive positions of the same wave front – will be parallel, and rays will be straight, for they are perpendicular to wave fronts.

The *time* between successive wave fronts, AB and $A'B'$ in Figure 4.10b, remains unchanged, so the wavelength must increase in the second rock, in proportion to the increase in velocity. From geometry,

$$\frac{v_1}{v_2} = \frac{\lambda_1}{\lambda_2} = \frac{BB'}{AA'} \qquad \text{Eq. 4.2}$$

By trigonometry,

$$\sin i_1 = \frac{BB'}{AB'} \quad \sin i_2 = \frac{AA'}{AB'} \qquad \text{Eq. 4.3}$$

Therefore

$$AB' = \frac{BB'}{\sin i_1} = \frac{AA'}{\sin i_2} \qquad \text{Eq. 4.4}$$

As BB' and AA' are in proportion to the velocities v_1 and v_2 (Eq. 4.2), this can be rearranged to give

$$\frac{\sin i_1}{v_1} = \frac{\sin i_2}{v_2} \qquad \text{Snell's Law} \qquad \text{Eq. 4.5}$$

Figure 4.14. Representing the Earth by many shells.

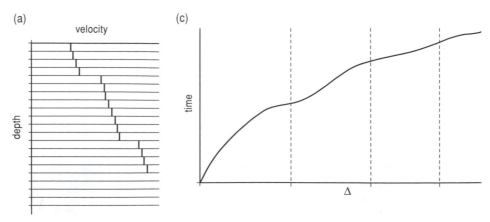

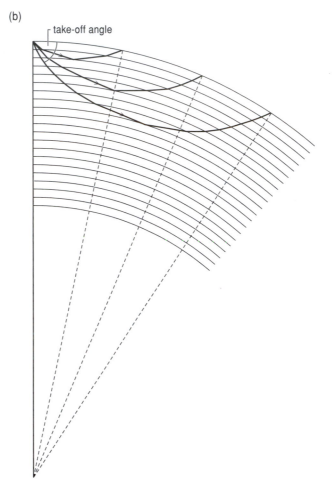

4.5 Seismic features of the Earth

4.5.1 Core and mantle

Figure 4.15a shows that seismic velocity generally increases with depth, except that about halfway to the centre there is an abrupt decrease. An abrupt increase or decrease is called a **velocity discontinuity,** and because these mark the boundary between bodies with different properties they are also called **interfaces.** This interface is a major feature, and *by definition* divides the **mantle** from the **core** (Fig. 4.15b). It is therefore called the **core–mantle boundary** (often abbreviated to **CMB**). To find out what it is, we use the fact that there are different types of seismic waves, with different velocities.

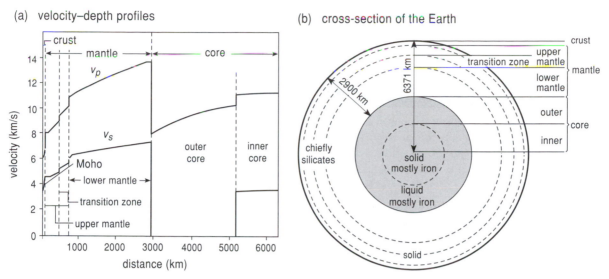

(a) velocity–depth profiles

(b) cross-section of the Earth

Figure 4.15 Velocity–depth curves for P- and S-waves, and the Earth's structure.

4.5.2 Longitudinal and transverse waves

Figure 4.1 showed waves along a spring and a rope. For the spring, the turns moved back and forth in the direction the wave travels; this is an example of a **longitudinal wave**. But for the rope, points oscillate at right angles to the direction of wave travel; this is a **transverse wave**. Seismic waves also exist in longitudinal and transverse forms; Figure 4.16 shows how an imagined column through rock is deformed as waves travel through it. If the column were marked into cubes, as shown to the right, the longitudinal wave would cause them to compress and then dilate as crests and troughs of the wave pass through, changing both their shape and size; but the transverse wave changes only their shape, making them lozenge shaped.

Longitudinal and transverse seismic waves are called **P-** and **S-waves** (for historical reasons); they may be remembered as **p**ressure or **p**ush–pull waves, and **s**hear or **s**hake waves. P-waves are essentially the same as sound waves, except that many of the frequencies recorded are too low to be heard by the human ear. Sound waves travel well through most rocks, usually better and more quickly than through air, which explains, for instance, why the Indians of North America used to put their ear to the ground to hear the approach of distant cavalry.

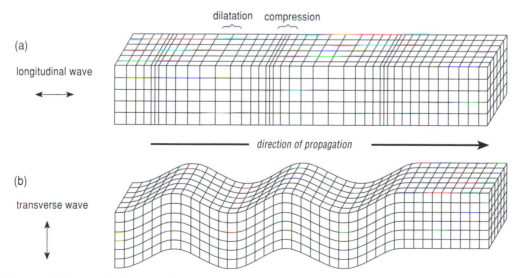

Figure 4.16 Longitudinal and transverse waves in rocks.

BOX 4.1 Elastic moduli and seismic velocities *(continued)*

(a) compression modulus

(b) shear modulus

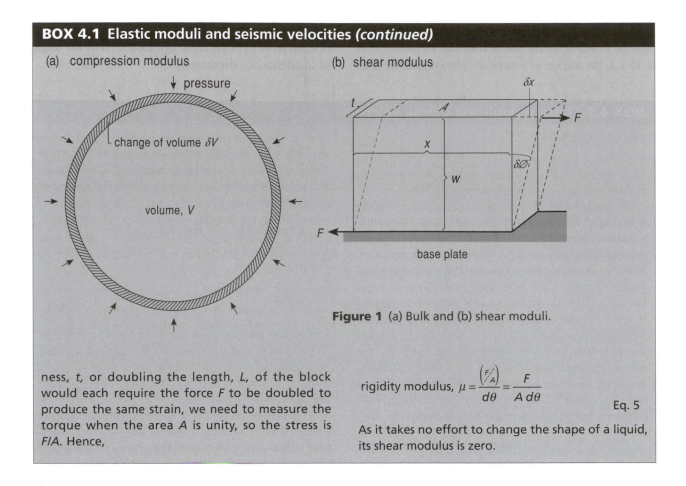

Figure 1 (a) Bulk and (b) shear moduli.

ness, *t*, or doubling the length, *L*, of the block would each require the force *F* to be doubled to produce the same strain, we need to measure the torque when the area *A* is unity, so the stress is *F/A*. Hence,

rigidity modulus, $\mu = \dfrac{\left(F/A\right)}{d\theta} = \dfrac{F}{A\,d\theta}$

Eq. 5

As it takes no effort to change the shape of a liquid, its shear modulus is zero.

4.5.3 The mantle–core difference

The relevance of P- and S-waves is that they reveal that the core–mantle boundary is a change from solid mantle to liquid core, for S-waves do not travel through the core, as will be explained further in Section 4.5.6. A change from solid mantle to liquid core also helps account for the value of v_p being less at the top of the core than at the base of the mantle (Fig. 4.15a), for v_p is reduced if μ is changed to zero (Eq. 2 of Box 4.1). However, the core being liquid does not rule out it being made of a different material as well; in fact, the next question is why a liquid should occur below a solid, deep within the Earth. From arguments involving meteorites and theories of formation of the Earth (see, e.g., Brown and Mussett, 1993) we believe the mantle is composed of crystalline silicates while the core is predominantly of molten iron, which settled to the centre of the Earth because of its high density; the iron has a lower melting temperature than the silicates, and the temperature at the core–mantle boundary is between the melting points of the two materials.

4.5.4 Other seismological features of the Earth

The shallowest significant feature on Figure 4.15a, just discernible at this scale, is a small jump at a depth of a few kilometres. This is the **Moho** (the generally used abbreviation for the discontinuity named after its discoverer Mohorovičíc), where the P-wave velocity increases abruptly from less than 7 km/sec to more than 7.6 km/sec. This *defines* the boundary between **crust** and mantle. The Moho exists over nearly the whole globe but varies in depth from 5 to 10 km (average 7) under the floors of the oceans to 70 km or more under the major mountain chains, with 40 km the average under continents. The term 'crust' suggests a solid layer over a liquid one, as in pie crust, but we know that crust and mantle are both solid, for S-waves travel through them.

About 100 km down is the **low-velocity zone** (**LVZ**), not a sharp change of velocity but a *decrease* over an interval before the general increase of velocity with depth resumes. Its lower boundary is chosen to be the depth where the velocity regains the value of the upper boundary (Fig. 4.19). The LVZ varies in thickness and in its velocity decrease, and it is not found at all under old continental cratons. Its cause is discussed in the next section.

The next important boundary is the 400-km discontinuity, where velocity increases abruptly. It is almost certainly due to a phase change, a reorganisation of the olivine and pyroxene – believed to be the dominant minerals at this depth – to more compact forms due to the increase of pressure with depth, just as graphite converts to diamond at sufficiently high pressures. Further down is the **660-km discontinuity,** whose nature is more controversial. Very probably it also is a phase change, to a yet more compact crystalline form, but there may be a small change of composition as well, also increasing the density. The region between 400 and 660 km is the **transition zone,** and it forms the lower part of the **upper mantle,** which is separated by the 660-km discontinuity from the **lower mantle.** There are probably other, smaller, discontinuities in the upper mantle.

The P-wave velocity increases smoothly with depth in the lower mantle, except for the bottommost 200 or so kilometres, where is it almost constant. This is called the D″ layer (after a former scheme for labelling the various layers of the Earth alphabetically). This may differ from the mantle above in having both higher temperatures and a somewhat different composition.

The core is divided into outer and inner parts by yet another discontinuity. The **inner core** is solid, probably mostly iron below its melting point at this depth. It exists within the liquid outer core, also predominantly of iron, mainly because the melting temperature of the **outer core** is lowered by the presence of other elements, such as sulphur.

4.5.5 Attenuation

The amplitude of seismic waves changes for two main reasons. One is that the wave front usually spreads out as it travels away from the source and, because the energy in it has to be shared over a greater area, the amplitude decreases. (Occasionally, energy is concentrated by reflection or refraction at interfaces and then the amplitude increases. If the waves come from a nearby earthquake, such concentration may increase the damage they do; see Section 5.10.1).

The second is when some of the wave energy is absorbed. This occurs if the rock is not fully elastic. A simple example is when waves enter unconsolidated sands: The sand grains move semi-independently, so the sand does not spring back to its original shape as a wave passes through. In loose sand the waves rapidly die away, but in partially consolidated sand their amplitude decreases progressively; this is called **attenuation.** We have seen that S-waves will not travel through liquids, though P-waves will; if a rock contains some liquid distributed through it S-waves will be noticeably attenuated, though the P-waves will be less affected. This is believed to be the reason for the low-velocity zone, for it is much more noticeable for S- than P-waves, and it occurs at a depth at which there is likely to be a few percent of partial melt; that is, between the crystalline grains of the rock is a few percent of liquid. Attenuation can therefore be used to map the presence of magma beneath volcanoes.

4.5.6 Ray paths in the Earth

The main ray paths through the Earth are summarised in Figure 4.20. A ray is named according to the parts of the Earth it travels through (the crust is ignored on this scale) and by whether it is a P- or an S-ray. A P-ray in the core is called K; thus a P-ray travelling successively through mantle, core, and again mantle is called PKP (sometimes abbreviated to P′). A reflection from the core is denoted by c, so a P-ray reflected back up from the core is PcP; if converted to an S-ray

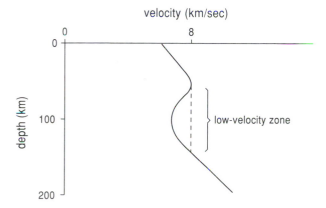

Figure 4.19 Low-velocity zone, LVZ.

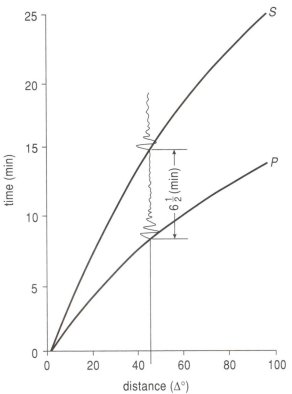

Figure 5.3 Distance of earthquake from difference of P and S arrival times.

The depth of the hypocentre can be found by measuring the difference in arrival of the P-ray – the direct ray – and the ray that reflects from the surface, pP (Fig. 5.5). If a series of earthquakes occurred at progressively greater depths below the same epicentre, the pP–P difference would increase. The differences have been tabled to give depths. (The P–S arrival-time differences are also affected, but there are tables that take this into account.)

5.3 Fault-plane solutions and stresses

5.3.1 Fault-plane solutions

As well as finding where and at what depth an earthquake occurred, it is also possible to make a **fault-plane solution,** which is deducing the orientation of the **fault plane** and the direction of the displacement in that plane. These are found from the directions of the first arrivals at a number of receivers encircling the epicentre. To understand how first motion at a seismometer depends on the orientation of the fault, first consider a peg driven firmly into the ground and struck by a hammer moving, say, northwards (Fig. 5.6a).

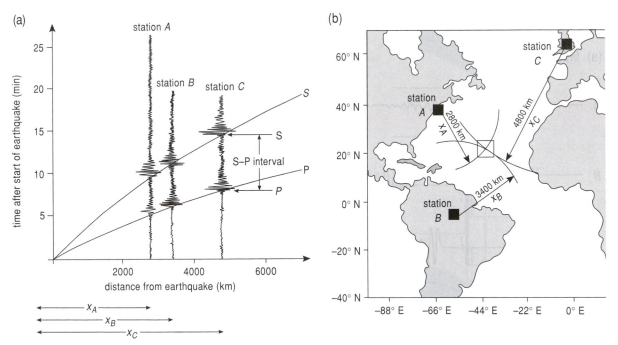

Figure 5.4 Location of an earthquake using P–S arrival-time differences.

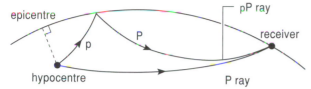

Figure 5.5 Ray paths from a deep earthquake.

Since P-wave oscillations are back and forth in the direction of propagation, there can be no P-waves to either the east or west, but they propagate with maximum amplitude to north and south. In other directions there will be P-waves with lesser amplitudes. Figure 5.6b shows, in plan view, relative amplitudes of the waves by the length of an arrow in the relevant direction; the tips of the arrows trace out two lobes. As the peg is struck towards the north, the *first arrival* or movement of the ground to the north of the peg is a **compression**, while to the south it is a rarefaction or **dilatation**, denoted respectively by + and − signs. S-waves, being transverse, are not propagated to north or south, but have progressively larger amplitudes as the direction approaches east or west, as summarised by Figure 5.6c; however, they are of little help in finding the fault plane.

We next extend these ideas to a fault displacement, which is more complex because the sides move in opposite directions. In the example of Figure 5.7a – a dextral strike-slip fault striking N–S – the amounts of *rebound* are shown by the half-arrows. On the eastern side of the fault the motion of the rebound is similar to that of the peg being struck towards the south, so that – in the eastern half – the P-wave first-motion pattern is the same as in the western half of Figure 5.6b. On the western side, with displacement in the opposite direction, the pattern is reversed.

What is recorded by receivers beyond the limits of the fault rupture? Exactly north and south, on the line of fault (Fig. 5.7a), the effects of the opposite displacements of the two sides cancel and there is no ground movement; nor is there any to east and west, for no waves are generated in these directions. In other directions, the effect of displacement of the further side of the fault is shielded by the fault, so the nearer side has the greater effect. Their combined effects are shown in Figure 5.7b, a *four*-lobed radiation pattern, with alternately compressive and dilatational first motions.

Because the radiation pattern is symmetrical about the N–S and E–W directions, it is impossible to tell from the radiation pattern alone which of these two directions is the fault plane and which a plane perpendicular to it, called the **auxiliary plane**; exactly the same patterns would result from an E–W sinistral fault (far side of fault moves to the left) as for the N–S dextral one (far side to the right) of the example. However, other considerations can usually be used to recognise which is the fault plane; these include (i) the fault plane may be revealed by a surface break; (ii) any aftershocks will be along the fault plane (Section 5.4); (iii) if several earthquakes in an area lie along a line, they are likely to be on a single fault plane rather than on several separate ones; (iv) the trend of faults in the area may be known from geological maps or photographs. Of course, receivers may not be so regularly distributed, and then it may only be possible to locate the fault and auxiliary planes approximately.

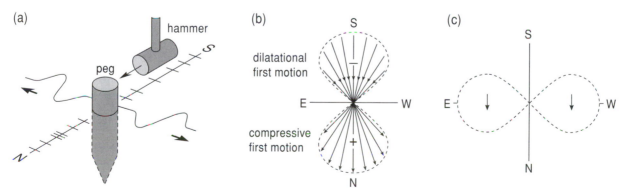

Figure 5.6 P- and S-waves around a struck peg.

BOX 5.1 Finding a fault-plane solution

Data needed

The epicentre and hypocentre of the earthquake (found as described in Section 5.2). Then for each receiver,

its bearing (azimuth) with respect to the epicentre,
its take-off angle (using a table that shows angles versus distance),
whether the sense of first motion of P arrivals is compressional or dilatational.

To plot the data

To illustrate how plotting is done, suppose the values of the above quantities for one receiver are azimuth 34°, take-off angle 46°, and dilatational.

(i) Place a piece of tracing paper over a pinpoint protruding through the centre of a Lambert (equal area) projection net (also known as a Schmidt net), and mark the position of north on it. Count clockwise round the net from the north an angle equal to the azimuth value (34°), and make a mark on the circumference. Rotate the tracing paper until this mark is at

north; count from the centre towards north an angle equal to the take-off angle (46°) as in Figure 1a.

(ii) Put a small circle at the position, solid for compressional first arrival, open for dilatational.

(iii) Repeat for each receiver.

Suppose the result is as in Figure 1b (some of the data are tabled at the end of the box).

Locating the fault and auxiliary planes

(iv) Rotate the tracing paper (Fig. 1c) to find an arc running from north to south of the stereo net (corresponding to a plane through the centre of the focal sphere) that separates compressional and dilatational points; the compressional and dilatational points will probably 'change sides' along the arc (as shown in Fig. 5.1d). There may be more than one arc that fits, depending on the distribution of points.

(v) Without moving the tracing paper, count 90° along the 'equator' of the stereo net from where the arc intersects it, passing through the centre of the net. Mark this point as P, for pole.

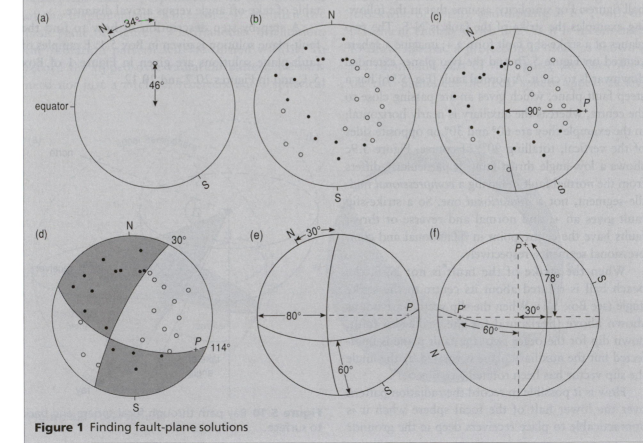

Figure 1 Finding fault-plane solutions

BOX 5.1 Finding a fault-plane solution *(continued)*

(vi) Rotate the tracing paper to find a second arc that separates compressional and dilatational points and *that also goes through P* (this ensures that its plane is perpendicular to the first plane). The two planes are shown in Figure 1d.

(vii) If you cannot find a second plane that separates the compressional and dilatational points, try with another arc that satisfies (iv).

If the points permit a wide choice of planes – because the points are few or poorly distributed – the planes are poorly defined.

Finding the values of azimuth and dip of the planes

(viii) When the two arcs have been found, position the tracing paper with north at the top. Read off the azimuth angles round the rim of the stereo net to the first end of each arc that you encounter (Fig. 1d) (the other end will be 180° further). These are the strike angles of the fault and auxiliary planes (you don't yet know which is which). In the example the azimuths are 30° and 114°.

(ix) Rotate the tracing paper until one of the arcs has its ends at north and south of the stereo net (Fig. 1e); read off the angle from its midpoint along the 'equator' to the edge of the net: this is the angle of dip of the plane (80°).

Repeat for the other arc – giving a dip of 60° in this example.

(x) Recognise whether the solution is for a predominantly strike-slip, normal, or reverse fault. This example is closest to a strike-slip fault but differs significantly from one.

(xii) If the solution is not approximately one of these three types of fault, you need to find the slip angle on the fault plane.

Guess that one arc is the fault plane; rotate the tracing paper until the ends of the *other arc* are at north and south (Fig. 1f); read off the angle from its midpoint to *P*, to give the slip angle (here 78°) on the guessed fault plane (it would be 90° for a pure strike-slip fault). Repeat for the other arc, after finding its pole P', as described above (it should lie on the first arc); it gives 60° for this example. In this example, the fault has either azimuth 30°, dip 80°, with a slip direction of 78° from the vertical, or 114°, 60°, and 60°.

(xiii) Determine which of the two planes is the fault plane, using other information as described in Section 5.3.1.

(*Alternatively*, the stress directions – which are unambiguous – may be deduced: mark on the two possible couples where the two planes intersect and combine them as explained in Section 5.3.2.)

Table 1 Selection of data from Figure 1b

Azimuth	Take-off angle	Sense*	Azimuth	Take-off angle	Sense*	Azimuth	Take-off angle	Sense*
14	41	C	134	74	C	218	62	D
24	49	D	144	51	C	234	65	D
49	45	D	177	54	C	266	54	D
76	23	D	178	37	C	278	49	C
134	54	D	212	39	C	348	50	C

*C: compressional; D: dilatational

5.3.2 The earthquake stress field and the double-couple mechanism

The directions of the stresses that cause earthquakes can also be deduced from the radiation pattern. The elastic rebound is not confined to just beside the fault rupture but extends out – in lessening amount – on either side. Squares drawn on the surface before the strain had begun to build up would have become lozenges (rhombuses) before the earthquake occurred (Fig. 5.11); these lozenges, of course, are just the tops of columns. To distort a square into a lozenge requires a shear couple (explained in Box 4.1); this could be produced, for example, by pulling on ropes attached to opposite faces (Fig. 5.12a), except that a single couple would rotate the rectangle (Fig. 5.12b).

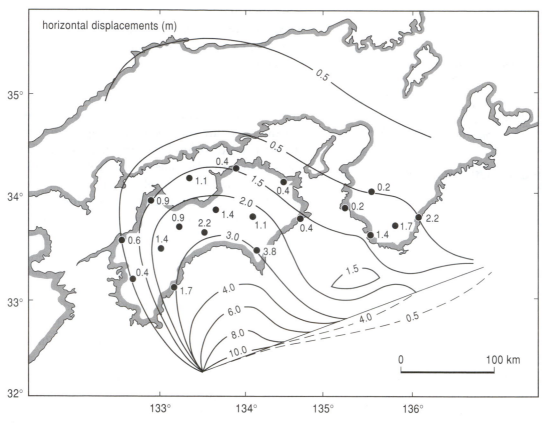

Figure 5.16 Horizontal displacements due to the Nankaido, Japan, earthquake of 1946.

Earthquakes have all sizes of rupture lengths up to hundreds of kilometres (see Section 5.6), widths up to tens of kilometres, and displacements up to tens of metres. The maximum strain varies far less, and has a value of roughly 10^{-4}, that is, the corner of a square in Figure 5.11 moves no more than about one ten-thousandth of the length of a side before there is failure. This is because rocks at depth do not vary greatly in their strength (surface rocks, such as alluvium, are often too weak to accumulate any significant strain, as mentioned above). Therefore, the energy of an earthquake depends mainly on the size of the strained volume – which can be millions of cubic kilometres – not on how much it is strained.

You may be wondering what is meant by the hypocentre of an earthquake, if a rupture can extend for hundreds of kilometres. When strain has built up to the point of rupture, there will be some point on the fault plane that is just a little weaker than the rest, and this is where rupture starts. This puts more strain on adjacent areas, which yield in turn, and failure rapidly spreads out until it reaches

parts of the fault plane where the strain is much less. As the rupture spreads more slowly than P- and S-waves, the first arrival at a distant receiver is of waves that originated at the initial point of rupture, and so that is where the hypocentre is located. Waves from other parts of the fault plane arrive a little later, and this is one reason why seismograms of earthquakes are complex.

5.5 Measures of earthquake size

5.5.1 Intensity: Severity of an earthquake at a locality

We need a way of measuring the size, or strength, of earthquakes. There are several ways, but first there is a fundamental distinction to grasp. Just as the term 'earthquake' may refer to either the effects at a locality or to the sudden displacement at the source, so the size can be a measure of the effects at a locality or of the disturbance at the source. We shall be concerned chiefly with the latter but consider the former first.

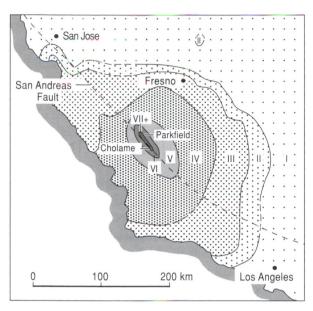

Figure 5.17 Isoseismals for Parkfield 1966 earthquake.

Intensity is a measure of the severity of an earthquake at a locality, judged by the effects produced, such as perceptible shaking, lamps swinging, destruction of masonry, and so on. The scale commonly used is the Modified Mercalli Scale of 1931, which has twelve categories, I to XII. For instance, intensity III includes, 'Felt indoors. Hanging objects swing. Vibration like passing of light lorries (trucks). Duration estimated. May not be recognised as an earthquake'. VI is 'Felt by all . . . many frightened and run outdoors . . . .'. Most alarming of all is XII: 'Damage nearly total. Large rock masses displaced. Line of sight and level distorted. Objects thrown in air' (the dedicated geophysicist continues to take observations!).

This may not sound as 'scientific' as, say, recording the local ground motion, but to do this would require having anticipated the earthquake and set up instruments. Instead, it allows the intensity at each locality to be estimated *after* the earthquake, by observing the damage and asking people what they felt and noticed (though it should be done as soon as possible, while memories are fresh and damage unrepaired). If enough data is collected the intensities can be contoured as **isoseismals** (Fig. 5.17). The zone of highest intensity indicates the position of the fault rupture, but only roughly, partly because data may be poor, but also because intensity is partly determined by the local geology; for instance, damage is less to buildings built on solid rock than on alluvium.

Measurement of intensity may not tell us much about the strength of the earthquake at its source, for a small but nearby earthquake may cause as much damage at a locality as a large but remote one. Besides, estimating intensity is time-consuming, not very precise, and limited to places where there are buildings to be disturbed and people to report their observations. A way of measuring the source strength of an earthquake is explained next.

5.5.2 Seismic moment: Size of the earthquake at source

The most commonly quoted measure of earthquake size is the Richter magnitude (explained in Section 5.8) but a later and better measure is the **seismic moment**, M_0. Just before a fault ruptures, the shear forces on either side of the fault (Fig. 5.18) exert a couple, whose size, or moment, equals the product of the shear forces and the perpendicular distance between them, that is, $2Fb$. The force F depends on the strain, the area of the rupture, A, and the rigidity modulus, μ (Box 4.1). The strain depends on the fault offset and the width of the strained volume, and so equals $d/2b$, where d is the average displacement; values are estimated as described in Section 5.4. These are combined as follows:

$$\text{moment of couple} = F \cdot 2b \qquad \text{Eq. 5.1}$$

As $F = \mu A \times$ strain, and strain $= \dfrac{d}{2b}$,

$$\text{moment of couple} = \mu A d = M_0 \qquad \text{Eq. 5.2}$$

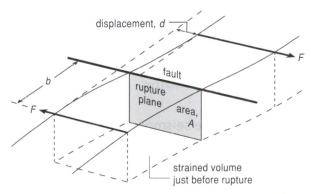

Figure 5.18 Seismic moment.

50 mm/year of the Pacific relative to the North American Plate (Section 20.5.2). On parts of it there occur large but infrequent earthquakes, notably those of 1857 (Fort Tejon) and 1906 (San Francisco), with ruptures extending for hundreds of kilometres along the fault. These parts of the fault experience negligible aseismic slippage, and also little seismicity between the large earthquakes, as illustrated by Figure 5.20; they are described as being 'locked'. Between these two ruptures is a length with mainly aseismic slip or creep, which has been constant over the 40 years of records, with a maximum movement of about 30 mm a year in the middle of the region; though there are frequent small earthquakes, they account for less than 10% of the slippage. Towards either end of this creeping section are transitional zones where significant earthquakes occur, such as the Parkfield earthquakes of moderate strength, which recur every few decades (one was the earthquake of Fig. 5.15).

The San Andreas Fault, though often described, is not typical. Most faults in continental crust and also those in oceanic crust (except subduction zones, Section 20.4.1) experience no more than a small proportion of aseismic slippage, so estimates using seismic moments will give good approximations of the total slippage.

5.7 Surface waves

In the previous chapter, P-waves and S-waves, which travel through the body of the Earth, were introduced. There are two other types of seismic waves (Fig. 5.21). Both are mainly confined to near the surface, and so are called **surface waves**, in contrast to P- and S-waves, which are termed **body waves**. Water waves are an example of a surface wave: A cork bobbing on the water reveals the amplitude of the waves, but a fish swimming down from the surface would find that the amplitude decreases rapidly (so one way to escape seasickness is to travel deep in a submarine). Similarly, the amplitude of seismic surface waves decreases rapidly with depth (Fig. 5.21) and this characterises surface waves. The amplitude of long waves decreases less rapidly than that of short ones, but in proportion to their wavelengths the decrease is the same; for instance, the amplitude of a 100-m wavelength wave would have decreased at a depth of 100 m by the same fraction as that of a 50-m wave at 50 m. P- and S-waves are called body waves because they can travel through the body of the Earth rather than only near the surface (their amplitudes do decrease, due to spreading out, reflections, and some absorption – as is also the case for surface waves – but not because of distance below the surface).

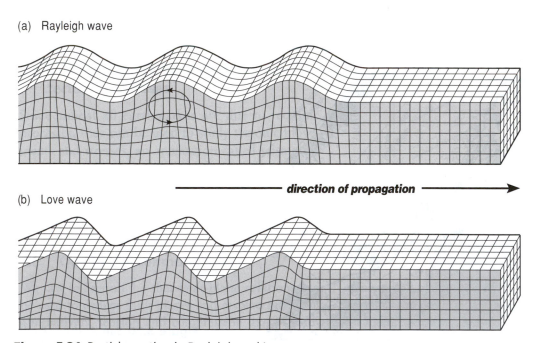

(a) Rayleigh wave

direction of propagation ⟶

(b) Love wave

Figure 5.21 Particle motion in Rayleigh and Love waves.

The two types of seismic surface waves are **Rayleigh waves** and – also named after their discoverer – **Love waves.** The main difference between them is that in Rayleigh waves particle motion is a vertical ellipse, whereas in Love waves it is horizontal and transverse (similar to horizontal S-waves) (Fig. 5.21). Therefore Love waves are not detected by a vertical component seismometer, whereas Rayleigh waves are detected by horizontal as well as vertical ones. In both cases, a horizontal component seismometer will only respond to the waves if it has its sensitive direction in the direction of particle motion. In Figure 5.22, the N–S instrument records by the far the largest Love wave amplitudes but the smallest Rayleigh amplitudes; the waves are therefore travelling approximately east or west.

Both types of wave are slower than P- and S-waves – as Figure 5.22 shows – and so are little used for measuring travel-times, but they are important in other branches of seismology. They are generated by most seismic sources and often have the largest amplitude, as Figure 5.22 illustrates; they are responsible for much earthquake damage (Section 5.10.1). They are also used to measure the size of earthquakes, as described in the following section.

5.8 Magnitude: Another measure of earthquake size

Seismic moment (Section 5.5.2) is the preferred measure of the size or strength of an earthquake at its source, but an earlier measure, **magnitude,** is still widely quoted. Magnitude was the first measure of the source strength of an earthquake, and the first magnitude scale was devised by Richter, an eminent seismologist, in 1935, and – with modifications (see Box 5.2) – is the one usually quoted by the media, as the **Richter magnitude.** Richter measured the amplitude in microns (millionths of a metre) of the largest oscillations – the surface waves – recorded by a particular type of seismometer, 100 km from the earthquake source. As the values have a very large

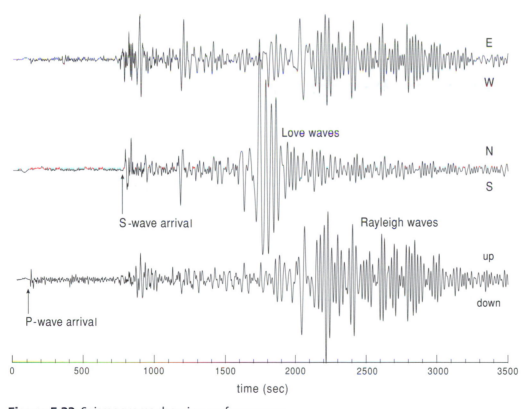

Figure 5.22 Seismograms showing surface waves.

5.9 Energies of earthquakes

Seismic waves carry energy, as is obvious from the damage they can cause. Waves are momentary deformations of rock, and if the energies of all such simultaneous deformations over the whole volume affected by the earthquake were added together, we would have the seismic energy of the earthquake. (This is less than the strain energy released by a fault rupture, for reasons given in Section 5.1.) This is a difficult sum to carry out but it has been estimated and related to the magnitude of the earthquake; the approximate relationship is shown in Figure 5.24.

An increase of 1 in magnitude means 10 times the amplitude but about 30 times the energy. The energy of some earthquakes far exceeds that of the largest nuclear bomb exploded, which is why great earthquakes are so destructive. The 1960 Chile earthquake released the equivalent of about 2500 megatonnes of high explosive.

Numbers of earthquakes and their total energy. The largest earthquakes, fortunately, occur less frequently than smaller ones; for instance, each year there are about 100,000 earthquakes of $M_L = 3$, compared to 20 of $M_L = 7$. For any region, or the Earth as a whole, a plot of numbers N of earthquakes greater than a given magnitude M (as a log) versus the magnitude gives an approximate straight line ($\log_{10} N = a - bM$), though the slopes (b) are not usually the same, as shown in Figure 5.25 (only earthquakes above a certain magnitude have been plotted, but smaller ones are even more numerous). There are many more small earthquakes than large ones, but as they have little energy the relatively few large earthquakes account for most of the seismic energy released. (Similarly for seismic moment; the 1960 Chile earthquake released a quarter of the total moment in the period 1904–1986.) The average annual energy release worldwide equals that of about 100 megatonnes of high explosive! But large as this is, it is small compared to some other forms of global energy, as explained in Chapter 17.4.1.

5.10 Earthquake damage and its mitigation

5.10.1 Causes of damage

It is hardly surprising that the huge energies released by large earthquakes produce enormous damage, but how is the damage actually caused? There are several mechanisms, both direct and indirect.

(a) $\log_{10}$ scale

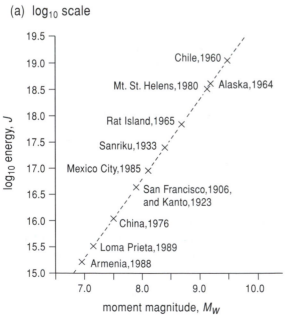

(b) linear scale

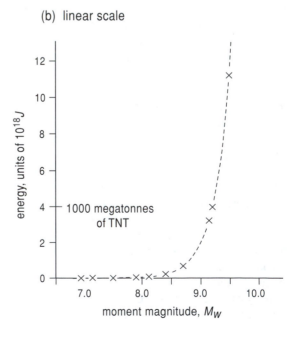

Figure 5.24 Energy and magnitude.

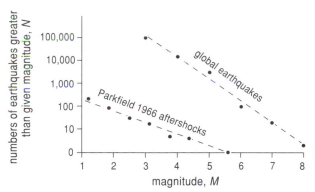

Figure 5.25 Numbers of earthquakes of different magnitudes.

Most obvious is the rupture, if it reaches the surface, for any house, road, pipe, and so on straddling the fault is likely to be ruptured. However, such damage is confined to the fault trace.

Much more extensive, and the chief cause of damage, is the shaking caused by the passage of seismic waves, particularly surface waves, for often these have the largest amplitude and continue the longest (Fig. 5.22). Aftershocks may cause extensive damage, for though their magnitudes are less than that of the main shock they may occur nearer to towns. Few traditional buildings are designed to be shaken, least of all from side to side, and many collapse; particularly at risk are masonry and adobe buildings, which lack cross bracing. More resistant are buildings with wood or steel frames, though buildings in earthquake-prone areas need to be specially designed to survive severe shaking. Some countries are zoned according to seismic risk and have building codes to match.

The amplitude of shaking depends not only upon the magnitude of the earthquake and its distance, but also on the local geology. When waves slow on entering rocks with a lesser seismic velocity, their wave trains shorten and therefore their amplitude increases; this commonly happens as waves approach the surface, for surface rocks often have lower velocities than deeper ones. In addition, unconsolidated rocks such as alluvium may be temporarily 'liquefied' by the shaking (just as sand and mud when rapidly shaken tend to flow like a thick liquid), and soil liquifaction has caused buildings to

topple or collapse. Another effect is that subsurface interfaces can reflect or refract the waves so that the seismic energy is partly focused into the area or, alternatively, deflected away.

Other mechanisms of destruction are indirect. The shaking of the ground may initiate an avalanche or landslip. In turn, these may dam a river valley, forming a lake; once the lake overflows the dam, it often cuts rapidly down through the unconsolidated jumble of rocks and soil, releasing the pent-up water to cause severe flooding downstream. The collapse of buildings containing fires and the rupturing of gas pipes may lead to widespread conflagrations, which, with the general chaos and damage to water pipes, may rage unchecked. Fire was a major cause of damage in the 1906 San Francisco earthquake and in the Kobe earthquake in Japan in 1995. People have even been known to set fire to their house on discovering that their insurance covers fires but not earthquakes!

Yet another form of damage may be suffered along coasts, from tsunamis (sometimes incorrectly called tidal waves – they have nothing to do with tides). These are generated when a mass of water is abruptly displaced, usually by movement of the sea floor below shallow seas, often by an earthquake, though also by volcanic eruptions and landslips. The waves travel across the deep oceans with a wavelength of perhaps 100 km but an amplitude of no more than a metre, so that a ship does not notice their passing, being slowly raised and lowered over a period of several minutes. But when the waves reach the shallow water near a coast they 'break', just as normal-sized waves do when they cause surf. Tsunamis may reach a height of 30 m or more, sweeping far inland, sometimes carrying ships to improbable places and devastating towns on their way. Around the Pacific, tsunamis are a sufficiently frequent hazard that the Seismic Sea Wave Warning System has been set up. When a large earthquake has been detected seismically, changes of sea level are checked locally to see if a tsunami has been generated; if it has, warnings are sent by radio to vulnerable shores. Though the waves travel at perhaps 800 km/hr, they still take several hours to cross an ocean, sufficient time for warning. (But the warning does not always save lives, for the curious may go to view the freak wave and be overwhelmed!)

of Love waves in the E–W direction. What is the approximate direction of the earthquake from the station?

9. Which of the following ways of measuring the size of an earthquake does not need an instrumental record:
   (i) Richter magnitude.
   (ii) M_W.
   (iii) Moment.
   (iv) Intensity.

13. When a certain locality experienced an earthquake with a Richter magnitude of 7 there was little damage. A newspaper reported that as this was only a little less than the largest known in the area, magnitude 8, citizens could be confident that recent work to make buildings, bridges, and so on safe had been successful. Why should they not be too confident?

10. Plot the data in the table, find the fault-plane solutions and deduce what type of fault was involved (strike-slip, etc.).

Azimuth	Take-off angle	Sense*	Azimuth	Take-off angle	Sense*	Azimuth	Take-off angle	Sense*
0	18	D	123	28	D	247	55	C
34	27	C	153	46	D	270	19	C
45	19	C	202	15	D	311	16	C
58	29	C	230	31	D	329	46	D
94	17	C	238	29	D	356	39	D

*C: compressional; D: dilatational

11. Plot the data in the table, find the fault-plane solutions and deduce what type of fault was involved (strike-slip, etc.).

Azimuth	Take-off angle	Sense*	Azimuth	Take-off angle	Sense*	Azimuth	Take-off angle	Sense*
46	74	C	138	67	D	259	64	D
64	38	C	140	37	D	317	58	D
89	43	D	192	54	D	318	78	C
104	68	C	198	64	C	334	51	C
128	76	C	243	71	C	351	22	D

*C: compressional; D: dilatational

12. Plot the data in the table and find the dip and strike of the possible fault planes and the slip angles.

Azimuth	Take-off angle	Sense*	Azimuth	Take-off angle	Sense*	Azimuth	Take-off angle	Sense*
9	70	D	128	85	D	276	32	C
26	69	D	150	48	D	276	68	C
41	21	C	206	63	D	298	50	D
62	66	C	226	37	D	310	81	D
86	45	C	237	22	C	335	82	D
			244	72	D	349	41	C
112	28	C	255	82	C			

*C: compressional; D: dilatational

Chapter 6

Refraction Seismology

Refraction seismology is a powerful and relatively cheap method for finding the depths to approximately horizontal seismic interfaces on all scales from site investigations to continental studies. It also yields the seismic velocities of the rocks between the interfaces, a useful diagnostic tool.

It exploits a particular case of refraction in which the refracted ray runs along an interface, while sending rays back up to the surface.

6.1 Critical refraction and head waves

A ray crossing an interface downwards into a layer with a higher velocity – a common situation – is refracted away from the normal, according to Snell's Law (Eq. 4.5). Consider rays progressively more oblique to the interface (Fig. 6.1): There must come an angle, the **critical angle**, i_c, where the refracted angle is exactly 90°; that is, the refracted ray travels along the interface (if the ray is yet more oblique, *all* the seismic energy is *reflected*, called total internal reflection). The value of i_c depends on the ratio of velocities either side of the interface:

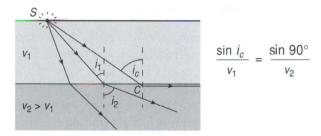

Figure 6.1 Critical angle of refraction.

$$\sin i_c = \frac{v_1}{v_2} \sin 90° = \frac{v_1}{v_2}$$

Eq. 6.1

What does a ray travelling along the interface mean: Is it in layer 1 or 2? This is best answered by considering wave fronts. As rays and wave fronts always intersect at right angles, a ray along the interface means waves perpendicular to the interface (Fig. 6.2a). A wave is due to oscillations of the rock, and as the layers are in contact, the rocks just above the interface must oscillate too. Therefore, as a wave front travels along just below the interface a disturbance matches it just above the interface. This in turn propagates waves – and therefore rays – up to the surface at the critical angle (Fig. 6.2b).

To understand why this is so, we use the concept of **Huygens's wavelets.**

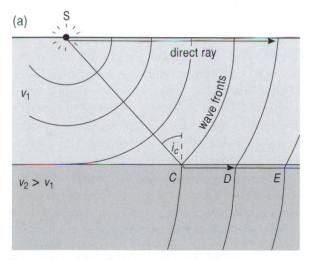

(a)

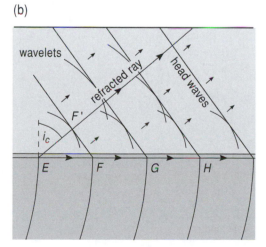

(b)

Figure 6.2 Wave fronts and head waves.

6.1.1 Huygens's wavelets

Huygens (a 17th-century scientist) realised that when a particle of a material oscillates it can be thought of as a tiny source of waves in its own right. Thus, every point on a single wavefront acts as a small source, generating waves, or wavelets as they are called. In Figure 6.3 the left-hand line represents a planar wave front travelling to the right at some instant. Wavelets are generated from all points along its length, and a few are shown. The solid half-circles are the crests of the wavelets after they have travelled for one wavelength (which is the same as that of the wave). At C_1, C_2, C_3, . . . these overlapping crests add, called reinforcement, so that they form a new crest. The succeeding troughs will have advanced half as far, shown by the dashed half-circles. At each of T_1, T_2, . . . there are two troughs and these will add, giving a trough between the two crests. But at Z_1, Z_2, . . . there is a crest and a trough, and these cancel perfectly. As there will be just as many troughs as crests along the left-hand line – assuming it is indefinitely long – then there will be perfect cancellation all along the line. Therefore, the crest has advanced from left to right. This is just what we have been using since waves were introduced in Sec-

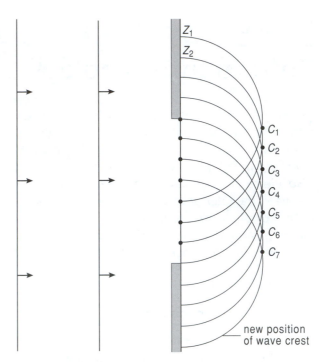

Figure 6.4 Diffraction into a shadow.

tion 4.1, which is why wavelets have not been introduced before. However, wavelets are helpful in some special cases, such as critical refraction and diffraction, the latter being considered first.

Diffraction. If waves pass through a gap, such as water waves entering a harbour mouth, it might be expected that once inside they would continue with the width of the gap, with a sharp-edged shadow to either side. But if this were so, at the crest ends there would be stationary particles next to ones moving with the full amplitude of the wave, which is not feasible. Considering wavelets originating in the gap shows that there is no such abrupt cutoff (Fig. 6.4). At C_1, C_2, . . . the crests will add as before, so the wave will continue as expected. But at Z_1 there is no trough to cancel the crest and so there is a crest there, though weaker because it is due to only one wavelet. The resulting wave crest follows the envelope of the curves.

This bending of waves into places that would be a shadow according to ray theory is called **diffraction**. It helps explain how we can hear around corners (there is often reflection as well). It will be met again in Section 6.10.1, and in some later chapters. Next, we return to the critically refracted wave of Figure 6.2.

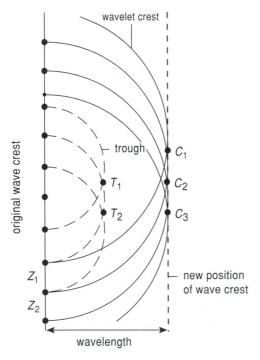

Figure 6.3 Huygens's wavelets.

6.1.2 Head waves

As a wave travels below the interface it generates wavelets (Fig. 6.2b). By the time a wavelet in the lower layer has travelled from *E* to *F*, a wavelet in the upper layer – travelling at velocity v_1 – has gone only a distance *EF′*. If we draw the wavelets produced by successive waves below the interface, at *E*, *F*, . . . , and add up their effects, the result is waves travelling up to the right (Fig. 6.2b). These are called **head waves**, and their angle i_{head} (which equals ∠*EFF′*) depends on the ratio of *EF′* to *EF*:

$$\sin i_{head} = \frac{EE'}{EF} = \frac{v_1}{v_2} = \sin i_c \qquad \text{Eq. 6.2}$$

This is the ratio of the velocities, and so is the value of the critical angle (Eq. 6.1): Therefore, head rays leave the interface at the critical angle.

The seismic refraction method depends upon timing the arrivals of head waves at receivers on the surface; the corresponding rays are usually called **refracted rays**.

6.2 The time-distance (*t–x*) diagram

Seismic energy can follow three main routes from the source to receivers (Fig. 6.5b): the refracted rays just described, the **direct ray** that travels just below the surface (shown in Fig. 6.2a), and **reflected rays**. They take different times to reach the receivers, as shown in the **time–distance**, or **t–x**, diagram (Fig. 6.5b).

The time it takes a direct ray to reach a receiver is simply the distance along the surface divided by the velocity, v_1, so its graph is a straight line from the origin. The slope of the line equals $1/v_1$, so the greater v_1 the shallower the slope. To find the travel-time to a receiver, follow straight up from the receiver on the lower diagram until you reach the line on the *t–x* diagram, then go left to read off the time on the vertical time axis.

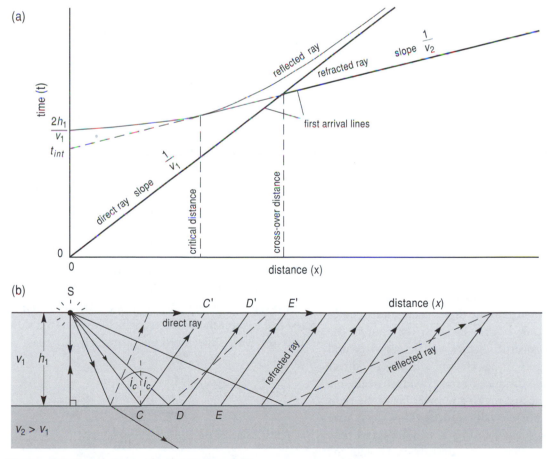

Figure 6.5 Travel-times of refracted and other rays.

Finding the time of a refracted ray is more complicated. Clearly, a refracted ray travels further than the direct ray, but part of its route is in the second layer, where it travels faster. If it travels far enough in the second layer it must overtake the direct ray (just as travelling to a distant destination via a motorway is quicker than using the shorter but slow road, even though you have to travel to and from the motorway on slow roads). All the refracted rays follow the critical ray SC down to the interface, and an equal distance up from the interface to the surface, CC', DD', EE', and so on, both at velocity v_1; to these must be added the connecting part in the second layer, CDE. . . . Therefore their total times will differ only by the different times they spend in the lower layer. For instance, ray $SCEE'$ takes longer than $SCDD'$ by time DE/v_2. As a consequence, the travel-time line for refracted rays is a straight line, with slope $1/v_2$; but it does not start at the origin, because the nearest point to the source it can reach is C', the **critical distance**. Nor does the travel-time line, extended back, reach the origin, because of the extra time taken travelling to and up from the interface.

The time for a refracted ray to reach a receiver is the time spent below the interface, plus the time it takes to go down to, and back up from, the interface. It works out to be

$$t = \frac{1}{v_2} \cdot x + 2h_1\sqrt{\frac{1}{v_1^2} - \frac{1}{v_2^2}}$$

time = slope · distance + intercept, t_{int} Eq. 6.3a

Equation 6.3a can also be written as

$$t = \frac{x \sin i_c}{v_1} + \frac{2h_1 \cos i_c}{v_1}$$
    Eq. 6.3b

This will be used later.

The third route to a receiver is by reflection. This obviously is further than the direct route and so takes longer. For a receiver very close to the source, the reflected ray goes vertically down to the interface and up again, taking $2h_1/v_1$, as shown, whereas the direct ray – having little distance to travel – takes negligible time. But the reflected ray to a receiver far to the right travels almost horizontally and so takes very little longer than the direct ray; this is why its travel-time curve approaches that for the direct ray. Though the reflected route is shorter than the refracted one it

always takes more time, except for the ray SCC', which can be regarded as either a reflected or a refracted ray; therefore the refracted and reflected times to the critical distance, C', are the same. Reflected rays are never first arrivals and are not considered further in this chapter. However, they form the basis of a most important prospecting method, reflection seismology, described in the next chapter.

Figure 6.5a describes what happens below ground, but we have to make deductions from surface measurements, in particular the times of first arrivals, as these are the easiest to recognise on a seismogram. The first arrivals for near distances are direct rays, but beyond a certain distance, the **crossover distance**, refracted rays are the fastest. Therefore, first arrival times for a range of distances trace out the two heavy lines of Figure 6.5b.

To get this information, a row of receivers – seismometers or geophones – is laid out in a line from the source, called the **shot point**. The resulting seismograms are shown in Figure 6.6a. The times of first arrival are 'picked' and their values are plotted against the distances of the corresponding receivers (receivers are often spaced more closely near the shot point, for reasons explained in Section 6.7). Then lines are drawn, for the direct and refracted travel-times (Fig. 6.6b).

These lines are used to deduce the depth to the interface and the velocities of both layers. First, the slopes of the direct and refracted lines are measured (this is easier if a convenient distance is used, as shown); their reciprocals yield the velocities of the top and lower layers:

Slope of direct line = 0.42 sec/250 m = 0.00168, so v_1 = 1/0.00168 = 595 m/s.
Slope of refracted line = 0.34 sec/500 m = 0.00068, so v_2 = 1/0.00068 = 1470 m/s.

These velocities can be useful for suggesting likely rock types (Table 6.1), but usually more useful is the depth to the interface, deduced as follows: Extend the refracted line to meet the time axis, which here has the value 0.28 sec. This is the intercept, t_{int}, given by Eq. 6.3:

$$\text{intercept, } t_{int} = 2h_1\sqrt{\frac{1}{v_1^2} - \frac{1}{v_2^2}};$$

$$0.28 \text{ sec} = 2h_1\sqrt{\frac{1}{595^2} - \frac{1}{1470^2}},$$

$$\text{so } h_1 = 91\,m$$

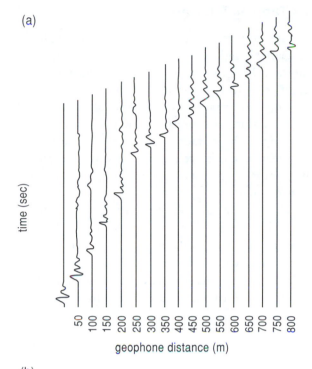

(a)

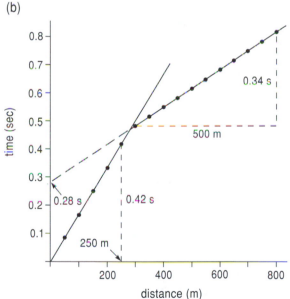

(b)

Figure 6.6 Making a time–distance diagram.

Table 6.1 Seismic velocities for rocks

Rock type	v_p (km/sec)
Unconsolidated sediments	
clay	1.0–2.5
sand, dry	0.2–1.0
sand, saturated	1.5–2.0
Sedimentary rocks	
anhydrite	6.0
chalk	2.1–4.5
coal	1.7–3.4
dolomite	4.0–7.0
limestone	3.9–6.2
shale	2.0–5.5
salt	4.6
sandstone	2.0–5.0
Igneous and metamorphic rocks	
basalt	5.3–6.5
granite	4.7–6.0
gabbro	6.5–7.0
slate	3.5–4.4
ultramafic rocks	7.5–8.5
Other	
air	0.3
natural gas	0.43
ice	3.4
water	1.4–1.5
oil	1.3–1.4

Ranges of velocities, which are from a variety of sources, are approximate.

If you have understood the above you have grasped the basis of refraction seismology. Obviously, not all situations are so simple: There may be several interfaces, and interfaces may be tilted or even undulating. All of these situations can be investigated using refraction seismology, as will be explained in the following sections.

6.3 Multiple layers

If there are several layers, each with a higher velocity than the one above – which is often the case – there will be critical rays for each interface, as shown in Figure 6.7. For each interface the critical angle depends only upon the velocities above and below it, but the ray paths down to an interface, SC_1, SC_2, SC_3, . . . , depend upon the thicknesses and velocities of all the layers above.

The t–x diagram for first arrivals is a series of straight lines, each with a less steep slope than the one to the left. As with a single interface, their slopes are the reciprocals of the velocities of the successive layers, so the velocities of the layers are easily found. In turn, these are used to calculate the critical angle for each interface, using Eq. 6.1. Depths are calculated from the intercept times but have to be done progressively. The depth, h_1, to the first interface, using Eq. 6.3b with $x = 0$, is

$$t_{\text{int}_1} = \frac{2h_1 \cos i_{c_1}}{v_1}$$

giving:

$$h_1 = \frac{v_1 t_{\text{int}_1}}{2 \cos i_{c_1}}$$

Eq. 6.4

Its value is put in the equation for the intercept time for the second interface, which can be shown to be

$$t_{\text{int}_2} = \frac{2h_1 \cos i_{c_1}}{v_1} + \frac{2h_2 \cos i_{c_2}}{v_2}$$

Eq. 6.5

which gives h_2. This is repeated for successive interfaces, the equation for the interface time increasing by one term each time.

6.4 Dipping interfaces

Tilting the interface does not change the value of the critical angle, but it rotates the ray diagram of Figure 6.3 by the angle of dip, α (Fig. 6.8). As a result, rays to successive receivers C', D', E', ... not only have to travel the additional distance CD, DE, ... along the interface, but also an extra distance up to the surface because the interface is getting deeper. Therefore the refracted line on the t–x diagram is steeper, its slope yielding a velocity less (slower) than v_2.

This presents a problem: There is no way to tell from Figure 6.8 that the refracted line is due to a dipping interface and not a horizontal interface over a layer with a slower seismic velocity equal to 1/slope. For this reason, velocities calculated from the slopes are called **apparent velocities**. However,

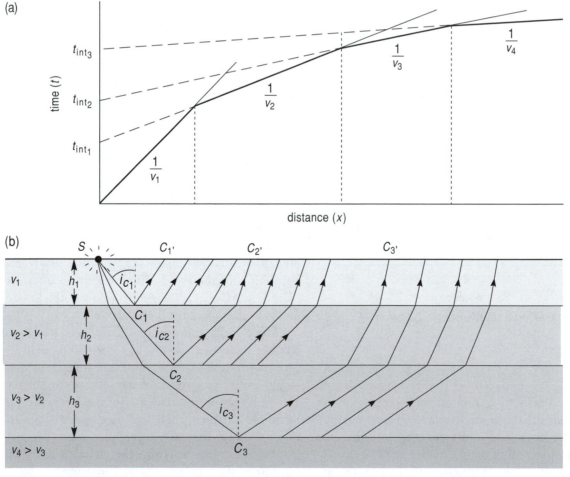

Figure 6.7 Multiple layers.

(a)

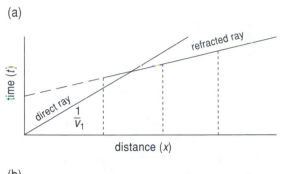

(b)

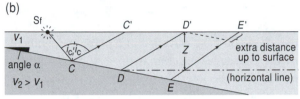

Figure 6.8 Dipping interface.

if the line is **reversed**, that is, repeated with the shot point at the right (Fig. 6.9), then successive head rays have *less* distance to travel up to the surface, and so the refracted line has a shallower slope, and appears faster. Therefore, *the presence of a dipping interface is shown by the **forward** and **reversed** refraction lines having different slopes* ('forward' and 'reverse' merely depend on which direction is shot first). If the interface has a small dip (less than about 5°), an approximate value of v_2 is found by averaging the forward and reverse slopes:

$$\frac{1}{v_2} \approx \frac{1}{2}\left(\text{slope}_{2_f} + \text{slope}_{2_r}\right)$$

Eq. 6.6

v_2 can then be used to find the critical angle, as usual (Eq. 6.1).

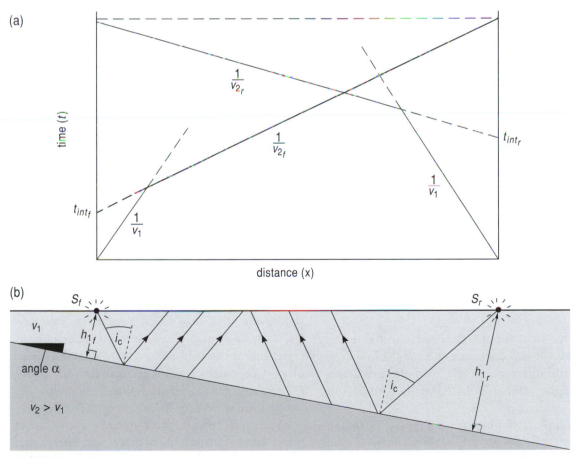

(a)

(b)

Figure 6.9 Reversed lines.

(a)

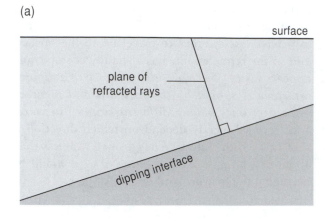

(b)

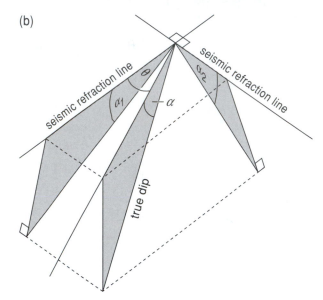

Figure 6.10 Dip in three dimensions.

The intercepts are different too, being less at the up-dip than the down-dip end. The depths to the interface are given by

$$h_{1_f} = \frac{t_{\text{int}_f} v_1}{2 \cos i_c} \qquad h_{1_r} = \frac{t_{\text{int}_r} v_1}{2 \cos i_c} \qquad \text{Eq. 6.7}$$

These are the distances perpendicular to the interface, not the vertical distances, but for shallow dips they are little different. (Exact formulas are given in some of the textbooks listed at the end of the chapter.)

In summary, t–x plots for a dipping interface compare with those for a horizontal interface as follows:

(i) The slope of the refraction line is less steep up dip, steeper down dip.
(ii) The intercept is less at the up-dip than the down-dip end.
(iii) The slopes for direct rays (first sections on t–x plot) are unchanged.

The true dip. It has been assumed so far that the seismic line is in the direction of the dip of the interface, but there is no reason why this should be so, unless it was known when the survey was set out. If the reversed lines are shot along strike, the rays recorded will have travelled in a plane perpendicular to the interface rather than in a vertical one (Fig. 6.10a), but the interface will appear to be horizontal. More generally, lines will be shot obliquely to the dip direction and give a value for the dip that is too low.

To find the true dip of the interface, two pairs of reversed lines at right angles can be used. Each is used to deduce a dip, α_1 and α_2 (Fig. 6.10b), from which the true dip, α, and its direction can be calculated from

$$\sin \alpha = \sqrt{\sin^2 \alpha_1 + \sin^2 \alpha_2} \qquad \text{Eq. 6.8a}$$

$$\cos \theta = \frac{\sin \alpha_1}{\sin \alpha} \qquad \text{Eq. 6.8b}$$

where θ is the angle between the dip direction and the seismic line that gave the dip component α_1.

6.5 Seismic velocities in rocks

The velocities found from travel-time diagrams give some indication of the types of rocks that form the layers, though rock types have a range of velocities. Velocities of some common rock types are given in Table 6.1.

In general, velocity increases with consolidation, so alluvium and loose sands have a very low velocity, cemented sandstones have a higher one, and crystalline rocks tend to have the highest velocities of all. Consolidation tends to increase with geological age, so a Palaeozoic sandstone, for example, may have a seismic velocity twice that of a Tertiary one.

Most rock types have a range of velocities, sometimes large, but in a particular area the range is often much less. Therefore, rock types can be identified more confidently if the velocities are measured in the area of interest – preferably in boreholes, for velocities measured on exposed rocks usually give a lower and more variable value because of weathering, opening of fractures under the reduced pressure, and pores not being fully saturated with water.

6.6 Hidden layers

There are two situations where a seismic interface is not revealed by a t–x refraction plot.

6.6.1 Hidden layer proper

It was explained in Section 6.2 that a refracted ray overtakes the direct ray, provided it travels a sufficient distance, but before that can happen it may be overtaken in turn by the refracted ray from the interface below. Figure 6.11a shows a 3-layer case where the second layer is revealed only by the short length of the line 'refracted ray 1' that is a first arrival. Suppose layer 2 were thinner: Then head rays from layer 3 would arrive earlier, displacing line 'refracted ray 2' downwards to the left, and no part of line 'refracted ray 1' would be a first arrival (Fig. 6.11b). Layer 2 is then **'hidden'**. Layer 2 would also be hidden if v_2 were decreased, or v_3 increased, for the first would displace crossover 1 to the right, while the second would displace crossover 2 to the

left. If the second layer is hidden, the t–x plot will be interpreted as two layers, one with velocity v_1 over a layer with velocity v_3. The depth to the top of the third layer, calculated in ignorance of the existence of layer 2, will be intermediate between the true depths to the tops of the second and third layers.

6.6.2 Low-velocity layer

If a layer has a *lower* velocity than the one above there can be no critical refraction, the rays being refracted *towards* the normal (Fig. 6.12). There is no refracted segment corresponding to the layer, so the t–x plot will be interpreted as a two-layer case. The calculated depth to the top of layer 3 will be exaggerated, because the slower velocity of layer 2 means that a ray takes longer to reach it, than if there really are two layers, v_1 over v_3.

The possibility of a hidden layer of either kind can be recognised only from independent information, particularly geological sections or borehole logs. Common low-velocity situations are sand below clay, sometimes sandstones below limestones, and most sedimentary rocks beneath a lava or sill (see velocities in Table 6.1). A hidden layer proper may be inferred, for instance, by comparing the seismic models with geological sections. Depending upon the information available, some correction may be possible.

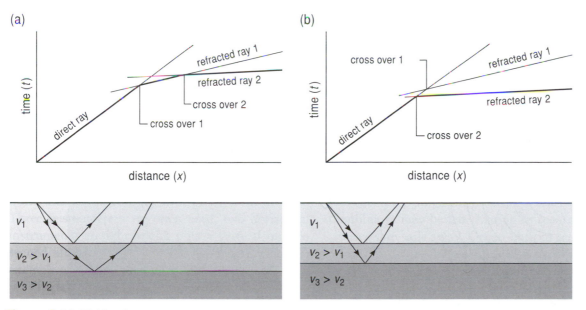

Figure 6.11 Hidden layer.

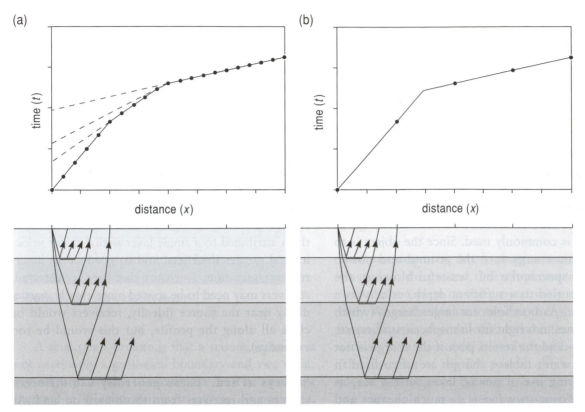

Figure 6.14 Effect of slow surface layers.

Having described how refraction surveys are carried out, we return to other layer geometries.

6.8 Undulating interfaces and delay times

If interfaces are not flat, a more sophisticated method of analysis is needed. Figure 6.15 compares the actual, undulating interface with a flat reference interface joining C_f to C_r. The forward and reverse refraction lines for this reference interface are shown dashed in the t–x diagram. As M, for instance, is closer to the surface than the reference interface, the actual travel time to M' plots below the reference line; conversely, that for N' is above it. The same argument applies to the reversed line, so the shapes of the refraction lines on the t–x plot show qualitatively how the interface differs from a flat surface. Note that the vertical separation of the forward and reversed refraction lines is about the same for the actual and reference interfaces.

These observations can be made more precise using the concept of delay times.

6.8.1 Delay times

The travel-time of a refracted ray is made up of three parts: travelling obliquely down to the interface, along the interface, and up to the receiver (Fig. 6.16a). But we can think of it as being made up in a different way (Fig. 6.16b): the time it takes to travel the distance between source and receiver, $S_v R_v$, just below the interface at velocity v_2, plus a term δ_S at the source end to equal the *extra* time it takes to go SC at velocity v_1 compared to going $S_v C$ at v_2, and similarly δ_R at the receiver end. In effect, we are pretending that the ray travels all the way at v_2, but there is a delay between the shot time and when the ray starts on its way, and another after it finishes before the signal is recorded by the receiver:

$$t_{SR} = \delta_S + \frac{SR}{v_2} + \delta_R \qquad \text{Eq. 6.9}$$

δ_S and δ_R are called **delay times** (or sometimes time terms).

(a)

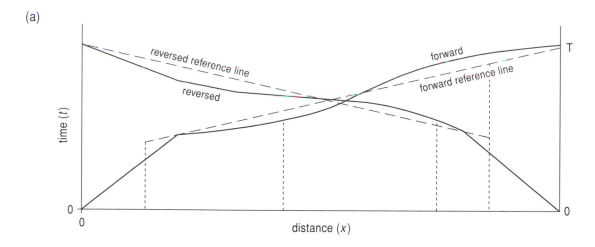

(b)

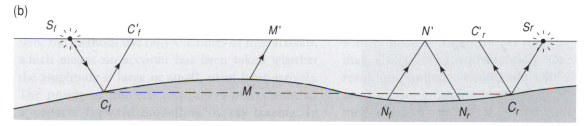

Figure 6.15 Undulating interface.

The delay time for a receiver is easily measured (Fig. 6.17a). The time, t_f, to go from one end to a receiver (path S_fCDR), and then on to the other end, t_r (path $REFS_r$), is longer than the time, t_{total}, to go from end to end (along S_fCDEFS_r) because of the extra times taken to travel from intercept to receiver, along DR and ER. As each of these extra times is just the delay time, we have

$$t_f + t_r = t_{total} + 2\delta_R \qquad \text{Eq. 6.10}$$

$$\delta_R = \frac{1}{2}\left(t_f + t_r - t_{total}\right) \qquad \text{Eq. 6.11}$$

As t_f, t_r, and t_{total} can all be read off the t–x diagram, δ_R can be calculated. This can be done for each receiver.

(a)

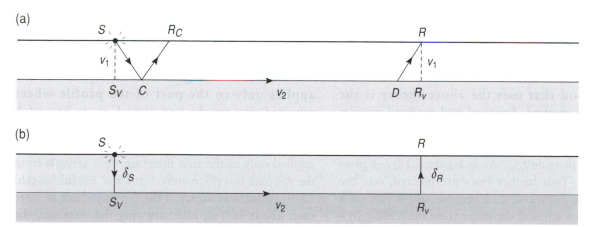

(b)

Figure 6.16 Two ways of treating travel-times.

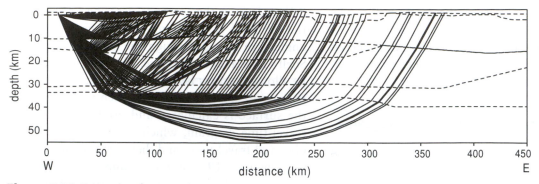

Figure 6.19 Example of ray tracing.

6.10 Detecting offsets in interfaces

Refraction seismology is mostly about measuring the depths to continuous, roughly horizontal interfaces, but it may be used to detect a fault if it offsets an interface. In Figure 6.20b a critically refracted ray travels just below the interface, producing head waves as usual along the interface from C to D. But what happens beyond the offset: Does the corner E cast a shadow preventing seismic energy reaching

EF . . . ? This is best answered by considering waves. F is not in shadow because waves diffract into it, as explained in Section 6.1.1, and therefore head waves originate along EF. . . . The t–x plot (Fig. 6.20a) has an offset of the refraction line: Little energy arrives in the interval D' to E', and when arrivals resume at E' they are later than those at D' not just because it is further from the source but also because E is further from the surface.

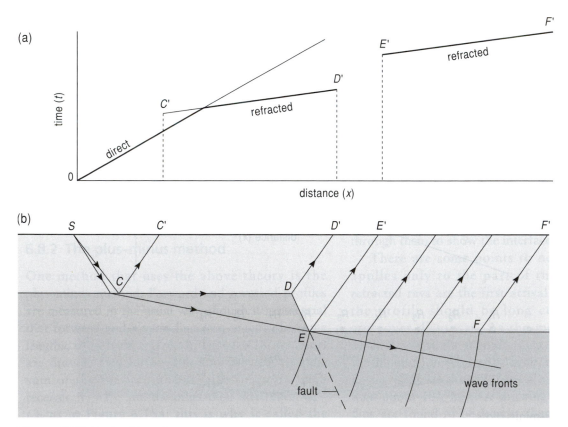

Figure 6.20 Faulted interface.

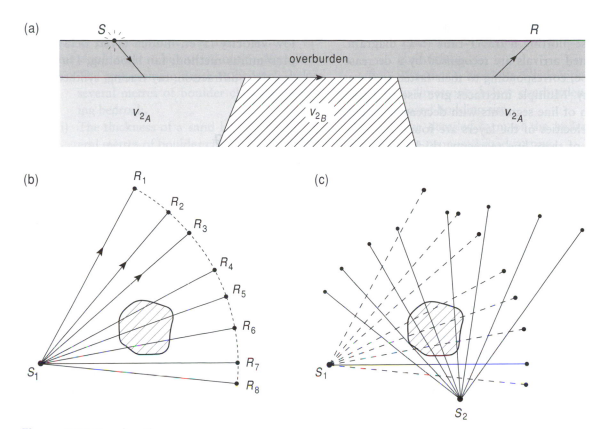

Figure 6.21 Fan shooting.

6.11 Fan shooting: Simple seismic tomography

This is a method that can detect steep-sided features even if they do not have a sharp boundary. Suppose the structure beneath the interface is as shown in Figure 6.21a, perhaps an intrusion that was truncated by erosion and then covered by sediments. Critically refracted rays travel just below the interface as usual, but those passing through the intrusion go faster and so have shorter travel-times. If receivers are arranged in an arc about the source, called **fan shooting** (Fig. 6.21b), the travel-times to R_4, R_5, and R_6 will be less than to the other receivers. This will reveal that *somewhere* along these ray paths is a body with higher seismic velocity. A set of travel-times using a different shot point will locate its position closely (6.21c). In practice, the receivers need not be all the same distance from the shot points and can be placed to record both shots without needing moving, though analysis is somewhat more complicated.

Fan shooting is an simple example of seismic tomography (Section 4.6). In the early days of oil prospecting in Texas it was used successfully to find salt domes because these were often associated with oil; the high seismic velocity of salt compared to most near-surface rocks made them fairly easy to locate.

Summary

1. Refraction seismic surveys are mainly used to detect roughly horizontal interfaces separating layers with different seismic velocities.
2. The seismic refraction method utilises head waves, which are generated by a critical ray that travels just below the interface. Head waves leave the interface at the critical angle.
3. Refracted rays arrive only beyond the critical distance but give first arrivals only beyond the crossover distance; up to the crossover distance the first arrivals are direct rays.

Chapter 7
Reflection Seismology

Seismic reflection is the most important tool we have for detailed imaging of approximately horizontal layering within the Earth, in three dimensions if required. It can reveal structural features such as folding and faulting. It is extensively used by the oil and gas industry to search for oil and gas fields, and to exploit them successfully.

An important branch of reflection seismology is seismic stratigraphy. On a regional scale it has been developed as sequence stratigraphy, revealing stratigraphic correlations that date different parts of a sedimentary succession, and even to establish eustatic sea-level changes. On a local scale, it can reveal lateral changes of lithology that may form stratigraphic oil traps.

Seismic reflection is a form of echo sounding, detecting 'echoes' from seismic interfaces. It can yield results that are the closest of any geophysical technique to a conventional geological section. However, there are several reasons why a seismic section must not be accepted uncritically as a geological section.

7.1 Seismic-reflection sections and their limitations

Reflection seismology, in essence, is echo or depth sounding. Carried out at sea, it is like conventional depth sounding, except that the echoes of interest come not from shoals of fish or the sea floor but from within the solid Earth.

It is most easily explained by describing a simple example (Fig. 7.1). A ship sails along emitting pulses of seismic energy, which travel downwards, to be partially reflected back up from the sea floor and from interfaces in the rocks, called **reflectors** (Fig. 7.1a). When a reflected pulse reaches the surface it is detected by seismic receivers. At the same instant as the pulse is emitted, a pen begins to move steadily across a roll of paper (Fig. 7.1b); it is connected to the receiver so that every reflected pulse produces a wiggle

on the trace, as shown. After a short interval, during which the ship has moved along, the process repeats, but as the paper is moving slowly along, each trace is slightly to one side of the previous one. The wiggles on the separate traces line up to show the interfaces.

As Figure 7.2 shows, the result, a **seismic section**, can give a very direct picture of the subsurface structure, but it is not a true vertical section, for several reasons:

(i) The vertical scale is not depth but time (usually the time to reach the reflector *and return:* the **two-way time, TWT**). Since velocity varies with depth, times cannot be easily converted into depths, as they can for measuring water depths.

(a)

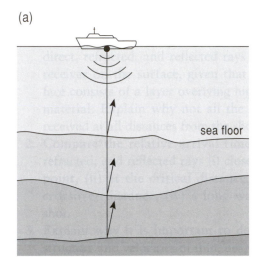

(b)

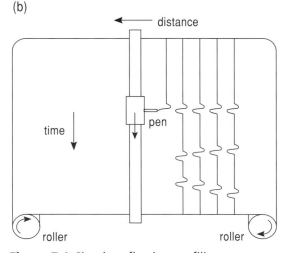

Figure 7.1 Simple-reflection profiling at sea.

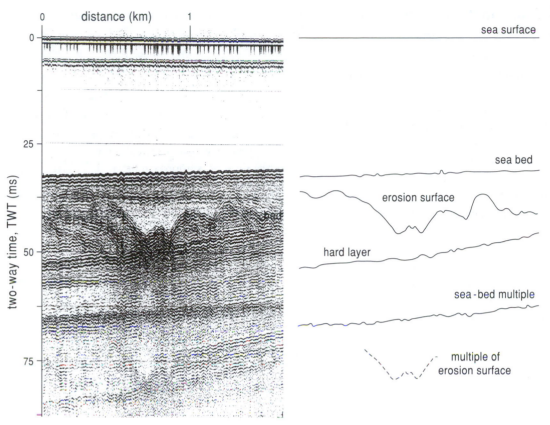

Figure 7.2 Seismic section of a buried channel.

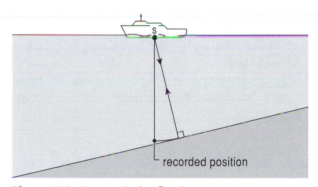

Figure 7.3 Nonvertical reflection.

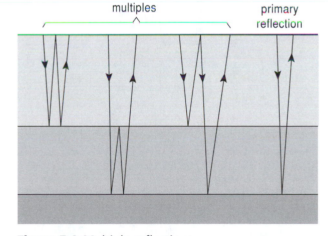

Figure 7.4 Multiple reflections.

(ii) Reflections may not come from directly below the source, since they reflect at right angles to the interface (Fig. 7.3), but the recording takes no account of this.

(iii) There may be multiple reflections in addition to the primary reflections (Fig. 7.4), and these produce spurious reflectors on the record, as shown in Figure 7.2.

(iv) Further reasons, which will be pointed out at appropriate places in the chapter.

How these difficulties can be recognised and largely overcome is explained in the sections that follow.

7.2 Velocity determination using normal moveout, NMO

To deduce the velocities – and hence depths – of the layers, we need a number of receivers along the line, offset to one or both sides of the shot point, so that most rays do not travel vertically (Fig. 7.5). The shortest reflected ray path (for a horizontal reflector) is the vertical one; the rays that reach receivers to either side travel progressively further, taking extra time, as shown by the curve (called a hyperbola) on the time-distance (t–x) diagram above. The *extra* time, Δt, depends on the velocity (as well as the thickness) of layer 1. It is calculated as follows: The vertical TWT, t_0, for the first reflector is just twice the depth divided by the seismic velocity, $2h_1/v_1$. The TWT, t, to reach offset receiver R is SAR/v_1, or $2SA/v_1$. SA is found by applying Pythagoras's theorem to right-angle triangle SVA:

$$t = \left(t_0 + \Delta t\right) = \frac{2}{v_1}\sqrt{h_1^2 + \left(\frac{x}{2}\right)^2} \qquad \text{Eq. 7.1}$$

Squaring both sides, and replacing $2h_1/v_1$ by t_0, we have

$$t^2 = \left(t_0 + \Delta t\right)^2 = t_0^2 + \frac{x^2}{v_1^2} \qquad \text{Eq. 7.2}$$

If the offsets are small compared to the thickness of the layer, so that the rays are never more than a small angle away from vertical, as is often the case, the extra time Δt is approximately

$$\Delta t = t - t_0 \approx \frac{x^2}{2v_1^2 t_0} \qquad \text{Eq. 7.3}$$

The later time of arrival of receivers offset from the source, for a *horizontal* reflector, is called **normal moveout, NMO**.

This equation can be rearranged to give v_1:

$$v_1 \approx \frac{x}{\sqrt{2\,t_0 \Delta t}} \qquad \text{Eq. 7.4}$$

Values of t_0, Δt, and x are measured from the t–x curve.

Multiple layers. The preceding formulas are true only for the topmost reflector. For deeper reflectors we have to allow for refractions through the interfaces above (Fig. 7.6). It turns out that Eq. 7.2 can still be used (with t_0 now the time for vertical reflection to a particular interface), but the velocity given by the equation is the root mean square velocity, v_{rms}:

$$v_{\text{rms}} = \sqrt{\frac{\left(v_1^2 \tau_1 + v_2^2 \tau_2 + \cdots\right)}{\left(\tau_1 + \tau_2 + \cdots\right)}} \qquad \text{Eq. 7.5}$$

where τ_1, τ_2, . . . are the one-way times spent in each layer (Fig. 7.6b). A special sort of average, v_{rms} is the velocity that a single layer, with thickness equal to the depth to the interface, would need to give the same TWT and NMO as observed.

However, to be able to deduce the thickness of each layer we need to know the velocity of each layer. These are found from the values of v_{rms} for successive interfaces. Firstly, v_1 is found using Eq. 7.2. For the second interface v_{rms} is

(a)

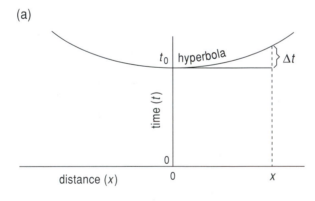

(b)

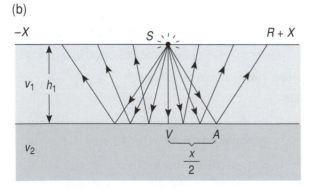

Figure 7.5 Offset receivers.

(a)

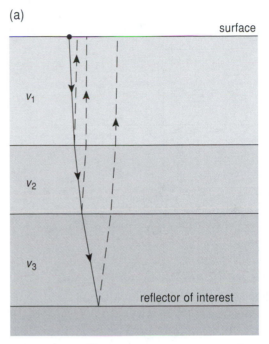

(b)

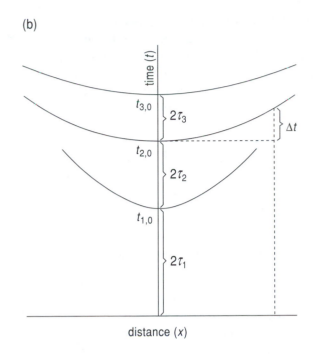

Figure 7.6 Offset ray paths with several layers.

$$v_{\mathrm{rms}} = \sqrt{\frac{v_1^2 \tau_1 + v_2^2 \tau_2}{(\tau_1 + \tau_2)}}$$

Eq. 7.6

$$h = v_{\mathrm{layer}} \left(\frac{\tau_B - \tau_T}{2} \right)$$

Eq. 7.8

As v_2 is the only value not known it can be calculated. Once v_2 is known, v_3 can be found from the expression for v_{rms} down to the third interface, and so on.

The velocity of any particular layer can be calculated using information for just the reflectors the layer lies between:

$$v_{\mathrm{layer}} = \sqrt{\frac{\left(v_{\mathrm{rms}.B}^2 t_n - v_{\mathrm{rms}.T}^2 t_{n-1} \right)}{t_B - t_T}} \quad \text{Dix's Formula}$$

Eq 7.7

where t_T and t_B are the TWTs to the top and base reflectors. This is known as Dix's Formula or Dix's Equation, and the velocity found is known as the **interval velocity.** Deducing the velocity of the various layers is called **velocity analysis.**

The thickness of a layer, h, is found by multiplying its velocity by the time taken to travel vertically through it, which is half the difference of the TWTs to its top and base interfaces:

7.3 Stacking

Often reflections are weak, particularly those from deep interfaces, for the downgoing pulse is weakened by spreading out and by losing energy to reflections at intermediate reflectors. They may then be hard to recognise because of the inevitable noise also present on the traces. Sections can be improved by stacking, the adding together of traces to improve the signal-to-noise ratio (Section 2.3). Section 6.7 described how repeated traces for a single receiver could be stacked; in reflection seismology the shot is not repeated but instead the records of the line of receivers are used. Before they can be stacked, each trace has to be corrected by subtracting its moveout, so changing the hyperbolas of Figure 7.6b into horizontal lines. But moveout is difficult to calculate when arrivals are hard to recognise. To surmount this impasse, finding velocities and stacking are done together. A value for v_{rms} is guessed and used to calculate the NMO at each receiver according to its distance from the shot point (using Eq. 7.3), and the times are displaced by the value of the NMO. If the velocity guessed was correct, the arrivals of the various traces will line up so that, when added together they produce a strong arrival; but if an incorrect velocity was guessed, the traces will not line up so well and then their total will be lower. Therefore, a range of velocities are used to calculate moveouts, and the one that results in the largest addition is used. This is repeated for each interface, for they have different values of v_{rms}. This method works despite noise on the traces, because the noise usually differs from trace to trace and so tends to cancel when stacked.

7.4 Dipping reflectors and migration

If a reflector is dipping, both its apparent position and dip on a seismic section are changed. This is most easily understood if we return to the single receiver system of Figure 7.1. A pulse sent out from S_1 (Fig. 7.7) reflects from P_1 and not from vertically below the shot point. But the recording device is just a pen moving across a roll of paper (or its electronic equivalent), so the reflected pulse is plotted as if it were vertically below S_1, at R'_1. Similarly for S_2, S_3, . . . , though the differences become smaller up dip. Consequently, the reflector appears to be *shallower* and with a *less steep* dip than is actually the case. This distortion cannot be recognised using only a single receiver, but a line of receivers gives rise to a hyperbola offset from the shot point (Fig. 7.8) because the shortest time is not for the ray that travels straight down from the shot point, but the one the returns vertically up from the interface. The hyperbola is displaced up dip by an amount $2h \sin \alpha$, where α is the dip of the interface (Fig. 7.8).

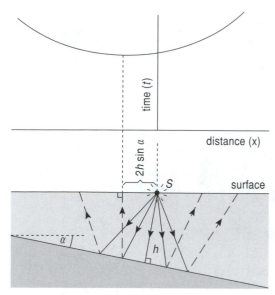

Figure 7.8 Moveout for dipping reflector.

If the reflector is curved, distortions may be more complex, for there may be more than one path between the reflector and a particular shot point. If the reflector is concave, as in a syncline, there will be three ray paths to the receiver, as shown in Figure 7.9a, and as they generally have different lengths – SP_1, SP_2, and SP_3 – they produce more than one arrival. The vertical ray will not be the shortest or the first to arrive if the bottom of the syncline is further from the shot point than the shoulders (i.e., the centre of curvature of the reflector is below the surface). As the shot point moves along, 1 to 10 (Fig. 7.9b), the shortest path switches from one side of the syncline to the other. The three arrival times for each of the transmitter positions 1 to 10 together produce a 'bow tie' on the section (Fig. 7.9c). If the centre of curvature is above the surface, a syncline produces a simple syncline in the section, though somewhat narrowed; for an anticline, the effect is to broaden it (Fig. 7.9d and e).

Correcting for the displacement of the position and shape of a reflector that is not horizontal is called **migration**. This is complicated and requires such large amounts of computer time that sometimes it is not carried out; in this case the section will be marked 'unmigrated'. If so, anyone interpreting the structure has to be aware of the possible distortions, as comparison of Figure 7.10a and b shows. These are discussed further in the next section.

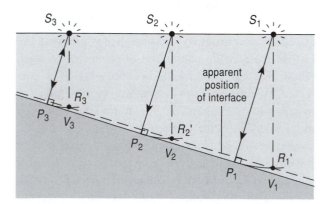

Figure 7.7 Displacement of a dipping reflector.

syncline

(a) ray paths, one shot point

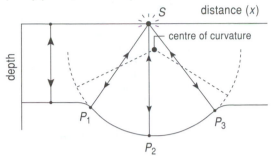

(b) ray paths, succession of shot points

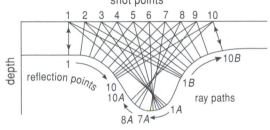

(c) section

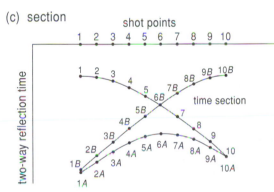

anticline

(d) ray paths

(e) section

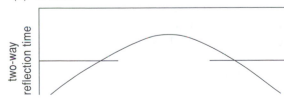

Figure 7.9 Distortion of synclines and anticlines.

a) unmigrated

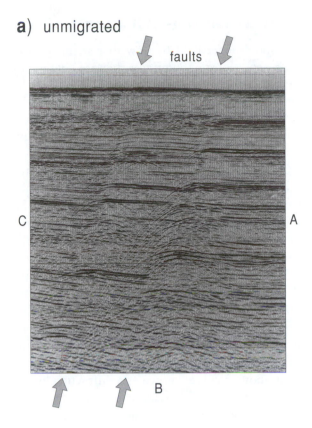

b) migrated

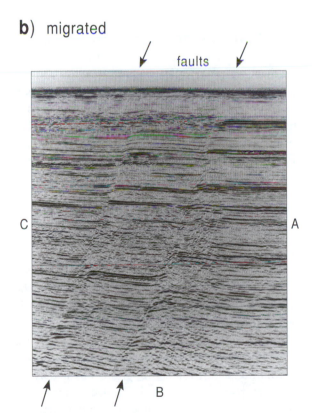

Figure 7.10 Unmigrated and migrated sections.

7.5 Faulted reflectors: Diffraction

You should understand by now that seismic reflection is about locating interfaces that are continuous and roughly horizontal. But an interface may not be continuous; for instance, it may have been offset by a fault. Such an abrupt end to a reflector produces diffracted waves, which are modelled using Huygens's wavelets, as explained in Section 6.1.1.

Consider first just a very short length of reflector (Fig. 7.11a). It acts as a point source; for example, when water waves encounter a pole sticking out of a pond, circular wave fronts travel outwards from the pole as if they were generated by the pole. As the source–receiver system approaches a buried reflector point it receives reflected waves from the point reflector, and these are recorded as if coming from below the transmitter–receiver position, at a distance equal to that of the reflector, as indicated by the arcs. Because this distance changes as the transmitter–receiver passes over the point, the resulting section shows a curve, known as a **diffraction hyperbola** (its amplitude decreases away from the centre of the hyperbola, because of the increasing distance of the reflector).

Turning to the stepped reflector (Fig. 7.11b), the Huygens's wavelets produce a horizontal, upward-moving reflected wave except near the ends, which act rather like point reflectors and produce diffraction hyperbolas. The overall result is that the two parts of the reflector are shown offset, but it is hard to tell exactly where the fault is located (Fig. 7.10a). Migration removes diffraction effects and reveals features more clearly (Fig. 7.10b).

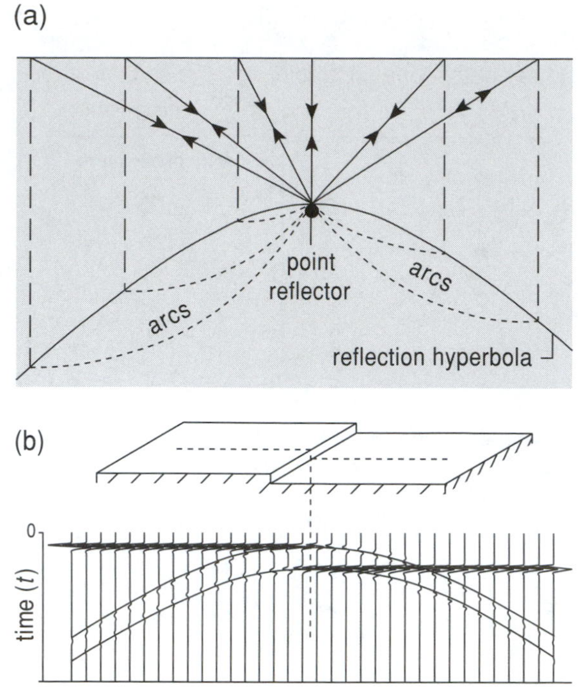

7.6 Multiple reflections

Some **multiple reflection** paths were shown in Figure 7.4, and some spurious arrivals were shown in Figure 7.2. The strongest multiple reflections are those with the fewest reflections and, of course, from the strongest reflectors. A common one is the double reflection from the sea floor (Fig. 7.2).

The positions of multiples on the seismogram can be anticipated from the positions of the primary reflectors higher up the seismogram, but it may be difficult to recognise a primary reflector that happens to lie 'beneath' a multiple (i.e., has nearly the same TWT as that of a multiple from a shallower

Figure 7.11 (a) Point and (b) stepped reflector.

interface). However, these can be distinguished because the moveout for the primary is less than for the multiple and so it stacks using a higher velocity.

We have now considered the three reasons given in Section 7.1 for why a 'raw' reflection seismic section (i.e., as recorded) is not a true geological section (additional reasons will be given later). Next, we see how a reflection seismic survey is carried out.

7.7 Carrying out a reflection survey

7.7.1 Data acquisition

As surveys on land and on water are carried out differently, they will be described separately as complete systems.

Marine surveys. Pulses can be generated by a variety of devices. Commonly used is an air gun, in which air at very high pressure (over 100 times atmospheric pressure) is released suddenly. Less widely used is the sparker, in which a high-voltage electrical charge is discharged from a metal electrode in the water, like a giant spark plug. Other devices use a pulse of high-pressure water (water gun), or explode a mixture of gas and oxygen (e.g., Aquapulse). Each of these sources has the ability to produce pulses that can be repeated at the required rate. The various types differ in the amplitudes of the pulses (dependent on their energy) and more particularly in the duration of the pulses they produce. The importance of a pulse's duration is explained in Section 7.8.2.

The reflected pulses – only P-waves, for S-waves cannot travel through water – are received by a line of hydrophones, which are essentially microphones that respond to the small changes of *pressure* produced by the arrival of a pulse, rather than to water movement. The hydrophones are usually mounted at regular intervals inside a **streamer** (Fig. 7.12), a flexible plastic tube filled with oil to transmit pulses from the water to the hydrophones. The streamer is towed behind a ship at a fixed depth below the surface. As the source has to be supplied with energy from the ship, it is placed in front of the streamer, and therefore the receivers are offset only to one side of the source. Commonly, several sources and many streamers are towed by a single ship, to cover an area more rapidly and thoroughly.

Land surveys. Explosive is probably the most common *impulsive* source used on land. Small charges (up to a few kilograms) are loaded into holes drilled through the top soil and weathered layer (typically a few metres thick), and each is fired in turn by an electrical detonator. Other sources include dropping a weight of 2 or 3 tonnes, or an adapted air gun.

The signals are received by geophones (described in Section 4.2). To improve the signal-to-noise ratio, each geophone may be replaced by a cluster of several, set out within an area only a few metres across, close enough for them to all to receive the signal from a reflector at practically the same time; their outputs are added together to increase the signal as well as partly cancelling any random noise, which differs between geophones. Moving the system along on land is much harder than on water, which makes land surveys about 10 times as expensive (the ratio for 3D surveying is even higher: See Section 7.9). A completely different kind of source, Vibroseis, will be described in Section 7.7.4.

Data recording. The output of each hydrophone or geophone (or cluster) is connected to an amplifier and forms one **channel** to the recorder, which nowadays records on magnetic tape.

The magnetic recording is usually digital (this is similar to that used in compact discs; the older 'wiggle trace' paper and magnetic tape recordings are called 'analog'). The signal is sampled at regular intervals, perhaps only a millisecond apart, and the height of signal at each interval is given as a number (Fig. 7.13). It is these numbers – in binary form, as used by computers – that are recorded and later changed back into the familiar wiggle trace. Digital recording both gives better records and make it easier to process them by computer.

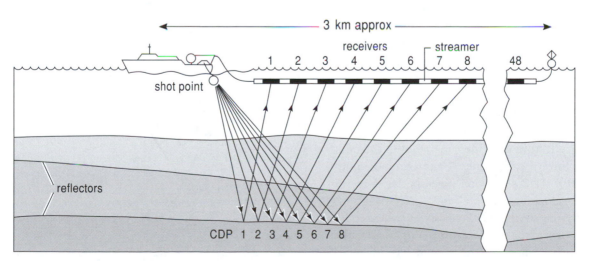

Figure 7.12 Seismic streamer for marine surveys.

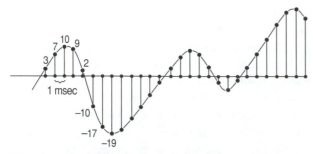

Figure 7.13 Digitisation of a seismogram trace.

Reflected rays are not the first arrivals, for they arrive after the direct ray (Section 6.2), and usually after the **ground roll,** the oscillations caused by surface waves (Section 5.7). Though surface waves travel more slowly than body waves, at small offsets from the source they arrive before reflected rays because of their shorter path, and then their large amplitude can make the reflected arrival difficult to recognise. Recording systems reduce the effects of direct rays and ground roll by using geophone arrays and filters but will not be described further.

7.7.2 Common-depth-point (CDP) stacking

Stacking was described in Section 7.3 as a way of adding together traces to deduce the velocities of layers from the moveouts and to improve the signal-to-noise ratio. The reflections used come from different parts of the interface (Fig. 7.14a); however, it is possible to stack using rays that have all reflected from the same part of an interface (Fig. 7.14b) by using pairs of shot points and receivers that are symmetrical about the reflector point, rays S_1R_3, S_2R_2, S_3R_1. Shots S_1, S_2, . . . are fired successively, so the traces have to be extracted from recordings of successive shots. The reflection point is called the **common depth point, CDP** (also known as the common reflection point, or – for horizontal reflectors with the shot point in the middle of the line of geophones – the common midpoint, CMP). CDP surveying gives better data for computing velocities and stacking.

To carry out marine CDP surveying, where both source and receiver move along (Fig. 7.12), the source is 'fired' whenever the system has advanced *half* the receiver spacing, so that reflections from a

particular common depth point due to successive shots are received by successive receivers. Thus, for shot position S in Figure 7.14c, R_1 receives the reflection from CDP_3, R_2 from CDP_2, . . . ; then when the shot has reached S', R_2 has advanced to R'_2 and receives the reflection from CDP_3, while R'_1 receives one from CDP_4. The traces are sorted by computer into gathers with the same common depth point. When CDP surveying is carried out on land, using a line of regularly spaced *fixed* receivers, the shot point is advanced a distance equal to the receiver spacing.

(a)

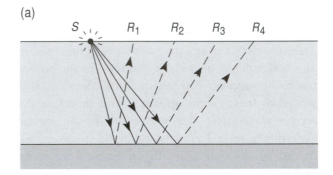

(b)

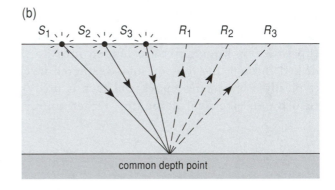

(c)

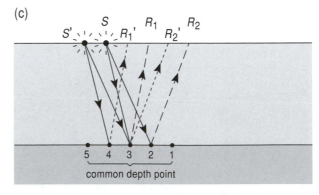

Figure 7.14 Common-depth-point stacking.

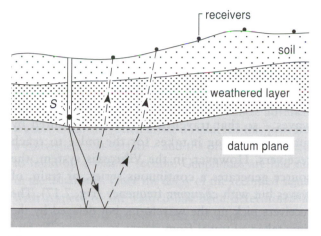

Figure 7.15 Static corrections.

The number of channels that are added together, or gathered, is termed the **fold** of the stacking; in Figure 7.14b it is threefold. The larger the number of receivers used the greater the fold, and 240-fold stacking is commonly used.

Velocity analysis, gathering, stacking, and migration are routinely applied to the 'raw' data. Additionally, in land surveys, any significant topography has to be corrected for, and allowance made for the effect of the topsoil and weathered layers; as pointed

out in Section 6.7, though they are usually thin, the time spent in them is not negligible because of their slow seismic velocities. These corrections are called statics, and the **static correction** corrects or reduces the data as if it had been obtained with shot and receivers on some convenient horizontal datum plane beneath the weathered layer (Fig. 7.15). If the variation of static correction along a profile time is not properly allowed for, time in the surface layers may be interpreted as spuriously thick layers at depth where the velocities are much higher.

7.7.3 Data display

After the data have been processed, a single trace is produced for each common depth point. These are seldom shown as simple wiggle traces because other forms show arrivals more clearly. The most common modification is to shade the peaks but not troughs of the wiggles, to give a '**variable area display**' (Fig. 7.16). Reflectors then show up as dark bands, as already displayed in Figure 7.10. Sometimes this idea is carried further, by shading the peaks more heavily the taller they are or only showing peaks and troughs over some value, so omitting minor wiggles which it is hoped are not important.

(a) wiggle trace   (b) variable area

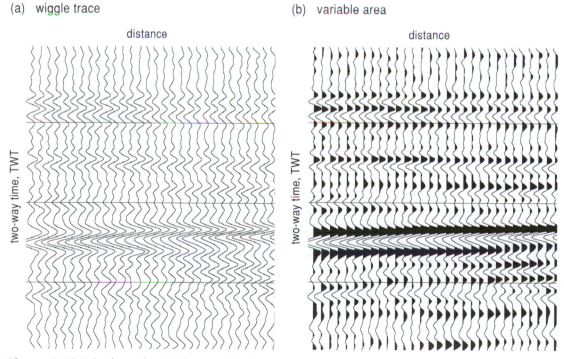

Figure 7.16 Wiggle and variable area displays.

This would give a very strong reflection, with 0.608^2 = 0.37 of the energy reflected.

Suppose instead that the rocks have these values: sandstone, v = 3.3 km/sec, ρ = 2.4 Mg/m³, and limestone, v = 3.0 km/sec, ρ = 2.64 Mg/m³. Then

$$R = \frac{(2.64 \times 3) - (2.40 \times 3.3)}{(2.64 \times 3) + (2.40 \times 3.3)} = 0$$

In this case, though there is a lithological boundary, there is no seismic reflector. Of course, acoustic impedances are unlikely to be identical, but they may be similar enough to produce only a weak reflection. (You may have realised that another choice of values could have given a negative reflection coefficient: This will be considered in Section 7.8.2.)

These calculations illustrate that the boundary between two lithologies need not be a detectable seismic interface, as was pointed out earlier. Conversely, a seismological interface need not be a geological boundary; an example is an oil–gas interface, discussed next.

Bright spot. The interface in a hydrocarbon reservoir between a gas layer and underlying oil or water produces a strong reflection, which appears on the section as a '**bright spot**'. It is also horizontal – a 'flat spot'. So the detection of a strong, horizontal reflector deep within sediments is strong evidence for the presence of gas (Fig. 7.20). Its detection is not always easy, though, because variability in the thickness or velocity of the layers above can decrease or increase the TWT to the interface, an effect called **pull-up** or **-down.**

7.8.2 Vertical resolution: The least separation at which interfaces can be distinguished

Suppose two interfaces could be brought progressively closer together, as shown by the series of diagrams in Figure 7.21a. The reflected pulses would overlap more and more, as shown by the 'combined' curves, until at some separation the pulses could not be distinguished, or resolved apart, and then the two interfaces would appear to be one. Figure 7.21a also shows that the shape of the combined pulse changes markedly with separation of the interfaces and may not look like the incident pulse.

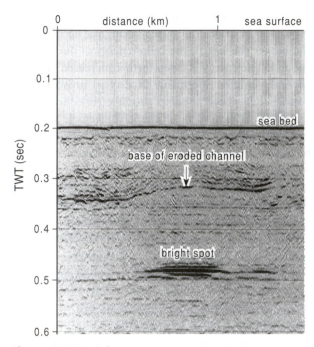

Figure 7.20 Bright spot.

Quite often the two interfaces are the two sides of a thin layer sandwiched within another lithology, such as a shale layer within sandstone. In this case *one* of the interfaces must have a positive reflection coefficient, R, and the other a negative one, because the values in Eq. 7.9 are exchanged. A negative value for R means that the reflected pulse is inverted compared to the incident wave, so one of the two reflected pulses is inverted compared to the other, as shown in Figure 7.21b (where the reflection from the lower interface is inverted). In this case, the closer the interfaces and the more the pulses overlap, the more completely they cancel, until there is no reflection at all! (This must be so, for bringing the interfaces together means shrinking the middle layer to nothing, resulting in a continuous medium, which obviously cannot produce reflections.) The shape of the combined reflection pulse depends on the actual shape of the pulse produced by the source, which is likely to be more complicated than the ones shown.

This addition or cancellation of waves or pulses (introduced in Section 6.1.1) is called **interference**, and it is termed constructive if the total is larger than either alone (Fig. 7.21a), destructive if the total is less (Fig. 7.21b).

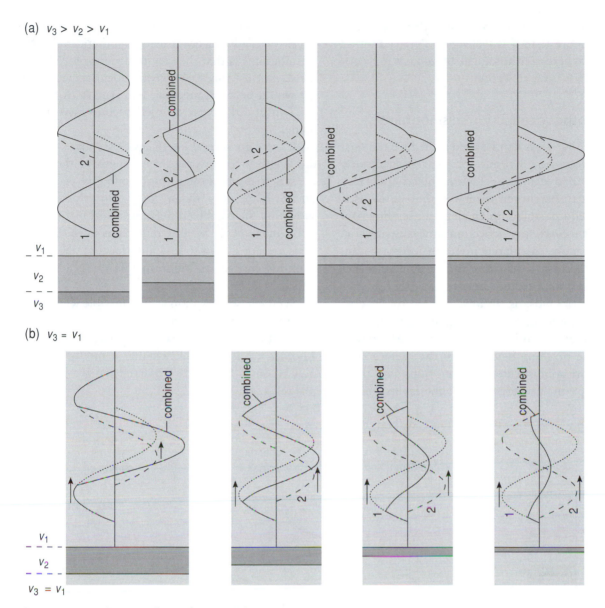

Figure 7.21 Reflections from close interfaces.

In practice, two pulses are difficult to distinguish when they are less than about a half of a wavelength apart. Since the pulse reflected from the lower interface has to travel further by twice the separation of the interfaces, interfaces cease to be resolvable when they are less than half a wavelength apart; that is, $\delta h < \lambda/4$.

Because **vertical resolution** depends upon the wavelength of the pulse, it can be improved by using a shorter pulse, which partly depends on the seismic source used; for instance, sparkers produce a shorter pulse than air guns. However, very short pulses are

more rapidly attenuated by absorption (Section 4.5.5), so they may not reach deeper reflectors. Often, a compromise has to be made between resolution and depth of penetration. The same problem is met in ground-penetrating radar (GPR), and an example is given in Section 14.8.2.

Another situation where there may be no reflection is when an interface is a gradual change of velocities and densities extending over more than about half a wavelength, rather than being abrupt. We can think of the transition as made of many thin layers with reflections from each interface; the many reflected pulses overlap and cancel if they are spread out over more than a wavelength. This is another situation where a lithological boundary may not show up by seismic reflection.

7.8.3 Synthetic reflection seismograms

The preceding ideas can be used to deduce the seismogram that would result for a given succession and to improve interpretation. Firstly, a pulse shape has to be chosen to match the source used (it is usually more complex than those shown in this book). Next, the reflection and transmission coefficients are calculated for each interface (Eq. 7.9); the necessary densities and seismic velocities are usually found from borehole logs (see Chapter 18). Then the pulse is 'followed down the layers', its energy lessening by spreading and by reflection at each interface; this is repeated for each reflection of the pulse on its return to the surface (Fig. 7.22). More sophisticated computations also allow for absorption of energy within the layers, multi-

ple reflections, and the effects of diffraction from edges. The advantage of using synthetic reflection seismograms is that they try to account for all of a trace, not just times of arrivals. Ideally, every wiggle would be accounted for, but even with powerful computers some simplifications have to be made.

If there are several interfaces closer together than the length of the average wavelength of the pulse, the reflected pulses often combine to give peaks that do not coincide with any of the interfaces, as is shown by the combined pulses trace of Figure 7.22. Thus, many of the weaker reflections in seismic sections are not due to specific interfaces at all. Understanding how a seismogram relates to a section improves its interpretation.

Synthetic seismograms – only feasible when there is good borehole control – are used to improve interpretation of seismic sections between boreholes. Another use is to check whether a multiple reflection accounts for all the observed signal or hides a real reflection.

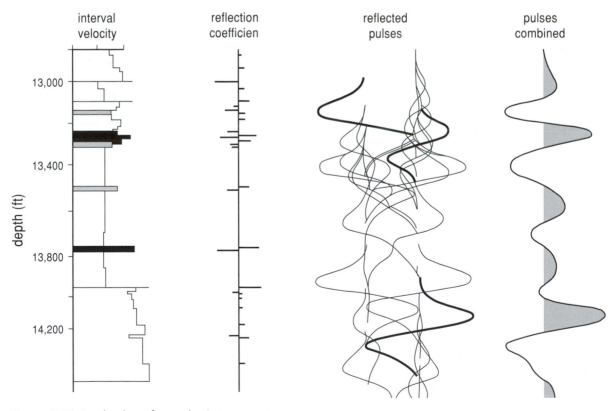

Figure 7.22 Production of a synthetic trace.

7.9 Three-dimensional (3D) surveying

Seismic surveying along a line as described so far allows interfaces to be followed only in the plane of the resulting section, whereas we often also wish to know the form of structures perpendicular to the section. A further limitation is that migration can be carried out only in the plane of the section, which neglects possible reflections from dipping reflectors outside the section. Such reflections are called **sideswipe**. For instance, if the section were taken along the *axis* of the syncline in Figure 7.9a, additional reflections would be received from the limbs on either side but would not be eliminated by migration.

These limitations can be surmounted by shooting a full **three-dimensional, or 3D, survey**, ideally with receivers occupying the points of a regular grid on the surface (Fig. 7.23), as little as 50 m apart, with shots being fired at all grid points in turn. On land, this requires many shot holes and large arrays of geophones. It is simpler at sea, for several sources and a number of streamers – up to 20 or more – are towed in parallel to cover a swathe of sea floor.

A CDP gather comprises pairs of shot points and receivers from all around the CDP, not just along a survey line (Fig. 7.23), where suffixes identify source-receiver pairs with the same CDP. Stacking, migration, and so on follow the same principles as for sections, though obviously they are more complicated.

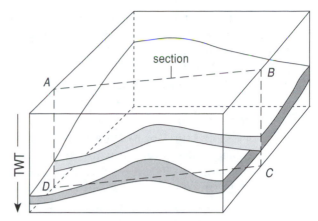

Figure 7.24 Three-dimensional block of data.

We can think of the reflectors revealed by this technique as being embedded in a block (Fig. 7.24). If there are many reflectors, this way of showing the results would be confusing; instead, the block can be 'sliced' with the aid of a computer, with colour normally used to emphasise features. Vertical slices (e.g., *ABCD* in Fig. 7.24) are similar to sections made in the usual way, except that migration will also have corrected for sideswipe reflections and their direction can be chosen independent of the direction the survey progressed. Horizontal sections can also be produced; these are called 'time slices' because the vertical axis is TWT, not depth. Alternatively, interfaces can be picked out. More sophisticated processing can reveal properties of the rock, such as its porosity, which is important, for it determines how much hydrocarbon a rock can hold and how easily it will release it (Chapters 18 and 22). Plate 6 shows a block of the subsurface from part of the Forties oil field in the North Sea, with the layers revealed by their porosity; in this case, higher porosity reflects a higher proportion of sand compared to shale. The low-porosity layer acts as a barrier to upward migration, so any oil will be just below it. Plate 7 shows a single layer from this part of the block. Its highest-porosity parts are due to a sandy channel present when the layer formed and forms a stratigraphic trap. This sort of information is essential when deciding where to drill extraction wells. A case study of the Forties field is described in Section 22.6.

A development of 3D surveying is **time-lapse** modelling (sometimes called **4D modelling**). By

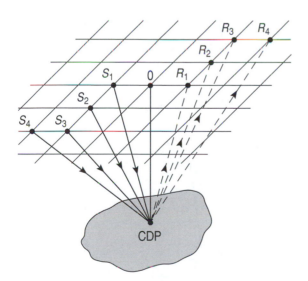

Figure 7.23 Part of a 3D survey grid and gather.

repeating surveys at intervals, the extraction of hydrocarbons can be followed and unextracted pockets of oil detected. An example is given in Section 22.6.2 and Plate 8.

Though 3D surveying is much more expensive, both for data acquisition and reduction, than linear surveying, it can pay for itself in the increased understanding of the structures of hydrocarbon reservoirs that it gives and in the precision with which it permits holes to be drilled in often intricate structures. By the mid-1990s 3D surveying had become the norm for marine surveys, and it is increasingly being used on land, though this costs up to 50 times as much as the same area of marine surveying.

7.10 Reflection seismology and the search for hydrocarbons

By far the biggest application of reflection seismology is in the hydrocarbon industry, for which the method has been developed, though it has other applications.

7.10.1 The formation of hydrocarbon traps

The formation of hydrocarbon reservoirs requires the following sequence: (i) Organic matter – usually remains of minute plants and animals – is buried in a **source rock** that protects them from being destroyed by oxidation; this is often within clays in a sedimentary basin. (ii) The organic matter is changed by bacterial action operating at temperatures in the range 100 to 200°C – due to the burial (Section 17.3.2) – into droplets of oil or gas. (iii) The droplets are squeezed out of the source rock by its consolidation. Being lighter than the water, which is the dominant fluid in the sediments, they tend to move up, but also often sideways, perhaps due to deformation of the basin. (iv) The hydrocarbon is prevented from leaking to the surface, where it would be lost, by an impervious **cap rock,** such as shale. (v) To be extractable, the hydrocarbon has to be in a **reservoir rock** that is porous and permeable, usually a sandstone or a carbonate. (vi) To be commercially useful, it has to be concentrated into a small volume. This requires a **trap,** which can be of several kinds.

Figure 7.25 Two types of structural trap.

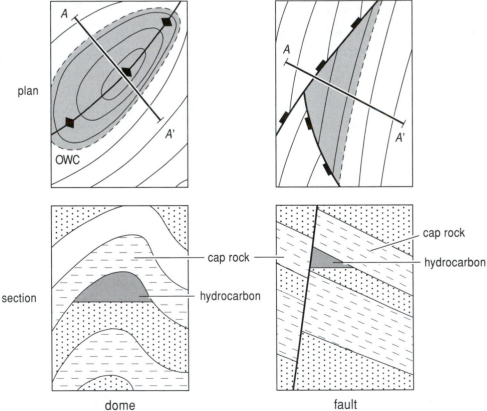

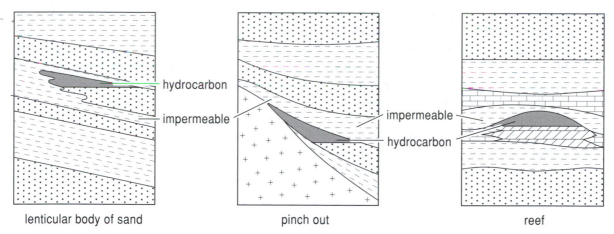

Figure 7.26 Three types of stratigraphic trap.

Structural traps result from tectonic processes, which produce folds, domes, faults, and so on. Figure 7.25 shows domal and fault traps, in plan and section. **Stratigraphic traps** are formed by lithological variation in the strata at the time of deposition, such as a lens of permeable and porous sandstone, or a carbonate reef, surrounded by impermeable rocks (Fig. 7.26). **Combination traps** have both structural and stratigraphic features, such as where low density salt is squeezed upwards to form a salt dome, both tilting strata and so causing hydrocarbons to concentrate upwards, and also blocking off their escape (Fig. 7.27). To find these traps seismically requires recognising how they will appear on a seismic section.

7.10.2 The recognition of hydrocarbon traps

Assuming that investigation indicates that traps filled with hydrocarbons may exist, how can they be located by seismic reflection? Easiest to recognise are structural ones (Fig. 7.25), where the tilting and/or termination of the reflectors reveals their presence. Stratigraphic traps which have tilted reservoir rocks terminating upwards in an unconformity are also fairly easy to spot, but other types are generally more difficult to detect, and other clues have to be pressed into service. The presence of a gas–oil or gas–water interface can give a particularly strong and horizontal reflector (bright and flat spot) (Fig. 7.20), as mentioned in Section 7.8.1, and also as the upper surface of a gas layer has a negative reflection

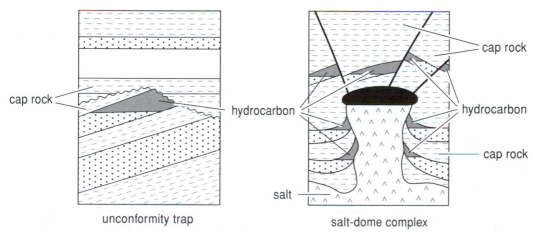

Figure 7.27 Two types of combination trap.

coefficient, it may be possible to recognise it from the inversion of the trace. Carbonate reefs also often produce a fairly strong reflector, which is often higher over the reef because reefs are often formed on a local topographic high and also generally are compacted less than other sediments following burial. The different velocities of reservoir and cap rocks may lead to a recognisable difference in interval velocity (Section 7.2) as well as a reflection. Other clues can come from more subtle changes in the character of the seismic section, which depends on a combination of the value of the reflection coefficient, the shape of the wiggles on the traces, and so on, and how these change laterally. Particular types of traps, such as sand lenses and river channels, can show up when contour diagrams are made, which is best done using 3D surveying. A model based upon 3D data is shown in Figure 22.14.

Because traps now being exploited are often smaller and harder to recognise than previously, seismic surveys must be of high-quality (closely spaced stations, high resolution sources, and so on), with static corrections and migration carried out precisely. Case histories of the finding and development of gas and oil fields are given in Chapter 22.

7.11 Sequence stratigraphy

Sequence stratigraphy (also called seismic sequence analysis and seismic facies analysis) grew out of seismic stratigraphy, which is the building up of a stratigraphy using seismic sections. As seismic data may cover large parts of a sedimentary basin, this provides a scale of investigation not possible from surface outcrops and a continuity not provided by boreholes unless they are very closely spaced. It can also allow recognition of global sea-level changes and so can provide a dating tool, valuable when palaeontological dating is not possible because of the absence of suitable fossils. It can also reveal stratigraphic hydrocarbon traps, described in the previous section. Sequence stratigraphy is possible because chronostratigraphic boundaries – surfaces formed within a negligible interval of time, such as between strata – are also often reflectors.

A stratigraphic sequence is a succession of strata with common genesis *bounded by unconformities*, or correlated unconformities. It is typically tens to hundreds of metres thick and spans an interval of

hundreds of thousands to a few millions of years. How a sequence is built up and terminated depends on the interplay of deposition and changes in sea level. The formation of an extensive sedimentary basin requires subsidence to continues for millions of years, but superimposed on this subsidence are variations in sea level due to local changes of subsidence rate or relatively rapid global changes in sea level, called eustatic changes (discussed shortly). What concerns us here are *relative* sea-level changes while the sediments are being deposited.

Consider deposition along a coastline where sediments are being supplied by a river, during a period when relative sea level is changing. If the sea level remains constant (Fig. 7.28a) sediment is deposited from the river where the flow first slackens pace, at the shelf edge. Each bed forms over and beyond the previous one, to give a prograding succession. The top end of each bed thins towards the land – causing it to disappear from the seismic section, for the reason given in Section 7.8.2 – this is a **toplap**; at its lower edge each bed downlaps onto the existing rocks. If the sea level rises steadily (Fig. 7.28b), there is deposition on top of earlier beds as well as beyond, producing successive near-horizontal layers; where they butt up against the sloping coastline there is onlap. Toplap grades into onlap as the sea level changes from being constant to rising, altering the balance between prograding and aggrading of sediment, as shown in Figure 7.28c and d, where the arrow shows the direction of growth of sediments.

An unconformity – which defines the boundary of a sequence – is formed when the sea level falls fast enough for erosion rather than deposition to occur. If the rate of fall is moderate, the sea retreats across the shelf truncating the *tops* of beds. If it is rapid, sea level falls to partway down the slope before the tops can be removed, then erosion begins to cut laterally. The eroded material, of course, is deposited elsewhere, supplementing the sediments being supplied by the river. Deposition and erosion may alternate cyclically if relative sea level fluctuates. Over a cycle, deposition varies from fluviatile facies, through coastal and offshore to basin facies though these cannot all be distinguished in seismic sections without the aid of borehole data. Beds that end – by lapping of various kinds or truncation – are termed discordant; they appear in seismic sections as reflectors that cease laterally and are important for working out the history of deposition and erosion. Figure 7.29 shows how a geologic section of lithostratic units – the arrangement of strata in space – converts into a chronostratigraphy – the arrangement of strata in time. Strata are numbered successively.

Sequence stratigraphy and eustasy. Sequence stratigraphy provides a record of the changes in local relative sea level. By identifying the types and volumes of the various facies within a sequence the amounts of rise and fall can be estimated, and their times can often be deduced from fossils taken from boreholes. It was found that the records from widely separated basins were similar, and this was attributed to global sea-level changes, rather than, say, changes in local subsidence rates. Some of the falls were rapid – up to 1 cm a year – and continued for thousands of years, and though some of these rapid falls could be attributed to global glaciations, others occurred when there were no glaciations, casting doubt on the reality of global changes. However, subsequent investigations have tended to confirm the reality of the global sea-level changes, though the mechanism is not known.

7.12 Shallow-reflection seismic surveys

Reflection seismology needs to be modified for interfaces less than a few hundred metres deep. This is for several reasons, including proportionately larger ground-roll and the likely need for higher resolution (Section 7.8.2), for shallow layers are often thin. High resolution requires a short though not powerful pulse; this can be provided by a small explosion, such as by a 'buffalo gun', which fires a shotgun cartridge in a steel tube placed in a hole drilled to about a metre's depth. The geophones and recorders have to be able to respond to the higher frequencies generated, typically about 400 Hz, compared to less than 100 Hz in hydrocarbon exploration. Processing of the results is similar to that used for normal depth surveys.

Shallow-reflection seismic surveying has been applied to small-scale geological problems, such as investigating glacial structures below the surface (Harris et al, 1997) or imaging a buried river channel (Brabham and McDonald, 1992). It has also suc-

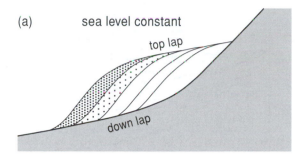

(a) sea level constant

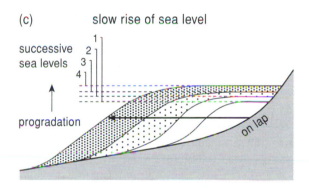

(b) sea level rising

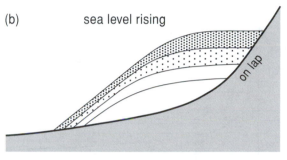

(c) slow rise of sea level

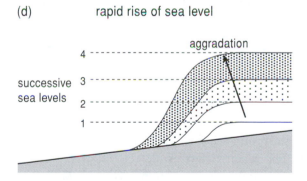

(d) rapid rise of sea level

Figure 7.28 Progradation and aggradation.

cessfully mapped lignite deposits within clay sequences in Northern Ireland (Hill, 1992). This last could not be done using refraction surveys, because the lignite and clay have closely the same *velocity*, so layers are hidden (Section 6.6). However, as lignite

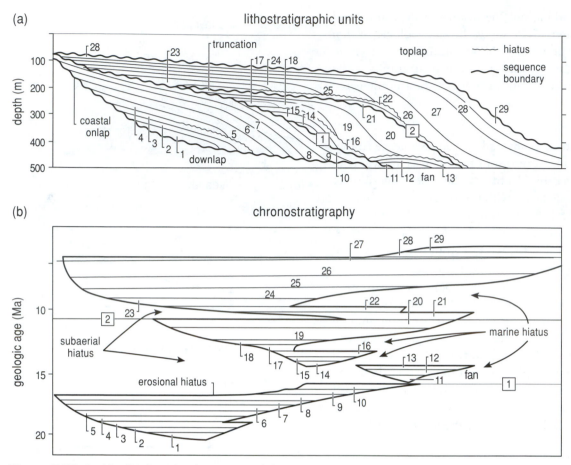

Figure 7.29 An idealised section in space and time.

has a considerably lower *density* than clay, the reflection coefficient (Section 7.8.1) is quite large, with a negative value for clay overlying lignite.

Summary

1. Reflection seismology is a form of echo or depth sounding for detecting interfaces below the ground surface, called reflectors.

2. The result of a seismic reflection survey – a reflection seismic section – is (with GPR, Section 14.8) the closest of any geophysical technique to a geological section. However, a 'raw' seismic section differs from a geological section for reasons that include these:
   (i) It gives the two-way time, TWT, rather than depth to the reflectors
   (ii) Reflections from a dipping reflector are displaced.

   (iii) Some apparent reflectors may be due to multiple reflections.
   (iv) Reflectors may not correspond to lithological boundaries, and vice versa.

3. Reflection surveying requires a source and one or more receivers. Impulsive sources for shipborne surveys are most commonly air guns or sparkers; on land usually explosives are used. Sources differ in energy and frequency (wavelength). Receivers for shipborne and land surveys are respectively hydrophones and geophones.

4. The Vibroseis source is nonimpulsive and 'sweeps' through a smoothly changing range of frequencies (or wavelengths). Its advantages are that (i) an intense source is not needed, (ii) it is quick to deploy, and (iii) it has a good signal-to-noise ratio. Therefore, it is much used in built-up areas, often along roads. It is mainly limited to use on land.

5. Velocities can be determined from the moveout. The velocity found is the rms velocity, v_{rms}, which is used to find interval velocities, using Dix's Formula, or Dix's Equation (Eq. 7.7).
6. Signals are stacked both to maximise the signal-to-noise ratio, and to find v_{rms}.
7. Migration is used to correct for the distortion in dip and depth resulting from dipping and curved reflectors, and to eliminate diffraction effects from sharply truncated reflectors.
8. The reflection and transmission coefficients of an interface depend on the difference of the acoustic impedances, ρv, on either side of the interface (Eq. 7.9). A negative reflection coefficient occurs when the acoustic impedance of the layer below an interface is less than that of the layer above it; it inverts the reflected pulse.
9. Interfaces are not resolved if they are less than about ¼ wavelength apart. Resolution is increased by using a source with a shorter pulse length (higher frequency), but at the expense of poorer depth of penetration. A transitional interface will not produce a reflection if the transition is significantly thicker than about ¼ wavelength. Many reflections on seismic sections are formed by interference of the reflections from several closely spaced interfaces.
10. Shallow-reflection surveys require high-frequency sources, receivers, and recorders to resolve shallow layers that are often thin. They find use in site investigation as well as for small-scale geological problems.
11. Synthetic seismograms, constructed using information on velocities and densities provided by borehole logs, can improve interpretation of sections away from the borehole.
12. Three-dimensional reflection surveys collect data from a grid. They permit migration perpendicular to a recording line as well as along it, eliminating sideswipe and allowing structures to followed in both directions. Results are presented as sections and time slices, and in other ways.
13. A major use of reflection seismic surveying is to find and exploit a variety of hydrocarbon traps.
14. Sequence stratigraphy allows extensive stratigraphies to be built up by recognising the various types of discordant endings of beds. The results are used to correlate successions within a sedimentary basin and to provide dates. On the largest scale they provide a record of eustatic sea level changes.
15. You should understand these terms: reflector, seismic section; two-way time (TWT); normal moveout (NMO), common depth point (CDP), channel, fold; rms and interval velocities; multiple reflections, 'bow tie', sideswipe, pull-up and -down, migration; variable area display, equalisation; acoustic impedance, reflection and transmission coefficients; vertical resolution, interference, diffraction hyperbola; three-dimensional surveying, time lapse; ground roll; static correction; bright spot; streamer; Vibroseis; source, reservoir and cap rocks; structural, stratigraphic, and combination traps, sequence stratigraphy.

Further reading

Reflection seismology is introduced in general books on applied geophysics, such as Kearey and Brooks (1991), Parasnis (1997), Reynolds (1997), and Robinson and Coruh (1988), and Doyle (1995) devotes several chapters to it; Telford et al. (1990) is more advanced. Among books entirely devoted to the subject, McQuillin et al. (1984) provides a good introduction though not of more recent developments, while Sheriff and Geldart (1995) is at a more advanced as well as more recent level.

Brown et al. (1992) contains a chapter on sequence stratigraphy, and Emery and Myers (1996) is devoted to the topic. Hill (1992) discusses a number of shallow-reflection case studies.

Problems

1. Why do marine seismic reflection surveys not record (a) S-waves? (b) refracted rays?
2. How does a migrated reflection seismic section differ from an unmigrated one? In what circumstances would they be the same?
3. How can a primary reflection be distinguished from a multiple one?
4. Will a migrated section correct for 'sideswipe'?
5. Is the dip of a reflector in an unmigrated seismic section more or less than its actual dip? Explain why with the aid of a sketch.

6. What is the main way in which the Vibroseis system differs from other data acquisition systems. Name two advantages that it has over other methods of land surveying.

7. What are the main purposes of stacking?

8. How can a reflection coefficient be negative? How can it be recognised?

9. How may synclines and anticlines appear in an unmigrated seismic section?

10. A succession consists of alternating sandstones and shales, with the top layer being sandstone. Calculate how the amplitude diminishes for reflections from each of the top four interfaces (ignore spreading of the wavefront and absorption), if the densities and velocities are as follows: sandstone 2 Mg/m³, 2.6 km/s; shale 2.3 Mg/m³, 2.8 km/s.

11. Seismic sections are not always what they appear. Explain how an apparent reflector may (a) have an incorrect slope, (b) may have an incorrect curvature, or (c) may not exist at all, while (d) three horizontal reflectors spaced equally one above the other may not be equally spaced, in reality.

12. What determines vertical resolution? Why does less than the required resolution sometimes have to be accepted?

13. Explain why a reflector on a seismic section need not correspond to a particular interface.

14. Why is a very strong horizontal reflection usually indicative of a gas–water interface? Why may a gas–water interface not always appear as a horizontal reflector?

15. A strong reflector that lies below several layers with different seismic velocities has the same TWT as the base of a single layer elsewhere. What do the total thicknesses above the two reflectors have in common?

16. Explain why a seismic interface may not be a lithological boundary, and vice versa. Give an example of each.

17. In what ways does shallow seismic land surveying differ from deep surveying?

18. A thin, horizontal layer of shale (v_P = 2.8 km/sec) lies within sandstone (v_P = 2.5 km/sec). What is the minimum thickness of shale that can be resolved in a Vibroseis survey? (Use an average frequency.)

19. Explain how (a) an interface may show up by seismic reflection but not seismic refraction, and (b) vice versa.

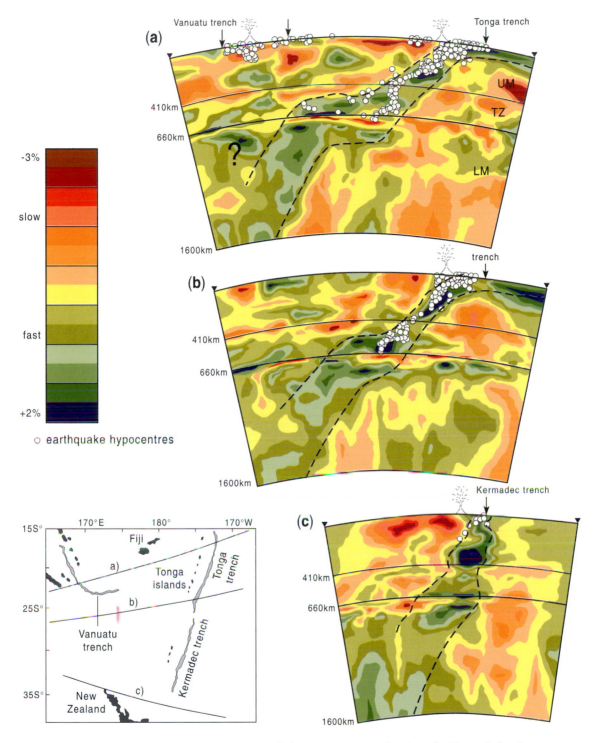

Plate 1. Three tomographic sections of the Tonga-Kermadec Trench. The subducting plate is revealed as a zone of higher velocity, believed to be because its temperature is lower than the surrounding mantle, and also – in the upper part of the plate – by earthquake epicentres shown by white dots. The plate is deflected by the 660-km discontinuity, the more so the faster the subduction rate. See Section 20.9.4 for details.

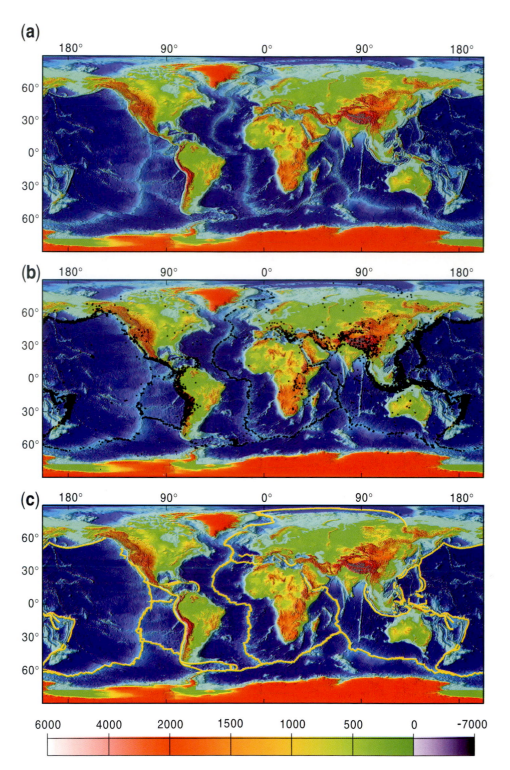

(a)

(b)

(c)

6000	4000	2000	1500	1000	500	0	-7000

Plate 2. Global topography, seismicity, and plates.
  (a) The topography of the continents and ocean floors; most major features are
      linear.
  (b) Earthquake epicentres have been added (for earthquakes with magnitude
      over 5, in the period 1980–90). Most occur in narrow belts that coincide with
      linear topographic features.
  (c) Boundaries of the largest plates have been added to the topographic map.

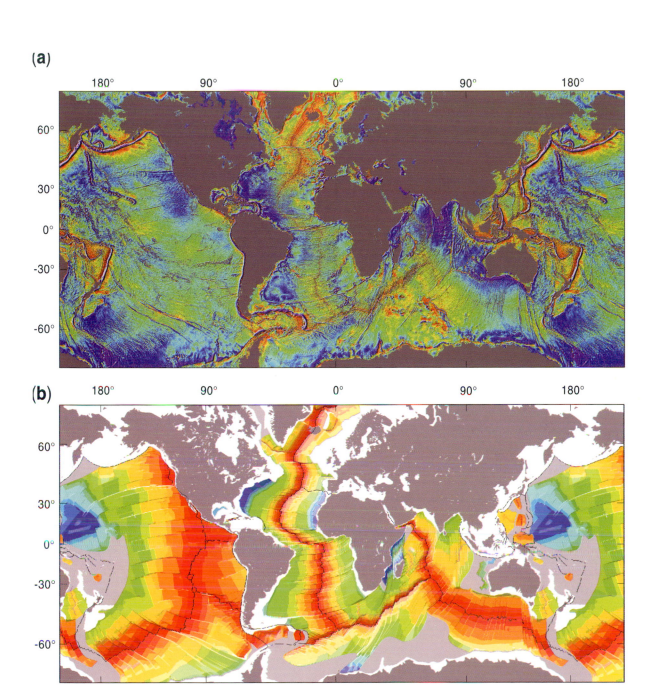

(a)

(b)

| Chron | 5 | 6 | 13 | 18 | 21 | 25 | 31 | 34 | M0 | M4 | M10 | M16 | M21 | M25 |

| 0.0 | 9.7 | 20.1 | 33.1 | 40.1 | 47.9 | 55.9 | 67.7 | 83.5 | 120.4 | 126.7 | 131.9 | 139.6 | 147.7 | 154.3 | 180.0 Ma |

Plate 3. The oceans: surface topography and ages of the ocean floor.
    (a) The topography after the effects of tides and waves have been allowed for. It reflects the boundaries of the tectonic plates, via their effects on the value of gravity, as explained in Section 9.5.1.
    (b) The ages of the ocean floor have been determined from magnetic anomalies ('sea-floor stripes'). This shows ages in terms of intervals between magnetic polarity chrons, as indicated in the key. See Section 20.2.1 for more details.

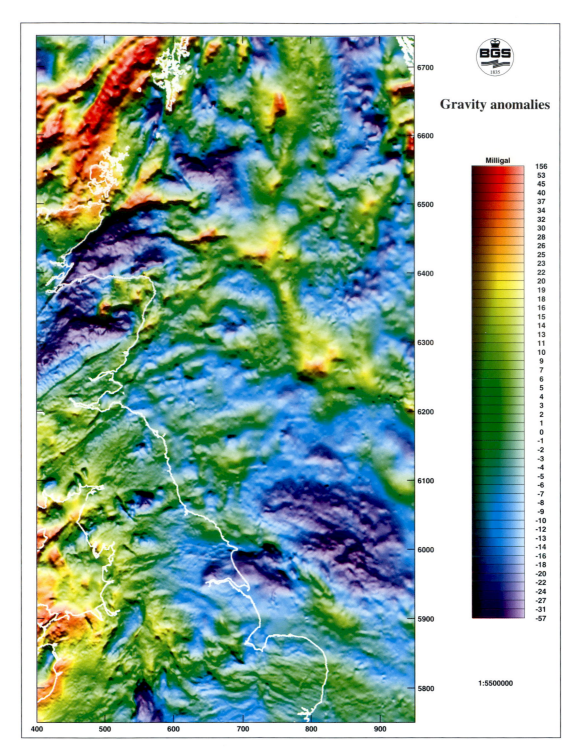

Gravity anomalies

Milligal

156	
53	
45	
40	
37	
34	
32	
30	
28	
26	
25	
23	
22	
20	
19	
18	
16	
15	
14	
13	
11	
10	
9	
7	
6	
5	
4	
3	
2	
1	
0	
-1	
-2	
-3	
-4	
-5	
-6	
-7	
-8	
-9	
-10	
-12	
-13	
-14	
-16	
-18	
-20	
-22	
-24	
-27	
-31	
-57	

1:5500000

Plate 4. Gravity anomaly map, in colour-shaded relief, of part of the U.K. and the North Sea. Bouguer anomaly for land areas, free air anomaly for sea areas. See Section 22.5 for its significance for hydrocarbon exploration.

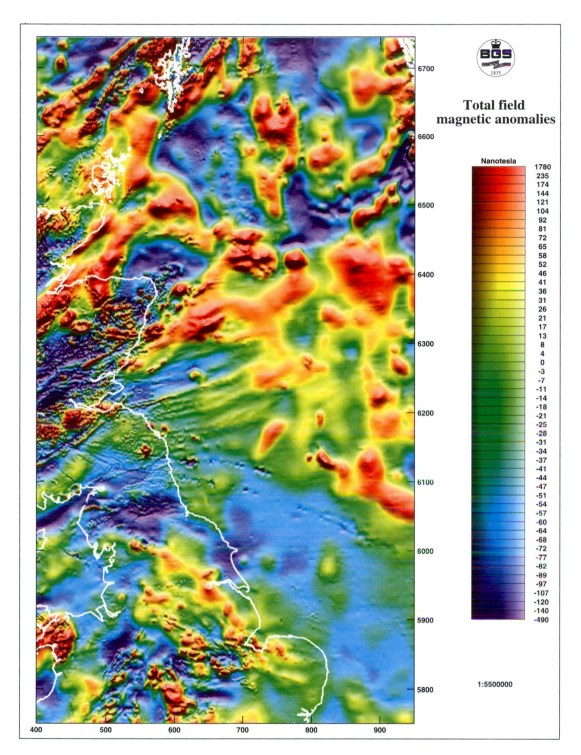

Plate 5. Magnetic anomaly map, in colour-shaded relief, of part of the U.K. and the North Sea. See Section 22.5 for its significance for hydrocarbon exploration.

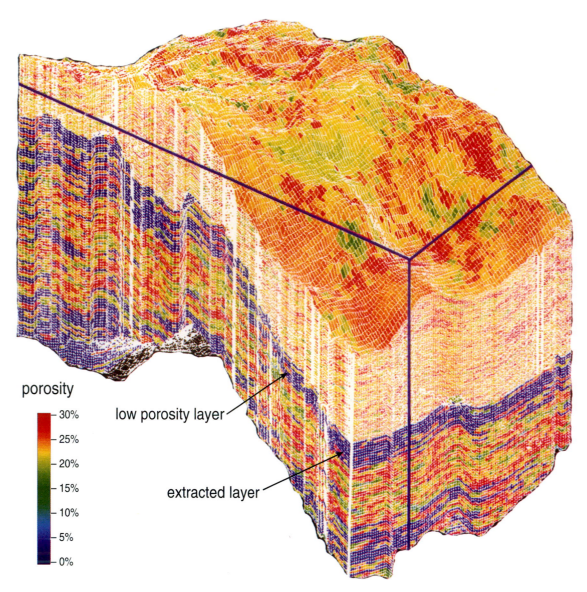

porosity

30%
25%
20%
15%
10%
5%
0%

low porosity layer

extracted layer

Plate 6. Results of 3D seismic reflection survey. This shows a model of part of the For-ties field in the North Sea, obtained by processing 3D seismic-reflection data. Varying porosity is shown by colour, which picks out reflectors. The 'extracted layer' is shown in Plate 7. See Section 7.9.4 and the case study in Section 22.6 for details.

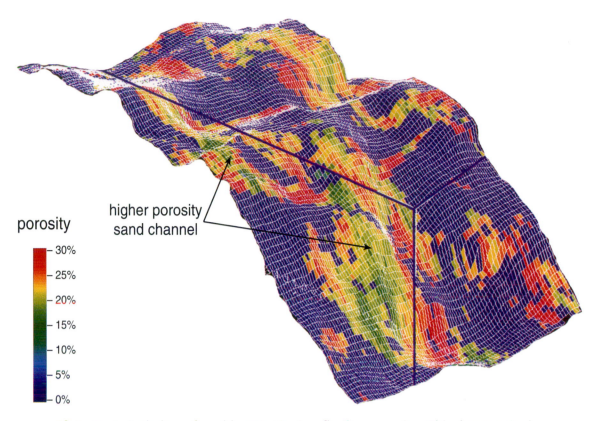

porosity

- 30%
- 25%
- 20%
- 15%
- 10%
- 5%
- 0%

higher porosity
sand channel

Plate 7. A single layer found by 3D seismic-reflection surveying. This shows a single layer from within Plate 6. A sandy channel shows up against shales by its high porosity. See Section 7.9.4 and the case study in Section 22.6 for details.

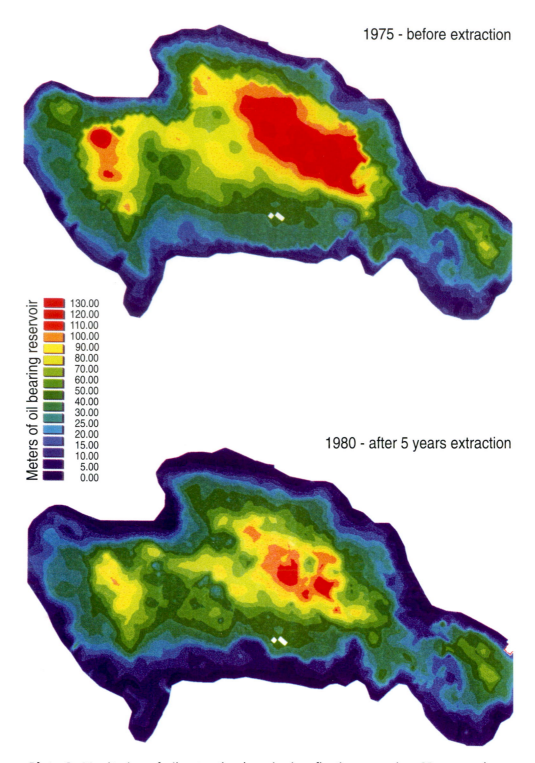

1975 - before extraction

Meters of oil bearing reservoir

130.00
120.00
110.00
100.00
90.00
80.00
70.00
60.00
50.00
40.00
30.00
25.00
20.00
15.00
10.00
5.00
0.00

1980 - after 5 years extraction

Plate 8. Monitoring of oil extraction by seismic-reflection surveying. 3D surveys have been used to monitor how the thickness of oil decreases due to extraction. See Section 22.6.2 for details.

SUBPART I.3

Gravity

'Gravity' is one of the most generally useful geophysical techniques, probably second only to seismology. It is used to detect lateral differences in the densities of subsurface rocks. It is therefore useful for finding buried bodies and structures, such as igneous intrusions and some faults, on scales that range from a few metres to tens of kilometres across. These uses are discussed in the first of the two chapters in this subpart, **Gravity on a small scale.**

But the Earth does not have the steady strength to support large lateral density differences – either on or below the surface, such as mountains and large intrusions – if they are more than a few tens of kilometres across; instead, these are supported by buoyancy, through the mechanism of isostasy in which the outer layer of the Earth floats on a yielding inner layer. Isostasy tends to reduce the size of gravity anomalies; however, large anomalies can exist if isostatic equilibrium has not yet been reached or if dynamic forces prevent equilibrium. Therefore the main use of gravity in **Large-scale gravity and isostasy** – the second of this pair of chapters – is to measure whether large-scale gravity anomalies exist, and if so, what prevents isostatic equilibrium.

Chapter 8

Gravity on a Small Scale

Gravity surveying, or prospecting, is a method for investigating subsurface bodies or structures that have an associated lateral density variation. These include ore bodies and intrusions whose density differs from that of the surrounding rocks, basins infilled with less dense rocks, faults if they offset rocks so that there are lateral density differences, and cavities.

Gravity surveys measure minute differences in the downward pull of gravity and so require extremely sensitive instruments. Gravity readings also depend on factors that have nothing to do with the subsurface rocks, and corrections for their effects are very important.

8.1 Newton's Law of Gravitation

We all feel the pull or attractive force of the Earth's gravity, particularly when we lift a heavy mass. This pull is due to the attraction of all the material of the Earth but depends most on the rocks below where we experience the pull. Therefore, by measuring how the pull varies from place to place we can make deductions about the subsurface. Evidently, we need to understand how subsurface rocks – and other factors – affect the pull of gravity at the Earth's surface.

Most people know the story about Sir Isaac Newton discovering gravitation after being hit by a falling apple. Of course, people did not need Newton to tell them that apples and other objects fall; what Newton had the genius to realise is that the familiar force that pulls objects towards the ground is also responsible for holding planets in their orbits around the Sun, and in fact that all bodies attract one another. He found that the attractive force (Fig. 8.1) between two small masses, m_1 and m_2, is proportional to the product of their masses (e.g., doubling either mass doubles the force, doubling both quadruples the force) and inversely proportional to the square of their separation, r (doubling their separation reduces the attraction to a quarter). These relationships are summed up by the equation

$$\text{attractive force, } F = G \frac{m_1 m_2}{r^2} \qquad \text{Newton's Law}$$

Eq. 8.1

Figure 8.1 Attraction between two small masses.

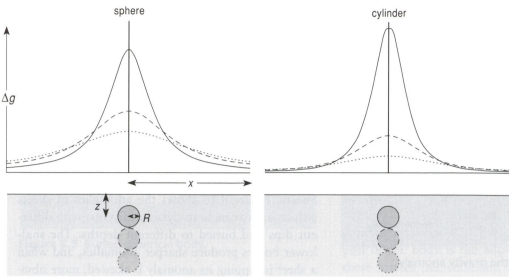

Figure 8.9 Anomalies of a sphere and a horizontal cylinder at different depths.

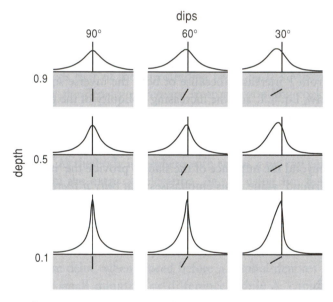

Figure 8.10 Anomalies of narrow sheets, or ribbons, at different depths and dips.

Though the slab produces no anomaly, its excess density does increase g everywhere, by an amount which can be calculated to be

$$\delta g = 2\pi G\, \Delta\rho t \qquad \text{Eq. 8.6}$$

where t is the thickness of the slab, and $\Delta\rho t$ is the excess mass of a square metre of the sheet. The depth to the slab does not matter, which is why it is not in the equation. This result will be used in Section 8.6.1.

However, a slab is detectable when it is not infinite, such as a 'half-slab' or 'half-sheet' (Fig. 8.12). Over the sheet far from its edge, at P_2, the situation will approximate to that of an infinite sheet and so g will be $2\pi G \cdot \Delta\rho \cdot t$ greater than at P_1, and there must be a transition between the two values, near the edge of the sheet. As with other bodies, the nearer the half-sheet to the surface, the sharper (i.e., shorter), is the transition.

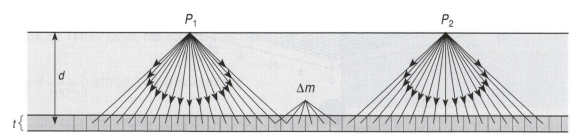

Figure 8.11 Horizontal sheet.

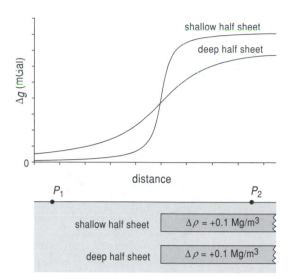

Figure 8.12 Horizontal half-sheet at two depths.

A half-sheet may seem unlikely but it can be produced by faulting. Figure 8.13a shows an example. Figure 8.13b is equivalent: It shows only the *lateral density differences,* for only these produce an anomaly. Each half-sheet produces an anomaly like those of Figure 8.12, and the total anomaly is their addition. In practice, the shallower half-sheets produce most of the anomaly, for the anomalies of the two halves of a deep faulted sheet on opposite sides of the fault will largely cancel; also a deep body produces a very broad anomaly that may not be noticed in a local survey (it will form part of the regional anomaly – see Section 8.6.2). Strike-slip faulting will not produce an anomaly if it offsets horizontal layers, but will if it offsets, say, parts of a vertical intrusion.

If a body has an irregular shape its anomaly is calculated by dividing it into simpler shapes, which approximate its shape, and calculating their combined anomaly, using a computer (e.g., Figure 8.17b).

8.5 Measuring gravity: Gravimeters

A gravimeter has to be able to measure differences much less than a millionth of the surface gravity, g. This can be done in three main ways. One way is to time the fall of an object; though this can be done very accurately in the laboratory it is not convenient for measurements in the field. A second way is to use a pendulum, since its period depends upon the value of g as well as upon its dimensions. This has been used but requires bulky apparatus and takes a considerable time. It has been largely superseded by the third way, which basically uses a spring balance, shown schematically in Figure 8.14a. The pull of gravity on the mass, m, extends the spring, so the spring has a slightly greater extension at places where g is larger. (A spring balance measures the *weight* of a body, the pull of gravity upon its mass, m, equal to mg. In contrast, a pair of scales compares *masses,* and if two masses are equal they will balance everywhere. Thus the mass of m is constant but its weight varies from place to place, as we know from the near-weightlessness of objects in space.) Modern gravity meters or **gravimeters** (see Box 8.1 for details) can detect changes in g as small as a hundredth of a mGal or about $10^{-8}\,g$, and some extra-sensitive instruments can do considerably better. (Such differences are much smaller than those measured in other branches of geophysics.)

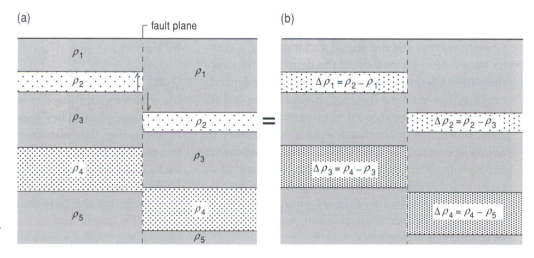

Figure 8.13 Layers offset by vertical faulting.

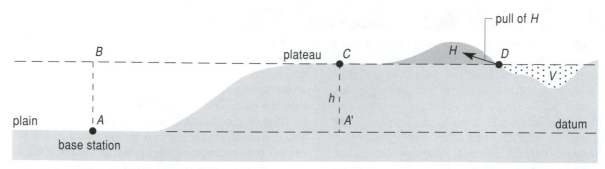

Figure 8.15 Topographic corrections.

Latitude correction. The Earth's rotation makes this correction necessary. Firstly, it tends to fling bodies away from its rotation axis, as water is flung off a spinning ball. This centrifugal force depends on the distance from the axis (as well as on the rate of rotation, which is essentially constant), being a maximum at the equator and decreasing smoothly to zero at the poles. Secondly, because of its centrifugal force, the Earth is slightly flattened, having an ellipsoidal section (Fig. 9.17). The Earth is said to have an *equatorial bulge* because the equatorial diameter is greater than the polar one, and this causes g to be less at the equator, because it is further from the centre of the Earth.

As both effects vary with the latitude, λ, of the gravity station, they can be combined in one formula (known as the International Gravity Formula):

$$g_\lambda = 978031.8(1 + 0.0053024 \sin^2 \lambda$$
$$- 0.0000059 \sin^2 2\lambda)\ \text{mGal} \qquad \text{Eq. 8.7}$$

For a survey extending no more than tens of kilometres the variation can be regarded as simply proportional to distance:

$$\delta g_{\text{lat}} = 0.812 \sin 2\lambda\ \text{mGal/km polewards} \qquad \text{Eq. 8.8}$$

where λ is the latitude where the survey is being made. Only the N–S distance matters, and as g increases towards the poles, the correction is *added* for every kilometre a station lies to the equator side of the base station, to cancel the decrease. For example, at the latitude of London the correction is about 0.8 mGal/km N–S; as a gravimeter can measure 0.01 mGal, a N–S movement of only 12 m produces a detectable difference.

Eötvös correction. This is needed only if gravity is measured on a moving vehicle such as a ship, and arises because the motion produces an centrifugal force, depending upon which way the vehicle is moving. The correction is:

$$\delta g_{\text{Eötvös}} = 4.040\,v \sin\alpha \cos\lambda$$
$$+ 0.001211\,v^2\ \text{mGal} \qquad \text{Eq. 8.9}$$

where *v* is the speed in kilometres per hour (kph), λ is the latitude, α is the direction of travel measured clockwise from north, because only E–W motion matters; the correction is positive for motion from east to west. For the latitude of Britain – roughly 55° N – the correction is about +2½ mGal for each kph in an east to west direction.

Topographic corrections. So far, we have supposed that the buried mass lies under a level plain, so that all stations are at the same altitude as the base station. When this is not so, there are three separate effects to be taken into account.

    i. *Free-air correction.* Firstly, imagine the gravimeter floats up in a balloon to B from the base station at A (Fig. 8.15). It is now further from the centre of the Earth and the reduced value of g is calculated using Equation 8.4 with R_E increased by the height risen; g decreases by 0.3086 mGal per meter rise. Therefore a gravimeter will respond to a rise of only a few cm!

    ii. *Bouguer correction.* Next, our balloonist floats horizontally from B to C, which is on a plateau (much wider than shown). At C there is an additional thickness, *h*, of rock below, compared to B, and its pull increases g. Assuming the plateau is wide, this rock can be approximated by an infinite sheet or slab and its effect on g calculated using Eq. 8.6:

$$\delta g_{\text{Bouguer}} = 2\pi G\rho h = 0.04192\ \rho h\ \text{mGal} \qquad \text{Eq. 8.10}$$

where ρ is measured in Mg/m^3 and h is in m.

As both the free-air and Bouguer corrections depend upon h they can be replaced by the combined elevation correction:

$$\delta g = h\left(0.3086 - 0.04192\ \rho\right)\text{mGal} \qquad \text{Eq. 8.11}$$

Here h is considered positive if above the reference level or datum (here assumed to be that of the base station). This correction effectively removes the rock above the datum and lowers the gravimeter to the datum. The free air correction is *added* for we know that going down increases g, but the Bouguer correction is *subtracted* because we have removed the pull of the intervening slab. The formula also works for stations below datum, provided h is given a negative value. For large-scale surveys the Bouguer correction is reduced to sea level so that surveys of adjacent areas can be compared, but for local surveys the base station is sufficient.

iii. *Terrain correction.* The Bouguer correction assumes an infinite slab, which is reasonable for a station on a wide plateau as at C, but not for a station situated as at D (Fig. 8.15). Imagine D was initially on the plateau and then a hill, H, was formed by bringing in material: Clearly, the mass of the mountain would give a pull sideways and partly *upwards*, so reducing g. Surprisingly, excavating valley V has the same effect, for it *removes* a *downward* pull. Therefore, the estimated pulls of hills above the station and of hollows below it are *added* to remove their effects.

The size of the terrain correction depends upon the detailed shape of the hills and hollows, particularly of the parts nearest the station, because of the inverse square law of the attraction. Methods are explained in more detail in the texts mentioned in Further reading.

When all the above corrections have been made, the corrected anomaly is called the **Bouguer anomaly** (not to be confused with the Bouguer *correction*, which is one of the corrections needed to calculate the Bouguer anomaly). In summary:

Bouguer anomaly = observed value of g + free-air correction – Bouguer correction + terrain correction – latitude correction + Eötvös correction

This assumes the station is above datum, and polewards of the base station; if either is not, the respective correction will have the opposite sign (it also assumes that instrumental readings have been converted to mGal and drift allowed for). If the terrain correction is small compared to the Bouguer anomaly, it is sometimes omitted because of the labour involved, in which case the result is sometimes called the 'simple Bouguer anomaly'.

The purpose of the Bouguer anomaly is to give the anomaly due to density variations of the geology below the datum, without the effects of topography and latitude. Most of these corrections can be made with a precision that is limited only by the care taken in measuring distances, heights, and so on, but Bouguer and terrain corrections depend upon the densities of rocks between the surface and the datum. If possible, these are estimated from samples, but even then sampling is likely to be limited to only a part of the rocks. There is the possibility of a change of lithology below the surface, particularly if the datum is far below the station; after all, the survey is looking for density variations and these may not be restricted to below the datum. Thus a degree of geological interpretation enters the last two corrections. How densities are found from samples or estimated in other ways is described in Section 8.7.

8.6.2 Residual and regional anomalies

A further stage of data reduction is needed to produce the anomaly we require. We have assumed that the only subsurface body present is the one we are interested in, but there may be others. The problem then is how to separate their anomalies. There is no foolproof way, but often there will be a difference in lateral scale of the various anomalies.

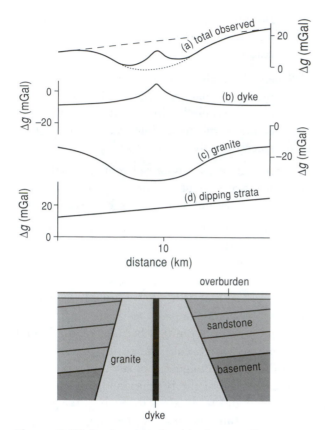

Figure 8.16 Regional and residual anomalies.

Figure 8.16 shows an area with dipping strata; if they have different densities they will produce a lateral variation in *g*. A granite has intruded the strata; its density is lower than the average density of the strata and so it produces a negative anomaly. A higher-density basaltic dyke intruding the granite produces a positive one. Above the section are shown the separate anomalies of the dyke, granite, and country rocks; their addition gives the anomaly we measure, shown at the top. If the objective is to locate the dyke, then the combined anomalies of the granite and strata provide a background value of *g* that varies laterally instead of being constant as we would like. To separate the dyke anomaly from the others, the anomaly of the rocks in the absence of the dyke has to be estimated; the *dotted line* shows the likely values. This is subtracted from the total anomaly to give the **residual anomaly**, which we hope is close to that of the dyke alone. The anomaly that is subtracted – due to the granite and other rocks – is called the **regional anomaly**; 'regional' does not necessarily mean that it extends over a geo-logical region, but just that it extends well beyond the body of interest.

If the objective were the granite then the regional is the anomaly due to the variation in the surrounding geology; this is estimated to be the *dashed line*. What had been the main cause of the regional for the dyke survey has become the anomaly of interest; conversely, the dyke anomaly is then just a nuisance. As it is smaller in lateral scale than the anomaly of interest it is sometimes referred to as noise (see Section 2.3), though noise will also include other small-scale effects such as variation in the thickness of overburden and imperfectly made corrections. In the example above, the regional anomalies were estimated by 'eyeballing'. Though this can be sufficient, the idea that the component anomalies have different lateral extents can be used in a more sophisticated way, using wavelength analysis or filters, as described in Chapter 3.

8.7 Planning and carrying out a gravity survey

The first step when planning a survey is to be clear about the objective, whether it is to detect the presence of a fault, the shape of a pluton, or whatever. Then an estimate should to be made of the size of the likely anomaly, using tabled values of rock densities unless actual ones are available; this determines the precision to which *g* needs to be measured and hence whether this is feasible. In turn, this determines how accurately the corrections need to be made; often the most important are the topographic corrections. For example, to measure an anomaly of about 10 mGal, *g* needs to be measured to a precision of at least 1 mGal; therefore, heights need to be known to about 30 cm (the free air correction is 0.3086 mGal/m, but the combined elevation correction is smaller than this; however, there will be other corrections, which will increase the error).

To find accurate heights of stations it may be necessary to carry out a topographic survey, though for regional survey with large expected anomalies, spot heights at bench marks may be adequate. Stations are often placed along roads, for ease of movement and because this is where most bench marks with surveyed heights are found. Global positioning systems (GPSs) using satellites are increasingly used, particularly in unsurveyed areas, though it is not yet possible to measure position much more precisely than a metre, unless sophisticated and time-consuming procedures are used.

The spring-balance type of gravimeter used for nearly all surveying measures only the *difference* of *g* between stations, one of which is usually the base station. This is sufficient to reveal a small-scale anomaly, but it does not allow results to be compared with those of independent surveys using a different base station. Therefore readings are usually converted to 'absolute' values by measuring the difference between the base station and a Standardisation Network station, where the value of *g* has been measured using special 'absolute' gravimeters. Topographic corrections are then usually made to sea level, rather than the base station datum, as mentioned earlier.

To allow the residual anomaly to be separated from the regional, the survey should extend well beyond the gravitational influence of the target body, on a line or grid, as appropriate to the shape of the target. The times of readings are recorded, with regular returns made to the base station, to allow the drift to be measured.

Densities are needed for Bouguer and terrain corrections as explained in Terrain correction in Section 8.6.1. Density is most accurately found by measuring a number of samples, preferably from unweathered exposures such as quarries, road cuttings, mines, or boreholes. If the rocks in situ are saturated with water, as is often the case, the samples will need to be saturated too. However, samples from outcrops tend to have a lower density than the same rock at depth, as was explained in Section 8.2. If the densities of rocks cannot be measured from samples, a variety of methods, which are described in more advanced books, can be used to estimate them, based on the form of the anomaly. If seismic data is available, another way is to use the empirically determined approximate Nafe–Drake relationship between density and seismic velocity (Nafe and Drake, 1957); an example is given in Section 21.4.

Marine surveys. Shipborne surveys differ in a number of ways from land ones. Firstly, the movement of the ship due to waves produces accelerations which are often far larger than the anomaly sought; these are reduced by averaging them over several minutes, with the help of a specially adapted meter, which is kept vertical on a stabilised platform. The Eötvös correction is needed to correct for the veloc-ity of the ship. But after all precautions, marine surveys are seldom accurate to better than 1 mGal (which can be attained using satellite gravity measurement, Section 9.5.1). Sea-bottom gravimeters, designed to operated remotely on the sea floor, have much higher precision but are slow to deploy. Station positions are found using modern navigational aids, particularly satellite positioning systems.

Topographic corrections differ from those on land. If a ship sails out from the coast *g* decreases because rock is being progressively replaced by less dense water. To correct for this and so allow the results of marine surveys in shallow water to match those on the adjacent land, the sea is 'filled in' using the Bouguer correction, but using $(\rho_{rock} - \rho_{water})$ in place of ρ_{rock}; *h* in Eq. 8.10 is the water depth and considered negative, so that the correction is *added*. For surveys in deep water there is little point in making this correction, so the Bouguer correction is omitted; the resulting anomaly is called the **free-air anomaly.**

Airborne surveys. These offer rapid coverage but the accelerations due to even slightly irregular motion of the plane reduce the accuracy to 5 to 20 mGal. However, accuracies of 2 mGal can be achieved by mounting the meter on a gyro-stabilised platform. In addition, sophisticated GPS (global positioning system) units monitor the plane's motion and can correct for any acceleration so accurately that flights made in smooth weather at constant barometric height agree with ground surveys to within 1 mGal. When combined with airborne magnetics (Section 11.1.3) airborne gravity is an attractive option, particularly for offshore areas with difficult access.

8.8 Modelling and interpretation

8.8.1 The inversion problem

Section 8.4 explained how the anomalies of known bodies can be calculated, but most geophysical surveys operate inversely: Having measured an anomaly, we wish to deduce the subsurface body that produced it. This is an example of the inversion problem (Section 2.4); for gravity, it means deducing a distribution of subsurface densities that can be interpreted as a body or structure. But there are several difficulties in doing so.

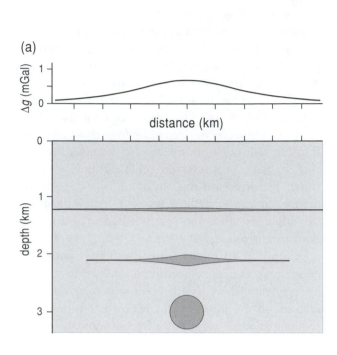

Figure 8.17 Different bodies giving identical anomalies.

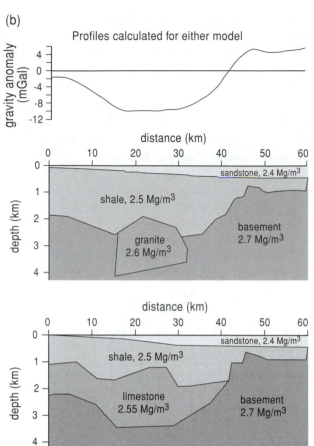

We saw in Section 8.4 that gravity anomalies of different bodies can look very similar – often a peak or trough tailing away to each side – even though the bodies that produced them have sharp edges and different shapes. In fact, bodies with different shapes can produce *exactly* the same anomalies. Figure 8.17a shows that a sphere and a lens-shaped body with the same density contrast with their surroundings can produce the same anomaly, while Figure 8.17b shows alternative models that match the observed anomaly of Figure 2.5 equally well. This is called **non-uniqueness:** Though it is possible to calculate uniquely the anomaly due to any specified body, it is not possible to deduce a unique body from an anomaly. This problem is additional to uncertainties arising from poor or sparse data, or interference from other anomalies (which make modelling even more difficult).

Another aspect of non-uniqueness is that anomalies depend only on density *differences* or contrasts: For instance, a sphere of density 2.6 Mg/m³ sur-

rounded by rocks of density 2.5 Mg/m³ produces the same anomaly as one of 2.4 Mg/m³ surrounded by rocks of 2.3 Mg/m³, and a half-slab to one side of fault with density higher than that of its surrounding produces the same anomaly as one to the other side with lower density (Fig. 8.18).

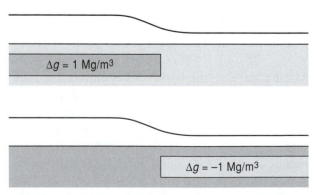

Figure 8.18 Two half-slabs with the same anomaly.

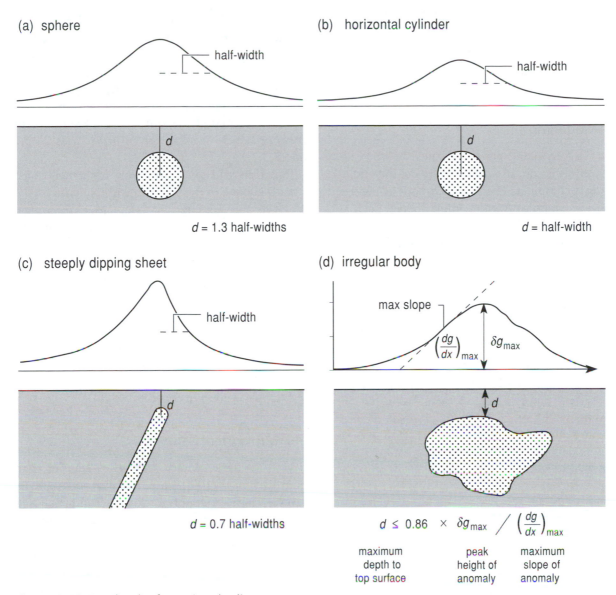

Figure 8.19 Depth rules for various bodies.

8.8.2 Depth rules

These limitations should not induce doubt of the usefulness of the gravity method for though many bodies may be compatible with an anomaly, most will be geologically implausible. Interpretation – as for all geophysical models – should be carried out in the context of all available knowledge.

8.8.2 Depth rules

Figures 8.9 and 8.10 show that the nearer a body is to the surface the narrower or sharper the anomaly it produces, and this can be used to deduce an upper limit to the depth of the body. The sharpness is mea-

sured as the **half-width:** *half* the width at *half* the peak height (Fig. 8.19a to c). The relation between half-width and depth to the body depends on the shape of the causative body; Figure 8.19 shows the rules for sphere, cylinder, and steeply dipping sheet. If the shape of the body is not known, or irregular, the gradient-amplitude method is used: the steepest slope of the anomaly, $(dg/dx)_{max}$, and the total height, δg_{max}, are measured and give a maximum depth to the top of the body as shown in Figure 8.19d. Another depth rule, Peter's half-slope method, is described in Section 11.3, and others can be found, for instance, in Telford et al. (1990).

Figure 8.20 Modelling a basin by slabs.

8.8.3 Modelling

If calculations using simple shapes do not give enough information, it is usual to model the anomaly using a computer. The modelled body is divided into a few components with simple shapes, such as polygons, with uniform densities (Fig. 8.17b). In practice, there is no point in refining a model beyond the point where the resulting changes in the calculated anomaly lie within the observational errors, or beyond its geological usefulness. As an example, a basin could be modelled by a succession of slabs (Fig. 8.20), with the number adjusted to give the required precision.

Most gravity modelling programmes for computers are 2½D (Section 2.4), that is, appropriate for bodies several times as long as their width; full 3D programmes are needed for more compact shapes, but require more computational effort. Examples of gravity modelling are given in Sections 21.3, 25.3.2, and 26.5.2.

8.9 Total excess mass

We saw earlier that modelling of gravity anomalies is limited by non-uniqueness, but the total **excess mass**, M_{excess}, of the body (or deficit mass for a body less dense than its surroundings) has no such limitation, and it can be shown that

$$M_{excess} = \frac{1}{2\pi G} \left(\text{volume of the anomaly} \right) \qquad \text{Eq. 8.12}$$

The volume is the volume 'under' the anomaly, which needs to be considered in 3D. In practice, the anomaly is divided into little columns whose volumes are added (e.g., Fig. 8.8). Each column should have an area small enough that there is little variation of g across it, so that its volume equals its area times its height.

The formula for the excess mass does not assume that the body has uniform density or require it to have any particular shape, nor does its

depth matter, but the contribution of the 'tails' around the anomaly needs to be measured carefully, for though their heights are small they can extend a long way. This requires that the anomaly has been correctly picked out from its surroundings, which in turn requires that the regional anomaly has been correctly estimated (Section 8.6.2). Examples of excess or deficit mass are given in Sections 24.4.1 and 27.3.2.

Equation 8.12 gives the *excess* (or deficit) mass of the body due to the difference of its density from that of its surroundings. To calculate its *total* mass – perhaps the mass of an ore – the densities of both body and its surroundings are needed (these may be available from borehole logs). Assuming the body and its surroundings each have a uniform density,

$$M_{body} = M_{excess} \left(\frac{\rho_{body}}{\rho_{body} - \rho_{surroundings}} \right)$$

$$\text{Eq. 8.13}$$

An example is given in Section 23.5.3.

8.10 Microgravity surveys

Surveys for targets with anomalies smaller than about 0.1 mGal (100 μGal) are called **microgravity surveys**. They are important for environmental and civil engineering surveys, such as checking sites for the presence of caves or mine workings.

Especially sensitive gravimeters, often called microgravimeters, are used, with stations often only a few metres apart on a grid, and heights measured to a precision of a few centimetres. Terrain corrections have to be carried out with great care and may include the effects of buildings. Examples are given in Sections 24.4, 27.3 and 27.4.

Summary

1. Gravity surveys measure variations of g, the acceleration due to gravity, over the surface of the Earth. These are produced by *lateral* variations in subsurface density and are used to investigate subsurface bodies and structures.
2. Gravity surveying depends on Newton's Law of Gravitation (Eq. 8.1).
3. Most rock densities fall in the range 1.5 to 3.0

Mg/m³ (1½ to 3 times the density of water). Density contrasts are often small, 0.1 Mg/m³ or less, though some metalliferous ores have a considerably greater contrast.

4. A gravity anomaly is the difference of g above or below its value in the surrounding area. The magnitude of an anomaly is often less than a millionth of the average value of g.

   Units of gravity are the mGal (10^{-5} m/s² ≈ 10^{-6} g) or the g.u. (10^{-6} m/s²); 1 mGal = 10 g.u. Gravimeters are extremely sensitive, able to measure changes in g of 1 part in 10^8 or smaller.

5. Field gravimeters measure *differences* in gravity between two stations. To determine the 'absolute' value of g at a station, it must be compared with the value at a Gravity Standardisation Network station.

6. Data reduction is very important and comprises instrumental calibration factor and drift, plus several corrections: tides, latitude, Eötvös (when the gravimeter is moving), free air, Bouguer, and terrain. Estimated densities – and geological judgement – are needed for the last two corrections. Densities of a several fresh samples of the rocks are better than tabled values.

7. The corrected data forms a Bouguer anomaly, corrected to sea level except for local surveys; a free air anomaly is used out at sea.

   The regional anomaly is removed to leave the residual Bouguer anomaly, which may be the anomaly due to the body or structure of interest. The regional anomaly itself may be of interest in the study of the deeper structure.

8. The spacing, direction (for elongated targets), and precision of gravity surveys are chosen to detect the likely size and shape of a target body; auxiliary measurements – particularly of heights – need to be made sufficiently precisely to allow corrections to match the precision of the readings.

9. Interpretation is either by approximating the target body to a simple geometrical shape or by modelling it with an assemblage of polygons, using a computer, in 2, 2½, or 3D, as appropriate to the structure and data. Depth can be estimated from the half-width or maximum slope of the anomaly profile.

10. Modelling is limited by non-uniqueness of shape, and because only density *differences* can

be detected (as well as by the quality of the data). Therefore geological, geophysical, and borehole constraints are needed to produce plausible models.

11. The total excess (or deficit) mass can be determined without theoretical limitation. The density of the country rock is needed to deduce the mass of the body.

12. Microgravity extends the method to smaller sized anomalies and is useful for investigating small bodies, including site investigations.

13. You should understand these terms: acceleration due to gravity, g, and G; gravimeter, mGal, g.u.; base station; drift, latitude, Eötvös, free air, Bouguer and terrain corrections; free-air and Bouguer anomalies; regional and residual anomalies; non-uniqueness; half-width; excess mass; microgravity.

Further reading

The basic theory of gravity prospecting is covered by chapters in each of Kearey and Brooks (1991), Parasnis (1997), Reynolds (1997), and Robinson and Coruh (1988), while Telford et al. (1990) covers it at a considerably more advanced level. Tsuboi (1983) is devoted exclusively to gravity and covers both chapters of this subpart, often at a fairly advanced mathematical level.

Problems

1. The value of g at the north pole, given g at the equator is 9.78, is m/sec².
   (i) 9.785, (ii) 9.805, (iii) 9.950, (iv) 9.832, (v) 10.005, (vi) 11.155 m/s².

2. A horizontal sill that extends well outside the survey area has a thickness of 30 m and density of 0.5 Mg/m³ in excess of the rocks it intrudes. Estimate the maximum depth at which it would be detectable using a gravimeter that can measure to 0.1 mGal.

3. An extensive dolerite sill was intruded at the interface between horizontal sandstones. Sketch the gravity profiles expected if the sill and beds have been displaced by:
   (a) A steeply dipping normal fault.
   (b) A shallow thrust fault.

levels, but it is simplest to choose the upper and lower levels to equal the highest and lowest parts of the blocks when these can be recognised. The lowest level of the blocks is called the compensation depth; below it, the density of the rocks is laterally uniform. It is at least 100 to 150 km deep, under continental areas, though much compensation takes place by variation in thickness of the crust, which is considerably less dense than the mantle below.

These two equations are used for solving simple isostasy problems, and we shall illustrate their use by calculating the effect of adding 2 km of ice on top of a continent (Fig. 9.6). The weight of ice causes the block to sink deeper, until a new equilibrum is reached (Fig. 9.6b). We regard the continent before and after the addition of the ice as two separate blocks, A and B. The base of block B is clearly the lowest part, so the lower level is placed here. Similarly, the upper level is chosen as the top of the ice, the highest point.

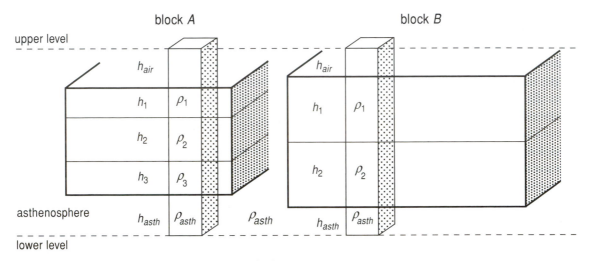

Figure 9.5 Columns through two layered blocks.

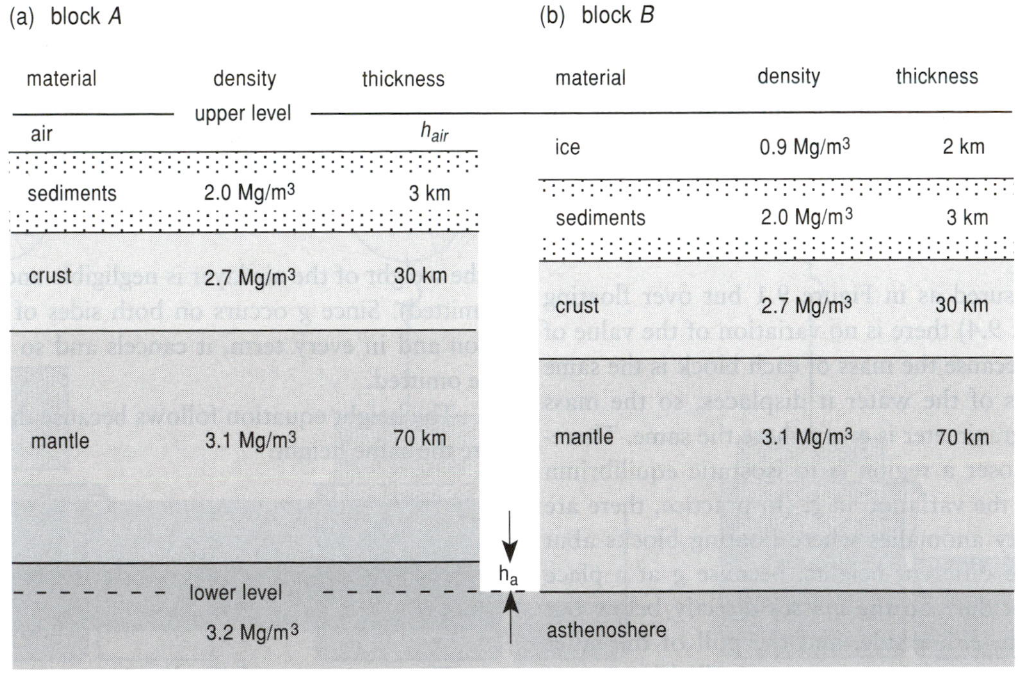

(a) block A

material	density	thickness
	upper level	
air		h_{air}
sediments	2.0 Mg/m³	3 km
crust	2.7 Mg/m³	30 km
mantle	3.1 Mg/m³	70 km
	lower level	
	3.2 Mg/m³	

(b) block B

material	density	thickness
ice	0.9 Mg/m³	2 km
sediments	2.0 Mg/m³	3 km
crust	2.7 Mg/m³	30 km
mantle	3.1 Mg/m³	70 km
asthenoshere		

h_a

Figure 9.6 Continent without and with ice sheet.

Applying the weight equation, Eq. 9.1, and putting in values where they are known gives

$$(3 \times 2.00) + (30 \times 2.70) + (70 \times 3.10) + (h_a \times 3.20)$$
$$= (2 \times 0.90) + (3 \times 2.00) + (30 \times 2.70) + (70 \times 3.10)$$

The symbol h_a is used for the one unknown quantity. All the terms cancel except one on each side, for most of the layers are the same for the two blocks, giving

$$h_a \times 3.20 = 2 \times 0.90$$

Therefore, $h_a = 1.8/3.2 = 0.5625$ km; that is, the weight of the ice caused the lithosphere to sink 0.5625 km.

As 2 km of ice was added but the lithosphere sank 0.5625 km, the top of the ice is only 1.4375 km above the original continental surface. Alternatively, the height equation (Eq. 9.2) can be used, which gives the same result.

$$h_{air} + 3 + 30 + 70 + 0.5625 = 2 + 3 + 30 + 70$$

(Conversely, melting of an ice sheet results in the lithosphere rising, as will be explained in Section 9.1.8.)

Because adding weight causes the lithosphere to sink, filling a large basin takes far more material than first expected, as the following example shows.

Suppose that on the previous continent there was an extensive lake 2 km deep in isostatic equilibrium (Fig. 9.7a). Gradually, material brought in by rivers replaces the water and completely fills the basin with new sediments (Fig. 9.7b). How thick will the new sediments be, if the area is in isostatic equilibrium? Since the top of the new sediments equals the former water level, this can be chosen as the upper level. Equating weights of columns (and omitting layers common to both columns) gives

$$(2 \times 1.00) + (h_a \times 3.20) = (h_s \times 1.80)$$

The old sedimentary layer, crust, and mantle have been omitted because they are common to both columns and so will cancel as in the previous example.

This equation cannot be solved immediately, because there are *two* unknown quantities, h_s and h_a (this h_a has a different value from the previous question). We use the height equation

$$2 + h_a = h_s$$

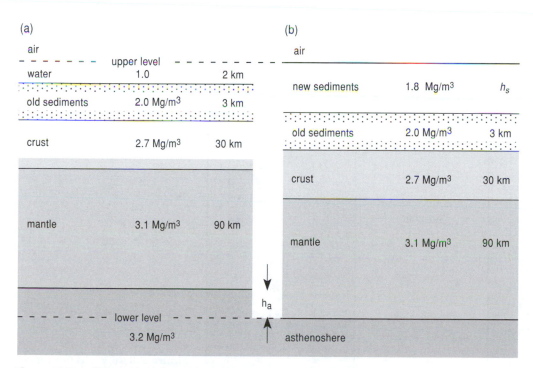

(a)

air

upper level — 1.0 — 2 km

water

old sediments 2.0 Mg/m3 3 km

crust 2.7 Mg/m3 30 km

mantle 3.1 Mg/m3 90 km

lower level

3.2 Mg/m3

(b)

air

new sediments 1.8 Mg/m3 h_s

old sediments 2.0 Mg/m3 3 km

crust 2.7 Mg/m3 30 km

mantle 3.1 Mg/m3 90 km

h_a

asthenoshere

Figure 9.7 Infilling of a lake with sediments.

(again omitting layers common to both columns). Substituting this expression for h_s into the preceding equation, we get

$$\left(2 \times 1.00\right) + \left(h_a \times 3.20\right) = \left(2 + h_a\right) \times 1.80$$
$$2 + 3.2 h_a = 3.60 + 1.8 h_a$$
$$h_a = 1.1429 \text{ km}$$

As $h_s = h_a + 2$, $h_s = 3.1429$ km. (You can check the values are correct by putting them in the weight equation and seeing that it balances.)

The filled basin is deeper than the original lake (and the increase can be greater, for thick sediments compact under their own weight). This effect helps explain how deep basins can form. Conversely, if material is removed by erosion the remaining material tends to rise, exposing it in turn to erosion, so that once deeply buried rocks can get exposed.

9.1.4 Airy and Pratt models of isostasy

We saw in Section 9.1 that mountains have low density roots, but have not yet considered the details. Two particular models have been proposed to account for isostasy. According to the **Airy model** (Fig. 9.8a), lithospheric blocks *all have the same density* but different thicknesses; the thicker blocks have their top surfaces higher and their lower surfaces deeper than thinner blocks, and so higher

ground is where the lithosphere is thicker. The weight equation, applied to the two columns, A and B, Eq. 9.1, simplifies to:

$$\left(h_{\text{lith}} \rho_{\text{lith}} + h_{\text{asth}} \rho_{\text{asth}}\right)_A = \left(h_{\text{lith}} \rho_{\text{lith}} + h_{\text{asth}} \rho_{\text{asth}}\right)_B \quad \text{Eq. 9.3}$$

The **Pratt model** (Fig. 9.8b) is more complex: Blocks all float *to the same depth* but have different densities; higher blocks are composed of less dense rocks, just as a sheet of plastic foam floats higher than a wooden plank. The weight equation becomes

$$\left(h_{\text{lith}} \rho_{\text{lith}}\right)_A = \left(h_{\text{lith}} \rho_{\text{lith}}\right)_B \quad \text{Eq. 9.4}$$

For both models, the height equation is unchanged.

Which model actually occurs? First, you need to realise that these are not the only possible models, just extreme ones. Blocks might differ in both thickness and density, in various combinations. In practice, seismic refraction surveys suggest that continental mountains are supported largely according to the Airy model, but with most of the increased thickness being in the crustal layer, which is buoyant, but there is also support due to lateral changes of density in both crust and uppermost mantle, whereas ocean ridges (Section 20.2.2) are close to the Pratt model. However, continents are higher than ocean floors due to both large density and thickness differences.

(a) Airy   (b) Pratt

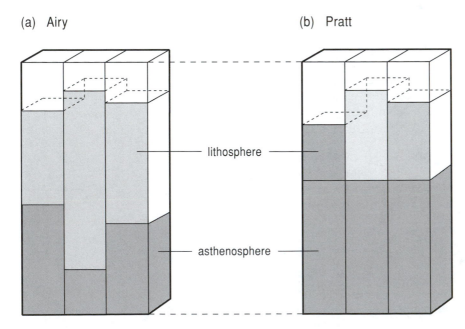

Figure 9.8 Airy and Pratt models of isostatic compensation.

9.1.5 Isostasy with regional compensation

The account of isostasy given so far is too simple. Both the Airy and Pratt models treat the lithosphere as being made of blocks that can float up and down independently, as material is added or eroded, but this would require enormous vertical faults, extending through the full thickness of the lithosphere (Fig. 9.9a), and these do not occur. Instead, the lithosphere is continuous and so has lateral strength, but when a large weight is added it flexes, depressing adjacent areas as well as the area beneath the load (Fig. 9.9b); this is **regional compensation**. Therefore, a better conceptual model than separate floating blocks of wood for representing the lithosphere is a sheet of plywood floating on a denser liquid (Fig. 9.10). Adding a small load has little effect (Fig. 9.10a), for it is supported almost entirely by the strength of the plywood with negligible bending; similarly, the weight of buildings or a small volcano is supported indefinitely by the lithosphere. But a large load (Fig. 9.10b) causes considerable bending. Its weight is still supported partly by the strength of the sheet, but also partly by buoyancy equal to the weight of liquid displaced by the depressed sheet.

(a) local compensation

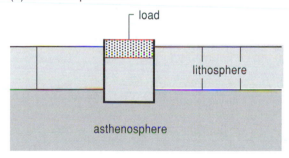

(b) regional compensation

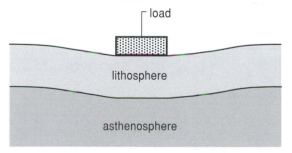

Figure 9.9 Lithosphere without and with lateral strength.

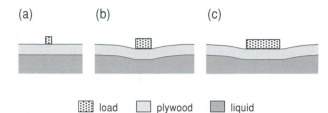

load   plywood   liquid

Figure 9.10 Flexural deformation with different-sized loads.

The sheet exerts a force only where it is curved, in trying to straighten itself. With a very extended load (Fig. 9.10c) – such as a large ice sheet or wide basin – the sheet will be flat except near the edges of the load and therefore the sheet exerts little force; the load is then supported mainly by buoyancy and so approximates to the simple isostasy model. In continental areas, which have a strong lithosphere, buoyancy is not usually the dominant support until loads are over a hundred kilometres across, but in some oceanic ones with a thin lithosphere it will be nearer 10 km.

In summary, small loads are supported by the strength of the lithosphere with negligible bending. Intermediate-sized loads are supported by a combination of the strength of the deformed lithosphere and buoyancy, with the deformation extending beyond the edges of the weight. The most extended weights – large mountains, ice sheets, and wide sedimentary basins – are supported largely by buoyancy. On this largest scale the simple Airy and Pratt models of compensation can be used. For loads of intermediate extent regional compensation is important. Some examples are given in Section 9.1.7.

9.1.6 The isostatic anomaly

In Section 9.1.2 we saw that if an area is in isostatic equilibrium g would be constant when measured in a balloon passing at constant height over the area. But as we measure g on the ground there will be variation due to any change in elevation, as explained in Section 8.6.1. Figure 9.11a shows a block floating in perfect isostatic equilibrium by the simple Airy mechanism, so gravity above the topography, along AA, is constant; but on the surface gravity at D will be less than at B because D is higher. However, if we apply the free-air correction to D (adding 0.3086 mGal times the height of D above B), they will be equal. Therefore, the free-air anomaly of an area in isostatic equilibrium will be uniform.

However, if we make the Bouguer correction as well, we effectively strip away the mass above the datum, and this will reduce the value of g. Therefore, the Bouguer anomaly of a plateau in isostatic equilibrium will be negative. This is why a Bouguer gravity map of, for instance, the British Isles – whose high land is partially isostatically compensated – shows generally negative values inland.

Conversely, if a high region has no isostatic compensation the Bouguer anomaly will be uniform but the free-air anomaly will be positive (Fig. 9.11b). A partially compensated area will have both a positive free-air and a negative Bouguer anomaly. The purpose of the isostatic anomaly is to measure how far there is isostatic compensation, and it equals the Bouguer anomaly minus the effect of the root.

The above considered the simple case of a very wide plateau, with compensation by the simple Airy mechanism. For more complicated situations, such as regional compensation, the free-air anomaly will not be uniform because compensation will not be locally perfect. However, it is possible to calculate, *according to a given compensation mechanism*, how

g of a fully compensated area would vary, and this is subtracted from the free-air anomaly to give the **isostatic anomaly**. It will also remove the edge anomalies (which occur when blocks with different heights abut), as shown in Figure 9.11. It is used to test the extent of isostatic compensation but does so only according to the mechanism assumed.

9.1.7 The evidence for isostasy

Now that we have introduced regional compensation, we shall look at some of the evidence that supports the idea of isostasy. The idea of isostasy arose from a topographic survey of northern India near the Himalayas. Surveying involves measuring angles from the vertical or horizontal, and these are defined using plumb lines or spirit levels. But the sideways pull of great mountains deflects the vertical and horizontal and so affects the angle measured. This deflection can be detected by comparing the vertical found using a plumb line with the direction to a star, which is not affected. A difference was found near the Himalayas, but only about a third that expected for their apparent mass. As other mountains had given the same result, it was realised that it was a general effect, and the Airy and Pratt models of isostasy were proposed to explain the apparently low mass of the mountains.

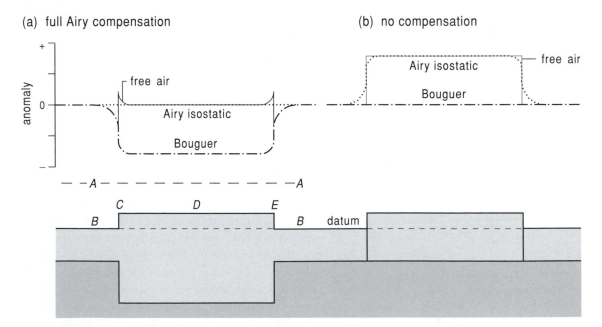

Figure 9.11 Anomalies with and without isostatic compensation.

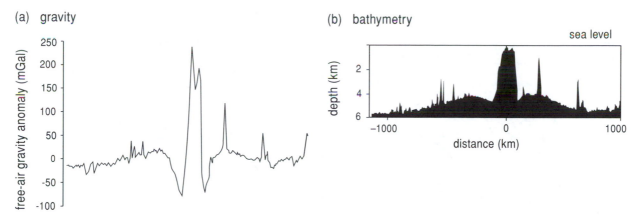

Figure 9.12 Bathymetry and gravity anomaly across Hawaiian Islands.

Evidence that the lithosphere sags under large loads (Fig. 9.9b) is provided by the Hawaiian islands. They have been 'built' on top of the oceanic lithosphere by volcanic eruptions, and in terms of their height from base to peak are the tallest mountains in the world, but with most of their mass below sea level. This weight has caused the ocean floor to subside and flex, so that it is deepest just beyond the foot of the islands (Fig. 9.12b), forming a 'moat' (this is superposed on a wider rise attributed to an uprising plume (Section 20.9.2) that has produced the volcanoes). The free-air gravity anomaly (free-air because it is measured at the sea surface, Section 8.7) is not level (Fig. 9.12a), the large troughs surrounding the peak showing that much of the support of the mountain is regional. Another example is that the surface of Greenland beneath the ice is depressed, as much as 250 m below sea level, while areas of northern Europe and America retain evidence of similar depressions, as described in the following section.

9.1.8 Isostatic rebound and the viscosity of the asthenosphere

We have not yet considered how long it takes for the lithosphere to regain equilibrium when a load is removed or added. If a large load – such as an ice sheet – is added, asthenosphere has to flow away to allow the lithosphere to subside (Fig. 9.13a and b); this can take thousands of years, for the asthenosphere is very viscous, like extremely thick oil. When the load is removed (e.g., the ice melts) the lithosphere slowly returns to its original shape by **isostatic rebound** (Fig. 9.13c and d).

We happen to live not long after the end of an extensive glaciation, which retreated from much of northern America and Europe roughly 10,000 years ago. The Great Lakes of North America lie in depressions produced by the weight of the ice. In Fennoscandia, a thick ice sheet once depressed the land, as the Greenland ice sheet still does. Surveys repeated over a period of years show that the area is rising (Fig. 9.14a), and most rapidly where the thickness of ice is likely to have been greatest, at the centre of the sheet. The total uplift since the end of glaciation has been found by measuring the height, above sea level, of beaches formed in the last glacial period; it is about 275 m, the rebound expected after the melting of an ice sheet about 1 km thick.

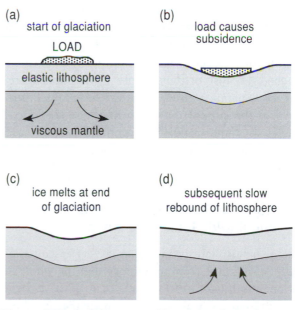

Figure 9.13 Subsidence and isostatic rebound.

(a) uplift

(b) gravity anomaly (Bouguer)

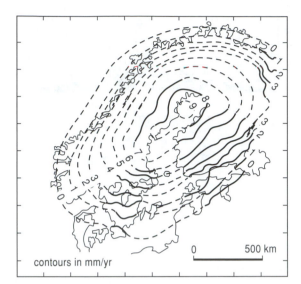 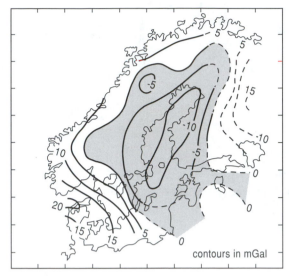

Figure 9.14 Uplift and gravity anomaly of Fennoscandia.

If rebound continues – perhaps with tilting, if the loading or unloading is uneven – it could have important long-term effects on harbours, or in off-setting locally the anticipated rise in sea level due to probable global warming, so how do we know whether equilibrium has been reached? As explained in Section 9.1.2, when isostatic compensation is perfect gravity (at a constant height above the land) will be uniform. Figure 9.14b shows that a large negative gravity anomaly exists around the Gulf of Bothnia, greatest where uplift is fastest, indicating a mass deficit, which is what we would expect for an area still rising by isostatic rebound. Of course, we should not expect perfect compensation to be reached, because of the strength of the lithosphere, but the anomaly is too large in extent for the lithosphere to be supporting much of the mass deficit. Therefore, Fennoscandia will continue to rise.

We consider next how the asthenosphere is able to deform.

9.2 How the mantle is both solid and liquid: Solid-state creep

The asthenosphere behaves like a very viscous liquid, yet we know from seismology that the whole of the mantle (apart from pockets of magma) is rigid enough to allow S-waves to propagate (Section 4.5).

How can the mantle, below the lithosphere, be both rigid and yielding?

These two behaviours are possible because they occur on different time scales. A ball of the material known variously as bouncing, or potty, putty slowly collapses to form a pool, like treacle, when left on a table (under the force of gravity steadily applied for some minutes), but it bounces elastically like a rubber ball if dropped onto the table (a force applied for a fraction of a second); if hit with a hammer, it shatters like glass (a large force applied very briefly). Pitch and toffee behave similarly, except that they take weeks or longer to deform under their own weight; similarly, glaciers flow slowly down valleys.

This dual behaviour is possible because of **solid-state creep**. There are several mechanisms of solid-state creep, but only one will be described, as an illustration. Most crystals are not perfect arrays of atoms, and one simple type of defect is a vacancy (Fig. 9.15). Random thermal vibrations with different energies continually occur in any material; occasionally, a sufficiently energetic one will cause an atom to break free from the bonds that hold it to adjacent atoms and move into the vacancy. In turn, another atom – or perhaps the original one – can jump into the position vacated so that in effect the vacancy wanders through the crystal. If the crystal is strained by an external stress, the vacancies will

(a)   (b)

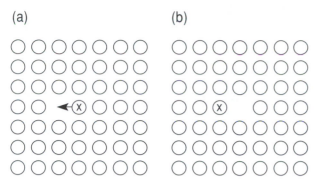

Figure 9.15 Vacancy diffusion.

tend to move to relieve the strain and so the material slowly deforms.

The essential difference of solid-state creep from the behaviour of a true liquid is that only a small proportion of the atoms are moving at any instant – whereas most are in a liquid – so the lattice remains intact. Therefore, if a stress is applied only briefly, as in the shaking associated with an S-wave, the material behaves as a rigid solid; but if the stress is applied steadily for a considerable time the jumps of atoms add up to a change of shape.

9.3 What is the lithosphere?

The idea of solid-state creep, coupled with temperature differences, explains most of the difference between the lithosphere and the asthenosphere (which do not correspond to any of the seismic layers described in Section 4.5). The vigour of lattice vibrations increases with temperature, so there is a temperature above which creep is significant (over a period of thousands of years), but below which the chance of atoms making lattice jumps is very small, so that the material behaves as a conventional solid. As temperature increases with depth in the Earth (Chapter 17), creep is significant below some depth, and this forms the boundary between lithosphere and asthenosphere. The boundary is not sharp, which is why it is not a seismic interface. It is deepest in areas with low geothermal gradient, such as cold, old continental areas (cratons), where it can exceed 200 km, and thinnest in areas of recent tectonic activity and young ocean floors, where it may be only a few km thick (see Chapter 17). The transition temperature also depends upon the material,

with acid rocks creeping at a lower temperature than basic ones; this partly determines the creep behaviour of the crustal part of the lithosphere, which has variable composition, but is much less important in the subcrustal part, which is more uniform. How thick the asthenosphere is is even less clear than for the lithosphere; it could be said to extend at least to the base of the upper mantle, but is usually taken to be the least viscous part underlying the lithosphere, roughly a hundred kilometres thick, corresponding roughly with the low-velocity zone (Section 4.5.4.).

9.4 Forces on the lithosphere

Because of isostatic compensation, we expect extensive loads to be supported mainly by buoyancy and therefore have only small gravity anomalies. However, large anomalies – large in extent and magnitude – do exist, for two possible reasons. One is that a load has been changed so recently – within the last few thousand years – that isostatic equilibrium has not yet been regained; the melting of ice sheets is an example. The second is that a large force is preventing isostatic equilibrium being reached.

The gravity anomalies with the largest magnitude are found near oceanic trenches (Section 20.4.1). These are not due to recent changes in loading, but are over subduction zones, places where an oceanic lithosphere is plunging into the mantle (Section 20.4.1); the large gravity anomalies are evidence of the forces that are pulling the plate into the mantle.

9.5 The shape of the Earth

9.5.1 Seeing the ocean floor in the ocean surface

The ocean surface is not perfectly level, even after allowance has been made for waves and tides, and Plate 3a shows its topography. Many readers will recognise a correspondence with many features of the ocean floors, particularly those associated with plate margins (compare with Plate 2c). Why does the ocean *surface* reveal the structure below? The link is through gravity anomalies.

Suppose a ball is resting on the Earth's surface at *B* (Fig. 9.16a), to one side of a buried sphere.

The ball will tend to move to the left because the sphere pulls it sideways as well as downwards. It will tend to move until it is directly over the sphere, at *A*, where the force is straight downwards. Though the force is too small to move a real ball, because of friction, it will move water, heaping it up over the sphere, until the water surface is everywhere perpendicular to the local pull of gravity (Fig. 9.16b), for then there is no longer a sideways pull to move it (such an equilibrium surface is called an gravity equipotential surface). Thus there is a small 'hill' where there is a positive gravity anomaly, a hollow for a negative one. A ship is unaffected by these slopes of the ocean surface, not because they are small, but because there is no sideways pull. The shape of the sea surface depends on the form of the gravity anomaly, but its calculation is beyond the scope of this book. Because even the largest gravity anomalies (over oceanic trenches) are only a fraction of 1% of the average value of *g*, the ocean surface deflections are no more than a few metres.

The shape of the ocean surface is measured using satellites that measure the distance to the ocean surface below them. This is done by timing the reflection of radar waves, in a form of echo sounding; in turn, the positions of the satellites are found by observation from ground-based stations. Great care is needed to eliminate the effects of tides and waves and other causes of error, but the average height of sea level can be measured to a precision of about 1 cm. The variations in sea level (Plate 3a) can be converted to provide a gravity map, accurate to about 1 mGal, which gives information about the structure of the ocean floors (Chapter 20).

9.5.2 The large-scale shape of the Earth

The first approximation to the shape of the Earth is a sphere; one with a radius of 6371 km has the same volume as the Earth. But this ignores the equatorial bulge, so any cross-section of the Earth through the poles is an ellipse rather than a circle (Fig. 9.17), though the difference is only a few kilometres. Therefore, a better approximation to the Earth's shape is an ellipsoid of revolution, the shape produced by rotating the ellipse about the Earth's axis. The one that best approximates the sea-level surface is called the **reference spheroid** and it has a polar radius of 6357 km and an equatorial one of 6378 km, giving a flattening of 1/298.

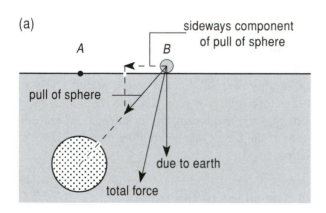

(a)

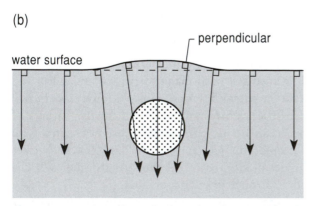

(b)

Figure 9.16 Distortion of a liquid surface by a buried mass.

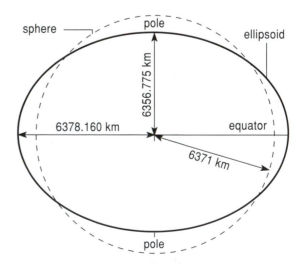

Figure 9.17 Shape of the Earth.

Even the reference spheroid is not an exact match to the sea surface, because of the effects of gravity anomalies, as described in the previous section. The actual mean sea level surface is called the **geoid** (for continental areas the geoid is the height that a sea level canal would have). The biggest differences between the reference spheroid and the geoid is no more than 80 m. Even so, they matter for exact surveying. Their cause is not well understood but must be due to extensive density differences deep in the Earth.

Summary

1. On the large scale – *lateral* density differences extending for tens of kilometres or more – the Earth behaves as having a solid lithosphere floating on a yielding and denser asthenosphere.

2. In the simplest model of isostatic adjustment, the lithosphere is treated as a series of blocks that float up or down in the asthenosphere, according to Archimedes' Principle, in response to changing loads. In the Airy model the blocks have the same density but different thicknesses; mountains are highest where the blocks are thickest and so go deepest, to form 'roots'. In the Pratt model, the blocks extend down to the same depth but have different densities; highest parts are where the blocks are least dense. The Airy model is appropriate for explaining major topographic variations on the continents, but the Pratt model accounts better for the shapes of ocean ridges. Continents are higher than ocean floors because of a combination of being lighter and thicker.

3. Isostatic adjustment of the simple block model follows Equations 9.1 and 9.2. [These are modified for the Airy and Pratt models, see Section 9.1.4].

4. The regional compensation model of isostasy takes account of the lateral, flexural strength of the lithosphere, which results in an extensive load being supported by an area that extents beyond its limits. Loads up to a few kilometres across are supported by the strength of the lithosphere with negligible deflection, intermediate-sized ones by both the strength of lithosphere and buoyancy, while the most extensive loads – upwards of tens of kilometres across, depending on the type of lithosphere – are supported mainly by buoyancy and this approximates to the simple isostasy model.

5. The mantle is able to behave as a liquid in isostatic adjustment, yet permits S-waves to propagate, because of solid-state creep. It has elastic rigidity when forces last only a short time, but deforms like a liquid under long-continued forces.

6. The boundary between the lithosphere and the asthenosphere is gradational and is where the temperature becomes high enough for the rate of solid-state creep to be significant. It also depends to some extent whether the rocks are acidic or basic. Its thickness varies from a few kilometres in young parts of the ocean floor to over 200 km in old continental areas (cratons).

7. Isostatic adjustment takes thousands of years because of the extremely large viscosity of the asthenosphere.

8. Isostatic compensation tends to eliminate gravity anomalies; therefore, gravity surveys reveal whether isostatic equilibrium exists. Large and extensive gravity anomalies exist only if the loading – such as an ice sheet – has changed too recently – within the past few thousand years – for equilibrium yet to have been regained, or because an internal force is preventing equlibrium.

9. The water surface is up to a few metres higher than average where there is a positive gravity anomaly, lower where there is a negative one. The ocean surface topography reveals major ocean bottom subsurface features, mostly reflecting plate tectonic structures.

10. The shape of the Earth is close to a reference spheroid with an elliptical polar cross-section. The geoid is the actual mean sea level surface (on land: sea level canals). It differs from the reference spheroid by no more than 80 m.

11. You should understand these terms: lithosphere, asthenosphere; isostasy, isostatic compensation, isostatic rebound, isostatic equilibrium; isostatic anomaly, Airy and Pratt models, regional compensation; solid-state creep; reference spheroid, geoid.

Further reading

Isostasy is explained in Tsuboi (1983) and in Fowler (1990), which also discuss the shape of the Earth.

Problems

1. Why does it seem a contradiction that the mantle can both transmit S-waves and rebound isostatically? Explain, on both an atomic and a bulk scale, how this apparent contradiction can be resolved.
2. How would you tell if an area is in isostatic equilibrium?
3. A large continental area covered with ice has a positive gravity anomaly. Which of the following might account for it?
   (i) The thickness of ice increased recently.
   (ii) The thickness of ice decreased recently.
   (iii) The thickness of ice increased several tens of thousands of years ago.
   (iv) The thickness of ice decreased several tens of thousands of years ago.
4. Explain how erosion of mountains can sometimes result in uplift of the peaks.
5. Melting of the ice in the arctic region would cause the sea level to
   (i) rise, (ii) fall, (iii) be unchanged.
6. If all the ice of Antarctica were to melt rapidly, would you expect (a) a thousand years later, (b) a million years later, that the shoreline around Antarctica would be, compared to the present, higher, lower, or the same?
7. If a continental area is in perfect isostatic equilibrium, which of the following are true?
   (i) A Bouguer anomaly map would show no variations.
   (ii) There are no lateral variations of density below the surface.
   (iii) No uplift or subsidence is occurring.
8. A wide block of wood 100 cm high and with density 0.72 Mg/m³ is floating in a liquid with density 0.96 Mg/m³.
   (a) Calculate how far the top of the block is above the surface.
   (b) How far would it be if 12 cm were removed from the base of the block?
   (c) How far would it be if 12 cm were removed from the top of the block?
9. A large area of continent consisting of 30 km of crust with average density 2.8 Mg/m³ and over 90 km of material with density 3.1 Mg/m³ is covered with ice (density 0.9 Mg/m³) 1.6 km thick, and is in isostatic equilibrium. Then the ice melts. By how much has the *rock surface* of

the continent changed when equilibrium has been regained? (The density of the asthenosphere is 3.2 Mg/m³.)
10. The crust of a continent contains a layer of salt, density 2.2 Mg/m³ and thickness 3 km, within sediments of density 2.4 Mg/m³. Over a period of time much of the salt is squeezed sideways, reducing its thickness to 1 km. This causes the surface of the continent to lower by how much? (Assume the asthenosphere has a density of 3.2 Mg/m³.)
11. A extensive area is intruded by three basaltic sills with uniform thicknesses of 30, 40, and 50 m. What is the change of the height of the surface after isostatic equilibrium has been restored? (Density of sill, 2.8 Mg/m³; density of asthenosphere, 3.2 Mg/m³.)
12. A sea, initially 2 km deep, over a long period of time fills to sea level with sediments. How deep are these sediments? (Use the following densities, in Mg/m³: water, 1; sediments, 2.4; asthenosphere, 3.2.) If the sediments had been denser, would the thickness have been greater or less?
13. How would you expect the depth to the asthenosphere below a continental area with high geothermal gradient (i.e., the temperature increases with depth more rapidly than beneath most areas) to compare with that below an area with a low gradient?
14. Why cannot the depth to the base of the lithosphere be measured using either seismic reflection or refraction surveys?
15. The uplift of former beaches around the Gulf of Bothnia is about 275 m. What thickness of ice would be needed to depress them back to sea level? (Density of ice, 0.9 Mg/m³; density of asthenosphere, 3.2 Mg/m³.)
16. The value of g at a place A is less than that at B. Which of the following might be the explanation?
   (i) A is at a higher latitude.
   (ii) A is at a higher elevation.
   (iii) A is underlain by lower-density rocks.
   (iv) A but not B was covered by a thick ice sheet a million years ago.
17. The value of g at a place varies with time due to which of the following?
   (i) Isostatic rebound.
   (ii) The topography of the continents.
   (iii) Lateral differences in the compositions of rocks.
   (iv) The Earth's rotation.

Magnetism

Compasses point approximately north-wards because the Earth has a magnetic field which aligns them; it is as if the Earth has a powerful magnet at its centre. Many rocks in the past gained a magnetisation in the direction of this field and retain that direction today. These past directions act like fossil compasses and can show if the rock has moved since it was formed. This has many applications, from following the movements of continents to the deformation of rocks in folding and even the angle through which a pebble has been rotated. The study and applications of fossil magnetism is called palaeomagnetism. The direction of the magnetic field has varied over time; in particular, it has reversed (i.e., inverted) many times. The reversal record in rocks provides a way of ordering and even dating them, leading to magnetostratigraphy.

Rocks become magnetised by various mechanisms, including cooling over a range of temperatures. This can be utilised to determine past temperatures, such as the heating caused by an intrusion. The magnetic properties of surface rocks and soils partly depend on their history; mineral magnetism exploits this to investigate, for example, the sources of sediments, rates of erosion, and the presence of volcanic ashfalls.

The first of the two chapters of this subpart, Palaeomagnetism and mineral magnetism, describes the above topics. The second chapter, Magnetic surveying, describes how measuring variations in the present-day magnetic field at the surface can be used to investigate the subsurface.

chapter 10

Palaeomagnetism and Mineral Magnetism

Palaeomagnetism utilises the fossil magnetism in rocks. One major application is to measure movements of rocks since they were magnetised, which can be due to plate movements or tectonic tilting. Another application is to measure the thermal history of a rock, such as reheating or the temperature of emplacement of a pyroclastic deposit. Successful applications require understanding the magnetic field of the Earth and how rocks become magnetised, and detecting whether this magnetisation has changed subsequently.

Mineral magnetism utilises the variation of the magnetic properties of rocks to study processes such as erosion and deposition. Magnetic fabric – when magnetic properties vary with direction in a rock – is used to deduce fluid flow when the rock formed, or subsequent deformation.

All the applications described in this chapter require access to the rocks, usually by collection of oriented material in the field.

10.1 The Earth's magnetic field, present, and past

10.1.1 Magnets and magnetic fields

Unlike the pull of gravity, you cannot feel magnetism, but its presence is revealed because it deflects a small suspended magnet, such as the needle of a magnetic compass. If a compass is placed in the vicinity of a stationary magnet and always moved along in the direction it is pointing, it traces out a path from one end of the magnet to the other (Fig. 10.1a). There are many such paths, starting at different points, and they crowd together near the ends of the magnet. The paths are called lines of magnetic field and they map out the magnetic influence or **magnetic field** of the magnet. They can be made more visible by sprinkling iron filings on a sheet placed over the magnet: the iron particles become tiny magnets, which join end to end along the field lines. Magnetic field is measured in Teslas, but for measuring the Earth's field this is too large a unit, so the **nanoTesla, nT,** which is 10^{-9} Teslas, is used instead.

10.1.2 The Earth's magnetic field at present

A compass points (approximately) northwards in normal use because the Earth has a magnetic field. This field – traced out by following a compass – is similar to that of the magnet in Figure 10.1a, greatly enlarged of course; it is *as if* there is a powerful bar magnet at the centre of the Earth. However, this magnet is not aligned exactly along the axis of rotation but is presently tilted by 11½° (Fig. 10.2a), and so **magnetic north,** as given by a compass generally differs from **geographic** or true north (geographic, or true, north and south poles are where the Earth's rotation axis intersects the surface). The places where the axis of the Earth's magnetic field meets the surface are called the north and south magnetic poles, and magnetic latitude and equator follow from them (Fig. 10.2b). The strength of the Earth's

magnetic field varies by a factor of about two over the surface, from roughly 30,000 nT at the magnetic equator to 60,000 nT near the magnetic poles. However, in this chapter we shall be mainly concerned with its direction, rather than its strength.

Figure 10.2a shows that the lines of magnetic field usually intersect the Earth's surface at an angle. Therefore, compass needles would dip up or down as well as point north, except that they are weighted to lie horizontally (as a result, a compass works properly only near the latitude it is balanced for). However, a magnetic needle pivoted with its axis horizontal so that it can rotate in a vertical circle aligned N–S (Fig. 10.3a) does indeed show the magnetic dip, called the **magnetic inclination.** The angle is called positive when the needle points down; so the magnetic inclination ranges from −90° at the magnetic south pole (pointing up), passing through zero at the magnetic equator, to +90° at the magnetic north pole (pointing down). The magnetic inclination, I, and magnetic latitude, λ, are related by

(a)

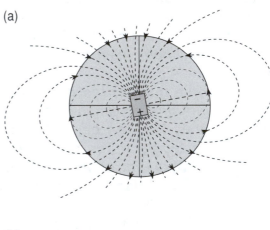

(b)

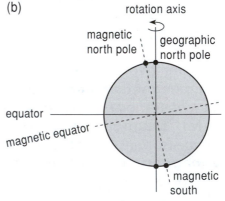

Figure 10.2 Magnetic field of the Earth.

(a)

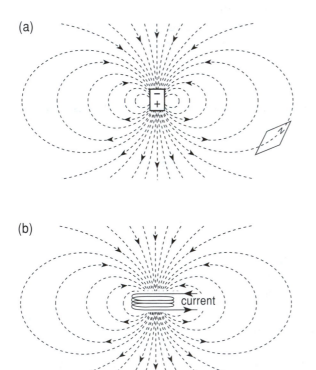

(b)

Figure 10.1 Magnetic field of a bar magnet and of a coil.

(a) dip circle

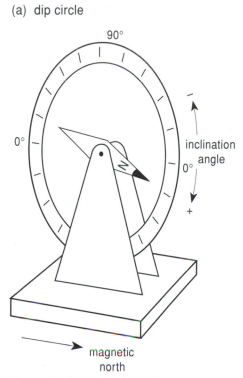

magnetic
north

(b) compass

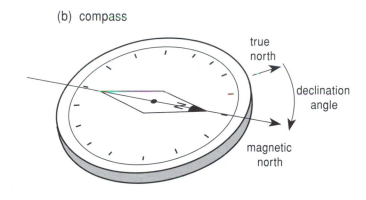

true
north

declination
angle

magnetic
north

Figure 10.3 Dip circle and compass.

$$. \tan I = 2 \tan \lambda$$
inclination latitude Eq. 10.1

This relation is shown graphically in Figure 10.4. Because the axis of the magnetic field is at an angle to the rotation axis, a compass usually only points approximately north; the difference from true north is the **magnetic declination** (Fig. 10.3b). Its value depends on one's position on the Earth's surface and is often very large near the geographic poles.

If you carried a dip circle and a compass you would know the direction of north (approximately) and your magnetic latitude (from Eq. 10.1), but you could not deduce your *longitude* because of the symmetry of the magnetic field about its axis.

We know that the Earth's magnetism is not actually due to a bar magnet, for reasons given later; instead, it is produced by electric currents in the liquid outer core. Figure 10.1b illustrates that a coil of wire carrying an electric current can produce the same magnetic field as a magnet. The currents in the Earth's core are more complex and are generated by flow of the outer core liquid in a complicated process called the geodynamo whose details

are only partly understood. Rather than refer incorrectly to a bar magnet, we shall call the source of the field a **magnetic dipole**, which just means that the source volume – magnet or coil – producing the magnetic field is far from where we are measuring its field; the Earth's magnetic dipole is within the core and so a long way from the surface. A dipole produces the simplest magnetic field and will be used in later sections.

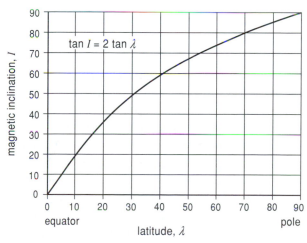

Figure 10.4 Magnetic inclination versus latitude.

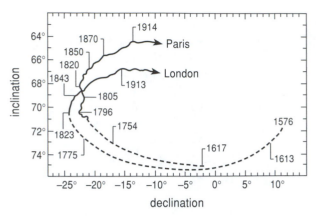

Figure 10.5 Secular variation in the past few hundred years.

10.1.3 The Earth's magnetic field in the past

We know from observatory records going back a few hundred years that the magnetic axis continuously changes direction; Figure 10.5 shows records for London and Paris, which differ because of their different locations. As a consequence, the correction to convert a compass reading to true north has to be changed every few years, and maps give both the declination and the rate of its change, the variation. Today the magnetic pole lies in northern Canada, but previously it has been in northern Europe and Asia. This slow, somewhat irregular, change in the direction of the field is called the **secular variation**. Changes before records were kept can be deduced from the palaeomagnetic record in rocks (described shortly), and these show that the magnetic axis wobbles about the rotation axis, somewhat irregularly, and takes about a couple of thousand years to complete a loop. Therefore, the direction when *averaged* over ten thousand years or more coincides with the direction of the rotation axis, and so with true north; this simplifies palaeomagnetic interpretations. In technical terms, we say that the Earth's field averaged over any period of several thousand years is close to that of a geocentric, axial dipole. When the *averaged* inclination is used, Eq. 10.1 and Figure 10.4 give the true palaeolatitude (i.e., the geographic latitude at the time of the rock's magnetisation).

However, there have been times when the magnetic poles have been interchanged. These **polarity reversals** – inversions of the field – have occurred at irregular intervals usually measured in millions of years (Fig. 10.21). In contrast, the time for the poles

to interchange is only a few thousand years; as this is short compared to most geological processes, it is rare to find rocks magnetised during the transition, and thus we can usually neglect the transition. We name the current sense of the field **normal** or **N polarity,** and the opposite **reversed** or **R polarity.**

In addition to reversals, there are excursions, in which the magnetic pole moves far from the geographic pole but returns within a few thousand years. They are significant only if they chance to be recorded, which is rare. The occurrence of reversals and excursions is one reason why the Earth's magnetic field cannot be due to a magnet. They complicate palaeomagnetic interpretation but can provide extra information, which is used in magnetostratigraphy (Section 10.5).

10.2 Palaeomagnetism

Many rocks retain a magnetism acquired long ago, often when they formed, and this is called **palaeomagnetism.** Its measurement, particularly of its direction, can provide different types of information, but first we need to understand how it is measured.

10.2.1 Measuring a palaeomagnetic direction

How is the past direction of the Earth's magnetic field found from rocks? As a simple example, suppose we wish to find the palaeomagnetic direction of a pile of Tertiary 'plateau' lavas, of the kind that cover large areas, such as the Columbia River Plateau in the United States, the Deccan of India, and most of Mull and Skye in northern Britain. Each lava was extruded and cooled in a few years and became magnetised; if the magnetism of the rock has been retained unchanged until the present, then each lava will give a 'instantaneous' record of the field direction where it was extruded and will not have averaged out secular variation. However, as lavas are typically extruded at intervals of several thousand years, the lava pile is likely to have taken hundreds of thousands of years to form, amply long enough to average out secular variation.

To find the direction of magnetisation of a lava, a sample is required. This may be a block, but usually it is more convenient to drill a short core, especially as this may penetrate into the unweathered interior. Before the block or core is detached it needs to be oriented: A line is drawn on it whose orientation relative to north (azimuth) and vertical (dip) is measured; north can be found by using a sun compass (i.e., taking a timed bearing on the sun), from a geographic bearing on a landmark, or – provided the rock is known not to be magnetic enough to deflect a compass – a magnetic bearing. If the rock has been tilted since its formation, this too has to be measured. Usually, six to eight samples are taken, separated by up to a few metres, to reduce errors due to orientation and other causes.

In the laboratory, short cylinders are cut from the cores or blocks and the magnetisations of these are measured using a magnetometer, specially designed to be sensitive enough to measure the usually weak magnetism of rocks. One common kind is a spinner magnetometer, in which the cylinder is spun, causing its magnetism to produce a tiny current in a nearby coil, in the way that electricity is generated by a dynamo, and this is measured by the instrument. This is repeated with the cylinder's axis in three mutually perpendicular directions, and the readings are combined to give the direction of magnetisation. This direction is with respect to the rock cylinder, but then it is calculated with respect to north and the vertical, using the orientation information of the core recorded in the field.

This measurement is repeated for cylinders from each of the cores taken from a single lava. We can make a check on our assumption that the magnetisation has not changed over time, for then their directions will be closely similar, as shown on Figure 10.6a. Usually, directions are displayed on a stereonet to give a **stereoplot** (Fig. 10.6b), which has the magnetic inclination ranging from vertical in the centre to horizontal at the circumference, while the declination is simply the angle around the circle clockwise from the top (north); Figure 10.6b shows the same directions as in Figure 10.6a. Downward magnetisations – positive inclination – plot on the lower hemisphere and are represented by open circles, directions above by solid ones. (An alternative way of showing directions, the vector component diagram, is described in Section 10.3.4.)

It is often convenient to replace the cluster of directions by an average, or mean, direction, plus an error (Fig. 10.7). We assume that sample directions are scattered randomly about the true direction, but with only a few samples it could happen, simply by chance, that they lie towards one side, giving an average away from the true direction; the larger the number of samples the less likely is such a bias, so the *average* direction becomes progressively closer to the true direction, and so its error shrinks.

(a)

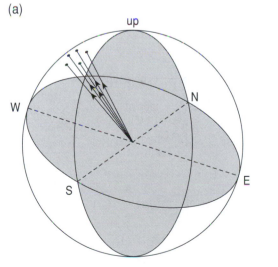

directions of magnetisation

(b)

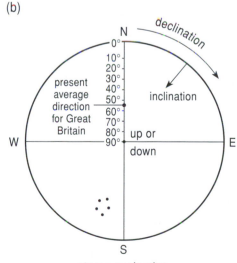

stereo projection

Figure 10.6 Palaeomagnetic directions.

(a)

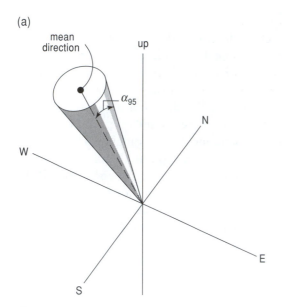

(b)

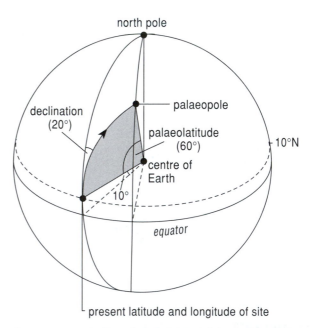

Figure 10.7 Mean direction and $\pm\alpha_{95}$ confidence angle.

The error used is the α_{95} confidence limit: A cone with this half-angle (Fig. 10.7) has a 95% probability of containing the true direction of magnetisation (*not* 95% of the sample directions), and is represented on the stereonet by an α_{95} **circle of confidence.** The average direction and the α_{95} error are calculated using special statistics of directions (see Further reading). Examples are given in Figure 10.19b.

10.2.2 Palaeopoles, palaeolatitudes, and rotations

As the inclination of the Earth's magnetic field depends upon latitude, rocks magnetised at the same time but at different latitudes have different magnetic directions. This makes it more difficult to recognise if, for example, two continents have moved apart. To get round this, the position of the north pole at the time of magnetisation is often calculated, as this is the same for rocks of the same age around the world *provided* they have not moved relative to one another since. For example, rocks from northern Canada and South America magnetised at the present day have quite different magnetic inclinations but the same magnetic pole (the present pole).

Suppose we drill cores from a group of lavas, formed hundreds of millions of years ago, which are now at latitude 10° N. Their palaeomagnetic directions are measured and averaged to give the direction of the axial field at the latitude of the rocks at the time

they were magnetised (we suppose that the rocks have not been tilted or rotated). The *direction* of north when the rocks were magnetised is given by the declination. Suppose it is 20° (east of north, unless stated otherwise) (Fig. 10.8). The *distance* of the north pole is calculated from the inclination, using Eq. 10.1. Suppose the inclination is +49°, which gives a palaeolatitude of 30° N; then the north pole was 60° (90° less the palaeolatitude) from the location of the rocks.

Figure 10.8 Finding the position of the palaeopole.

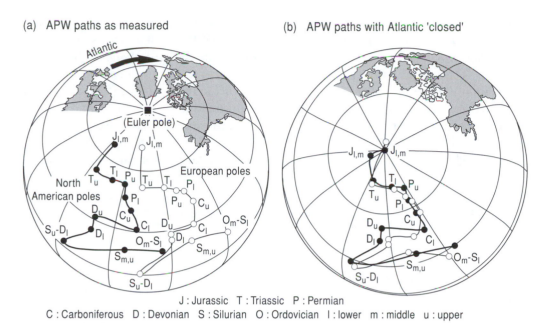

(a) APW paths as measured   (b) APW paths with Atlantic 'closed'

J : Jurassic   T : Triassic   P : Permian
C : Carboniferous   D : Devonian   S : Silurian   O : Ordovician   l : lower   m : middle   u : upper

Figure 10.9 Apparent polar wander curves for North American and Europe.

So the palaeopole is found by travelling 60° around the Earth along a great circle, starting from the present location of the rocks, in the direction of the declination, 20°.

We are not calculating the position of the palaeo north pole, which of course was at the north geographic pole, but where it was *relative to our rock sample,* so it is called the **apparent pole.** To understand this more clearly, think of a rock formed today at the equator: It has an magnetic inclination of zero. Suppose future continental movement takes it to the south pole: Its zero inclination would show that it had been formed at the equator, so its apparent pole would lie on the equator of that future time.

What can we learn about the movement of a rock? If the apparent palaeopole isn't at the present pole, the rock must have moved (assuming secular variation has been averaged out). If the declination is not due north the rock has rotated about a vertical axis. If the inclination does not correspond to its present latitude, the rock must either have moved north or south, or been tilted. Tilting can often be recognised and allowed for, in which case any remaining discrepancy is due to change of latitude, but because of the axial symmetry of a dipole field, *we cannot tell if the rock has changed its longitude;* for instance, recently formed rocks taken from

around the world along, say, latitude 30° N all have an inclination of +49° and declination 0°.

However, palaeolatitudes and rotations can be useful, even without knowing palaeolongitudes. For instance, the palaeolatitude can be used to predict the likely climate at some time in the past, for climate depends mainly on latitude. For example, in the Permian, Australia was further south, partly within the Antarctic circle, and so was much colder. An instance where rotation is important is shown in Figure 21.2.

In summary, the magnetisation of a single rock can often reveal its past latitude and whether it has been rotated, but not whether it has changed longitude. However, if we have a series of rocks of successive ages down to the present we may be able to deduce longitude as well, as the next section explains.

10.2.3 Apparent polar wander (APW) paths and relative continental movements

If a land mass has moved northwards or southwards over geological time, the palaeopoles of rocks of successive ages will change and trace out an **apparent polar wander (APW)** path (Fig. 10.9); it is called apparent polar wander because it is not the position of the Earth's pole that moves, but the continent. The APW paths can show if there has been *relative* movement between land masses, provided the APW paths cover the same time span. Figure 10.9 shows paths measured using rocks from the parts of Europe and North America shown.

The two APW paths for the period from the Ordovician to the Jurassic are not the same, but they have similar shapes. However, if the Atlantic is 'closed' by bringing together its coastlines, at the edge of the continental shelves, the paths coincide down to the Triassic but then diverge slightly. This is evidence that the two land masses were together until the Triassic. Note that if there were only one pole for each land mass many possible arrangements would bring the poles together, but with a long path there is only one. (In making such reconstructions, one land mass is usually treated as if it were stationary and the other moved.) After the Triassic the measured poles (only the Jurassic ones are shown) approach each other (Fig. 10.9a) until at the present day they coincide (at the north pole, of course) but on Figure 10.9b they would diverge. Thus, working back from the present, we can deduce past longitudes as well as latitudes.

APW paths can also show if what is now a single landmass formed from smaller parts. If so, the APW paths of the parts will be the same back until some past time, earlier than which they differ. Figure 10.10 shows the APW paths for Europe and Siberia as measured: They are the same back until the Triassic, but before this time the Siberia pole is to the west of the European pole, consistent with the two land masses coming together in the Triassic.

It is not always possible to detect relative movement. For example, if North America and Europe were to move apart without changing latitude or rotating (as if segments of the globe extending from the north to the south pole were being removed from the opposite side of the globe and inserted between the continents) their poles would remain at the north pole and so would not diverge. This occurs when the Euler pole of rotation – explained in Section 20.5.1 – is at a geographical pole. There are also, of course, uncertainties due to errors in measurement of the apparent poles.

10.3 The magnetism of rocks

The examples given so far have assumed that the rocks became magnetised when they formed and have retained that magnetisation unchanged to the present. Neither of these assumptions is always true, so we need to understand how rocks become magnetised and how to test if their magnetisation dates from their formation. This requires an understanding of how rocks are magnetic, which turns out to yield other useful information about rocks.

10.3.1 The atomic nature of magnetisation

Palaeomagnetism is possible because rocks have **remanent magnetisation,** the ability to retain magnetisation in the absence of a field or in the presence of a different magnetic field.

Ultimately, the source of the magnetism of a rock or magnet is magnetic atoms, each of which can be regarded as a minute bar magnet. The atoms of about half of the chemical elements are magnetic, but for most of these the directions of magnetisation of their atoms are randomly oriented (Fig. 10.11a). If such a substance is placed in a magnetic field, such as the Earth's, the magnetic atoms align with it to a very small degree (Fig. 10.11b), causing the material to become weakly magnetised, but this alignment and hence magnetisation disappears as soon as the field is removed, so this **paramagnetism** cannot produce a remanence.

In just a few materials the directions of the magnetisations of the magnetic atoms spontaneously align (Fig. 10.11c). Most common of such materials are iron or its compounds, so this magnetic behav-

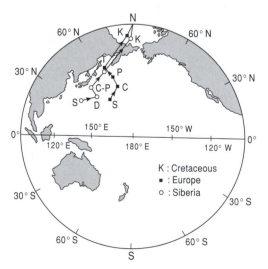

Figure 10.10 Apparent polar wander curves for Europe and Siberia. Letter key is given with Figure 10.9.

(a)   (b)   (c)

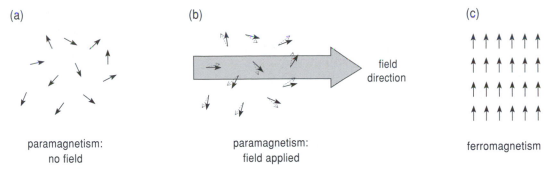

field
direction

paramagnetism: paramagnetism: ferromagnetism
no field   field applied

Figure 10.11 Paramagnetism and ferromagnetism.

iour is called **ferromagnetism.** (In this book 'ferro-magnetism' is used as a general term for the spontaneous alignments of atomic magnets. Sometimes the term is used in contrast to 'ferrimagnetism' and 'antiferromagnetism', which differ in how the atomic magnets are arranged, but these distinctions need not concern us.)

Most rocks contain some ferromagnetic minerals, compounds of iron. If the crystals, or grains, of ferromagnetic material are tiny, the directions of all the atomic magnets are aligned along one of a number of particular crystallographic directions, called easy axes, and the grains have a strong magnetisation for their size (Fig. 10.12). But a rock contains many such grains, and if their directions of magnetisation are randomly oriented, the rock as a whole has little remanence. However, if a magnetic field is applied to the rock the individual grain magnetisations will each tend to rotate into an easy axis closer to that of the field (Fig. 10.12). To do this they have to rotate through directions that are not easy, and this requires a relatively large magnetic field, but once they have rotated they will not easily change

back when the field is removed, so the rock has a remanence.

10.3.2 Magnetic domains

Only tiny grains have all their atomic magnets aligned in the same direction. Above a certain size (which ranges from 0.001 to 1 mm, depending on the magnetic mineral) the magnetisation spontaneously divides into small regions, called **magnetic domains** (Fig. 10.13), magnetised along different easy directions and separated from adjacent domains by a domain wall, within which the directions of the atomic magnets have intermediate directions (Fig. 10.13b). Subdivided grains are called **multidomain grains,** in contrast to **single-domain grains.** Domain walls tend to locate themselves where there are imperfections in the crystal, so it takes a significant field to move a wall to another imperfection. When the field is removed these imperfections tend to prevent the walls or domains returning to their earlier states, so the rock tends to retain a remanence. It is easier to change the magnetisation of multidomain grains than that of single-domain ones, because a wall will be moved by a smaller field than is needed to rotate the direction of all the magnetisation of a single grain; conversely, multidomain grains are less likely to retain their magnetisation unchanged until the present.

Because both single- and multidomain grains tend to return to some extent to a demagnetised configuration, the remanence after the field has been removed is less than the magnetisation while it is being applied. The part of the magnetisation that depends on the field is called the induced magnetisation and also is important, as will be explained in Section 10.6 and in Chapter 11.

no field   preferred   field applied
directions

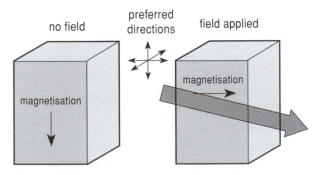

magnetisation

magnetisation

Figure 10.12 Single-domain grain.

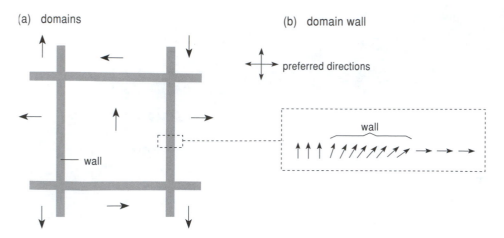

(a) domains

(b) domain wall

preferred directions

wall

wall

Figure 10.13 Domains and domain walls.

10.3.3 Curie and blocking temperatures

One way to magnetise a rock is by applying a magnetic field as just explained, but another way is by heating. If the temperature is slowly raised from room temperature, the random thermal oscillations associated with heat become more vigorous on average, so sooner or later one will chance to be strong enough to move a domain wall or even rotate the direction of magnetisation of a grain. *In the absence of a magnetic field*, such changes on average tend to randomise directions of magnetisation and so **demagnetise** a sample. Just as domain walls, or grains, require different magnetic fields to move them – depending on the size of the grains or the imperfections that 'pin' walls – so they need different temperatures. These temperatures are called **blocking temperatures,** T_b, and normally a rock has a range. Thus, raising the temperature results in progressive **thermal demagnetisation** of any initial remanence

(Fig. 10.14a). ('Blocking temperature' has similarities with 'closure temperature' in radiometric dating, as will be explained in Section 15.9.6.)

At a sufficiently high temperature, a second effect occurs: The individual atomic magnets cease to align with one another, and the spontaneous magnetisation necessary for ferromagnetism disappears. For a specific material, this happens at a characteristic temperature called the **Curie temperature,** or Curie point, T_C. Values for common ferromagnetic rocks and minerals are given in Table 10.1. The Curie temperature can be distinguished from the blocking temperature because it is higher and because above it the material has no significant magnetism even while a magnetic field is applied to it (Fig. 10.14b). (Above the Curie temperature the material becomes paramagnetic, and its weak magnetisation is responsible for the 'tail' in Fig. 10.14b.)

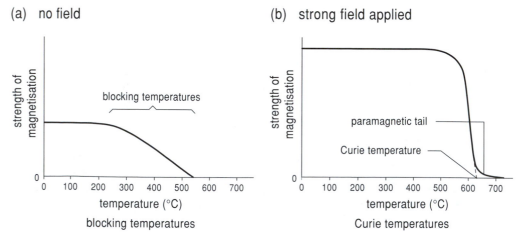

(a) no field

(b) strong field applied

blocking temperatures

paramagnetic tail

Curie temperature

blocking temperatures

Curie temperatures

Figure 10.14 Demagnetisation by increasing temperature.

The loss of magnetisation with increasing temperature is another reason why the Earth's magnetic field cannot be due to a magnet, for its interior is very hot and the outer core – being liquid – is above its melting point; no Curie temperature is as high as this. In fact, only the uppermost few tens of kilometres of the Earth are cool enough to have remanent magnetisation. Raising the temperature of a rock can destroy its remanence, but lowering it can allow a remanence to form, as explained next.

10.3.4 Thermal remanent magnetisation (TRM)

When a magma cools, it first solidifies, which includes the formation of grains of magnetic materials. With further cooling it passes through the Curie temperature as the atomic magnets within each grain align spontaneously to form one or more magnetic domains. The directions of magnetisation of the domains keep changing, or domain walls in multidomain grains keep moving, because the material is above its blocking temperatures, but on average there will be a small preponderance of magnetisation in the direction of the Earth's field. Then, as the rock cools through its range of blocking temperatures, first one, then another domain, or domain wall, will become fixed or 'block', and a net magnetisation is 'frozen' in. This is an effective way of magnetising a rock, for the resulting **thermal remanent magnetisation (TRM)** has a strength much larger than would result if the same field had been applied to the cool rock, and it often persists through geological time. It is the way that igneous rocks usually become

magnetised, and metamorphic rocks may also become magnetised in this way.

Measuring reheating temperatures. Sometimes an igneous rock is subsequently reheated, perhaps by an intrusion nearby. If it is heated above its highest blocking temperature, all its original remanence (termed **primary remanence**) will be demagnetised so that, when it cools again, it remagnetises in the Earth's field at the time. (This could lead to wrong deductions if the remanence is thought to date from the formation of the rock, so rocks needs to be examined for possible reheating.) If the temperature does not rise sufficiently to destroy all the primary remanence, some primary remanence is retained, but a new, **secondary remanence** is added as it cools. The **natural remanent magnetisation (NRM)**, which is the remanence of a rock as sampled regardless of how it was magnetised, is then a mix of the two remanences. It may be possible, in the laboratory, to retrieve the direction of the primary remanence.

Figure 10.15a and b show how part of the primary remanence *(AD)* is removed as the rock is heated, and a secondary remanence is added on cooling *(DA′)*, so that the new direction is *OA′* (Fig. 10.15b). If the rock is progressively reheated in the laboratory (in the absence of a magnetic field, even the Earth's, to inhibit further remagnetisation) this begins to remove the secondary remanence first (Fig. 10.15c), and the remaining remanence changes direction *(OB, OC, . . .)* until all the secondary remanence has been demagnetised *(D)*; thereafter, only the primary remanence remains, so a further rise of temperature causes the magnetisation to decrease but not change direction *(OE, OF, . . .)*.

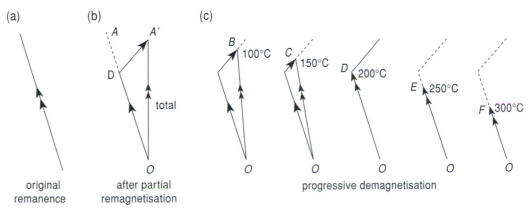

(a) original remanence

(b) after partial remagnetisation

(c) progressive demagnetisation

Figure 10.15 Partial remagnetisation by reheating.

In this example, the reheating temperature was 200°C, and above this the direction is primary.

Figure 10.15 shows the two directions of magnetisation in the plane of the paper, but we need to know what these directions are in 3D space. Figure 10.16a shows the directions relative to – in this example – up, east and south. A perspective diagram is inconvenient in practice, so the directions are projected onto two planes, here up–east and east–south, which are then displayed as in Figure 10.16b; the east axis is common to both parts. This is called a **vector component diagram** (or Zijderveld diagram); an example is given in Figure 11.21. Alternatively, a stereo plot could be used (Fig. 10.16c), but this does not show how the magnetisation changes in strength, only in direction.

One common cause of reheating is an intrusion. Reheating will reach its highest temperature close to the intrusion, at R (Fig. 10.17), and will be progressively less further away, until, at C, there will be no reheating at all. Figure 10.18 shows results for a Palaeozoic lava baked by a 6.2-m-wide dyke of early Tertiary age. D is the direction of magnetisation of the dyke and L that of the lava where it has not been

heated by the dyke; their directions are roughly antiparallel. A lava sample close to the contact (0.75 m from it) has a direction fairly close to D (Fig. 10.18), but thermal demagnetisation causes it to change, moving into the upper hemisphere towards L. Apart from changes at low temperatures, probably due to removal of a VRM (see Section 10.3.6), changes are small until about 515°C, after which the direction of the remaining magnetisation changes rapidly towards the original direction; this shows that the lava was heated to about 515°C. Samples further away from the contact were heated to lower temperatures (see McClelland Brown, 1981, for further details). A similar study showed that a kimberlite – an unusual type of intrusion from great depths that sometimes contains diamonds – probably was intruded at a temperature no higher than 350°C (McFadden, 1977).

Sometimes this type of analysis can give the temperature of emplacement of pyroclastic rocks. If a block is emplaced while still moderately hot, its subsequent cooling gives it a secondary magnetisation in the direction of the Earth's field of the time, whereas

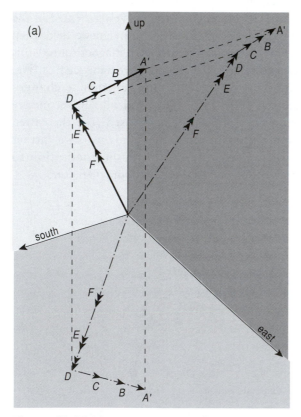
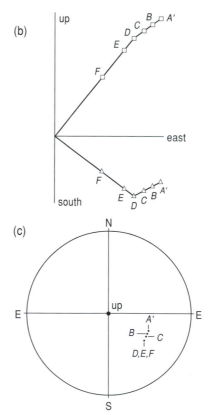

Figure 10.16 Vector component diagram.

any magnetisation acquired before emplacement is unlikely to have this direction; laboratory heating can recover the emplacement temperature.

There are other ways rocks become magnetised, but before describing them we shall introduce the most common magnetic minerals, for different minerals tend to be associated with particular magnetisation mechanisms.

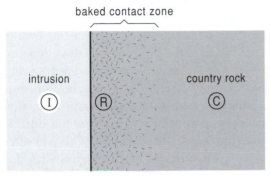

Figure 10.17 Lava reheated by dyke.

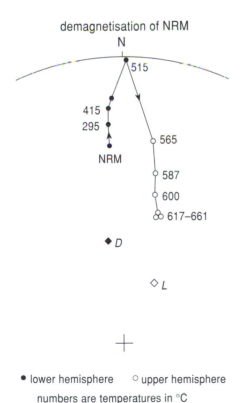

Figure 10.18 Thermal demagnetisation of a reheated lava.

10.3.5 Magnetic minerals

The most familiar magnetic material is iron. It is responsible for the magnetism of rocks, but combined with oxygen or sulphur, not as metallic iron. Other compounds of iron are common but are paramagnetic, and so do not contribute to the remanence.

Table 10.1 lists some magnetic properties of the most common magnetic minerals. The magnetisation given is the saturation remanence, the maximum remanence that the mineral can have, measured after applying and removing a very strong magnetic field. This is easy to measure, but as it differs from the magnetisation acquired in the Earth's field – being much larger – it is best used for comparing minerals; in any case, the magnetisation of a rock depends on the amount of a mineral present as well as on its type.

The magnetic mineral with the greatest remanence is **magnetite** (the iron is often partly replaced by titanium – forming a continuous solid solution series of titanomagnetites – which reduces its remanence). Another common mineral, but magnetically much weaker, is **haematite** (it forms a solid solution series of titanohaematites). It has a higher proportion of oxygen than does magnetite and so tends to form under oxidising conditions, such as weathering, and it is responsible (often in a hydrated form) for the rusty colour on rock surfaces and of water leaching from mines. **Maghaemite** forms by low temperature oxidation of magnetite; though it has the same chemical composition as hematite, it retains the crystal structure of magnetite and has a fairly high magnetisation; it is important in soils and sometimes is responsible for the magnetism of archaeological sites (Section 28.3.1). The other magnetic minerals listed in the table will not be discussed.

How magnetic a rock is depends on the compound formed as well as the concentration of iron present, as the great difference of saturation remanence of magnetite and hematite illustrates. In some igneous rocks, such as basalt, most of the iron is present in nonferromagnetic minerals such as olivine and pyroxene, and it is the small amount of magnetite also usually present that gives the rock its relatively large magnetisation. A further factor that determines remanence is grain size, for multidomain grains have a lower remanence than single-domain grains. Grain size can vary within a single rock mass; for example, chilled margins of igneous rocks are fine grained and may have a higher intensity of magnetisation that the interior.

Magnetic grains in sediments and those formed by weathering do not usually have a thermal remanence (TRM), and other ways of magnetising rocks are described next.

10.3.6 Mechanisms that magnetise rocks at ambient temperature

Many rocks, particularly sediments, become magnetised without being heated. One way, described in Section 10.3.2, is simply to apply and then remove a magnetic field. This is called isothermal remanent magnetisation, or IRM (i.e., magnetisation at constant temperature, though it is only one of several constant temperature mechanisms). It is a convenient method in the laboratory, for a strong field can be produced by electric currents, but it is not how rocks become magnetised (except occasionally by lightning strikes, when strong electric currents flow into the ground).

Chemical remanent magnetisation (CRM) takes place when a magnetic crystal forms in a magnetic field. This can happen by the chemical alteration of a nonmagnetic iron mineral into a magnetic one, as by weathering, or by precipating iron oxide (usually haematite) from water percolating through a rock. The latter is an important magnetising mechanism in sandstones, which also helps cement the sand grains together, so forming a 'red bed'. If the crystals grow above a certain size, domain walls form spontaneously and they cease to be single-domain grains. CRM is an effective way of magnetisation because, as with TRM, it does not involve changing the preexisting magnetisation of the grains. However, as the mineral usually formed in this way is haematite, which is

magnetically weak (Table 10.1), in practice it leads only to weak, though measurable, magnetisation.

If existing magnetised grains are deposited – together with other rock erosion products – to form a water-lain sediment, they tend to align their magnetisations with the field, like tiny compass needles, as they settle through the water. This **detrital** or **depositional remanent magnetisation (DRM)** is affected by any turbulence in the flow, which tends to randomise the orientations of the grains, while grains that have a flattened shape tend to align horizontally as they settle through the water, just as a dropped sheet of paper tends to land flat on the floor. Consequently, the direction of DRM may not align closely with the inclination of the Earth's field.

Some types of bacteria living in soils and near-surface sediments can produce grains of magnetite or maghaemite, particularly when there are organic compounds on which they can live. This can alter the magnetic properties of archaeological sites, for instance.

A further mechanism operates in rocks that have some blocking temperatures only a little above the ambient temperature. Though thermal fluctuations have an average energy that depends on temperature, they always have a range, and there will be occasional ones with much higher than average energy. Over a long period of time these cause domains with blocking temperatures that are not too far above the ambient temperature to be remagnetised in the direction of the field at the time. This slow, partial remagnetisation in the direction of the field is called **viscous remanent magnetisation (VRM)**, because the magnetisation changes direction sluggishly, like a compass needle in very thick oil. Some amount of VRM is common and is removed by heating to 100 to 220°C. An example is shown in Figure 11.21.

Table 10.1 Magnetic minerals in rocks

Mineral	Chemical formula	Saturation remanence (kA/m)	Curie temperature (°C)	Susceptibility* (rationalised SI units)
magnetite	Fe_3O_4	5–50†	585	0.07–20
haematite	Fe_2O_3	1	675	0.0004–0.038
maghaemite	Fe_2O_3	80–85	c. 740	
goethite	$FeO \cdot OH$	≤1	c. 120	
pyrrhotite	c. Fe_7O_8	1–20	c. 300	0.001–6.3
(iron)	Fe		780	0.2

*Defined in Section 10.6
†All ranges, which are from various sources, are approximate.

It should now be clear that a rock may not have acquired all its NRM when it formed, so it is important to test for this.

10.4 Testing when the remanence was acquired

There are two types of test, laboratory and field.

10.4.1 Laboratory tests

It was explained in Sections 10.3.3 and 10.3.4 how the magnetisation associated with lower blocking temperatures could be selectively removed by progressively heating a sample in the laboratory, in zero magnetic field, by the process called thermal demagnetisation. The method is rather slow to carry out because usually the sample has to be cooled before its remaining remanence is measured, and the heating–cooling–measuring cycle has to be repeated for progressively higher temperature. Further, the heating may change the nature of the magnetic minerals and hence their remanence.

Alternating-field demagnetisation ('a.f. demagnetisation') avoids some of these difficulties by using magnetic fields, rather than temperature, to demagnetise samples. A magnetic field of some value – the peak value – is applied, and this moves some domain walls or rotates some single-domain grains and then is switched off. To prevent this simply remagnetising the sample in the direction of the applied field, the field is then applied in the opposite direction but with a slightly smaller value so that most but not all of the domain walls or grains will be now be magnetised in the opposite direction. By alternating the applied field in this way while slowly reducing its value to zero, the magnetisations will end up divided equally into opposite directions and so cancel. To improve this cancellation the sample is usually 'tumbled' in space so that the magnetisations end up with random rather than just opposite directions. Only some walls or domains will have been demagnetised, depending on the value of the peak field, so the procedure is repeated with progressively larger peak fields. The magnetisation of the sample is measured each time the alternating field has been reduced to zero, corresponding to when the sample has cooled in thermal demagnetisation (Section 10.3.4).

Successful thermal or alternating field demagneti-

sation depends upon the secondary remanence being more easily demagnetised than the primary. This is usually true for thermal and viscous remanences, but often not for chemical remanence, for the magnetic material formed is often haematite, which is very hard to demagnetise, particularly using alternating-field demagnetisation. Demagnetisation by lasers or microwaves, which may select grains by type, is being developed but is not in general use.

Thermal or alternating-field demagnetisation – often called magnetic cleaning or washing as it removes the unwanted 'dirty' magnetic components – is routinely used but, because of its limitations, field tests are also used when possible.

10.4.2 Field tests

Fold test. Suppose a rock layer – a sediment or a lava – has been folded. If its magnetisation was acquired *before* the folding the directions of sites A and B, on opposite limbs (Fig. 10.19a), will not be the same, but they will be the same if they were magnetised *after* the folding. Therefore, if a tilt correction is made – effectively straightening the fold – the directions will group more closely if they pre-date the folding, less closely if they post-date it. This test may be applied to other forms of tectonic tilting. Figure 10.19b shows directions from four sites in the Himalayas that have been tilted by Late Eocene (c. 38 Ma) folding; each site has a different average direction, shown by their α_{95} circles of confidence being quite separate, though there is some overlap of individual sample directions between groups. After tilt corrections, the directions form a single group, making it very probable that the rocks were magnetised before the folding. Their average inclination after tilt correction is 55°, close to the 57° expected for an axial dipole field at the latitude of the sites, about 38°.

Conglomerate test. If a stably magnetised rock is eroded into clasts that are rounded during transport by water and then deposited to form a conglomerate (Fig. 10.20), the directions of magnetisation of the pebbles will be random, but if they were magnetised after deposition their present directions will be the same. Therefore, scattered directions show that the magnetisation pre-dates erosion, which makes it likely that the source rock of the pebbles retains its primary direction.

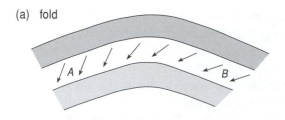

(a) fold

(b) fold test

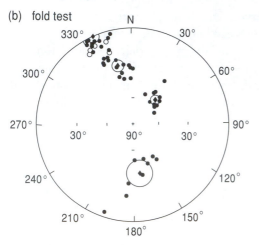

before tilt correction

○ reverse directions that have been inverted

after tilt correction

Figure 10.19 Fold test.

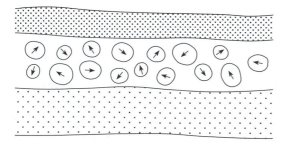

Figure 10.20 Conglomerate test.

Baked contact test. This test depends on intrusions remagnetising the adjacent country rock, as shown in Figure 10.17. Close to a large intrusion, *R*, the rock is heated to a high temperature and is completely remagnetised as described in Section 10.3.4, and so has the same direction of remanence as the intrusion, *I*, whereas rock far from the intrusion, *C*, retains its original direction (with, between, a zone of partial remagnetisation). If the intrusion and country rock have different ages, they are likely to have different directions of magnetisation. Then, if samples at *I* and *R* have the same direction but that of *C* is different, it is likely that *C*'s magnetisation pre-dates the intrusion; if they are all the same, it is likely they have all been remagnetised subsequent to the intrusion. If the direction at *R* agrees with that at *I* but not that at *C*, the directions at *I* and *R* probably record the field direction at the time of the intrusion.

These various tests only show if the magnetisation pre- or post-dates some geological event, such as folding or intrusion; if the former, it does not proves that it dates from the formation of the rock, though it gives it support.

10.5 Magnetostratigraphy

It was explained in Section 10.1.3 that the direction of the Earth's magnetic field is not constant – because of secular variation on a timescale of a few thousand years, and excursions and reversals, which occur at intervals of hundreds of thousands or millions of years. These changes of direction may leave their record in the rocks, and can be used to establish a stratigraphic order – a magnetic stratigraphy or **magnetostratigraphy** – or even to date rocks. Most useful are reversals, for they are global, occur abruptly compared to most geological processes, and are usually easy to recognise.

10.5.1 The magnetic polarity timescale

The times of reversals are found in principle by measuring the polarities of rocks with known ages, preferably dated radiometrically (Chapter 15). To ensure that no reversals are missed, and for other reasons, this needs to be done on a near-continuous succession of lavas; in practice, rocks of the ocean floor are used, for these are formed at a nearly steady rate at oceanic ridges, as is explained in Section 20.2.1. There are few such lavas older than about 160 Ma (million years) old, which is why the **polarity timescale** only extends back to this age. How the timescale has been worked out is explained in more detail in Box 10.1.

A current timescale is shown in Figure 10.21; the prefix C stands for chron, which is a significant interval of one polarity. These are numbered (with additions and deletions) up to the very long N inter-

nal which lasted for much of the Cretaceous; intervals earlier than this are numbered afresh, prefixed with M (for Mesozoic, see Fig. 15.19). Reversed intervals – unnumbered in the figure – are indicated by a suffix r and refer to the interval preceding the N interval with the same number (e.g., C21r preceded C21N). The *times* of the older reversals may have errors of up to about 1 Ma, but the *durations* of polarity intervals are known to a fraction of this.

BOX 10.1 Deducing the magnetic polarity timescale

The times of the last few reversals were found using many *isolated* lavas, by measuring both their magnetic polarity and their radiometric age (Chapter 15). Figure 1a is a schematic figure that shows invented dates of N and R lavas for the past few million years; the dot is the measured value, and its error is shown by the length of the bar through it. Because of the errors of the dates there is some apparent overlap in time of N and R lavas, but it is easy to recognise that the field reversed about 0.7, 1.6, 1.9, and 2.4 Ma ago. However, errors tend to be larger the older the lava, so it becomes progressively harder to recognise when and even if reversals occurred, especially as some older polarity intervals were very short. In practice, it is not possible to extend the polarity timescale back before about 5 Ma ago because the errors become larger than the durations of the polarity intervals.

Suppose, instead of isolated lavas, we had a *continuous* succession of lavas extruded at short intervals. This would provide a record of all reversals occurring while the succession formed; Figure 1b shows schematically how the lava polarities might appear. If, additionally, the lavas were extruded at regular intervals, their ages could be deduced by dividing the age of the oldest lava by the number of lavas. In this example, the oldest lava is about 10 Ma old, giving a lava interval of about 200,000 years, and hence the times of each reversal as shown. Though these times have errors, deriving from the error on the date of the oldest lava, we do not have the problem of the dates of N and R lavas overlapping, and all reversals can be detected and dated.

Though no such suitable pile of lavas is known, the equivalent exists on the ocean floors, which form nearly continuously at spreading ridges (Section 20.2.1). This allows the polarity time scale to be

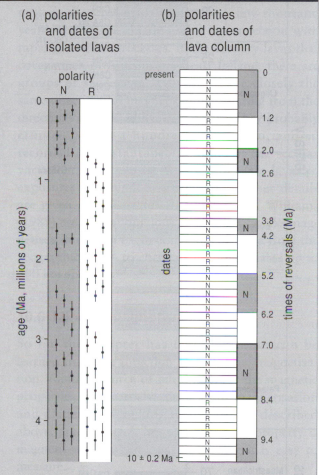

Figure 1 Establishing the times of reversals.

extended back to about 160 Ma. Continuously deposited sediments can also provide a record of polarity changes, but often the rate of deposition is so slow that a single sample may embrace a long period so that times of reversals are poorly defined. It is also often difficult to date sediments (Section 15.10).

$$M_i = \chi \times H$$
$$\text{induced magnetization} = \text{susceptibility} \times \text{field} \quad \text{Eq. 10.2}$$

The value of the susceptibility, χ (pronounced 'kai'), of a rock depends on both the type of magnetic mineral (Table 11.1) and its concentration (and also on the system of units used for the other quantities in the equation, though susceptibility itself has no units). If magnetite is present, it is likely to dominate the susceptibility (as well as the remanence). Susceptibility is measured by applying a known magnetic field to a sample and measuring the increased magnetism of the sample by the extra magnetic field it produces. Measurements can be made in the field or in the laboratory, the latter being more sensitive.

One example of using susceptibility is on glacial varves, which are formed of coarser particles carried by spring melt waters followed by finer particles as water flow slackens, in an annual cycle. The percentage of detrital magnetic minerals varies with the grain size and so also reflects the annual cycle; therefore, measuring susceptibility down a core drilled from the deposits can provide a convenient way of counting varves. An extension to this idea is to identify the contributions of different parts of a catchment area, when these supply different magnetic minerals, or to measure changes in supply over time, perhaps due to floods or to changes in erosion rates. Another cause of susceptibility change is deposition of volcanic ash, providing either a record of eruptions (see Section 24.3.1 for an example) or a stratigraphic marker, for ash can fall over a large area.

In China, large areas are covered by hundreds of metres thickness of loess, wind-blown dust, dating back some millions of years. Palaeosol horizons (fossil soils) within the loess record wetter climates, in some cases due to glacial cycles. These differences are reflected in differences in susceptibility and other magnetic properties, which provide a rapid and quantitative method for studying climatic changes on land, complementing marine studies using oxygen isotope ratios. In archaeological surveys, susceptibility is sometimes measured in the field to discover whether a particular area was once inhabited (Sections 28.3.1 and 28.4.2).

A more active role can be taken when studying fluviatile transport of suitable sediments. Material is removed from a river and made much more magnetic by heating, paramagnetic minerals being oxidised to ferromagnetic ones, which then acquire a TRM on cooling. The material is returned to the river upstream, and its progress downstream detected by the increase in susceptibility. Results are likely to be more valid than when using introduced materials, for these may travel at rates different from those of material found in the river.

10.7 Magnetic fabric: Susceptibility anisotropy

Some rocks are magnetically anisotropic, so the value of their susceptibility varies with the direction in which the field is applied. Though magnetic grains are often individually anisotropic, they also need to be at least partially aligned throughout a rock for it to be anisotropic. Susceptibility is measured in perpendicular directions and the results expressed as mutually perpendicular directions of maximum, minimum, and intermediate susceptibility (these in turn can be used to define a susceptibility ellipsoid, similar in concept to the strain ellipsoid). Anisotropy of susceptibility is often called **magnetic fabric.**

Alignment may be due to flow. When a sediment is deposited, magnetic grains may be aligned, and the resulting magnetic fabric can reveal the direction of the water flow. This has been used to determine palaeoslopes in a core of deep sea mudstone, which in turn indicated the direction from which the sediment came (Sayre and Hailwood, 1985). Magma flow is another aligning mechanism. In the dykes of the giant Mackenzie radial dyke swarm of Canada (Fig. 10.22a) the direction of maximum susceptibility is vertical within 500 km of the centre but horizontal beyond this distance (Fig. 10.22b). Maximum susceptibility is believed to be along the direction of flow, suggesting that magma ascended vertically in a column 1000 km across but flowed horizontally away from the centre of the swarm when it neared the surface. The swarm has been attributed to a plume, a column of magma that ascends from deep in the mantle, and then spreads out horizontally (Section 20.9.2). Further details are given in Ernst and Baragar (1992).

Another major cause of magnetic fabric is strain, grains being aligned by the deformation.

(a) Mackenzie dyke swarm

(b) stereo plots of susceptibility directions

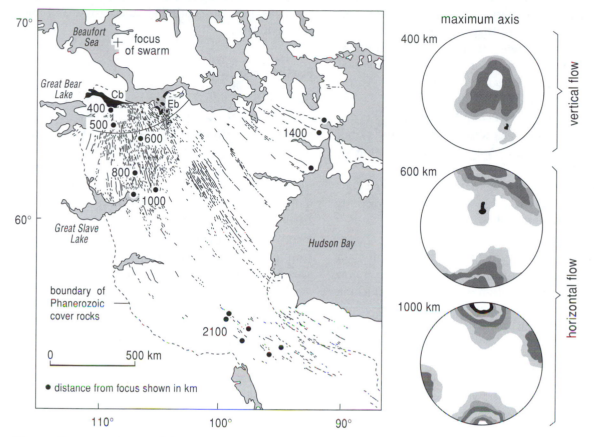

Figure 10.22 Mackenzie dyke swarm and magnetic fabric directions.

Summary

1. The Earth's magnetic field approximates to that of a geocentric dipole oblique to the rotation axis. Its direction varies – over hundreds of years – producing secular variation of the field. Its average over some tens of thousands of years is an *axial*, geocentric dipole. The Earth's magnetic field is mostly due to electric currents in the core, not to a giant magnet.

2. The direction of the magnetic field at a locality is given by its declination, D (positive east of north), and its inclination, I (positive downwards). The time-averaged inclination, I, is related to latitude, λ, by $\tan I = 2 \tan \lambda$.

3. The past magnetic direction recorded in rocks can reveal if a rock has changed latitude or been rotated about a vertical or horizontal axis, but it does not give the past longitude. However, past continental positions and relative movements can sometimes be deduced from apparent polar wander (APW) curves.

4. On a period of roughly millions of years the field reverses, taking only a few thousand years to do so. The record of reversals forms the magnetic polarity time scale, which is known well for the past 160 Ma. There are also excursions, during which the direction deviates by tens of degrees from the geographic axis but returns to its original polarity.

   These changes of direction are utilised in magnetostratigraphy for correlating and dating successions.

5. Palaeomagnetism is fossil magnetism, acquired in the past. It is possible because *some* iron compounds – oxides and sulphides – are ferromagnetic; that is, their atomic magnets sponta-

neously align. The most important magnetic minerals are magnetite and haematite.

All but small ferromagnetic grains spontaneously divide into magnetic domains. A sample has a remanence (remanent magnetisation) when – in the absence of a field – the volume of domains, or grains, magnetised in some direction exceeds that in all others.

6. The natural remanence (NRM) of a rock is due to one or more of thermal, chemical, detrital, and viscous remanence (TRM, CRM, DRM, and VRM).

7. Field tests (fold, conglomerate, and baked contact) can show whether remanence pre- or post-dates the event (folding, etc.).

8. A blocking temperature, T_b, is the temperature at which thermal fluctuations cause a domain wall to move or the magnetisation of a grain to rotate; rocks usually have a range of blocking temperatures. At a higher temperature, the Curie temperature, T_C, the ferromagnetism ceases.

9. Progressive laboratory demagnetisation (magnetic washing), by heating or alternating magnetic fields, may show if the remanence has more than one component. Sometimes a reheating temperature can be deduced.

10. Mineral magnetism utilises the magnetic properties of rocks, particularly magnetic susceptibility, and it is used for a variety of purposes, from establishing stratigraphic horizons to monitoring erosion.

11. Anisotropy of magnetic susceptibility – magnetic fabric – can be used to deduce flow directions in sediments or magmas or study strain in rocks.

12. You should understand these terms: magnetic field, magnetic dipole; magnetic and geographic north, declination and inclination; secular variation, excursion, polarity reversal, N and R polarities; apparent pole, apparent polar wander (APW); stereo and vector component plots, α_{95} circle of confidence; paramagnetism, ferromagnetism; remanent magnetisation, natural (NRM), thermal (TRM), chemical (CRM), deposition or detrital (DRM), and viscous (VRM) magnetisations; Curie and blocking temperatures; magnetic domain, single- and multidomain grains; magnetite and haematite; thermal and alternating-field demagnetisation;

primary and secondary remanence; polarity timescale, magnetostratigraphy; induced magnetisation, susceptibility, anisotropy, and magnetic fabric.

Further reading

Butler (1992) and Tarling (1983) provide general accounts of palaeomagnetism. Piper (1987) is mainly concerned with continental movements but describes the underlying palaeomagnetic methods and theory. Hailwood (1989) is a short account of magnetostratigraphy, while Opdyke and Channell (1996) is more extended and starts with a short account of the Earth's field and processes of magnetisation in sediments. Thompson and Oldfield (1986) describe environmental uses of mineral magnetism, with an account of the magnetism that underlies it.

Problems

1. The declination of a lava is 8°. Give at least two possible explanations for why it is not zero.

2. What are true north, geographic north, magnetic north?

3. Explain what these test for and how they differ: (a) Fold test, (b) Conglomerate test, (c) Baked contact test.

4. You wish to travel due north; you have only a magnetic compass to provide a bearing, but you know that for your location the magnetic inclination is +5° and the declination is 8°. Then, with respect to the direction of your compass needle, which of the following should you choose? (i) 5° East, (ii) 5° West, (iii) 8° East, (iv) 8° West.

5. When a particular sample was thermally demagnetised, its intensity increased up to 250°C, after which it decreased. Why was this?

6. The Earth's magnetic field is similar to the field that would be produced by a huge bar magnet at its centre. Give two reasons why it cannot be a magnet.

7. Which of the following relations between blocking temperature, T_b, and Curie temperature, T_C, is correct?
   (i) T_b is always greater than T_C.
   (ii) T_b is always less than T_C.
   (iii) T_b may be less or greater than T_C.
   (iv) T_b is less than or equal to T_C.

(v) T_b is equal to or greater than T_C.

8. If you wished to identify the magnetic mineral in a rock, would it be more useful to measure its Curie or blocking temperature?

9. Two rocks have the same iron content, yet their remanent magnetisations differ in strength. Give at least three reasons why the strength of the remanence of a rock does not depend just on its iron content.

10. The declination and inclination of rocks from the western end of an island a few kilometres long are 20° and +30°, whereas the declination and inclination of rocks from the eastern end are 35° and +30°. What explanation can you offer?

11. An American geologist, finding on arrival in New Zealand that his compass does not rotate freely, blames the airline for damage in transit. But there could be another explanation.

12. The equation $\tan I = 2 \tan \lambda$ is used to calculate palaeolatitudes. What assumptions are made in this equation? How valid are they likely to be?

13. A rock magnetised when it was 20° S of the equator has drifted to 20° N today. Its inclination differs from that of its present position by which of the following angles?
(i) 36°, (ii) 40°, (iii) 56°, (iv) 72°.

14. A small area of a continent has rocks of a range of ages. Their palaeomagnetic directions were measured, giving the data below (dates are accurate to about ±4 Ma, palaeomagnetic directions to ±2°). Plot the palaeolatitudes against time. What can you deduce about the continent?

Age (Ma)	Inclination (degrees)	Age (Ma)	Inclination (degrees)
300	−54	140	−38
250	−56	80	0
190	−55	60	+10
160	−50	present	+43

15. A small area of a continent has rocks of a range of ages. Their palaeomagnetic directions were measured, giving the data below (dates are

accurate to about ±4 Ma, palaeomagnetic directions to ±2°. Calculate palaeolatitudes against time. What can you deduce about the movement of the continent?

Age (Ma)	Declination (degrees)	Inclination (degrees)	Age (Ma)	Declination (degrees)	Inclination (degrees)
present	0	+36	100	10	−38
25	6	+26	140	190	+61
40	10	+19	160	10	−69
65	10	−6			

16. A sandstone has been cut by a 2-m-wide basaltic dyke. For each of the following results, is it likely that either the magnetisations of dyke or of the sandstone are primary?
(a) Samples of sandstone from within a few tens of centimetres of the dyke have the same direction of magnetisation as the dyke, but not of the sandstone several metres from the contact.
(b) All the samples of dyke and sandstone have the same direction.

17. A lava is cut by a dyke, both of late Jurassic age. Samples were taken from two sites in the dyke and also from the lava remote from the dyke. How do you account for the results, tabled below?

	Lava			Dyke	
Site	Declination (degrees)	Inclination (degrees)	Site	Declination (degrees)	Inclination (degrees)
1	341	+24	1	157	−24
2	337	+21	2	159	−22

18. What can you deduce from the information given about the age or duration of eruption of a succession of lavas, for each of the following two cases?
(a) They are of Palaeocene age and display the polarities succession N, R, N.
(b) They are of Campanian age, and all the lavas are normally magnetised except for a few at the base.

chapter 11

Magnetic Surveying

Magnetic surveying is useful in a wide range of problems, from planetary studies on a scale of thousands of kilometres using satellites to ones of a few metres over archaeological sites, and including mapping geological boundaries and faults, finding some types of ores, and locating buried pipes and old mine workings.

Magnetic surveying depends on the target producing a magnetic anomaly by locally modifying the Earth's magnetic field. The relationship of the anomaly to its source is much more complex than for gravity, for as well as depending on the source's shape and magnetic properties, it also depends on its orientation, the latitude at which the anomaly occurs, and – if it has a remanent magnetisation, as is usual – upon on its history.

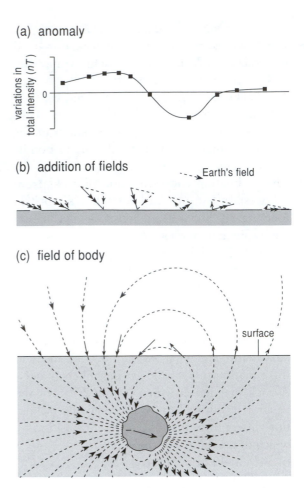

Figure 11.1 Magnetic field of a buried dipole.

11.1 Magnetic surveying

11.1.1 Anomaly of a buried magnet

To illustrate the idea of magnetic surveying, suppose a short but powerful dipole, such as a magnet, were buried (Fig. 11.1c). It produces a magnetic field, shown by the dashed lines, whose directions and magnitudes at the surface are shown by the arrows. Also present is the Earth's field, which over a small area is close to uniform. The actual field at the Earth's surface is found by adding the two fields together, taking into account their directions as well as strengths – that is, by vector addition (Fig. 11.1b). At some locations the field of the dipole is opposite in direction to the Earth's field, so the total field – shown by the double arrow – is less, giving a negative anomaly, while in others it is greater, giving a positive anomaly (Fig. 11.1a). In most real examples, the field produced is much less than that of the Earth's, so to detect the anomaly requires careful measurements using sensitive instruments.

This illustration shows that the anomaly of a buried magnetic body depends on both the direction of magnetisation of the body and on the Earth's field, which in turn depends upon the magnetic latitude (Section 10.1.2). This is one reason why magnetic anomalies are more complex than gravity ones.

The *measured* anomaly also partly depends on the type of magnetometer used, so magnetometers are discussed next.

11.1.2 Magnetometers

Magnetometers used for magnetic surveying differ from the types used to measure the magnetisation of rock samples in the laboratory because they need to measure the strength of the field rather than the strength of magnetisation of a sample, and done, whereas fluxgates give a continuous reading, which can result in quicker surveying when readings are taken at short intervals. Both types can measure the field to 1 nT or less, which is more than sufficient for most anomalies. In practice, proton magnetometers are used for most ground surveys, except that fluxgates are usually chosen for detailed gradiometer surveys (Section 11.7) and in boreholes (Section 18.6.2), for a variety of reasons. Both types are used in aerial surveys. How these instruments function is explained in Box 11.1. (A third type, the caesium vapour magnetometer, is more expensive but is sometimes used when higher sensitivity is needed, but will not be described further.)

11.1.3 Data acquisition

Ground magnetic surveys. The operator carrying the sensor should be free of magnetic objects, including keys, knives, zips, internal surgical pins, some types of coins, and fittings on boots, and – for the field geologist – not forgetting the compass. It is advisable to check for 'personal magnetism' by keeping the magnetometer still and taking readings with the user facing successively north, east, . . . and seeing that any differences are small compared to those expected in the survey.

The sensor is often mounted on a pole, with height selectable up to about 2 m. Raising the sensor makes it less affected by shallow magnetic bodies, particularly small ferrous objects, which are common in our throwaway society, though, as a result, source bodies will be less sharply defined. (Section 11.7, on gradiometry, considers this topic further.) However, this will not eliminate the unwanted effects of motor vehicles, railway lines, steel gates, barbed wire fences, and so on, so they should be passed at a distance, if possible, and a note made. Such objects generally give rise to sharp (i.e., very localised) anomalies (e.g., Fig. 11.16), so any such anomaly should be treated with suspicion, even if no ferrous object has been noticed. Readings are taken along a profile or sometimes over a grid, as described in Chapter 2, sometimes at closer intervals when the field is varying rapidly, to record the shape of the anomaly in more detail. Further details of field procedure are given in Milsom (1996).

Figure 11.2 Proton magnetometer.

they have to be portable. There are two main types. The proton precession magnetometer, usually abbreviated to **proton magnetometer** (Fig. 11.2) measures the *total strength of the magnetic field* but not its direction and so shows a **total field anomaly** (also called a total intensity anomaly), while the **fluxgate magnetometer** measures the *component of the field* along the axis of the sensor; for land surveying the sensor is usually aligned vertically for this is easy to do, giving a **vertical component anomaly,** but in aircraft, ships, and satellites it is kept aligned along the field and so measures the total field and also gives its direction. The sensor of a proton precession magnetometer need not be oriented (except roughly), and its readings do not 'drift' with time, in contrast to that of a fluxgate magnetometer, which has to be kept aligned and does drift. However, proton precession magnetometers take a few seconds to make a reading and are best kept stationary while this is being

BOX 11.1 Magnetometers: How they work

Proton precession magnetometer. The nucleus at the centre of a hydrogen atom is simply a proton, which acts as a tiny magnet and aligns with any magnetic field. Any liquid rich in hydrogen – such as water – could be used, but a liquid hydrocarbon such as kerosene is chosen because it will not freeze in cold weather. It is contained in a 'bottle' within a coil of wire (Fig. 1a). When a button is pressed to measure the magnetic field, the following sequence of operations is initiated: A current ('polarising current') flows in the coil and produces a field much larger than the Earth's (Fig. 1b), and the protons align along it. Then the current is switched off and the protons realign with the Earth's field, but they do not realign directly; each proton is spinning like a top, and – like a top, or gyroscope – this causes it to wobble, called precession, about the direction of field for a few seconds (Fig. 1c). The frequency of precession depends only on the strength of the field, so a measurement of the frequency can be converted to give the value of the field.

All the protons within the bottle precess together – provided the field is uniform throughout the bottle – and collectively they produce a tiny current in the now de-energised coil (by electromagnetic induction, described in Section 14.1.1). The frequency is accurately measured by suitable electronics, and further circuitry converts the frequency into the value of the field, which is displayed.

The requirements for a valid reading are these: (i) The coil has to be roughly aligned so that the direction of its field is at a large angle to the field to be measured, so that the protons precess before becoming realigned with the Earth's field. (ii) The field to be measured should be uniform throughout the bottle, or else the protons will precess at different frequencies, giving incorrect readings (a nonuniform field can occur near small, strongly magnetic bodies such as ferrous objects).

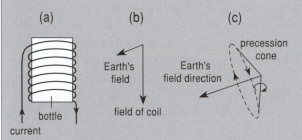

(a)   (b)   (c)

Figure 1 Principle of the proton precession magnetometer.

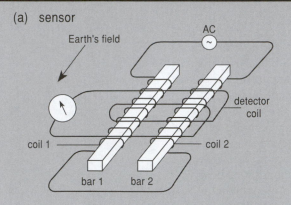

(a)   sensor

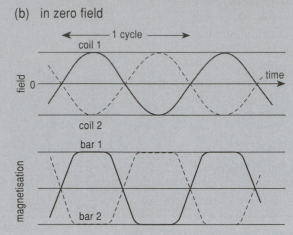

(b)   in zero field

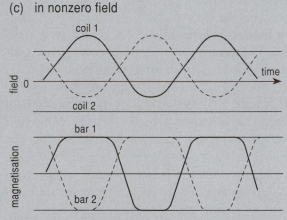

(c)   in nonzero field

Figure 2 Principle of the fluxgate magnetometer.

Fluxgate magnetometer. The sensor has two identical bars of magnetic material around which are wound magnetising coils, 1 and 2 (Fig. 2a), which are identical except that one is wound clockwise and the other counterclockwise. Around both of them is wound the detector coil. Alternating current flows through the magnetising coils to

BOX 11.1 *(Continued)*

produce a changing magnetic field, which in turn induces currents in the detector coil by electromagnetic induction. Because the coils are wound in opposite senses their fields are opposite in direction at any instant and together cancel, so no current is actually induced.

The bars are made of a special magnetic material that becomes magnetised to saturation (i.e., all domain magnetisations are aligned in the same direction) in a field less than that produced by the magnetisation coils; therefore, the bars have saturation magnetisation for much of each half-cycle of the alternating current (Fig. 2b). As the two bars are identical, their magnetisations do not upset the exact cancellation at the detector coil, but if there is an external field, such as the Earth's, parallel to the axes of the coils, as shown, it will add to, then subtract from, the field of the magnetising coil as the current alternates. Now the fields experienced by the two bars, though opposite, are no longer equal (Fig. 2c); in one half-cycle the bar in which the field of the coil and the Earth add reaches saturation sooner than the other; in the following half-cycle the other bar reaches saturation first. During the parts of the cycle when the bars are not equally magnetised there is a net field within the detector coil, which therefore has a current induced in it, and this is used to measure the strength of the Earth's field in the direction of the bars (in practice, a null method is used, in which the Earth's field is cancelled by passing a current through the detector coil and this is calibrated in terms of field). To work correctly the instrument is adjusted to give zero when there is no external field; if it does not keep in balance, the readings 'drift' with time. Only the component of the Earth's field parallel to the axes of the bars is measured. Therefore, if the sensor is turned the reading will change.

Figure 11.3 shows how a total field profile reveals two major geological contacts, between metagabbro (altered gabbro) and altered ultrabasics (this is actually the Moho, brought to the surface by obduction of sea floor, Section 20.8), and then schists that have been thrust over the ultrabasics. The variation of roughly 3000 nT is a very large anomaly; for comparison, the strength of the Earth's field at the latitude of Unst (about 60° N) is about 50,000 nT. Traversing from east to west, the strength of the field begins to rise some distance from the contact, because magnetic field extends far from a body, and returns to low values on crossing the second contact. Magnetic surveying therefore offers a convenient way of mapping the contacts. The variation within the ultrabasics (due to inhomogeneity and varying amounts of alteration of the rock) illustrates how the character of the anomaly can sometimes be used as a preliminary identification of lithology, useful in the interpretation of aeromagnetic maps.

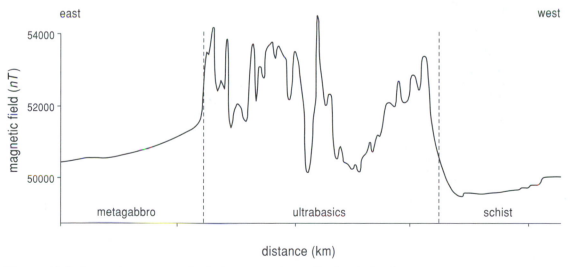

Figure 11.3 Anomaly across geological contacts, Unst, Shetland Islands.

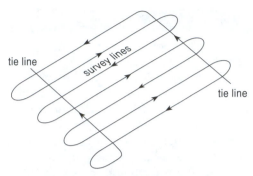

Figure 11.4 Aeromagnetic survey flight lines.

Aeromagnetic surveys. Magnetic surveying is one of the few geophysical techniques that can be effectively carried out from the air. The magnetometer sensor is either mounted on a boom extending from the plane or put in a 'bird' towed behind on a cable to reduce the effect of any magnetism of the plane. The plane usually flies along parallel lines at constant height above the ground surface, so far as this is feasible, with occasional tie lines at right angles to check for errors by comparing readings where lines cross (Fig. 11.4).

Aeromagnetic surveying has the advantage that large – and perhaps inaccessible – areas can be surveyed rapidly and more cheaply than by land-based surveys, but because of the height of the plane (and perhaps also the wide spacings of the lines), it cannot provide the detail of a ground survey. However, this may be an advantage if a regional survey of the deeper rocks is required, preferably without the effects of near-surface rocks (see Section 11.6.3). If more detail is required, ground surveys can follow.

An example of an aeromagnetic survey is shown in Figure 11.5a, for an area of Ayrshire, in southwest Scotland. It was known to have a variety of rock types and be intersected by several faults, but poor exposure had limited mapping. A low-resolution survey (ground clearance of 305 m, 2 km flight-line spacings) had shown magnetic anomalies due to belts of serpentinite, but with insufficient detail to identify individual rock units. A later survey, at only 60 m ground clearance and flight spacing of 250 m, provided much higher detail of part of the area. The anomalies have been emphasised by artificial illumination (Section 2.6) from the north, which picks out both northeast- and northwest-trending features. The boundaries between individual formations can be delineated in detail for the first time, permitting a map to be drawn (Fig. 11.5b). More information is given in Floyd and Kimbell (1995). Other examples of aeromagnetic surveys are given in Sections 22.5 and 23.5.1.

(a) aeromagnetic map

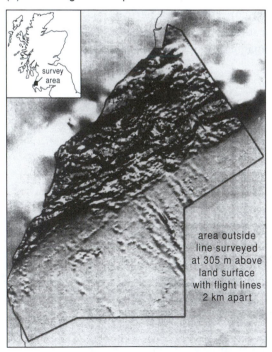

(b) geological map

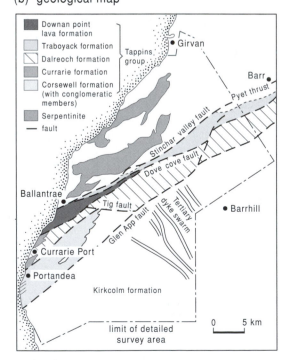

Figure 11.5 Aeromagnetic map of an area of southwest Scotland.

Shipborne surveys. Magnetic surveys are also easily carried out using a ship (in fact, the fluxgate magnetometer was originally developed to detect submarines). The procedure is similar to aeromagnetic surveying, with the sensor being towed some distance behind the ship at a constant depth below the surface; the device, appropriately, is called a fish.

11.1.4 Data reduction

Corrections needed for magnetic surveys are fewer than for gravity surveys and generally are of less importance.

Diurnal variations. The strength of the Earth's field changes over the course of a day, called **diurnal variation.** Variation seldom reaches as much as 100 nT over 24 hrs, but during occasional magnetic 'storms' it can exceed 1000 nT. As the diurnal variation changes smoothly (unlike storms), it can be measured by returning periodically to the base station, in the manner of drift correction of a gravimeter (Section 8.6.1); alternatively, a magnetometer that records values at regular intervals can be left at the base station and is more convenient. Figure 11.6 shows an example. To correct survey readings for diurnal variation, first some time is adopted as a reference – here it's the time the recording began – and then diurnal changes from this time are read off at the times of survey readings and added to (or subtracted from) the survey readings.

If the anomaly is large or the time spent recording it is short, correcting for diurnal variation may make little difference to the anomaly, but recording it provides a check that the instrument is functioning correctly. It will also detect if a magnetic storm is occurring; these can last for hours or days, greatly increasing the variation and sometimes rendering a survey valueless.

Magnetic field decreases with altitude but usually by too small an amount to matter (the decrease ranges from about 0.01 to 0.03 nT/m, depending on latitude), nor are readings reduced to a common datum, such as sea level. Because proton precession magnetometers give the actual strength of the magnetic field (unlike most gravimeters), there is no need to link surveys to stations of a regional network.

Subtracting the Earth's global field. Unlike gravity, the magnetic field varies markedly over the Earth's surface, being roughly twice as strong at the Earth's magnetic poles as the equator. This variation does not matter for surveys of small areas, but is corrected for on regional ones. The field subtracted is the IGRF – International Geomagnetic Reference Field. (This matches the actual field measured at observatories around the Earth by the fields of a series of fictitious magnets at the Earth's centre, known as a spherical harmonic series; a fairly good fit is provided by a single dipole (Section 10.1), but progressively smaller magnets with various orientations are used to improve the match to about 1 nT on average, though without matching local anomalies. The IGRF is considerably more complicated than the International Gravity Formula used for latitude corrections in gravity surveys. It is revised every five years to take account of secular variation, with the latest revision in 2000.

11.2 Anomalies of some simply shaped bodies

The purpose of this section is to show how anomalies relate to the shape, orientation, and latitude of the bodies that produce them. An example was given in Section 11.1.1; this showed that we need to know both the direction of the magnetic field produced by the body and its strength. We therefore need to understand how magnetic fields are produced.

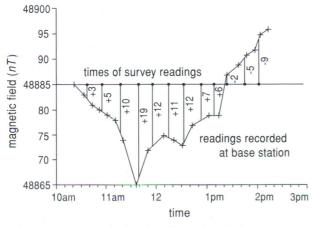

Figure 11.6 Correcting for diurnal variation.

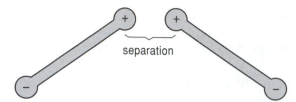

Figure 11.7 Magnets with isolated poles.

11.2.1 Magnetic poles and fields

A magnetic body can be thought of as being made of **magnetic poles**. Poles always occur as positive and negative pairs, but magnets can be made so long and thin that the effect of a single pole can be isolated (Fig. 11.7). Poles repel each other if they are both positive or both negative but attract if of opposite polarity. In either case, the force is proportional to the inverse square of their separation, like gravitational attraction between small masses (Section 8.1). Magnetic field is measured by the force – and its direction – acting on a standard positive pole (standard poles are defined in terms of coils of wire carrying an electric current, but will not be explained further.) It is measured in units of nanoTesla (nT), as explained in Section 10.1.1.

When calculating the magnetic field due to a body we usually have to add the fields of one or more pairs of opposite poles. The simplest body is a dipole.

11.2.2 The field of a dipole

A dipole consists of just a pair of poles whose separation, $2l$, is much smaller than their distance, r, from where the field is being measured (Fig. 11.8). The field of each pole is found separately and then added. On the axis of the dipole, at the north magnetic pole, P_N, the fields due to the two poles are in exactly opposite directions, but that of the negative pole, which is closer, is the greater, so the total – shown by the double arrow – is towards the dipole; at P_S it is away. At E, on the magnetic equator, the forces are equal but not quite opposite and together produce a force antiparallel to the dipole. At some general point X the forces are neither quite equal nor opposite, and they add as shown. The strength of the field at P_N and P_S is twice that at E, which is why the Earth's field at the poles is roughly twice as strong as that at the equator, as mentioned earlier. Expressions for the field strength and direction at other latitudes can be found in some of the books given in Further reading.

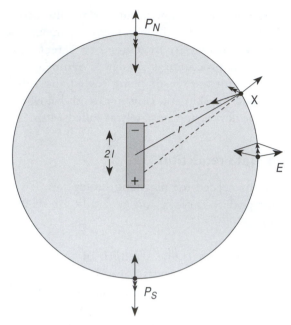

Figure 11.8 Deducing the field of a dipole.

A compass needle – which also has a pair of poles, one near each end – points northwards because its north-seeking pole (which is a positive pole) is attracted in the direction of the field while its other pole is pulled in the opposite direction.

11.2.3 Anomaly of a dipole, or small body

In the previous section we deduced the magnetic field of a dipole; to deduce the anomaly it produces we also need to take account of its orientation and the Earth's field, as was explained in Section 11.1. To simplify matters, assume that the dipole is aligned in the direction of the Earth's field (this is equivalent to assuming it has only induced magnetisation, as will be explained in Section 11.4); even so, its anomaly varies with latitude because the Earth's field varies with latitude.

Figure 11.9 shows the total field anomaly at three magnetic latitudes (from $\tan I = 2 \tan \lambda$, the inclination at latitude 27° N is 45°, downwards). The profiles are taken in the S–N direction (at the pole, profiles in all directions are the same). At the equator the anomaly is symmetrical, with a trough between two peaks (Fig. 11.9a): Immediately above the dipole at A the field is opposite to the Earth's field, so the total intensity is *reduced*, while at B and C it *increases* the total field, though by a lesser amount, because of the greater distance from the dipole.

(a) equator (b) latitude 27° N (c) north pole

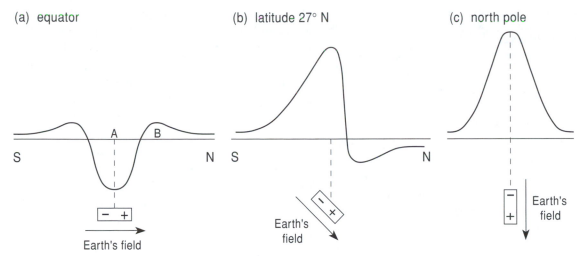

Figure 11.9 Anomaly of a dipole at different latitudes.

The anomaly at the pole is also symmetrical (Fig. 11.9c), but for the inclined dipole (Fig. 11.9b) there is a larger positive to the south than negative to the north; in the southern hemisphere the anomalies are inverted compared to those in the northern hemisphere at the same latitude. So the shape – and size – of the anomaly depends markedly on latitude. This dependence of magnetic anomalies upon latitude increases the difficulty of interpreting them; a partial way around this will be described in Section 11.6.1. (The anomalies found by a horizontal or vertical component fluxgate magnetometer would differ from the total field anomalies shown.)

11.2.4 Anomaly of a sphere

The field of a uniformly magnetised sphere – approximating to, for example, a compact ore body – is the same as that of a dipole at its centre (regardless of whether it is deep compared to its radius). So we already know how a S–N profile over its centre appears (Fig. 11.9). Figure 11.10b shows anomaly contours over the Marmora ore body in Canada, found by an aeromagnetic survey, while Figure 11.10a shows that an S–N profile through it is similar to that of a buried sphere at the same latitude.

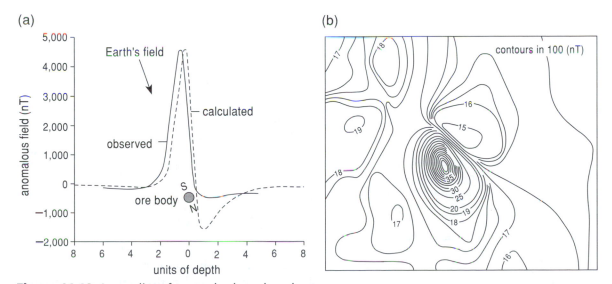

Figure 11.10 Anomalies of an ore body and a sphere.

11.2.5 Anomaly of a vertical sheet

The anomalies of extended bodies – ones whose dimensions are not small compared to their depth – can be built up from arrays of dipole magnets, but first we need to understand how magnets behave when divided or joined.

Positive and negative poles always occur in pairs, as mentioned above; if a magnet is cut in two, poles form at the new ends (Fig. 11.11). This is because the magnet is made up of a vast number of atomic dipoles, and because in any little volume there are as many positive as negative poles their fields cancel; only near the ends are there more of one sort of pole than the other, and these net poles produce the field outside the magnet. Conversely, if magnets are put end to end, *with a positive next to a negative pole*, they form a single magnet, so far as their external field appears. Therefore, when considering the field produced by an extended body we need consider only the net poles that form near its surfaces. Which faces the poles form on depends on how the body is oriented with respect to the Earth's field (assuming magnetisation is along the Earth's field). Figure 11.12 shows the poles formed on a long, vertical, sheet – approximating, for example, a dyke or ore vein – that is thin (i.e., its thickness is small compared to the depth to its top surface) at the same three latitudes as in Figure 11.9, striking E–W and also N–S.

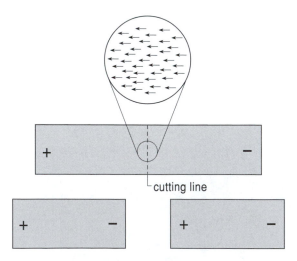

cutting line

new magnets made by division

Figure 11.11 Dividing a magnet.

Sheet striking E–W. In Figure 11.12a the field is horizontal and poles are formed on opposite vertical faces of the sheet, like dipoles stacked one above the other. Each dipole produces a field like that in Figure 11.9a, except that the deeper the dipole, the lower and broader its anomaly. Together, they add to an anomaly somewhat broader than that of the top one alone, as shown above it.

At the north magnetic pole, where the field is vertically downwards, poles are formed along the upper and lower edges of the sheet (Fig. 11.12c). If the sheet extends to great depths, the field of the lower poles will be negligible because of their distance from the surface, so only the top edge need be considered. It produces a symmetrical, positive anomaly. At intermediate latitudes poles form on top, base, and sides (Fig. 11.12b), resulting in the anomaly shown. An example that approximates to this is shown in Figure 11.16, profiles of a dyke striking roughly E–W, at a latitude of about 55° N.

Sheet striking N–S. In this case, poles form only on the edges. At the magnetic equator, these are the vertical ends (Fig. 11.12d); if the sheet is long with its ends remote from the survey area, these will produce little field, so, surprisingly, it would not be detectable magnetically! At the magnetic north pole, the poles form only along the top and base (Fig. 11.12f); this is the same as for an E–W striking sheet, because at the poles strike direction has no significance. At intermediate latitudes, poles along the top dominate (Fig. 11.12e), producing a symmetrical though reduced anomaly.

Of course, rock masses are often more complex than the simple bodies considered above; then their anomalies are usually interpreted using computer modelling, as described below (Section 11.5), but understanding simple anomalies allows some immediate insight into the subsurface structure.

11.3 Depth of the body

It is a general rule, as with gravity surveying, that the shallower a body, the sharper is its anomaly (the shorter its main wavelength), and this can be used

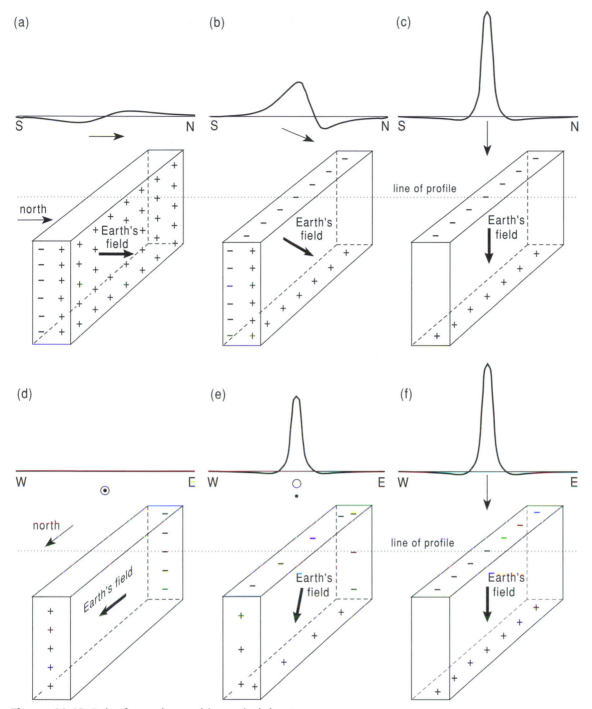

Figure 11.12 Poles formed on a thin vertical sheet.

to deduce its depth. However, the shapes of anomalies also depend on the direction of magnetisation, and often have both positive and negative parts. Figure 11.13a shows anomalies of a magnetic dipole or sphere, at three depths. As well as becoming smaller with depth they also become wider, which is made clearer in Figure 11.13b, where the peak values have been adjusted to be the same. However, the peak-to-trough distance in each case equals the depth to the dipole, so – for this particular case of a dipole dipping at 45° – it is easy to deduce its depth.

(a) actual profiles

(b) adjusted profiles

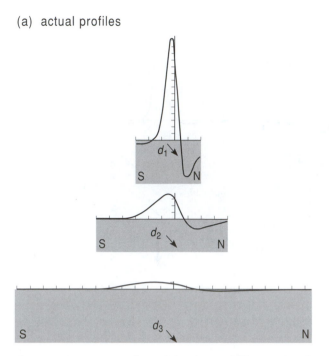
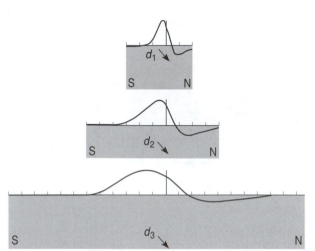

Figure 11.13 Anomalies of dipoles at different depths.

However, we need a depth rule of more general use. There are several, one of which is Peter's half-slope method. The slope of the steepest part of the anomaly is drawn (Fig. 11.14, line 1); then a line half as steep is drawn (line 2). Next, positions A and B, either side of the steepest part, are found with the slope of line 2, and their horizontal separation, S, measured. The depth, d, to the top of the body is given approximately by depth, $d \sim 1.6\ S$. This relationship works best for dykes or sheets aligned N–S (near-symmetric anomalies), and at high latitudes, and the approximation

deteriorates the more the situation departs from this. For a 3D anomaly (i.e., measured over a grid) a profile is drawn through the peak, crossing the narrowest part of the anomaly, to give the sharpest profile.

11.4 Remanent and induced magnetisations

The magnetisation of rocks has two parts, as mentioned in Section 10.3. Induced magnetisation (Section 10.6) exists only while a magnetic field exists and is aligned in the direction of the field (unless the rock is magnetically anisotropic, Section 10.7). It is due to paramagnetic as well as ferromagnetic minerals. As explained in Section 10.6, the strength of magnetisation, M_i, is proportional to the strength of the field, H, and to its magnetic susceptibility (Table 11.1). Remanent magnetisation, M_r, in contrast, can persist for millions of years, largely irrespective of the direction of the magnetic field; it may therefore have a direction very different from that of the present day, either because the field has changed (e.g., reversed) or because the rock has moved. The total magnetisation, M_{total}, of a rock is the addition of the induced and remanent magnetisations, taking into account their directions (Fig. 11.15).

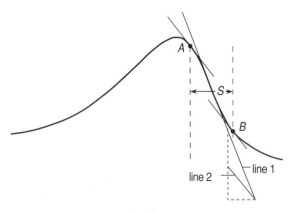

Figure 11.14 Peter's half-slope method for estimating depth.

Table 11.1 Magnetic susceptibilities of common rocks

Rock type	Susceptibility (rationalised SI units)
Sediments	
chalk	c. 0
limestone	0.00001–0.025*
salt	–0.00001
sandstone	0–0.2
shale	0.0006–0.02
Igneous and metamorphic	
basalt	0.0005–0.18
gabbro	0.0008–0.08
gneiss	0–0.003
granite	0.00002–0.05
peridotite	0.09–0.2
rhyolite	0.0002–0.04
serpentinite	0.003–0.08
slate	0–004
Other	
water, ice	–0.000009

*Ranges, which are from several sources, are approximate.

The presence of remanence makes interpreting magnetic anomalies more uncertain because its direction is generally unknown – unlike that of induced magnetisation. If possible, its value should be determined by measuring samples of the rock, as described in Section 10.2.1.

The ratio of remanence to induced magnetisation, M_r/M_i, is called the **Königsberger ratio, Q**. Q may vary within a single body, further complicating interpretation. Though the induced magnetisation usually dominates in old rocks and sediments, the remanence may be large enough to be significant. Sometimes it is both large enough and with a direction sufficiently different from the Earth's field that the total magnetisation has a direction markedly dif-

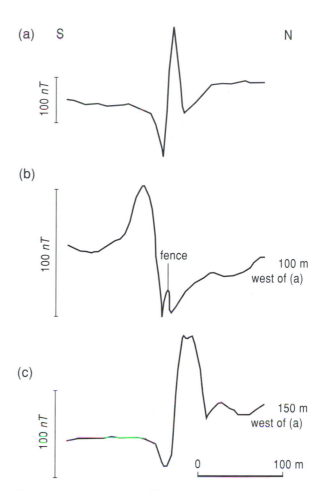

Figure 11.16 Three profiles across a dyke.

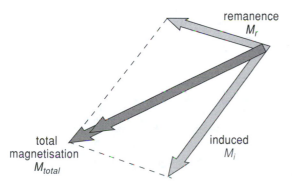

Figure 11.15 Addition of induced and remanent magnetisations.

ferent from the Earth's field; this is commonly found in the Tertiary basaltic lavas of Scotland and Ireland, for instance, which often have a strong, reversed magnetisation.

As an illustration, Figure 11.16 shows three profiles across an early Tertiary basaltic dyke in northwest England. It is vertical and strikes roughly E–W, and profile (b) is as expected for induced magnetisation for such a dyke in the northern hemisphere (see Fig. 11.12b), that is, asymmetric with the peak to the south of the trough, but profiles (a) and (c) – no more than 100 m away – each has its peak to the north. Sampling has shown that the dyke has reverse remanence; evidently, the remanent exceeds the induced magnetisation (so Q exceeds 1) at profiles (a) and (c), but is less at profile (b). However, such rapid variation in Q is uncommon. Another survey involving bodies with significant remanent magnetisation is described in Section 11.8.

11.5 Computer modelling

Section 11.2 described the anomalies produced by various simply shaped bodies, and so allows simple interpretations. Often, though, bodies have more complicated shapes, or they have both induced and remanent magnetisations at an angle, and then forward modelling (guessing a model, computing its anomaly, and comparing it with the observed one, etc.) is a more useful approach. An additional difficulty is the problem of non-uniqueness already encountered in gravity (Section 8.8.1), so it is particularly important that models be constrained by all available information and the remanent magnetisation and susceptibilities of samples measured if possible. 2D, 2½D, or even 3D models are used, as appropriate to the problem, though 3D modelling allowing for both induced and remanent magnetisations with a complicated geometry is not usually possible. An example of 2D modelling is given in Section 11.8.

11.6 More advanced processing of data

11.6.1 Reduction to the pole

One reason interpreting magnetic anomalies is more difficult than interpreting gravity ones is that they vary with the latitude and strike of the causative body (e.g., Fig. 11.12). Anomalies are simplest for bodies at the magnetic pole, where the field is vertical; if the body is symmetrical, its anomaly is symmetrical and independent of strike (e.g., Fig. 11.12c and f) and similar to its gravity anomaly (assuming the magnetisation is induced and uniform, and the density is uniform, respectively). To transpose an anomaly as if its causative body were at the pole, but without knowing what the body is, requires methods beyond the level of this book. This '**reduction to the pole**' is most accurately achieved if the body is at high latitudes, where the field is already steep.

11.6.2 Pseudogravity

This is used to compare magnetic and gravity surveys of the same area. After a body has been

deduced that accounts for the magnetic anomaly, it is used to calculate a gravity anomaly by assigning it a plausible density. If the **pseudogravity** anomaly matches the actual gravity anomaly, the same body is probably producing both the gravity and magnetic anomalies; if it doesn't, there are different causative bodies, or their magnetisations and densities are not uniform.

11.6.3 Upward and downward continuation

These are mathematical operations that approximately recalculate the anomaly as if had been measured in a survey at a greater, or lesser, height than was actually done. **Upward continuation** gives much the same result as would an aeromagnetic survey and so can be used to decrease the relative size of the anomalies of shallow, near-surface bodies compared to those of deeper ones without the expense of flying a plane. As example, consider the anomalies of two dipping dipoles, one (M) buried at 10 m, and the other, which is 8 times as large (8M), at 20 m (Fig. 11.17). The anomaly at ground level of the smaller dipole will be as tall as that of the larger one (the field of a dipole decreases as the cube of the distance from it), but 100 m up – though both anomalies will be much smaller – that of the larger

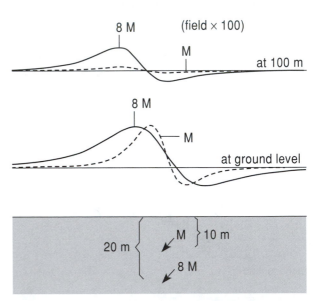

Figure 11.17 Anomalies at different heights of two dipoles.

dipole will be five times as large as the other's ($8/140^3$, compared to $1/120^3$). Though the anomalies of extended bodies do not decrease so rapidly as those of dipoles, the anomalies of small shallow bodies still decrease much more rapidly than those of deep bodies. Filtering – described in Section 3.2 – can have a similar effect.

Upward continuation can also be used to see if the body causing a surface anomaly is also responsible for the aeromagnetic one, or whether there are other bodies as well. **Downward continuation** is the converse process and can be used to emphasise the anomalies of near-surface bodies, but it needs to be used with caution because unwanted anomalies, such as those due to ferrous constructions, will also be emphasised.

Upward and downward continuation can also be performed on gravity measurements, but the vertical change above a small body is less rapid than for small magnetic bodies because the latter have both positive and negative poles whose fields partially cancel (the magnetic field of a sphere – a dipole – decreases as the cube of the distance, whereas its gravitational pull decreases as the square).

11.7 Magnetic gradiometry

It was just explained that downward continuation can be used to emphasise anomalies, particularly of shallow bodies, but that it has limitations. If possible, it is usually better to emphasise the wanted anomaly when carrying out the survey, and for shallow bodies this can be done by placing the magnetometer sensor close to the ground. A further step is to use a **gradiometer,** which has two sensors, a fixed distance apart, usually one above the other, and record the *difference* of their readings. The result is given either as the difference in nT or as the gradient of the field, which is the difference divided by the separation of the sensors, and is given in nT/m. As explained in the previous section, the closer a body is to the surface, the greater the difference between the fields it produces at the two sensors and so the greater the gradient, whereas deep bodies produce only a small gradient. An advantage of using a gra-

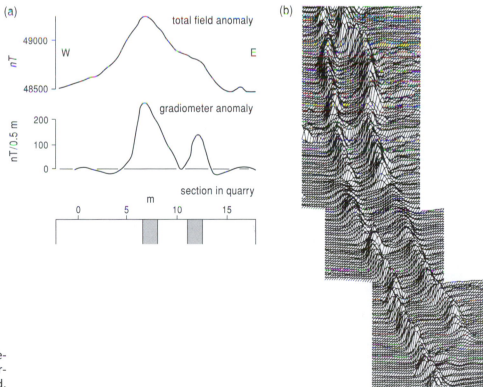

Figure 11.18 Gradiometer survey of the Butterton dyke, England.

diometer is that the Earth's field is the same at the two sensors and so completely cancels; for this reason there is no need to correct for diurnal variation. Instruments may be proton precession or fluxgate magnetometers; the latter is the lighter instrument and gives more rapid readings, but is susceptible to drift and needs careful orientation. A sensor separation of a half to one metre is usual.

Gradiometers may reveal geological boundaries more clearly than a magnetometer when the overburden is thin. Figure 11.18a compares total field and gradiometer profiles of a dolerite dyke; it appears to be a single body in the magnetometer profile, but the gradiometer profile reveals it to be double, as its confirmed by its exposure in a nearby quarry. (The anomalies are fairly symmetric, as expected for a dyke striking only about 25° from N–S, and positive as expected for induced magnetisation.) Figure 11.18b shows results in stacked profiles of a gridded gradiometer survey, with readings taken every ½ m along lines a meter apart. It reveals great detail, including small offsets of the dyke.

As gradiometers detect shallow targets, they are widely used in archaeological surveys, as described in Section 28.3.1, and examples are given in Sections 28.4.2 and 28.5.2. They are also useful for searching for buried ferrous pipes, or to find old shafts and wells if these have been infilled with magnetic material; an example is given in Section 27.2.3.

11.8 The Blairgowrie magnetic anomaly: A case study

This example illustrates many of the points made in this subpart, including palaeomagnetic measurements. On an aeromagnetic map, the Blairgowrie anomaly shows up as one of the largest in Scotland, reaching a positive value of about 600 nT (Fig. 11.19a). It is elongated and lies along the Highland Boundary Fault Zone (HBFZ), a major fault (Fig. 11.19b); near Blairgowrie the fault has several parallel branches, with faulted blocks between.

To provide more detail, a 21-km-long traverse was surveyed at ground level, along convenient roads roughly perpendicular to the fault (Fig. 11.19c). Stations were every 100 m, and readings were taken some metres from the road where possible, to reduce the effects of ferrous fences, vehicles, and so on. The positions of the stations were projected on to the straight line *AB* perpendicular to the fault. The diurnal variation was allowed for by

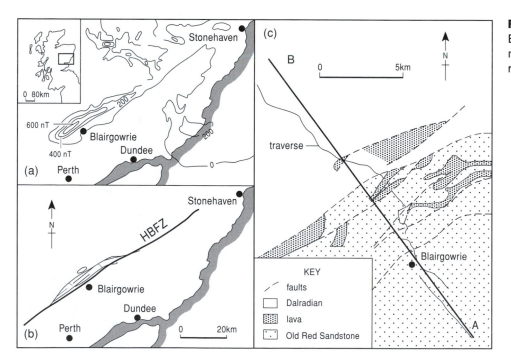

Figure 11.19
Blairgowrie aeromagnetic anomaly in relation to geology.

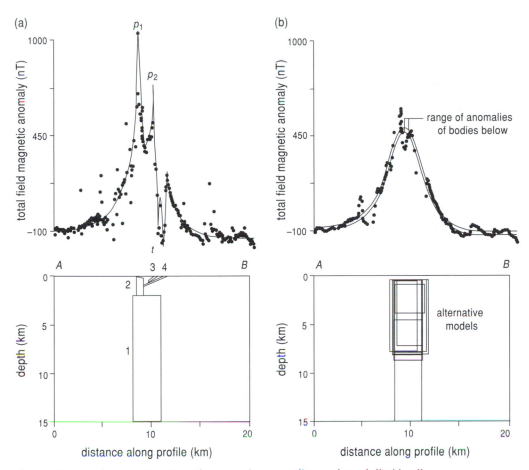

Figure 11.20 Blairgowrie ground magnetic anomalies and modelled bodies.

subtracting the changes at a base station in Blairgowrie. The anomaly, shown in Figure 11.20a, has narrow peaks and troughs (notably p_1, p_2, t) not seen on the aeromagnetic anomaly, though some of the scatter of readings is probably noise due to ferrous objects. It was realised that the narrow features must be due to small, shallow bodies, whereas the aeromagnetic anomaly was due to a deeper and larger body.

As the anomaly is so elongated, it was modelled using a 2D program (i.e., assuming a uniform cross-section), which permitted independent induced and remanent magnetisations to be specified. The first step was to model the deeper body (body 1), so the sharp peaks and troughs were removed and a short-wavelength (low-pass) filter used to remove noise (Section 3.2). The remaining anomaly (Fig. 11.20b) was matched by a vertical,

rectangular body about 3 km wide, roughly coinciding with one of the fault blocks. Its top was nearly 2 km below the surface, but, as is usual when modelling narrow bodies, the position of its lower edge was not well constrained; though the best match was for a body extending down to 15 km, bodies half or twice as deep fitted nearly as well (Fig. 11.20b).

To model the shallower bodies, the calculated anomaly of the deep body – which acts as a regional anomaly with respect to those of the smaller bodies – was removed, and then the remaining anomalies were matched by forward modelling. Peak p_1 was modelled by vertical sheet 2 (Fig. 11.20a); peak p_2 was modelled by gently dipping sheet 3 (a lava) with normal magnetisation, and trough t by another lava, 4, but with reversed magnetisation.

To confirm these magnetisations, the three shal-

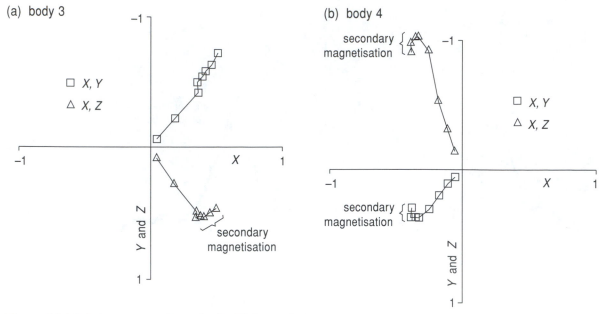

Figure 11.21 Palaeomagnetic results for Blairgowrie lavas.

low bodies were sampled and their remanent directions determined after magnetic 'cleaning' using thermal demagnetisation. Figure 11.21 gives the results for the two lavas, shown on a vector component diagram (Section 10.3.4). After small secondary magnetisations – probably VRM – had been removed from their NRMs, their remanent directions were found to be close to those deduced by the modelling. Their Königsberger ratios exceeded 1 (1.7 for body 3 and 1.4 for body 4), which is why body 4, with reverse remanence, produces the anomaly of a reversely magnetised body.

The deep body (body 1) is probably serpentinite, judging from an occurrence of this rock type along the fault and its known strong magnetism. Body 2 is probably also serpentinite, in a small uplifted block, while bodies 3 and 4 are andesitic lavas. Further details are given in Farquharson and Thompson (1992).

Summary

1. Magnetic surveying depends upon subsurface rocks having magnetisations that differ from those of their surroundings, so producing anomalies in the surface magnetic field.

2. Magnetic surveying can be carried out on the ground, from aircraft, and from ships. Commonly used instruments are the proton (precession) magnetometer, which measures the total field, and the fluxgate magnetometer, which measures a component of the field; for ground surveys this component is usually the vertical one, but in ships and aircraft it is along the field and so effectively measures the total field and its direction. Gradiometers use a pair of sensors, usually fluxgates.

3. Corrections are far fewer and usually less important than for gravity surveys; the chief one is the diurnal variation. Regional surveys have the IGRF (International Geomagnetic Reference Field) subtracted, to remove the large latitudinal variation of the Earth's field.

4. The magnetism of rocks is likely to be a combination of induced magnetisation (usually in the direction of the present field) plus remanent magnetisation, which can have a direction far from that of the present field.

5. Magnetic anomalies are more complex than gravity ones, for they depend on the orienta-

tion and latitude of the body, as well as its geometry and magnetic properties, and often have both positive and negative parts. The presence of significant remanence further complicates the relationship between anomaly and body.

6. Anomalies can be transformed to a standard form, independent of latitude and strike, by 'reduction to the pole'.

7. Gradiometer surveys are useful for emphasising the anomalies of near-surface bodies. They automatically remove the Earth's field and are not affected by diurnal variation.

8. Upward/downward continuation is a mathematical technique that transforms an anomaly approximately as if it had been measured at a different height.

9. Depths to bodies can be estimated using depth rules, such as Peter's half-slope method.

10. Pseudogravity is used to see if the magnetic causative body is the same as the gravity one, but assumptions need to be made.

11. You should understand these terms: magnetic pole, Königsberger ratio, Q; diurnal variation, IGRF; total field and vertical component anomalies; proton precession and fluxgate magnetometers, gradiometer; reduction to the pole, upward and downward continuation, pseudogravity.

Further reading

Magnetic surveying is described in several textbooks, including Kearey and Brooks (1991), Milsom (1996), Parasnis (1997), Reynolds (1997), Robinson and Coruh (1988), and Telford et al. (1990), of which the last is the most advanced.

Problems

1. A vertical dyke has a roughly E–W strike. Sketch the total field anomaly profile of the dyke for each of the following latitudes, assuming there is only induced magnetisation: (a) 0°, (b) 50° N, (c) 85° S. Make clear which way your profiles have been drawn.

2. Describe how you would carry out a magnetic survey to locate the position, width, and depth of a basic dyke, about 10 m across, running approximately E–W across an area of England.

3. What are the advantages and disadvantages of aeromagnetic surveys in comparison to ground-level ones?

4. Sketch the anomaly of a buried sphere with induced magnetisation at the south magnetic pole.

5. What is the advantage of a magnetic anomaly being 'reduced to the pole'?

6. In each of the following magnetic surveys, how important would it be to measure the diurnal variation?
   (a) A survey to locate the position and extent of a granite intrusion.
   (b) Traverses to follow a basaltic dyke across country.
   (c) A survey to locate a normal fault in a sandstone.

7. Compare the advantages and disadvantages of proton precession and fluxgate magnetometers when carrying out ground surveys.

8. The field at the equator, compared to that at the pole, is
   (i) 4 times. (ii) Twice. (iii) The same. (iv) Half. (v) One quarter.

9. The magnetic field one Earth radius above the north pole compared to the field at the pole is:
   (i) 8 times. (ii) 4 times. (iii) Twice. (iv) Half. (v) A quarter. (vi) An eighth.

10. How is the aeromagnetic map of a sedimentary basin likely to differ from that over old crystalline basement rocks?

11. Discuss whether magnetic surveying could be helpful in the following cases, and, if so, how surveys would be carried out.
    (a) To map dipping beds of sandstone, shale, and limestone beneath thin glacial deposits.
    (b) To determine whether an extensive lava flow is below a given area.
    (c) To locate infilled sinkholes in limestone.
    (d) To locate natural cavities in limestone.
    (e) To locate a thrusted contact between schists and gabbro.
    (f) Regional mapping of a heavily wooded area where sandstones, serpentinites, and schists often have faulted contacts.

12. Discuss whether a magnetic gradiometer would probably be superior to a magnetometer when surveying in the following situations:
    (a) To detect the boundaries of a buried granite.
    (b) To locate a steel water-pipe.
    (c) To locate the faulted margin of a lava beneath a metre or two of soil.
    (d) To locate the faulted margin of a sill.

13. The Königsberger ratio of a roughly N–S–trending basic dyke that was intruded into sandstones and conglomerates varies along its length from about 0.8 to 1.3. What effect does this have upon its anomaly?

14. Plate 5 is described as 'total field magnetic anomalies'. What does 'total' mean?

Electrical

Most of the methods of this subpart have been particularly associated with mineral prospecting, but they are also used for a wide variety of other purposes, which include hydrogeological and ground contamination surveys, site investigation, and archaeological surveying. The methods exploit the differences of various electrical properties of rocks and minerals.

Resistivity surveying investigates variations of electrical resistance (or conductivity, the inverse of resistivity), by causing an electrical current to flow through the ground, using wires connected to it.

Electromagnetic (e-m) surveying responds to much the same targets as resistivity surveying but usually does not use wires connected to the ground. This allows it to be used in aerial surveying and where the very high resistivity of surface layers prevents the resistivity method being used. Ground-penetrating radar (GPR) records radar waves – a type of electromagnetic wave – reflected from interfaces and gives a more direct image of the subsurface – rather like reflection seismology – than other electrical methods, but is usually limited to the top few metres of the earth. Magnetotelluric (MT) surveying can reach much deeper than the previous methods – to depths of tens or even hundreds of kilometres – by utilising natural currents in the ground, using a combination of e-m and resistivity methods.

Self-potential (SP) is mainly used for prospecting for massive ores and relies on the natural production of electricity by them. Induced polarisation (IP) is also widely used for mineral prospecting, particularly for disseminated ores, which are hard to detect using other electrical methods. It depends on the ability of the ores to store a small amount of electricity when a current is passed into the ground, which is released after the current is switched off.

These methods have been grouped into three successive chapters: **Resistivity methods** (Chapter 12), **Induced polarisation and self-potential** (Chapter 13), and the others in Chapter 14, **Electromagnetic methods**.

chapter 12

Resistivity Methods

The resistivity of rocks usually depends upon the amount of groundwater present and on the amount of salts dissolved in it, but it is also decreased by the presence of many ore minerals and by high temperatures

The main uses of resistivity surveying are therefore for mapping the presence of rocks of differing porosities, particularly in connection with hydrogeology for detecting aquifers and contamination, and for mineral prospecting, but other uses include investigating saline and other types of pollution, archaeological surveying, and detecting hot rocks.

Resistivity surveying investigates the subsurface by passing electrical current through it by means of electrodes pushed into the ground. Traditionally, techniques have either been designed to determine the vertical structure of a layered earth, as vertical electrical sounding, VES, or lateral variation, as electrical profiling; however, more sophisticated electrical imaging methods are being increasingly used when there are both lateral and vertical variations.

12.1 Basic electrical quantities

Matter is made of atoms, which can be conveniently visualised as a small, positively charged nucleus encircled by negatively charged electrons (the naming of charges as positive and negative was arbitrary but is long established). Usually, the amounts of positive and negative charges are equal, so they balance to give electrical neutrality; only when there is imbalance does a body have a charge and its electrical properties become apparent. In resistivity surveying, we are concerned with the movement, or *flow*, of charges through rocks, rather than imbalances as such.

Electric charge flows round a circuit (i.e., a closed path or loop). Figure 12.1 compares an electrical circuit with water flowing through pipes. Just as water flow is a current, so flow of charge is an **electric current** (by convention, current is considered to flow from positive (+ve) to negative (−ve), though in wires current is due to electrons moving the other way). Electrical current is measured in amperes, usually abbreviated to **amps;** an amp is not an actual amount of unbalanced electric charge existing at some instant but the amount that *passes* any point in the circuit in 1 sec; charges pass round the circuit many times, just as the water circulates repeatedly around a central heating system. And just as a pressure difference between ends of the piping is needed to make water flow, so, to make electric current flow, an electrical pressure difference is needed; it is called **potential difference** (abbreviated to **p.d.**), is often given the symbol *V* (for voltage), and is measured in **volts;** thus a 1.5-volt battery produces a p.d. of 1½ volts (sometimes p.d. is called voltage difference, or voltage).

For most materials, including most rocks, the current through a piece of the material increases in proportion to the p.d. (Fig. 12.2): For example, doubling the p.d. doubles the current. Figure 12.2a shows how these could be measured: The voltmeter measures the p.d. across the ends of the piece, while the ammeter

measures the current flowing through it (the voltmeter is designed so that a negligible fraction of the current flows through it, and the parts are connected by wires with small resistance, so the effects of the connecting circuitry are minimised). This simple proportion of current to p.d. is called **Ohm's Law,** and it is the basic relation of this chapter. The amount of current flowing when the p.d. is 1 volt is called the **resistance** of the piece; it also equals the slope of Figure 12.2b. Resistance is measured in **ohms** (given the symbol Ω). As an equation, Ohm's Law is:

$$\frac{\text{potential or voltage difference (volts)}}{\text{current (amps)}} = \frac{V}{I}$$
$$= \text{resistance } R \text{ (ohms)} \qquad \text{Ohm's Law} \qquad \text{Eq. 12.1}$$

The value of resistance depends upon both the material and its shape; a wire of copper has less resistance than one of lead with the same dimensions, while a long thin wire has greater resistance than a short, fat one of the same material: doubling the length doubles the resistance, but doubling the area of cross-section halves it, just as doubling the cross-sectional area of a pipe would double the rate of water flow through it. For a uniform bar (Fig. 12.3) these effects can be summarised as follows:

$$\text{resistance, } R = \text{resistivity } (\rho)$$
$$\times \frac{\text{length}}{\text{area of cross-section}} \qquad \text{Eq. 12.2a}$$
$$\text{resistivity, } \rho = \text{resistance}$$
$$\times \frac{\text{area of cross-section}}{\text{length}} \qquad \text{Eq. 12.2b}$$

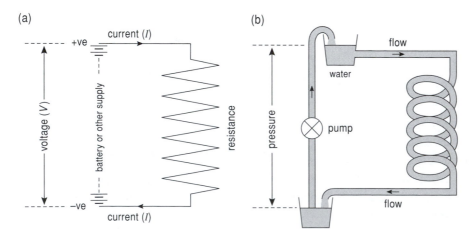

(a) (b)

Figure 12.1 Analogy between electricity and water.

Figure 12.2 Measuring resistance.

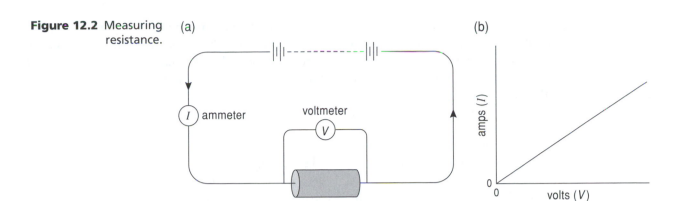

(a)

ammeter

voltmeter

resistance

(b)

amps (*I*)

volts (*V*)

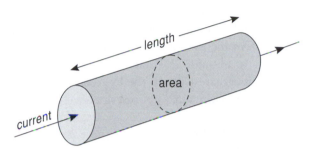

length

area

current

Figure 12.3 Resistance of a wire depends partly on its dimensions.

The **resistivity** (given the symbol ρ, 'rho') characterises the material independent of its shape; it is measured in ohm-m. (The inverse of resistivity, 1/ρ, called **conductivity** and given the symbol σ, 'sigma', is also used, particularly in Chapter 14.) Resistivity is the quantity investigated by resistivity surveying.

12.2 Resistivity surveying

12.2.1 Resistivities of rocks and minerals

Everyday uses of electricity, such as house wiring, use good conductors (very low resistivity), such as copper wires, to lead the current where it is wanted, and insulators (very high resistivity) enclosing the wire to prevent the current leaking elsewhere. Though very good conductors and insulators occur naturally – for example, silver and quartz respectively (Table 12.1) – most rocks have resistivities between (this does not make them semiconductors, which have special electrical properties). Nevertheless, they still have a wide range of resistivities, which is utilised in resistivity surveys.

Table 12.1 Resistivities of some rocks and minerals

Rocks, minerals, ores	Resistivity (ohm-m)
Sediments	
chalk	50–150*
clay	1–100
gravel	100–5000
limestone	$50–10^7$
marl	1–100
quartzite	$10–10^8$
shale	10–1000
sand	500–5000
sandstone	$1–10^8$
Igneous and metamorphic rocks	
basalt	$10–10^7$
gabbro	$1000–10^6$
granite	$100–10^6$
marble	$100–10^8$
schist	$10–10^4$
slate	$100–10^7$
Minerals and ores	
silver	1.6×10^{-8}
graphite, massive ore	$10^{-4}–10^{-3}$
galena (PbS)	$10^{-3}–10^2$
magnetite ore	$1–10^5$
sphalerite (ZnS)	$10^3–10^6$
pyrite	1×100
chalcopyrite	$1 \times 10^{-5}– 0.3$
quartz	$10^{10}–2 \times 10^{14}$
rock salt	$10–10^{13}$
Waters and effect of water and salt content	
pure water	1×10^6
natural waters	$1–10^3$
sea water	0.2
20% salt	5×10^{-2}
granite, 0% water	10^{10}
granite, 0.19% water	1×10^6
granite, 0.31% water	4×10^3

*Values or ranges, which have come from several sources, are only approximate.

Most minerals that form rocks, such as quartz, feldspar, mica, and olivine, are good insulators, but rocks have pores and cracks that generally contain at least a film of water on their surfaces and often are full. As groundwater has a fairly low resistivity, the resistivity of the rock depends chiefly upon its porosity and how saturated the pores are with water. It also depends upon how conductive the water is: Pure water is quite a good insulator, but natural waters contain dissolved salts of various kinds, usually derived from the weathering of rocks; the salts dissociate into positive and negative ions, which are atoms or groups of atoms that have lost or gained one or more electrons – for example, common salt dissociates into sodium ions, Na^+, and chloride ions, Cl^- – and these can move through the water, in opposite directions, and so form a current (Fig. 12.4a). This is called **ionic conduction** and differs from **electronic conduction**, where current is due only to electrons (Fig. 12.4b); electronic conduction occurs in metals and some ores. The salinity of pore water is usually small, but can be high if, for example, it is contaminated with seawater or because – in hot, dry countries – evaporation has concentrated the salts. Because of these various factors, the resistivities of most rock types have a large range, which makes it difficult to identify a lithology from its measured resistivity.

The resistivity of porous, water-bearing sediments, termed the formation resistivity, P_ρ, can be calculated approximately, knowing the porosity ϕ (the fraction of the sediment that is pore space), the water saturation S_w (fraction of the pore space filled with water), and the resistivity of the water in the pores ρ_w, using **Archie's Law,** which was deduced empirically:

$$\rho_t = a\,\rho_w\,\phi^{-m}\,s_w^{-n} = a\frac{\rho_W}{\phi^m\,S_w^n} \qquad \text{Archie's Law Eq. 12.3}$$

where a, m, and n are 'constants' whose values are determined from laboratory and field measurements (a is in the range 0.5 to 2.5, n is usually about 2 if the water saturation is more than about 0.3 but can increase greatly for lesser saturations, while m depends upon the cementation of the grains, which tends to increase with the age of the rock, with m increasing from about 1.3 in Tertiary sediments to about 2 in Palaeozoic ones). The formula has its greatest use in the hydrocarbon industry, in connection with geophysical well-logging (Box 18.1).

The formula does not hold if there is a significant amount of clay in the rock, for its fine particles trap a film of electrolyte around them except when it is very dry. As a result, clay is one common rock with low resistivity (usually in the range 10 to 100 ohm-m). Some ore minerals and also graphite have low resistivities, but the resistivity of the bulk ore also depends on the concentration of ore grains and the extent to which these are interconnected. When they are unconnected, as in disseminated ores, the bulk resistivity is high, and they are not revealed by resistivity surveys (instead, the induced polarization method is used, Section 13.1).

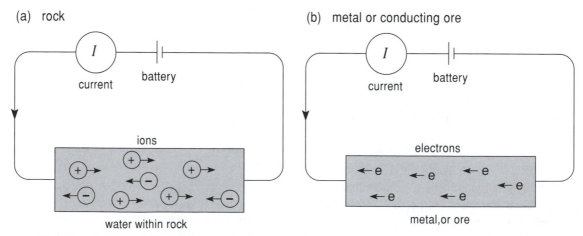

Figure 12.4 Conduction of electricity in metals and rocks.

Resistivity and temperature. Resistivity generally decreases as the temperature rises. This needs to be allowed for in well logging, when deducing porosity and other properties of rocks (Chapter 18) but is not usually otherwise important. However, it permits very hot rocks to be detected and so is useful for locating potential geothermal areas and the presence of magma. An example of the latter is given in Section 21.7.

12.2.2 How electricity flows through rocks

Section 12.1.1 described how to measure the resistance and hence resistivity of, say, a rod of material in the laboratory, but we need to adapt the method to measure the resistivity of the subsurface using measurements made at the surface. Electrical connections are made through **electrodes**, metal rods pushed a few centimetres into the ground. Two current electrodes are used, C_1 and C_2, current flowing in at one and out at the other (Fig. 12.5). You might expect the current to flow by the most direct route, just below the surface, from one electrode to the other, but would this mean just the topmost metre, or 1 millimetre, or how thick a layer? A thin layer has a large resistance, as Eq. 12.2 shows, so this would not be the easiest path. Instead, the **current paths** spread out, both downwards and sideways, though there is a higher concentration of current near the surface (and close to the electrodes) as indicated by the closeness of the current paths. In uniform ground, only about 30% of the current penetrates below a depth equal to the separation of the current electrodes.

12.2.3 The need for four electrodes

In Figure 12.2, the potential difference is measured between the ends of the resistance, where the current enters and leaves, but this is not done in resistivity surveys because there is usually a large and unknown extra resistance between each electrode and the ground. Rather than try to measure these contact resistances, the p.d. is measured between two other electrodes (Fig. 12.6); though these also have contact resistances, this does not matter because the voltmeter is designed to draw negligible current through them, and so – applying Ohm's Law – if the current is negligible the potential difference across the contact resistances is negligible too. The p.d. measured between the potential electrodes is given the symbol ΔV.

The power supply (usually run from batteries except for very large-scale surveys, where generators are used), current meter, and voltmeter are usually combined into a single instrument, a **resistivity meter** (Fig. 12.6), which often displays the ratio $\Delta V/I$ rather than the quantities separately. It is connected to the electrodes by wires whose resistances are very small. The voltage applied to the current electrodes is typically about 100 volts, but the current is only milliamps or less (sufficient to give a shock but not be dangerous) and ΔV ranges from volts to millivolts. Because ions can accumulate at electrodes and affect readings, they are dispersed by reversing the current flow a few times a second, though this is not apparent in the readings.

We shall label the current electrodes C, C and the potential ones P, P. However, A, B and M, N are also commonly used for the two pairs.

(a) section

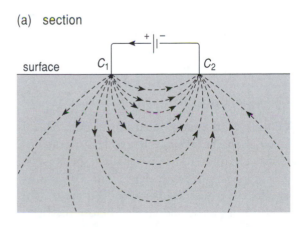

(b) plan

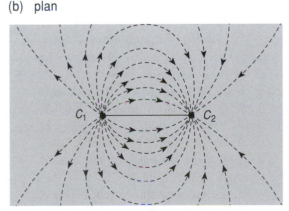

Figure 12.5 Current paths between electrodes.

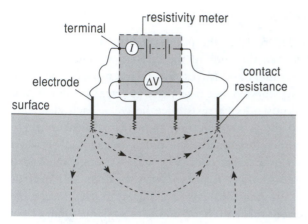

Figure 12.6 Use of four electrodes.

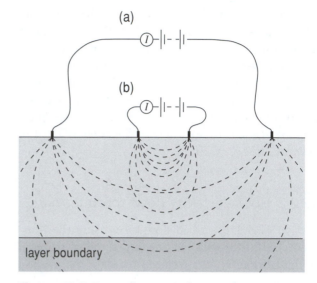

Figure 12.7 Expanding array for VES.

Having four rather than two electrodes has advantages, for the arrangement of the electrodes in a **resistivity array** can be chosen that is most suitable for the type of investigation being made. One common type of investigation is how resistivity varies with depth, called vertical electric sounding.

12.3 Vertical electric sounding, VES: Measuring layered structures

12.3.1 The basic concept

Vertical electrical sounding (VES) – also called depth sounding or sometimes electrical drilling – is used when the subsurface approximates to a series of horizontal layers, each with a uniform but different resistivity. The essence of VES is *to expand the electrode array from a fixed centre* (i.e., to increase at least some of the interelectrode spacings). Suppose we start with the current electrode separation being much less than the thickness of the top layer (Fig. 12.7b). Though some current spreads down into all layers, nearly all will be in the top layer, so the resistivities of lower layers have negligible effect on the current paths or, therefore, on the readings. This is no longer true when the separation has been expanded to be comparable to, or larger than, the depth to the second layer (Fig. 12.7a), and then the presence of the second layer will be detected.

We need to understand how the current lines are affected by the change of resistivity.

12.3.2 Refraction of current paths

Within a uniform layer the current paths are smooth curves (as shown in Fig. 12.5), but they bend, or refract, as they cross an interface separating different resistivities. They refract towards the normal when crossing into rock with higher resistivity (Fig. 12.8a), and conversely in rock with lower resistivity (Fig. 12.8b). (There are similarities with refraction of seismic rays, but the angles of current lines are related by $\rho_1 \tan \theta_1 = \rho_2 \tan \theta_2$, not sine as in Snell's Law, and seismic rays refract away from the normal when crossing into rock with higher seismic velocity.) Because refraction changes the distribution of current in a layered subsurface, compared to uniform ground (Fig. 12.9), the ratio $\Delta V/I$ is also changed, which makes it possible to measure the change of resistivity with depth.

12.3.3 Apparent resistivity

In a VES survey, the ratio of current to p.d., $\Delta V/I$, is measured with increasing electrode separations. This ratio changes for two reasons, because of changes of resistivity with depth but also simply because the electrodes are being moved apart. This second effect has to be allowed for before the first – the object of the survey – can be deduced.

(a)

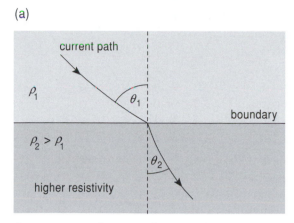

(b)

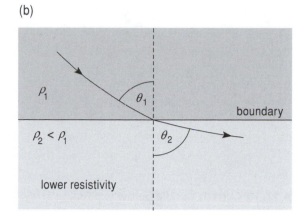

Figure 12.8 Refraction of current flow lines.

As the current travels through the ground, the current paths diverge from one current electrode before converging on the other (Fig. 12.10). The figure shows how a bundle of adjacent current paths define a shape like a pointed banana, with no current crossing the 'skin' of the banana. The resistance of the bundle is proportional to its length, but inversely proportional to its cross-sectional area (Eq. 12.2a). Suppose the separation of the current electrode is doubled; in uniform ground, the diagram simply scales up, with the banana becoming twice as long but with four times the area of cross-section; the quadrupling of the cross-sectional area outweighs the doubling of length, and so the resistance $\Delta V/I$ *halves*, the opposite of what you might expect. To allow for this effect the ratio $\Delta V/I$ is multiplied by a geometrical factor that depends on the electrode separation,

$$\rho_a = \text{geometrical factor} \times \frac{\Delta V}{I} \qquad \text{Eq. 12.4}$$

The factor is such that, for a uniform subsurface, ρ_a remains constant as the separation is changed and equals the resistivity of the ground.

The factor depends on the particular array being used. For instance, in the commonly used **Wenner** (pronounced 'Venner') **array,** the four electrodes are equally spaced, with this separation referred to as a (Fig. 12.11), and the geometrical factor is $2\pi a$, so $\rho_a = 2\pi a\,\Delta V/I$. In uniform ground, doubling a would halve the value of $\Delta V/I$, but the value of ρ_a remains the same. The geometrical factors for some other arrays are given on Figure 12.17.

If the subsurface is layered, the value of ρ_a depends on the resistivities of the various layers. The value calculated as above is called the **apparent resistivity, ρ_a,** and is the resistivity that uniform ground giving the same $\Delta V/I$ with the same electrode separations would have. As the array is expanded, ρ_a changes because more current flows in deeper layers, and this variation is used to deduce the electrical layering.

(a)   (b)

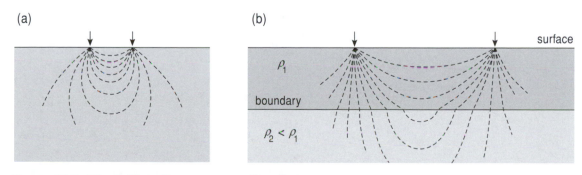

Figure 12.9 Effect of interface on current flow lines.

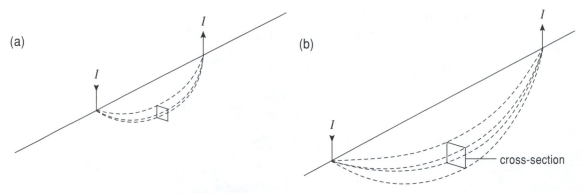

Figure 12.10 Current bundle below expanding array.

12.3.4 Carrying out a Wenner VES survey

To carry out a Wenner VES survey, two measuring tapes are laid out end to end (Fig. 12.11a) and four electrodes are pushed into the ground at equal intervals, symmetrically about the junction of the tape measures; they are connected to the resistivity meter, and readings are taken. The electrode separation is then increased (Fig. 12.11b, c), not equally but progressively, for the same increments at wide spacings would produce little change in the readings. Spacings of 0.1, 0.15, 0.2, 0.3, 0.4, 0.6 and 0.8, then 1, 1½, 2, 3, 4, 6, 8, . . . m (i.e., doubling spacings of 0.1, 0.2, 0.4, . . . and 0.15, 0.3, 0.6 . . . interleaved) are convenient values that will give approximately equally spaced points on the plot to be used (Fig. 12.14a), and the results are tabled. Expansion is stopped when the current is judged to have penetrated deep enough, roughly half the separation of the current (outer) electrodes. To expand large arrays quickly, several persons are needed to move the electrodes and operate the resistivity meter.

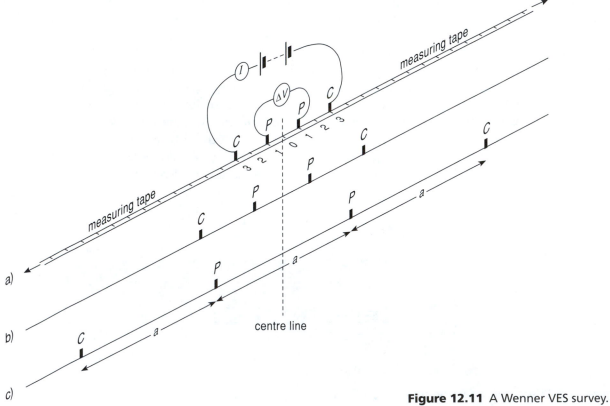

Figure 12.11 A Wenner VES survey.

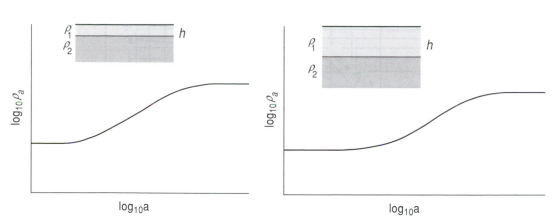

Figure 12.12 Plots for two thicknesses of top layer.

Apparent resistivity is calculated for each spacing, using $\rho_a = 2\pi a\ \Delta V/I$, and a graph is plotted of $\log_{10}\rho$ versus $\log_{10}a$ (logarithmic graph paper can be used to save working out the logs of the values – see Fig. 12.13). Logs are used because they allow the wide range of values to be plotted with roughly equal spacing. Figure 12.14a shows a simple plot.

12.3.5 Modelling the data

The curves of Figure 12.12 are both for two-layer cases that have the same resistivities but different thickness of the upper layer. How do we deduce the thicknesses and resistivities of the layers? The apparent resistivity is the same for the first few electrode spacings, which means that little current is penetrating down to the second layer. The apparent resistivity is therefore the actual resistivity of the top layer at this location.

The plots are also flat at separations that are large compared to the thickness of the top layer, because then most current lines travel for most of their length through the lower layer and the earth approximates to uniform ground with the resistivity of the lower layer.

This leaves only the thickness of the upper layer to be found. Figure 12.12b differs from Figure 12.12a in having a longer left-hand level part, because the electrodes have to be further apart before significant current penetrates into the lower layer. However, it is not easy to deduce the thickness from the horizontal length because the length also depends upon the values of the resistivities, being shorter the more the resistivities differ. In practice, the thickness and the resistivities are found by comparing the actual plot with master curves calculated for different values of thickness and resistivity (Fig. 12.13). To reduce the number of master curves needed (which would be vast for just these three variables) the electrode separation is plotted as a *ratio* to the thickness of the top layer, a/h_1, and the apparent resistivity as a *ratio* to the resistivity of the top layer, ρ_a/ρ_1, with the different curves labelled by the value of the *ratio* of the resistivities (some master curves are labelled by the value of $(\rho_1 - \rho_2)/(\rho_1 + \rho_2)$, the resistivity constant called k); of course, we don't know these ratios to start with. To use the master curves the apparent resistivity versus electrode spacing plot *must be on the same scale,* so that a factor of 10 (a difference of 1 in log values, see Fig. 12.14) is the same distance for both plot and master curves. Either the results are plotted on tracing paper or the master curves are on transparent overlays for easy comparison.

The plotted points are slid over the master curves (keeping the axes of the two figures parallel) until they lie along a single one of the curves; for Figure 12.14a this is the master curve of Figure 12.13, labelled $\rho_a/\rho_1 = 6$. Though the measured curve does not become horizontal, at its left end it can be extended using master curve 6; the horizontal part of master curve 6 (labelled $\rho_a/\rho_1 = 1$) cuts the y axis at a log value of 1.27, giving 18.9 ohm-m for the value of ρ_1. Since the matching master curve has $\rho_2/\rho_1 = 6$, ρ_2 is 6 times as large, or 113 ohm-m. So resistivity values can be deduced even though the measured curve does not become horizontal: All that is needed is sufficient length to match to a particular master curve.

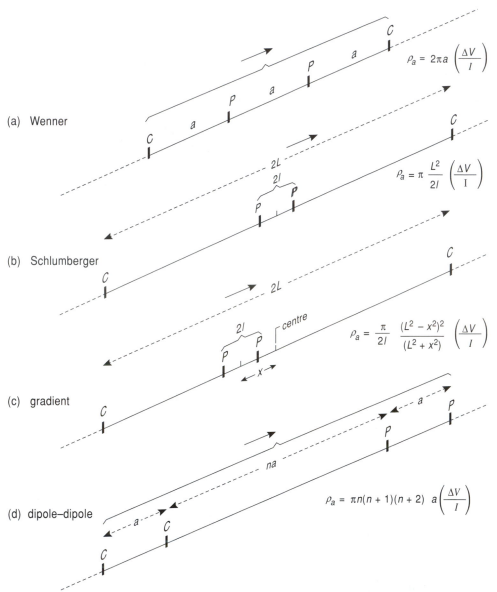

Figure 12.17 Various resistivity arrays.

(a) Wenner

$$\rho_a = 2\pi a \left(\frac{\Delta V}{I} \right)$$

(b) Schlumberger

$$\rho_a = \pi \frac{L^2}{2l} \left(\frac{\Delta V}{I} \right)$$

(c) gradient

$$\rho_a = \frac{\pi}{2l} \frac{(L^2 - x^2)^2}{(L^2 + x^2)} \left(\frac{\Delta V}{I} \right)$$

(d) dipole–dipole

$$\rho_a = \pi n(n+1)(n+2)\; a \left(\frac{\Delta V}{I} \right)$$

The Schlumberger array. The **Schlumberger array** differs from the Wenner array in having the P electrodes much closer together (Fig. 12.17b), though still placed symmetrically about the centre of the array. Readings are taken with only the current, C, electrodes being moved progressively and symmetrically apart. Moving only the C electrodes has two advantages: There are fewer electrodes to move, and with the P electrodes fixed the readings are less affected by any lateral variations that may exist. However, when expansion causes the value of ΔV to become so small that it cannot be measured precisely, the P electrodes are moved much further apart, while keeping the C electrodes fixed; then further readings are taken by

expanding the C electrodes using the new P electrode positions. Though the values of ρ_a for the two separations of the P electrodes but the same C-electrode separation should be the same, in practice they may be different (perhaps because of lateral variation), giving an offset of the ρ_a–current electrode spacing curve, (Fig. 12.18), and each section is adjusted to join smoothly to the preceding one, as shown. Master curves and modelling programs similar to those for the Wenner array are available for the Schlumberger array, and are used in the same way.

Examples of the use of the Schlumberger array are given in Sections 23.5.3 and 26.5.2. The two other arrays of Figure 12.17 are discussed in Section 12.4.

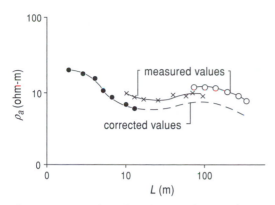

Figure 12.18 Plot of Schlumberger results.

The BGS Offset Wenner array system. The Wenner and Schlumberger arrays require several persons to move the electrodes, and one development is to set out many electrodes initially and select the required ones using switches near the resistivity meter.

One such system is the BGS Offset Wenner array system. Two special cables, with predetermined connection points, are laid out in opposite directions (Fig. 12.19a), and an electrode is connected to each of these points, with an additional one at the centre point, before any readings are taken; no tape measures are needed. The cables are connected to a resistivity meter via a switch-box, which has two switches: One selects a set of five equally-spaced electrodes – which always includes the centre electrode (Fig. 12.19b) – while the other selects combinations of four electrodes from these five, and con-

nects them to the four terminals of the resistivity meter. Five combinations are used (more are possible) (Fig. 12.19c). First consider the D_1 and D_2 combinations. Both are simple Wenner arrays (i.e., four equally-spaced electrodes) but are offset to either side of the centre by half the electrode spacing; this is why it is called an offset array (BGS stands for Barker Geophysical Services). If apparent resistivity values for D_1 and D_2 differ there is lateral variation or inhomogeneity; they can be averaged when modelling, but if the difference is large (say, over 30%) simple models should be treated with caution. Use of the other three combinations has several advantages, the main one being that they allow errors to be calculated. Including the comparison of D_1 and D_2, three measures of error are offered; two are measures of lateral inhomogeneity, while the third (that the value of $A - C$ should equal that of B) is a test that the apparatus is functioning correctly. Modelling is as for the simple Wenner array, except that the errors of the data points can be used to guide the fit. An example that used the Offset Wenner array is given in Section 26.7.2.

12.3.7 Limitations of vertical electrical sounding

How deep and how detailed a structure can be determined? In theory, if the subsurface truly consists of a series of uniform horizontal layers, then all can be determined. But reality is usually less simple.

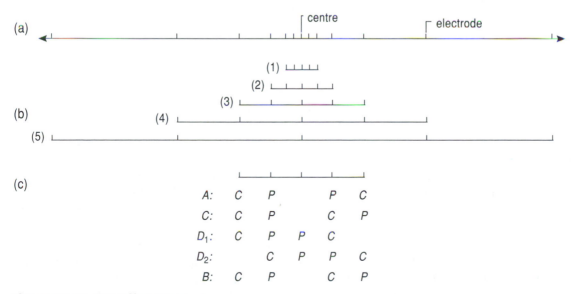

Figure 12.19 BGS Offset Wenner array.

Maximum detectable depth (depth of investigation). This chiefly depends on the maximum separation of the current electrodes (Fig. 12.7) but also partly depends on the resistivity difference of the layers; Figure 12.13 shows that the larger the contrast of resistivities between layers, the sooner the apparent resistivity begins to change as spacing is increased. It is not realistic to give simple rules for the maximum detectable depth of a layer because of the huge range of possible electrical structures and variety of arrays, but half the current electrode separation can be used as a guide to the maximum depth of penetration. Another approach is given shortly.

Penetration may be limited in practice if values of ΔV become so small – because of the geometrical effect of expansion or low-resistivity layers at depth – that they become difficult to measure accurately (this occurred in the survey described in Section 26.6.2). Accuracy can be improved by increasing the current or by averaging several readings, but not by a large factor.

Detail of structure. Within the depth of investigation, there are limits to how thin a layer can be detected (i.e., resolved). What is important is not the layer's actual thickness, but its thickness compared to its depth. For example, a metre-thick layer below a metre of overburden would be easily detected, assuming a significant resistivity contrast, but not one at 20 m. When a layer cannot be detected because of its combination of thinness and lack of resistivity contrast, it is said to be suppressed.

It is not easy to give rules for what is detectable; instead, it is better to construct relevant models. For instance, to check if a layer of clay within the top 15 m of sand in some proposed sand quarry would be detectable, apparent resistivity curves could be computed for sand with and without such a layer (using plausible values for resistivities) to see if the differences would be detectable. A similar approach could be used to see if, say, the known bedrock of an area is within the depth of investigation, provided plausible values can be assigned to the various layers. There should be at least as many electrode spacings used as layers modelled. If not, a simpler model will fit the data (Fig. 12.20).

Other factors. Modelling assumes uniform layers, whereas they may grade vertically one into another. In principle, this could be modelled by treating

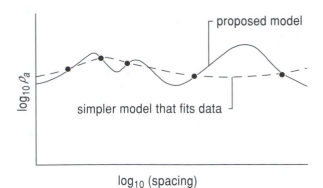

Figure 12.20 Overdetailed model.

them as a series of layers with slightly different resistivities, but this would result in too many layers to be practicable.

Layers may vary laterally in thickness or resistivity, which can be detected by carrying out surveys in different parts of the area. If the variation is not too large it may be possible to follow layers laterally and map the changes (Fig. 12.21), but if the differences are large, electrical imaging is a better method (Section 12.5).

Sometimes layers are **anisotropic,** so their resistivities differ with direction. In shales, slates, and schists, resistivity perpendicular to the lamination can be several times that parallel to it, and though the difference in sediments is usually less it can still be significant. If beds of these rocks dip, apparent resistivities will depend on the angle of the survey lines to the strike. Often, surveys are made at right angles to check for this. Other limitations are explained in Box 12.1.

12.4. Resistivity profiling: Detecting lateral variations

12.4.1 Introduction

Vertical electrical sounding (VES) investigates how resistivity varies with depth, assuming layers with no lateral variation; in contrast, **electrical profiling** investigates lateral changes, such as the presence of a mineral vein or a geological boundary. Whereas VES 'expands' an array symmetrically about a point, in profiling some or all the electrodes are moved laterally with fixed spacing. Covering an area by parallel traverses, giving a grid of readings, is called resistivity mapping; an example of this is given in Section 28.4.2.

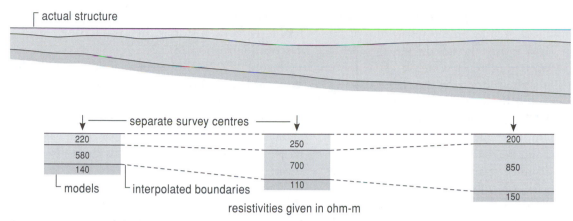

Figure 12.21 Varying and dipping layers.

BOX 12.1 Further limitations on VES modelling

Principle of equivalence. Because of refraction, current flows almost vertically through a layer if its resistivity is much *higher* than those of the layers that sandwich it (Fig. 12.8). If the layer is thin, it can be shown that replacing the layer by another with the same *product* of $t\rho$ has a negligible effect on readings. For instance, in Figure 1, halving the thickness t_2 of layer 2 but doubling its resistivity has negligible effect, for halving the length of the current tube through the layer compensates for doubling of its resistivity, and, as the layer is already thin, making it thinner has little effect upon the proportions of the figure. Therefore the ρ_a curve can be modelled equally well by different layers that have the same resistivity-thickness product, provided they are thin and with a high resistivity. If a thin layer has *lower* resistivity, current paths tend to flow in the layer, and then layers that have the same *ratio* of t/ρ

(doubling the thickness compensates for doubling the resistivity) will appear the same.

Anisotropy. As pointed out in Section 12.3.7, anisotropy can be detected by measuring resistivity along perpendicular lines if laminations are tilted but not if they are horizontal. However, horizontal laminations will still distort the lines of current flow and this causes models that assume no anisotropy to give a greater thickness but a lesser resistivity to the anisotropic layer than is actually the case. Anisotropy should be borne in mind, particularly if laminated rocks are likely to be present. These limitations apply in addition to the modelling assumption that the layers are uniform in thickness and resistivity, and near to horizontal, but has taken no account of reading errors, which also limit the accuracy of models.

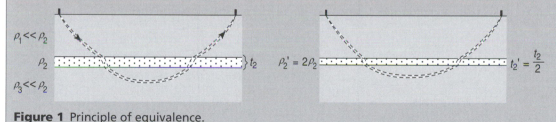

Figure 1 Principle of equivalence.

12.4.2 Some arrays for profiling

As with VES, a number of arrays can be used for profiling, but only a few will be described, to illustrate the range and some of their advantages and disadvantages. The simplest target is a vertical boundary between two resistivities, such as two lithologies offset by a vertical fault or the contact of a large intrusion.

Far to either side of the contact, the apparent resistivity is constant and equal to the resistivity of the rock below (modified by any overburden), but the transition between them can be complicated because, as the array crosses the boundary, different combinations of the electrodes are in each rock. The anomaly produced by the Wenner array is particularly complicated (Fig.

12.22a), with its detailed shape depending on the electrode separation as well as on the resistivities of the rocks. If the array is used 'broadside' (i.e., parallel to the boundary), the transition is a simple curve, for at any station all the electrodes are in the same rock.

A simpler anomaly is produced by the **gradient array** (Fig. 12.17c) as shown in Figure 12.22a. As with the Schlumberger array, the current electrodes are not moved and the potential electrodes are kept a small, fixed distance apart, but the potential electrodes are moved laterally over the target, though they are not allowed to approach close to the current electrodes; for example, the current electrodes may be 1 km apart, the potential electrodes only 20 m, and do not approach closer than 200 m to the current electrodes. This array can be adapted to survey an area by moving the potential electrodes out of the line joining the current electrodes (in which case the expression used to calculate the apparent resistivity is more complicated)

The **dipole–dipole array** (Fig. 12.17d) also produces a simple anomaly (Fig. 12.22a); the two current and the two potential electrodes are moved as pairs with the same fixed separation *a*. Often, only the potential electrodes are moved, either along a line or over a grid. The meter that measures the potential difference can be quite separate from the power supply and current meter; if so, this greatly reduces the length of wires that have to be moved, a particular advantage if the pairs of electrodes need to be widely separated to give deep penetration. This array is useful for surveying an area, because of its ease of use and good resolution of detail.

The twin array, in which a current and a potential electrode are moved together, with the other pair fixed, is very popular in archaeology and is described in Section 28.3.2.

Figure 12.22b shows anomalies over a vertical sheet, such as an ore vein. The anomalies, of course, are more complicated than for a single interface, but again the Wenner array gives the most complicated anomaly, its shape depending in part on the width of its electrode spacing compared to the width of the sheet.

Once an ore has been shown to exist in potentially economic concentration, by its presence in one or more boreholes, its extent may be explored using the **mise-à-la-masse method**. One current electrode is placed in contact with the ore, down a borehole, while the other makes contact with the surface beyond the likely extent of the ore; one potential electrode is fixed on the surface above the known body, so the only electrode that moves is the second potential electrode. This roving electrode is moved over the surface, or down boreholes, and the apparent resistance between the two potential electrodes ($\Delta V/I$) is measured and contoured to indicate the extent of the body. Figure 12.23 shows results for a nickel sulphide orebody, which had been located using ground magnetic surveys. After the near-surface ore had been worked out, an exploratory borehole found a small body at depth but its extent was not known, so potentials were measured down a number of subsequent boreholes. The resulting contours outline the body; because all the contours reach the surface they show that the body is continuous (had it consisted of separate bodies the contours would have closed around them).

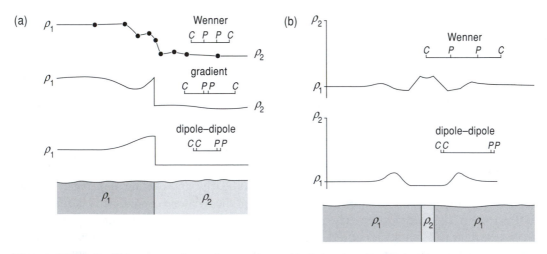

Figure 12.22 Profiles across a boundary and a vertical sheet, using various arrays.

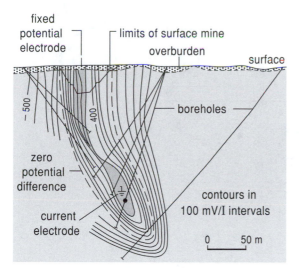

Figure 12.23 Mise-à-la-masse method, Tekkälä nickel sulphide ore body, Finland.

The mise-à-la-masse method is particularly useful in conjunction with a drilling program to evaluate the extent of a body: As additional boreholes are drilled the roving electrode can be placed in them to extend the survey in the third dimension ahead of the drilling. An example is given in Section 23.5.3.

12.5 Electrical imaging

Obviously, resistivity may vary both vertically and horizontally, so neither profiling nor VES gives accurate results. To investigate such electrical struc-

tures, arrays need to be both expanded and moved laterally.

If the body is very elongated horizontally – such as a vein or fault – its electrical structure can be investigated using a **pseudosection**. The survey, in effect, takes repeated profiles along the same traverse crossing the body, but with a range of electrode separations. Figure 12.24a shows, by the solid lines, one pair of electrode positions for a dipole–dipole array, here with $n = 3$ (Fig. 12.17d). The measured value of apparent resistivity is plotted at the intersection of 45° lines drawn down from the midpoint of each pair, shown by the solid dot. When the whole array is moved one spacing at a time to the right, the intersection points move an equal distance, giving values along the horizontal row, $n = 3$. Repeating with different values of n give rows at different depths. When the survey is complete, the values are plotted and contoured (Fig. 12.24b), showing roughly how resistivity varies laterally and vertically, like a section. However, the plotting depths are rather arbitrary and the variation of resistivity within the section is not accurate, which is why it is called a pseudosection. A common distortion is a small area of low resistivity, such as the massive sulphide in Figure 12.24c giving rise to an inverted V, or 'pant legs', as seen above it in Figure 12.24b. The Wenner array can also be used, with values plotted below the midpoint at a depth equal to the electrode spacing a (Fig. 12.24d).

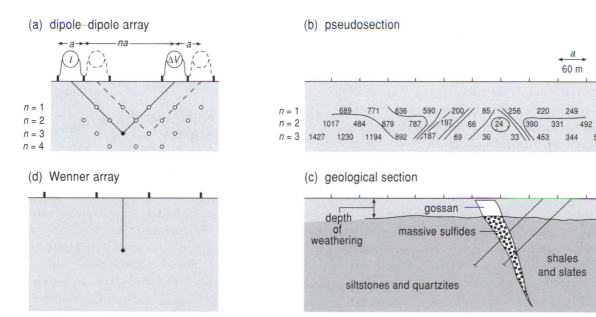

Figure 12.24 Pseudosection surveying.

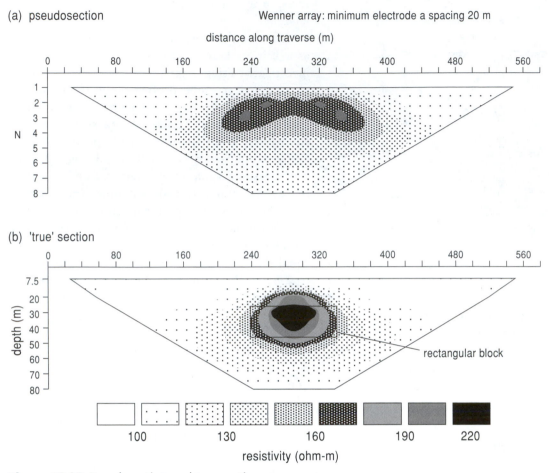

Figure 12.25 Pseudosection and true section.

In the past few years it has become possible to invert a pseudosection into a 'true' resistivity section (using tomographic theory beyond the scope of this book), and this is called **electrical imaging**. Figure 12.25 shows a contoured pseudosection and the true section derived from it, for a test section with a rectangular area of 500 ohm-m surrounded by 100 ohm-m. In the 'true' section, the 'pant legs' have been replaced by an oval area much closer to the size of the target; however, as with other tomographic methods, sharp boundaries have been 'blurred', and though the resistivity is higher than in the pseudosection it is less than the true value.

Making the measurements for a section can be greatly speeded up by inserting electrodes at each of the positions to be used and connecting them to the resistivity meter via a special cable; switches allow sets of electrodes to be selected for each spacing and lateral position.

Electrical imaging is finding increasing applications in groundwater studies, including contamination, and for investigating engineering and archaeological sites. Examples and more about the method can be found in Loke and Barker (1995, 1996).

12.6 Designing and interpreting a resistivity survey

12.6.1 Choosing a resistivity array

The choice of array and its dimensions largely depend upon the target: its size, depth, and resistivity contrast with its surroundings. Electrode spacing must be large enough to achieve penetration to the target, but the larger the spacing the poorer the resolution, both laterally and vertically, so it may not be possible to detect a small body at depth. In principle, any of the arrays described may be used for sounding or profiling, but in practice some are more suitable than others, as will be illustrated by some applications.

An important use of resistivity surveying is in the search for groundwater, for the water table in porous rock may form a horizontal electrical interface. The objective is to deduce the depths and resistivities of layers as precisely as possible. The Schlumberger and Wenner arrays are used, but the former has the advantage that its smaller separation of the potential electrodes reduces noise due to ground currents (from industrial and telluric sources, Section 14.7), which may limit the useful depth of penetration. That only the *P* electrodes in the Schlumberger method need to be moved, compared to all four in a Wenner array, is less of an advantage now that there is increasing use of systems with multicore cables with electrodes 'moved' by switching (e.g., the Offset Wenner array). In areas of dipping strata (or anisotropy) the results will depend on the orientation of the array, so it is usual to repeat with different orientations to detect the effect.

In mining, industrial, and engineering investigations, where the lateral variation is of most interest, profiles – rather than soundings – are more usual. Precision is less important, for interpretation is usually qualitative. Figure 12.26 illustrates the general situation, with an overburden containing lenses of sand or gravel and varying thickness over basement with steeply dipping mineralised zones. If the interest is in the sand and gravel lenses for building materials the equally-spaced Wenner array, with spacing equal to its electrode separation, is convenient. The choice of a suitable spacing may require some experimentation with different values, preferably in conjunction with a logged borehole to relate electrical layers to geological ones. Penetration needs to be sufficient to reach the lenses, but not so

deep that they form only a small fraction of the volume 'looked at' and so produce small anomalies with poor resolution; spacing is often chosen to be between two and four times the thickness of the overburden. But if the target is the basement mineralisation, the spacing will need to be up to 40 times the depth to the basement. With such a large separation, inhomogeneities in the overburden will be averaged out (in mineral surveys, resistivity is usually only one of several methods used, and is often subordinate to IP, described in the next chapter).

Supposing the depth to basement is needed along the route of a proposed road, it may be sufficient to measure a profile using electrode spacings 5 to 10 times the depth, provided the overburden has a strong resistivity contrast with the basement. If there are layers within the overburden and contrasts are small, it would be better to use a series of depth soundings, with the array aligned along strike if its direction is known. (For this purpose, resistivity would probably be used to interpolate between boreholes or seismic refraction surveys, which, being more expensive, would be used more selectively.)

12.6.2 Geological interpretation

The purpose of most electrical surveys (as with many other geophysical surveys) is interpretation in terms of geology. The values of resistivity provide only a rough indication of what the lithologies are (Table 12.1). Even within a single formation the resistivity often varies by as much as 30%. Further, electrical interfaces may not be geological boundaries, because they may be the water table or due to abrupt changes in salinity or clay content; conversely, a geological boundary may not be detectable if it separates layers with small electrical contrast. Geological interpretation must therefore be based, as usual, on a sound knowledge of the geology as revealed by borehole data, together with detailed study of the survey results and their variation with position.

Summary

1. Current, *I*, is caused to flow through a material by the 'pressure' of a potential difference, *V*, applied between two places. Current and potential difference are proportional to each other

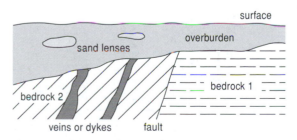

Figure 12.26 Possible profiling targets.

(Ohm's Law), and their ratio, *V/I*, is the resistance. Resistance depends upon (i) the geometry of a body, and (ii) its electrical resistivity, which is a fundamental physical property of a material.

2. Most rock minerals are insulators, but the presence of water in interconnected pores and cracks makes the rocks ionically conducting. The resistivity of the water, in turn, depends on its content of salts. Because these factors can vary so much, rock types have wide and overlapping ranges of the values of resistivities. For sediments, Archie's Law relates resistivity to porosity and other parameters. A few minerals, chiefly those of some ores and graphite, are electronic conductors and conduct without the presence of water.

3. To measure the resistivity of the ground, an array of two current electrodes and two potential ones is used, and the ratio $\Delta V/I$ is measured. Apparent resistivity is found by multiplying this ratio by a geometrical factor, which depends upon the array being used. It is the resistivity of uniform ground that would give the same ratio $\Delta V/I$ for the same electrode spacings.

4. The electrical structure of approximately uniform, horizontal layers is found using vertical electric sounding, VES, in which expansion of an array – or part of an array – results in the apparent resistivity changing as the current penetrates to deeper and deeper layers. Commonly used arrays for this purpose are the Wenner (including the BGS Offset version) and Schlumberger arrays. The electrical structure is found by plotting apparent resistivity versus electrode spacing (on logarithmic scales) and then comparing the result with master curves or with curves generated using a computer.

5. In principle, a uniformly layered structure can be fully determined; in practice, there are limits, depending on layer thicknesses, resistivities, and depths; the number of readings; and other factors. Also, layers may not be uniform in either thickness or resistivities, and may be anisotropic.

6. Lateral variation of resistivity is detected by profiling, moving an array – or part of it – along a line, or over an area. Values of apparent resistivity are profiled or contoured. Arrays used include Wenner, gradient, and dipole–dipole, which have various merits and limitations.

7. Variations of resistivity both laterally and in depth can be investigated by a combination of expansion and traversing of arrays, with results presented as a pseudosection or, better, by full electrical imaging.

8. You should understand these terms: current, current path, potential difference, resistance, resistivity, apparent resistivity, amp, volt and ohm; ionic and electronic conduction; electrode, Wenner, Schlumberger, dipole–dipole, and gradient arrays; resistivity meter; vertical electrical sounding (VES); electrical anisotropy; profiling, mise-à-la-masse, pseudosection, and electrical imaging.

Further reading

Resistivity is treated in most textbooks on applied or exploration geophysics, of which Parasnis (1997) has a larger amount than most, while Telford et al. (1990) goes into more of the underlying physics; Milsom (1996) concentrates on field procedures. Other useful books include Kearey and Brooks (1991), Reynolds (1997), and Robinson and Coruh (1988). Advanced methods for modelling sections so far are only described in Reynolds (1997) and in scientific journals.

Problems

1. Distinguish the geological situations for which the following forms of surveying are most appropriate: VES, profiling, imaging.
2. In modelling VES results, what assumptions are normally made about the subsurface?
3. Plot the data given below for the 5 VES surveys and deduce the minimum number of layers that exist in the subsurface. Why is it the minimum number?

Spacing (m)	Readings (ohms)				
	(a)	(b)	(c)	(d)	(e)
0.5	76	88	36	25	261
1	51	39	15	13	150
2	34	21	5.9	7.2	74
4	21	12.9	2.7	4.35	37
8	10.4	6.3	1.27	2.1	21
16	2.95	2.0	0.59	0.70	10.2
32	0.46	0.32	0.26	0.23	2.8
64	0.15	0.12	0.11	0.10	0.40

4. A VES survey using a Wenner array is carried out where beds, successively downwards, have

the following thicknesses and resistivities: 1 m, 20 ohm-m; 5 m, 400 ohm-m; 30 m, 100 ohm-m. Sketch roughly how the results would plot; you should label and number the axes and give rough values.

5. You are to carry out a survey whose purpose is to find out whether a 4 m-wide galena ore vein, seen in a known outcrop to strike roughly east-west, continues westwards beneath overburden 2 to 3 m thick, and if so, to locate its position. Describe how you would set about it, using a Wenner array. Why might you prefer to use another array, and what might it be?

6. A reading is taken at a particular place using a Wenner array and yields an apparent resistivity value of 100 ohm-m when using an *a* spacing of 5 m.
   (a) What is the simplest electrical structure that would give that reading.
   (b) If the same value was also obtained for spacings of 0.2, 2 and 20 m, what can you say about the subsurface?

7. Classify the following as likely to have resistivities greater or less than 100 ohm-m: clay, massive galena ore, disseminated pyrites, coastal sand below sea level, clean gravel beneath the water table, shale, dry sand, sphalerite ore, granite.

8. Consider if a resistivity survey would be appropriate to help solve the following problems. For each case where it would be appropriate, what form should the survey take?
   (a) To locate a disseminated ore of galena.
   (b) To locate a massive ore of sphalerite.
   (c) To measure the depth to the water table in sands.
   (d) To determine if there is saline water in the lower part of an aquifer that extends roughly from 10 to 70 m.
   (e) To measure the thickness of clay overlying sand saturated with fresh water.
   (f) To locate a strike-slip fault in crystalline bedrock, beneath dry sand.
   (g) To locate saline water 'pools' beneath a few metres of dry sand.
   (h) To detect if the thin soil covering a limestone area covers sinkholes.

9. Why is a pseudosection so called?

10. (a) Sketch how current flows between two electrodes 5 m apart inserted into a very thick uniform sands.
    (b) How would it be modified if the sand below 5 m were replaced by rock with lower resistivity?

11. Why are four electrodes used in electrical surveying?

12. Describe the electrode arrangements for the following arrays: Wenner, Schlumberger, gradient, and dipole–dipole. What are their relative advantages and disadvantages?

chapter 13

Induced Polarisation and Self-Potential

These two methods are mainly used to prospect for conductive ores. Self-potential (or spontaneous potential), SP, depends on small potentials or voltages being naturally produced mainly by some massive ores. Induced polarisation, IP, in contrast, depends upon a small amount of electric charge being stored in an ore when a current is passed through it, to be released and measured when the current is switched off. IP is significant only for disseminated ores but can often be used to locate massive ores as these are commonly surrounded by disseminated ore. For both methods, potentials can also arise in other, usually unwanted, ways.

Both methods require electrodes and wires to make contact with the ground.

13.1 Induced polarisation, IP

13.1.1 What induced polarisation is

Induced potential is a potential difference that sometimes exists briefly after the current in a resistivity array has been switched off. It arises from the presence of small particles of conductor in rocks, so it is used to detect **disseminated ores,** which are composed of discrete particles of conducting minerals, in a nonconducting matrix.

In rocks other than ores, current is conducted by positive and negative ions (Section 12.2.1) moving through the groundwater, often in tiny channels formed of interconnecting pores (Fig. 13.1). If a channel is blocked by a grain that is insulating no current can flow through it, but if the grain is conducting electrons can pass through, though ions cannot. Negative ions reaching the blockage will lose electrons and become electrically neutral; the electrons pass rapidly through the grain to its other side, where they are available to combine with positive ions coming to the blockage from the opposite direction. Because the exchange of electrons to and from ions is relatively slow there is a 'queue' or

buildup of ions on each side of the grain, forming a small accumulation of charge (Section 12.1); the rock behaves similarly to an electrical capacitor. When the current is switched off these charges disperse through the pore water and briefly produce a small current; as a result, the potential difference between the potential electrodes does not fall immediately to zero, as it would in a rock without ore grains. This buildup of ions is called **electrode polarisation.** It is significant only for disseminated ores, which is fortunate as the rock as a whole is not a good conductor, and so these ores are hard to detect using resistivity surveying. **Massive ores,** in which the conducting grains are in contact with one another, also often respond to IP surveying because they are commonly surrounded by disseminated ore.

Induced polarisation is developed by many ores of copper, iron, cobalt, and other metals, including most sulphides (though not sphalerite – zinc sulphide – which is not a conductor) and some oxides (e.g., magnetite and cassiterite – tin oxide) but also by deposits of graphite, which seldom have commercial value.

There is a second mechanism that produces unwanted IP. It is due to clay particles along the sides of the narrow pores. Negative charges that normally exist on their surfaces are in contact with the pore water, and these tend to attract positive ions when an electric current flows past (Fig. 13.2). These small accumulations of charge also disperse and produce a small current when the current is switched off. This **membrane polarisation,** fortunately, is generally only a fraction of the potential due to electrode polarisation but may act as background 'noise' when surveying for ores (Fig. 13.2).

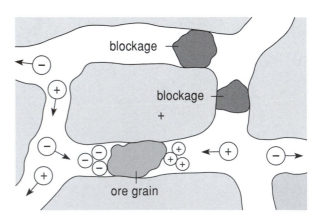

Figure 13.1 Microscopic pore channels in rocks.

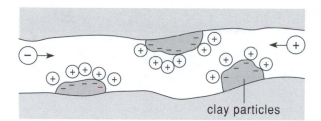

Figure 13.2 Membrane polarisation.

13.1.2 Carrying out an IP survey

An IP survey is similar to a resistivity one for it has to inject current into the ground and measure potential differences. A four-electrode array is used, as with resistivity surveys (Section 12.2.3), but the potential electrodes need to be nonpolarising (Section 13.2.2) to measure the small potential differences (p.d.'s). As with resistivity, the current direction is reversed at intervals to average out any steady potential present (mainly due to self-potential, Section 13.2, or telluric currents, Section 14.7). Commonly used arrays are the gradient and dipole–dipole arrays (Fig. 12.17).

To take a reading, the current is switched on for a couple of seconds. This is long enough to allow the charge at the grains to build up to a steady value, and so the current and potential difference become constant. Then it is switched off for an equal interval (Fig. 13.3). The residual p.d. is much smaller than the figure suggests; only a few thousandths of that when the current is on. It is not possible to measure it immediately the current is switched off, because the applied current itself needs a short time to drop to zero (due to the way equipment operates). The dying away of the p.d. is not measured continuously; instead, values are taken during a number of separate intervals each only a few milliseconds long, V_1, V_2, V_3, . . . ; these show the shape of the decay curve. Repeated readings are averaged to provide a more consistent value.

The area under the decay curve between the first and last interval (Fig. 13.3 detail) is calculated, and this area, divided by the potential difference just before the current was switched off, V_c, is called the **chargeability, M;**

$$\text{chargeability, } M = \frac{I}{V_c}\left(\text{area under curve}\right)$$

Eq. 13.1

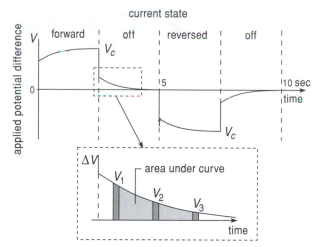

Figure 13.3 Measuring the decay of the induced potential.

and is measured in milliseconds. It is proportional to the total charge stored in the ground and so is a measure of the concentration of disseminated ore in the vicinity of the electrode array. Alternative measures of the IP effect are described in Box 13.1.

Surveys are usually profiles rather than depth soundings, but it is common to produce a pseudosection by using a dipole–dipole or other array with different separations of the pairs of electrodes, as described in Section 12.5. Occasionally, readings are taken over a grid and contoured.

As the induced potential is much smaller than the voltage between the P electrodes while the current is flowing (which is the value measured in resistivity surveying), much larger currents are used than for resistivity, up to 10 amps (compared to a fraction of an amp for most resistivity surveys); to provide this current all but the smallest systems require a portable generator rather than batteries. To inject large currents into the ground requires large voltages, and these are potentially lethal, so great care has to be taken to avoid receiving shocks and to prevent livestock electrocuting themselves by chewing the insulated cables, which they like to do. Apparent resistivity is also usually calculated, using the steady values of current and potential reached while the current is on.

In the above, the IP is measured using measurements at different times and is called a time-domain measurement; an alternative way, which makes measurements at different frequencies and so is called a frequency domain measurement, is described in Box 13.1.

BOX 13.1 Frequency-domain measurements: Frequency effect and metal factor

When a steady potential difference is applied to the current electrodes to inject current into the ground, the current reaches a maximum value almost at once but then begins to decrease as charge accumulates around the ore particles and opposes the current flow. When the charge stops accumulating the current has the value that would be reached if there were no ore particles, which is the value used when calculating apparent resistivity (Section 12.3.3). Therefore, as the current decreases the apparent resistivity increases from an initial value to a steady value, and the *increase* is a measure of the accumulated charge and hence of the amount of disseminated mineralisation.

To measure the change of apparent resistivity, first a rapidly alternating p.d. is applied to the current electrodes, to inject an alternating current into the ground. It alternates so rapidly (several times a second) that there is little time for charge to accumulate before the current reverses, and so the measured apparent resistivity, ρ_{ac}, is close to the initial value. Next, a steady applied p.d. is used to measure the steady value, ρ_{dc}. The two values are used to calculate the frequency effect, FE:

$$FE = \frac{\left(\rho_{dc} - \rho_{ac}\right)}{\rho_{ac}} \qquad \text{Eq. 1}$$

If there is no induced polarisation, there is no difference of the apparent resistivities and FE is zero.

Sometimes the FE value is multiplied by 100 to give the percentage frequency effect, PFE. In practice, ρ_{dc} is not measured with a steady current but with a current that alternates every few seconds, which is slow enough to allow the charge to accumulate to close to the steady value, but fast enough to average out natural ground currents and avoid other difficulties of steady-current readings. Nonpolarising electrodes are therefore not needed.

Metal factor The calculated value for FE depends partly upon the value of the apparent resistivity of the rock hosting the ore; for example, if the resistivity of the pore water in a ore-bearing rock were halved by influx of saline water, ρ_{dc} would halve and so FE would increase without a change in the mineralisation. Therefore, a better measure of mineralisation is the increase in *conductivity*, σ, due to the mineralisation, expressed as the metal factor, MF,

$$MF = 200,000\,\pi\left(\sigma_{ac} - \sigma_{dc}\right)$$
$$= 200,000\,\pi\,\frac{\left(\rho_{dc} - \rho_{ac}\right)}{\left(\rho_{dc}\,\rho_{ac}\right)}$$
$$= 200,000\,\pi\,\frac{FE}{\rho_{dc}} \qquad \text{Eq. 2}$$

The factor of 200,000π is inserted to convert small fractional values into more convenient numbers; the three expressions are exactly equivalent.

13.1.3 Data reduction and display

There is little data reduction or interpretation for IP, often no more than plotting the values of the chargeability (or frequency effect, etc., see Box 13.1) as a profile or on a grid and selecting obvious anomalies for drilling. However, when a range of electrode spacings has been used, a pseudosection is plotted as described for resistivity (Section 12.5), to give a better idea of how the ore concentration varies with depth, allowing a more informed choice of drilling locations, directions, and depths. Figure 13.4 shows pseudosections for both chargeability and resistivity; the former gives a much clearer anomaly, though with the inverted V or 'pant-leg' shape characteristic of a concentrated target, men-

tioned in Section 12.5. The ore was confirmed by drilling, as described in Section 23.5, which also gives further information about the IP surveys of this body.

13.2 Self-potential, SP

13.2.1 What self-potential is

Self-potential (SP) (also called spontaneous polarisation) is a naturally occurring potential difference between points in the ground. SPs are particularly associated with sulphide and some other types of ores, though they are produced by a number of mechanisms.

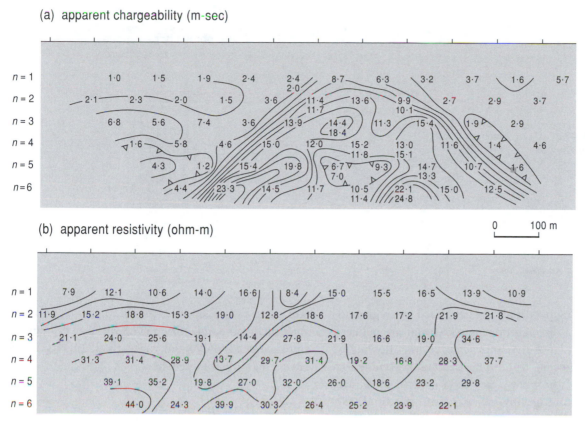

(a) apparent chargeability (m-sec)

(b) apparent resistivity (ohm-m)

0   100 m

Figure 13.4 IP and resistivity pseudosections, Elura.

Mineral potentials are associated with ores that conduct electronically rather than ionically (see Section 12.2.1), such as most sulphide ores (but not sphalerite, zinc sulphide) and magnetite; graphite, which is common but usually in worthless concentrations, can also give large values and arouse false hopes of a valuable ore discovery. It is not certain how the SP of an ore is generated, but most likely it is due to oxidation of the part of the ore body above the water table. In practice, we can regard an ore body as acting like a buried battery, nearly always with its negative terminal at the upper end (Fig. 13.5a), causing current to flow through the ground as shown. These produce potential differences, just as when a current flows in a resistivity survey. Since current flows from higher to lower potential the *lowest* potentials on the surface will be roughly above the ore (Fig. 13.5b). Anomalies are therefore mainly negative measured with respect to an electrode in **barren ground** (i.e., far from the ore body).

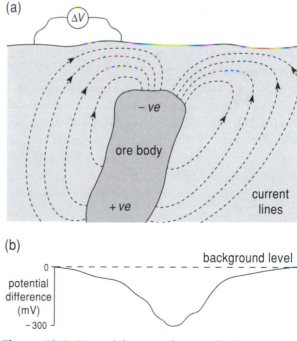

Figure 13.5 Potentials around an ore body.

(a) survey layout

(b) porous pot electrode

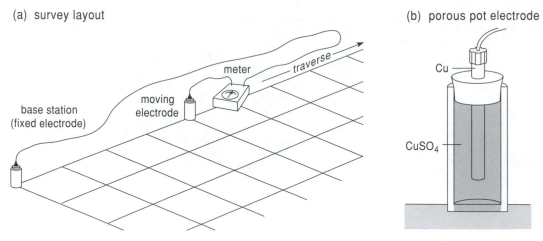

Figure 13.6 SP surveying.

Streaming potentials are generated when groundwater flows through porous rocks and are believed to be due to positive and negative ions flowing at different speeds, with some becoming attached to the sides of the pores, causing a charge imbalance. Higher ground usually has a positive potential with respect to lower ground, and the difference can exceed 1 volt. Streaming potentials usually are of no value, though they have been used to detect the positions of leaks in dams. Potential differences can also be due to concentration differences between adjacent formations such as clay and sandstone, and this is made use of in well logging (Section 18.5.2). Self-potentials can also be due to magnetotelluric currents (Section 14.7); these are usually small and fluctuating and form 'noise'.

Ores can usually be distinguished from other sources of SP because they produce much greater potential *gradients*, with potentials differing by more than tens of millivolts over distances of only tens of metres, compared to at most a few millivolts for other sources. The potential differences between different parts of an area over an ore sometimes exceed a volt.

You may be wondering why such ore bodies are not being used as power supplies for lighting homes and powering factories. There are two reasons: Firstly, the potential difference is seldom as much a volt, much less than the hundreds of volts used in the mains supply. Secondly, the ore supplies only a tiny current, too small even to power a torch, and any attempt to draw more causes the potential dif-

ference to fall. Therefore, the SP meter has to measure small potential differences without drawing a significant current from the ground (it is said to have a high input impedance).

13.2.2 SP surveying

In essence, an SP survey just measures the potential difference (voltage) between points on the surface (Fig. 13.6a). The instrumentation is small and light because – unlike resistivity and IP surveys – it does not have to inject current into the ground, but simply measures existing potentials – which range from a few millivolts to over a volt – without drawing more than a tiny current. The metal electrodes used for resistivity surveys are unsuitable for SP surveys, for polarisation occurs at their surface where conduction changes from ionic to electronic (Section 12.2.1); ions accumulate and so build up a potential comparable with the one to be measured (in resistivity surveys this accumulation is prevented by reversing the current a few times a second). Instead, **nonpolarising electrodes** are used, in which the end of a metal rod is immersed in a solution of one of its salts (Fig. 13.6b). The base of the container is porous (such as unglazed porcelain) and, once solution has soaked through it, allows electrical connection to the ground without polarisation. Though there is a contact potential between the electrode and the solution surrounding it, the two electrodes tend to drive current around the circuit in opposite directions and so cancel, provided the electrodes produce exactly the same potential. To ensure this, the solution is kept saturated by putting in more salt than will dissolve. Commonly, electrodes have a copper electrode immersed in saturated copper sulphate solution. The electrolyte leaks slowly out, so the level should be checked daily. Copper sulphate is mildly poisonous and stains, so gloves should be worn and reserves of copper sulphate crystals and solution should be kept safely.

At the start of a survey, the electrodes are checked by placing them close together and seeing that the reading is no more than a few millivolts. In damp weather, the electrodes will usually make electrical contact as soon as they are placed on the ground, but in dry weather it may be necessary to cut a small hole through turf or dig through dry soil to damp ground below. Readings can be made on bare rock if a pad soaked in copper sulphate is placed under the electrodes.

Surveys can be made in various ways. In potential surveys one electrode is put at some convenient base station and the second electrode plus meter are moved along a profile, or over a grid, as shown in Figure 13.6a. When the extent of the cable is reached, the last station can be used as a secondary base station and further readings taken with respect to it. The value of the secondary station relative to the base station is added to each reading; for example, if a station has a potential of +47 mV with respect to secondary base station SB1, which has a potential with respect to the base station of –167 mV, the potential of the station is +47 – 167 = –120 mV.

Another way is to find a pair of positions where the potential difference is some convenient value – say 50 mV – and then, keeping one electrode at the base station, locate a succession of positions where the other electrode maintains this difference, so tracing out the 50-mV contour; then repeat it for 100 mV, and so on. This is called equipotential line mapping but is not easy to carry out in rough terrain.

A third way, which avoids having a long connecting wire – difficult to move in rough terrain – is gradient surveying. The electrodes are kept a fixed distance apart, say 20 m, and advanced this distance by two persons, so that the rear electrode moves up to the position previously occupied by the front one. The p.d.'s of each step are added cumula-

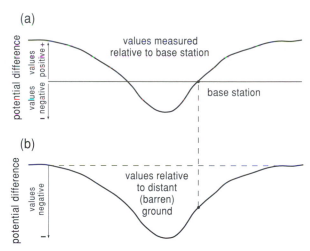

Figure 13.7 Dependence of anomaly on position of base station.

tively. It is best to work around loops, checking that potentials add up to about zero for a complete loop.

Data reduction. Though the anomaly is usually negative with respect to barren ground away from ore, the base station may not be in barren ground, so readings can be positive or negative with respect to it (Fig. 13.7a). When the barren background readings have been identified by their higher and approximately constant readings, the zero line can be adjusted to match (Fig. 13.7b). Figure 13.8a shows a contoured SP anomaly drawn from profiles, one of which is shown (Fig. 13.8b). The body is a shallow, massive pyrrhotite (iron sulphide) ore in the form of steeply dipping sheets.

Interpretation. There is little formal interpretation of SP anomalies, particularly as the cause of mineral potentials is not well understood. Anomalies larger than about 100 mV usually are due to ores, though not necessarily of commercial importance. The position of the lowest potential indicates the approximate centre of the ore body, though an asymmetric anomaly shows that either the body is dipping in the direction where contour lines are further apart or the concentration decreases less rapidly in this direction. Wide anomalies reflect extensive bodies. The half-width of the anomaly (i.e., half the width at half the minimum value of the anomaly) gives a rough estimate of depth; this is unlikely to be more

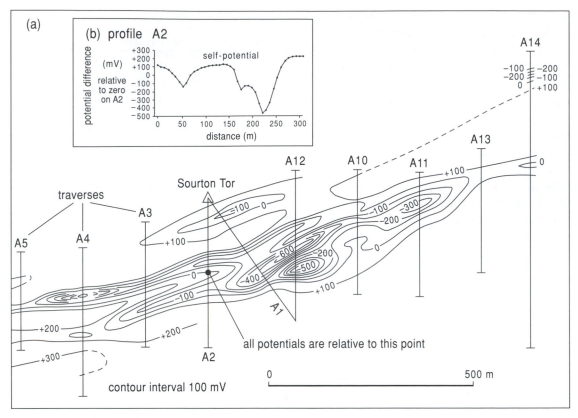

Figure 13.8 SP anomaly over Sourton Tor, Dartmoor.

than 30 m, as deeper ore bodies are not usually detectable by SP.

Summary

1. Induced polarisation (IP) depends upon charge accumulation on minute ore grains in pores when current is passed through the ground. It is measured after the current is switched off, as a small potential difference that dies away within a few seconds.

2. IP is developed by most disseminated ores, particularly sulphides (but not sphalerite), some oxides and – unfortunately – graphite.

3. IP equipment is very similar to that used for resistivity except that the current and voltage used are much larger. Gradient and dipole–dipole arrays are commonly used. Apparent resistivity is usually measured as well.

4. An IP result is expressed as chargeability (or, for measurements in the frequency domain, as frequency effect, FE or PFE, or metal factor, MF,

Box 13.1). Anomalies may be displayed in profile or by contouring; a pseudosection can be drawn if the dipole–dipole array has been used with different separations.

5. Self-potential, SP (spontaneous potential), is a small voltage (millivolts to about a volt) generated by conducting ores, including most – but not all – sulphides, some oxides, plus graphite, which is usually worthless. Smaller potential differences can be generated by a variety of processes, including streaming potential, which are usually unwanted and makes the background (zero) potential harder to recognise.

6. SP anomalies are usually negative with respect to barren ground, with the lowest potential (trough) over the ore body. Contoured results show the approximate extent of the ore. Ores below 30 m are unlikely to be detected.

7. SP is measured using of a special meter that draws very little current, connected to nonpolarising electrodes.

8. You should understand these terms: charge;

induced and spontaneous potentials; electrode and membrane polarisation; chargeability; mineral and streaming potentials; nonpolarising electrode; massive and disseminated ores, barren ground; time and frequency domains.

Further reading

IP and SP are discussed in the standard textbooks of exploration or applied geophysics, such as Kearey and Brooks (1991), Reynolds (1997), Robinson and Coruh (1988), Parasnis (1997), and Telford et al. (1990), with the last two being more advanced. Milsom (1996) concentrates on field procedures.

Problems

[ ] denotes problems that draw on material in a box.

1. Explain why IP, but not resistivity, is a suitable method for detecting disseminated ores. Explain why IP, as well as resistivity, is often successful for detecting massive sulphide bodies.
2. Which of the following gives the approximate position of the centre of an ore body when using an SP survey?
   (i) Below the maximum value.
   (ii) Below the lowest value.
   (iii) Midway between maximum and minimum values.
   (iv) Where the values are changing most rapidly.
   (v) Where the value is zero.
3. In which of the following are dangerous voltages and currents used?
   (i) IP, (ii) SP, (iii) Resistivity?
[4. Why are nonpolarising electrodes not needed when frequency effect is measured?]
5. How does the performance of a nonpolarising electrode differ from that of metal one? Describe, with a labelled sketch, the essential

features of a nonpolarising electrode. How would you ensure in an SP survey that the electrodes do not produce spurious readings?
6. Why is a resistivity survey often carried out in conjunction with an IP one?
7. Distinguish between electrode and membrane polarisations. Which one is associated with ore bodies?
8. Select from the list the electrode combinations used in (a) resistivity surveys, (b) IP surveys, (c) SP surveys.
   (i) 2 C and 2 P electrodes.
   (ii) 2 C electrodes.
   (iii) 2 P electrodes.
   (iv) 1 C and 1 P electrodes.
9. Select from the list the types of electrodes used in (a) resistivity surveys, (b) IP surveys, (c) SP surveys.
   (i) Metal.
   (ii) Nonpolarising.
   (iii) Metal C electrodes, nonpolarising P electrodes.
   (iv) Metal P electrodes, nonpolarising C electrodes.
10. How does a resistivity survey in the vicinity of ore bodies avoid measuring self-potentials?
11. A typical value of the SP of a massive sulphide ore is:
    (i) A few millivolts.
    (ii) A few tens of millivolts.
    (iii) A few hundred millivolts.
    (iv) A few volts.
    (v) A few tens of volts.
    (vi) A few hundred volts.
12. Compare and contrast the apparatus used for IP, SP, and resistivity surveys.
[13. If the fresh water in the pores of a disseminated lead sulphide ore were replaced by saline water, how would this affect the values of the frequency effect and metal factor?]

chapter 14

Electromagnetic Methods

Most electromagnetic (e-m) methods of surveying are used for targets similar to those of resistivity surveys, because both respond to variations in the resistivity (or conductivity) of the subsurface. The main difference is that e-m methods 'induce' current flows in the subsurface, usually without using electrodes. Many e–m methods can therefore be used in aerial as well as ground surveys.

E-m methods are particularly useful for ground surveys where the surface layer has a very high resistivity – such as dry sand or frozen ground – which prevents resistivity electrodes making electrical connection with more conductive layers below; conversely, a very conductive surface layer limits penetration more severely for e-m methods than it does for resistivity ones. A further limitation of e-m surveying is that generally it maps the subsurface less precisely than resistivity surveying. Smaller e-m instruments are quick to use on the ground because there are no electrodes and wires to set out.

Magnetotelluric, MT, surveying relies on naturally induced currents and can investigate down to tens, or even hundreds, of kilometres. Ground-penetrating radar, GPR, operates quite differently, by reflecting radar waves from subhorizontal interfaces, and so has similarities with seismic reflection, except that the discontinuities

are of electrical rather than seismic properties. Like seismic reflection, it can provide high-resolution sections, but penetration is limited to a few metres, which limits its use to shallow targets, which include engineering, hydrogeological, and archaeological as well as some geological ones.

14.1 Basic concepts

14.1.1 Electromagnetic induction

Electromagnetic (e-m) methods are mainly used, like resistivity methods, to investigate variations in subsurface resistivity, but they depend on different physical principles, the most important of which is **electromagnetic induction**. Figure 14.1a shows schematically one e-m system. A coil (with many turns of wire) forms a **transmitter** to which a power supply delivers an alternating current. This produces a magnetic field around it – the **primary field** – both above and below ground, including through the target. As the primary field is alternating – that is, regularly and smoothly reversing direction (Fig. 14.1b) – it produces a **magnetic field** through the target that changes in the same way. When a magnetic field passing through an electrical conductor (such as an ore body) *changes*, it causes, or induces, an alternating current to flow in the conductor. This induced current in turn produces its own alternating magnetic field – the **secondary field** – which can be detected at the surface, so revealing the presence of the ore. The usual way of detecting this secondary field is by the current it induces in a **receiver** coil. Thus one way to detect the target is to traverse the transmitter and receiver coils along the surface: When they are in the vicinity of the target the receiver shows a signal due

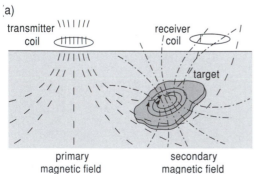
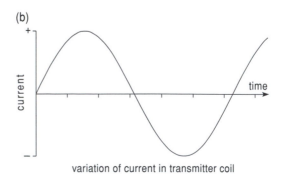
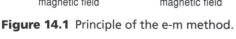

Figure 14.1 Principle of the e-m method.

to the current induced in it by the field of the target. The same principle is used in metal detectors, though these penetrate only tens of centimetres. In e-m surveying 'conductivity' is used rather than 'resistivity'. **Conductivity, σ,** is just the reciprocal of resistivity: $\sigma = 1/\rho$, and it is measured in Siemens/m, S/m.

14.1.2 Factors that affect the signal

To be able to interpret the results of an e-m survey, we need to know how the signal at the receiver depends upon the material, shape, and depth of the target, and also upon the design and positions of the transmitter and receiver coils, which roughly correspond to the role of the current and potential electrodes in a resistivity array. To simplify matters, treat the target as if it were a loop of wire (Fig. 14.2). The size of the current induced in it by the field of the transmitter, at any instant, depends on (i) the number of lines of magnetic field through the loop, (ii) the *rate of change* of this number, and (iii) the material of the loop, being greater the greater its conductivity, which is why the method is used for detecting conductive bodies. The number of lines through the loop is called the **magnetic flux**, and it depends upon (i) the strength of the magnetic field at the loop (Fig. 14.2a), (ii) the area of the loop (Fig. 14.2b), and (iii) the angle of the loop to the field (Fig. 14.2c). The equations underlying electromagnetic surveying are given in Box 14.1.

The field through the loop due to the transmitter depends upon the size and number of turns of the transmitter coil, how much current is flowing through it, and also how far it is from the target. The field in the receiver, due to the current induced in the target, depends upon how many turns it has and how far it is from the target. Of these various factors, some depend on the instrument – the combination of the transmitter, the receiver, and their separation – and some on the target – its size, its shape, its depth, and how conductive it is. The deeper and smaller the target the weaker the magnetic field it produces at the receiver, and if it is sheetlike its orientation will affect the flux. To detect weak targets the transmitter can be made more powerful (e.g., larger coil or current), and the receiver more sensitive (e.g., more turns), but there are practical limits, such as size and weight. The transmitter current can be made to alternate more rapidly, to increase the *rate of change* of flux, but this can limit the depth of penetration of the magnetic

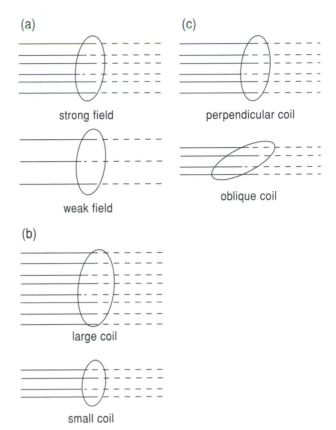

Figure 14.2 Flux through a coil.

field, as will be explained in Section 14.4.2. Frequencies of alternation range from hundreds to tens of thousands of Hertz (cycles/sec). Many of these factors are selected by the manufacturer of the instrument, but more usually the user has some choice of transmitter–receiver distance or frequency or both.

Next, we consider some different types of e-m systems.

14.2 Some e-m systems

14.2.1 Moving transmitter–plus–receiver system (Slingram)

Many types of instrument are used in e-m surveying, and many have different variants, and only a selection is described to indicate of the range of design and scope of application. This type of instrument – often known by the Swedish name **Slingram** – is like the arrangement shown in Figure 14.1. It has the transmitter and receiver connected by a cable and their separation kept constant as they are moved together along a traverse. As Figure 14.3 shows, the

BOX 14.1 Equations for e-m induction

To explain what determines the current induced in the target and hence what signal it produces requires a number of steps. Firstly, the target will be replaced by a loop as being more easily defined (Fig. 1). The number of field lines due to the transmitter passing through the target coil is the magnetic flux, ϕ. It depends (see Fig. 14.2) on the magnetic field, on its angle to the coil and the area of the coil, and on the number of turns if there are more than one:

$$\text{flux, } \phi = \text{magnetic field} \times \cos\theta \times \text{area} \times \text{number of turns} \qquad \text{Eq. 1}$$

When the flux through the coil *changes*, it produces a voltage (potential difference) around the loop, called the electromotive force (emf), ε; ε is all around the loop, but if a break were made in the loop this voltage would exist across the gap. Its value equals the rate of change of flux:

$$\text{e.m.f., } \varepsilon = -\left(\frac{d\phi}{dt}\right) \qquad \begin{array}{l}\text{Faraday's Law of}\\ \text{Electromagnetic Induction}\end{array} \text{ Eq. 2}$$

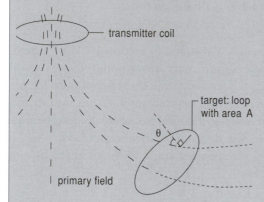

Figure 1 Target represented by a loop.

($d\phi/dt$ means rate of change of ϕ with time). Because the rate of change of flux is greatest when the field is changing from forward to reversed or vice versa (Fig. 14.1b), the induced current is a maximum when the transmitter current and field are zero, and vice versa (the negative sign is not important here; it shows that the induced current lags the transmitter current, as explained in Section 14.6). The rate of change increases with the frequency of alternation of the current.

This induced emf tends to make a current flow in the loop, but Ohm's Law by itself is not adequate for calculating the current in this situation because the current flowing around the loop produces its own magnetic flux through the loop (see, e.g., the transmitter coil in Fig. 14.1a). As it is alternating the changing flux produces an emf, which tends to decrease the rate of current *change*. This self-induction acts like a resistance but one that increases as the frequency, f, is increased. How much self-produced flux the target loop produces depends on the current in it, but also on the geometry of the loop: its size, shape, and number of turns; the ratio of flux to current is called the self-inductance, L, of the loop. The larger the self-inductance, the smaller is the current. The loop also has an ordinary resistance, R (Section 12.1), and the various quantities are related by

$$\varepsilon = I\sqrt{R^2 + \left(2\pi fL\right)^2} \qquad \text{Eq. 3}$$

compare with:

$$V = IR = I\sqrt{R^2} \qquad \text{Ohm's Law} \qquad \text{(Eq. 12.1)}$$

If f, the frequency, is zero they are the same.

magnetic field of the transmitter passes through the receiver coil (as well as through the target), so the magnetic field through the receiver has two sources: the primary field of the transmitter and the secondary field produced by the target, usually considerably the smaller of the two. To annul the current induced by the primary field, the cable from the transmitter supplies a current to the receiver that cancels the induced current; cancellation is exact only if the transmitter and receiver coils are kept at the same separation and orientation to one another. Smaller instruments of

this type have the transmitter and receiver coils held rigidly in a frame (Fig. 14.4a and b); for larger instruments, with coils 10 m or more apart, a frame is not practicable, and the coils have to be carefully positioned at each station (Fig. 14.4c).

The size of the secondary field is usually given as a percentage of the primary field. Alternatively, some instruments are calibrated to read apparent conductivity, that is, the conductivity that a uniform subsurface giving the same reading would have (this is the inverse of apparent resistivity, Section 12.3.3).

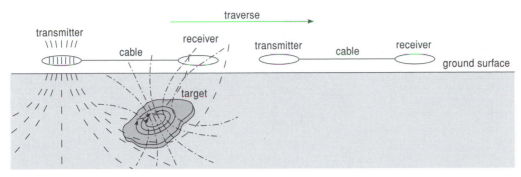

Figure 14.3 Moving source and receiver.

(a) Geonics EM31 instrument

(b) EM31 schematic layout

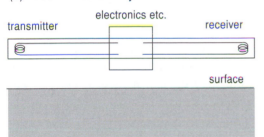

(c) Geonics EM34 instrument

Figure 14.4 Slingram systems.

If the target is a vertical, thin sheet, such as an ore vein, its anomaly is as shown in Figure 14.5a. There is no signal when the transmitter is directly over the target (Fig. 14.5b) because its magnetic field does not pass *through* the target, from one side to the other (i.e., there is no flux through the coil). When the receiver coil is over the sheet there is no signal either, because the secondary field at the receiver, being horizontal, does not pass *through* it (Fig. 14.5d). The signal is largest when the midpoint of the coils is over the sheet (Fig. 14.5c); it is negative because the primary and secondary fields through the receiver are in opposite directions at any instant. If the sheet is dipping, the profile is asymmetric (Fig. 14.5e).

To respond to deeper targets, both the coil separation and transmitter current are increased (the frequency may also need to be reduced, see Section 14.4.2). By repeating a profile with a range of coil separations it is possible to detect variations of conductivity both vertically and laterally, and a pseudo-section can be plotted, as for resistivity and IP arrays (Section 12.5).

As the coils are not connected to the ground they can be used in aerial surveys, either mounted on the ends of the wings of a aeroplane or with the transmitter and receiver carried in a 'bird' hung below a helicopter. This offers several advantages, which include the speed and low cost with which a survey can be carried out, particularly when large, rugged areas are to be covered, and the ability to carry out other aerial surveys – such as magnetic and radiometric ones – at the same time with little extra cost. The reduction in signal due to the extra distance from the target can be partly compensated by using large coils and currents; the large power supply this needs is carried by the aircraft. An alternative is to use a transient e-m (TEM) system (Section 14.3).

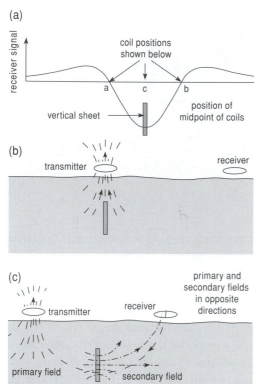

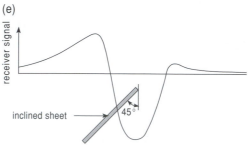

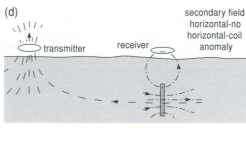

Figure 14.5 Profiles over sheets.

(a) apparent resistivity

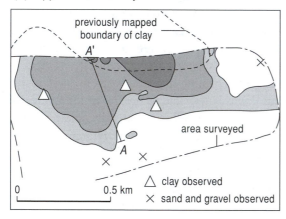

(b) profiles with two coil separations

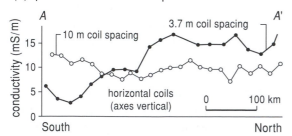

Figure 14.6 Geological mapping of an area of East Anglia, England.

Figure 14.6 shows an example of geological mapping using Slingram instruments. Boulder clay overlies sands and gravels interbedded with clays, which in turn rest on sands. The area is difficult to map in the conventional way for there are few exposures and the soil cover does not correspond simply to the underlying deposits, and even closely spaced hand-augered holes are hard to interpret. However, it is suitable for conductivity surveying because the proportion of clay is revealed by its much higher conductivity compared to that of sand or gravel (Table 12.1). Traverses 200 to 400 m apart enabled apparent conductivity to be contoured (Fig. 14.6a) and showed high conductivities (darkest areas) south of what had previously been thought to be the limit of the boulder clay. Two instruments were used, with coil separations of 3.7 and 10 m (Geonics EM31 and EM34 instruments, Fig. 14.4a and c). Figure 14.6b shows profiles for the two instruments along traverse *AA′* of Figure 14.6a. The shallower-penetrating instrument gives higher apparent conductivities to the north, consistent with a near-surface clay layer, while the lower value for the deeper-penetrating instrument indicates that the clay is not thick here, but there is more clay at depth in the south. (More details of this and other examples are given in Zalasiewicz et al., 1985.) A Slingram survey for saline water is included in a case study in Section 26.6.

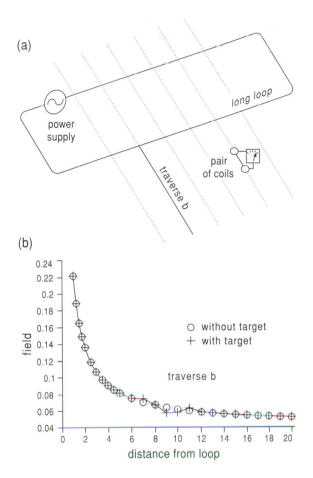

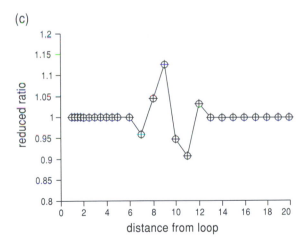

Figure 14.7 Turam system.

14.2.2 Turam system

The Slingram system is limited in the size of transmitter coil and current that can be used, since it has to be moved for each station. More powerful systems use a very large stationary transmitter coil or wire laid out on the ground, and only the receiver is moved, often on a grid of stations.

One such is the **Turam** system (Fig. 14.7a). The transmitter coil is usually a narrow rectangular loop with the long side typically 1 to 2 km long – though loops over 10 km long have been used – and roughly parallel to the expected strike, while the receiver consists of two portable coils connected together by a cable and kept a fixed distance of between 10 and 50 m apart. As the receiver coils pass over a conductive body the secondary field will generally be different for the two coils and so the variation in the ratio of the induced currents in them is measured to indicate the position of the body. However, there will be a difference even in the absence of a target, for the primary field decreases with distance from the transmitter and so differs at the two receiver coils (Fig. 14.7b); this difference is subtracted to give the 'reduced ratio', whose value is recorded as being at the midpoint of the receiving coils (Fig. 14.7c). The advantages of this arrangement are not only to have a very large transmitter coil to give deep penetration, but also to avoid having a cable between transmitter and receiver, which would be very cumbersome with the large surveys often made using this system.

Figure 14.8 shows the results of a survey over a massive base metal (copper–lead–nickel) sulphide with 10 millions tons of mineable ore, which drilling has shown subcrops beneath several metres thickness of overburden, and dips down to the west at about 45°. The stacked profiles of reduced ratios show clear anomalies, which correspond closely to the known subcrop.

14.3 Transient electromagnetic, TEM, systems

14.3.1 The basic concept

One of the problems with e-m surveying is cancelling exactly the current induced in the receiver by the primary field of the transmitter; when the signal from the target is weak any spurious signal due to imperfect cancellation can swamp it. This is particularly a problem for aerial surveys where the receiving coil is often towed behind the aircraft – to give a large transmitter–receiver distance – and may move relative to the transmitter coil, and the secondary

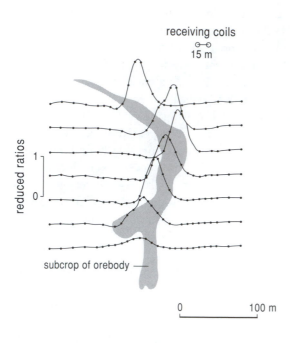

receiving coils
o—o
15 m

reduced ratios

subcrop of orebody —

0 _____ 100 m

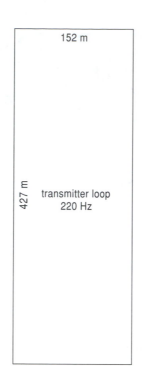

152 m

427 m

transmitter loop
220 Hz

Figure 14.8 Turam survey of the Woodlawn, Australia, massive sulphide ore body.

field is small because of the distance from the target. One way around this is to use **transient electromagnetic (TEM) surveying,** in which the receiver signal is measured only after the transmitter has been switched off. This will be explained in connection with a particular system.

14.3.2 The INPUT system

There are a number of TEM systems in use, but only INPUT (INduced PUlse Transient), one of the most successful systems, will be described. As the alternating transmitter current (Fig. 14.1b) changes from forward to reverse direction of flow the current is zero, but it is *changing* at its most rapid rate. Therefore, the current induced in the target will be a maximum, as will the secondary field and signal it produces. In the INPUT system the transmitter is switched off at this point (Fig. 14.9a), so the current induced in the target will become zero. But the current does not become zero instantaneously because the field produced by the target passes through the target and so produces a flux in it; as this flux decreases it in turn induces a current, which is in the same sense as the current induced by the primary field, and this slows the rate of decrease (if the target were perfectly conducting the current would continue indefinitely). As a result, the current does not cease abruptly but continues as a transient current

that dies away in a fraction of a second. (Another way of regarding this is that establishing a magnetic field takes energy, derived from the electric current, and this is returned when the field is switched off.)

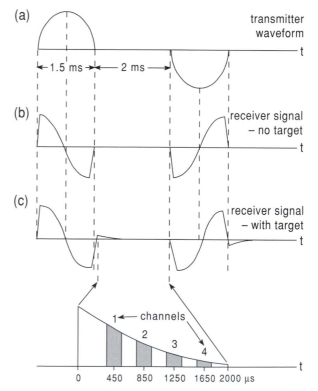

(a) transmitter waveform

← 1.5 ms → ← 2 ms →

(b) receiver signal – no target

(c) receiver signal – with target

1 ← channels
2
3
4

0   450   850   1250 1650 2000 μs

Figure 14.9 INPUT transmitter waveform and receiver signal.

Figure 14.9b shows the receiver signal in the absence of a conducting target; at the end of the half-cycle it goes rapidly to zero (but not instantaneously, because of the same effect occurring in the transmitter–receiver system, though this can be minimised by careful design). When there is a conducting target the transient current causes the signal to persist significantly longer (Fig. 14.9c).

The transient signal is sampled as shown (rather as with IP, Fig. 13.3). Each interval is called a channel and recorded separately. If there is a shallow body that is a good conductor the transient is relatively long lived and so the values of all channels are relatively large, but the poorer the conductor the quicker the values decrease. If there is a deep conductor there is little signal in the early channels, for it takes time for the field to propagate up to the surface, so the signal grows before decreasing. The INPUT system can detect targets to a depth of about 300 m below the aircraft.

Figure 14.10 shows two of a number of profiles that led to the discovery of the Detour zinc–copper–silver deposit in Quebec, Canada. They were taken in a N–S direction, to be perpendicular to the geological strike of the area. The transmitter was carried on the aircraft 120 m above the ground surface, with the receiver towed about 100 m behind and 67 m below. Each profile shows six channels, the first at the bottom, and the scale bar is 1000 ppm (parts per million, so 1000 ppm = 1/10%) of the transmitter field. Both show a clear anomaly, but for the first profile it appears only on the first three channels, whilst it extends to four or even five on the second, indicating a richer ore here. As the first channel gives the largest anomaly the ore cannot be deep, and drilling showed it to be only a few tens of metres. Another example of the use of TEM is included in a groundwater case study, in Section 26.5.

14.4 Electromagnetic waves

14.4.1 Wavelengths

When an electric current through a wire or coil is switched on, it produces a magnetic field in the space around the wire, but it is not established everywhere at the same instant; instead, it first occurs near the wire and then extends away, though extremely rapidly. If the current is alternating (ac), flowing first one way then the other, these alternations of the magnetic field also spread out, as a series of waves, just as waves spread out from an oscillating seismic source. These are electromagnetic (e-m) waves and are of the same kind as radio waves, X-rays, γ rays (Section 16.1), and light, the main difference between them being their wavelengths.

As explained in Section 4.1, wavelength is related to frequency by the relation: velocity = frequency × wavelength ($v = f \times \lambda$). The waves travel in air or space at about 300,000 km/sec (the speed of light) so that for a frequency of 1000 Hz – a typical frequency in e-m equipment – the wavelength is about 300 km. Waves travel more slowly in the ground, so there peaks are closer together, rather as

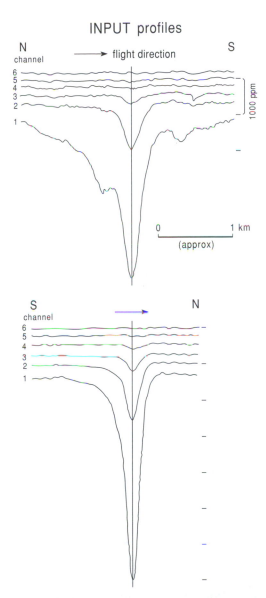

Figure 14.10 Two sets of profiles, Detour deposit.

cars close up as traffic slows; therefore, as the equation shows, wavelengths decrease in proportion.

In most e-m surveying, such as described above, the wavelengths are so much longer than the extent of the survey that, at any instant, the field is effectively the same over the area and need only be regarded as varying with time. However, at very high frequencies this is not true, and the wave behaviour becomes more important. This is most evident in ground-penetrating radar, GPR (Section 14.8). Of more immediate importance is that the wavelength partly determines the depth to which e-m surveys can investigate, through its effect on absorption, similar to that of seismic waves (Section 4.5.5).

14.4.2 Absorption and attenuation of e-m waves

Waves become weaker, or attenuate, as they propagate, for two reasons. As they spread out their energy is shared over an enlarging wave front, which is why radio reception becomes poorer further from the transmitter; secondly, they can be absorbed.

E-m waves can travel indefinitely through space or the atmosphere but are progressively absorbed as soon as they travel through a conductor – which all rocks are to some extent – and more rapidly the higher its conductivity. This is why radios and cellular phones do not work in a tunnel. For rocks of a given conductivity, amplitude decreases by the same factor for each wavelength travelled, leading to exponential decrease, as shown in Figure 14.11.

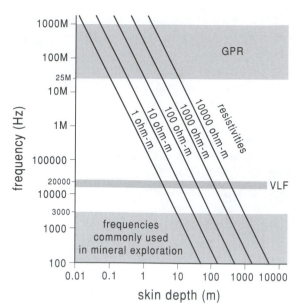

Figure 14.12 Skin depths for different frequencies and resistivities.

As the amplitude never becomes zero, a simple absorption distance cannot be given; instead, the **skin depth** is used, the distance in which the amplitude drops to $1/e$, which is about a third (e is the exponential, 2.718 . . .); at twice the depth it is reduced to $1/e^2$, and so on. The skin depth is given by the formula

$$\text{skin depth, metres} = 500\sqrt{\frac{1}{\sigma f}} = 500\sqrt{\frac{\lambda}{\sigma v}}$$
$$= 500\sqrt{\frac{\rho}{f}} = 500\sqrt{\frac{\rho\lambda}{v}} \quad \text{Eq. 14.1}$$

where λ is the wavelength, f the frequency, v the velocity, σ the conductivity, and ρ the resistivity. Skin depths are shallower for both higher frequencies and higher conductivities (lower resistivities). Figure 14.12 shows how they are related: To find the skin depth first select the frequency on the left-hand axis, then move horizontally to reach the appropriate sloping resistivity line, and finally drop down vertically to the skin depth axis. The figure also shows the frequency ranges used in different e-m surveying equipment.

Because of absorption, lower frequencies – as well as a more powerful transmitter – may be needed to reach deep targets, but if the overburden is very conductive it may be impossible to use e-m to explore below it. Conversely, e-m surveying can work well for

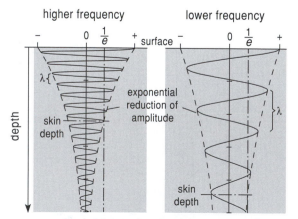

Figure 14.11 Amplitude decreasing with depth due to absorption, at two frequencies.

detecting conductive bodies beneath a highly resistive overburden such as dry soils, which make resistivity surveying difficult. For instance, when choosing a route for a buried metal pipeline through a desert it may be necessary to check that there is no saline groundwater – which would cause corrosion – beneath the dry surface; a survey with a Slingram system along the proposed route would do this quickly.

14.5 VLF (very-low-frequency) method

14.5.1 Basic concepts

This method does not have its own transmitter, for it makes use of very powerful radio transmitters that are used for communicating with submarines. A transmitter for this purpose has an aerial that is a vertical wire, up and down which a large current alternates (Fig. 14.13), producing magnetic field lines that are horizontal circles. These circles expand away from the aerial as waves travelling at the speed of light.

The frequencies used are around 20,000 Hz (20 kHz), which is a **very low frequency, VLF,** for radio, though it is quite high for e-m surveying. (The transmissions are for communicating with submerged submarines; because the sea is quite conductive, due to its salt, it rapidly absorbs radio waves, so very low frequencies are used to give the deepest penetration.)

A long way from the transmitter – hundreds or thousands of kilometres – the curvature of the wave fronts is so slight that they are effectively flat, vertical surfaces, so far as a survey of a few kilometres is concerned. The magnetic field is horizontal in the plane of the wave fronts (e-m waves are transverse – like seismic S-waves, Section 4.5.2).

Suppose the target is a buried vertical sheet of conducting material, such as an ore vein; to get the maximum flux linkage and hence signal, the magnetic field should be perpendicular to the sheet, which therefore should strike approximately towards the transmitter (Fig. 14.13). The survey traverse, as usual, should be roughly at right angles to the strike of the sheet. Though you can't choose the strike of the sheet, there is usually a choice of transmitter directions, for several nations keep in touch with their submarines secretly cruising the oceans. If the strike of conductors is not known, surveys should be carried out in turn with two transmitters in directions roughly perpendicular to one another, to ensure that all conductors will be detected.

To see how an anomaly is produced, think of the sheet as a rectangular, vertical loop of wire, which has a current induced in it (Fig. 14.14). The top part of the coil, being much closer to the receiver, contributes most of the signal, so the lower part can be neglected. The secondary field it produces has circular lines of magnetic field around a horizontal axis (Fig. 14.14b; compare with the transmitter of Fig. 14.13).

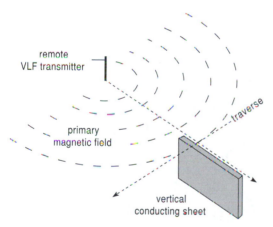

Figure 14.13 Transmitter of VLF radio waves.

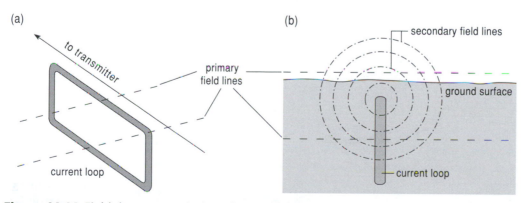

Figure 14.14 Field due to a vertical conducting sheet.

It is not possible to cancel the effect of the field due to the transmitter, as in the Slingram system, so the receiver detects the combined fields of transmitter and target. The combined field is usually tilted, whereas the primary field is horizontal, and so the tilt is used to detect the presence of the buried sheet.

14.5.2 Carrying out a VLF survey

The instrument is a radio receiver tuned to receive the particular transmitter selected; because a transmitter does not have to be provided, the instrument is light and compact (Fig. 14.15a and b). It has a receiver coil in its handle to detect the alternating magnetic field. It gives out a single audible (and irritating) note, whose loudness is used to adjust the instrument. First, it is swung around in a circle with the receiver coil in a *horizontal plane* (Fig. 14.15c): There will be two antiparallel directions of maximum signal, when the handle is aligned along the field, which then is passing through the receiver coil; the direction of the survey line should be roughly along one of these maximum directions. Next, it is held upright to the eye, with the reference coil aligned along the survey direction (Fig. 14.15d) and rocked in the *vertical plane* to find the *minimum* signal (generally not zero). In the absence of a conductive target the receiver coil will be vertical, but it will be tilted near a target. The tilt angle is read from a scale seen in the eyepiece. As Figure 14.16a shows, the instrument is *tilted up* (in VLF surveying this is considered a positive dip angle) approaching the sheet, is horizontal above the sheet, and is *tilted down* as the sheet is left behind (for this reason it is known – together with some other methods not described in this book – as a tilt-angle e-m method). This is true regardless of which way one goes along the survey line, so parallel traverses should be surveyed with the observer *always facing the same direction* to avoid having some anomalies inverted with respect to the others. If the ground is not level, the magnetic field – in the absence of a target – tends to be parallel to the surface, giving a positive dip uphill and negative downhill; this effect should not be mistaken for a conducting body.

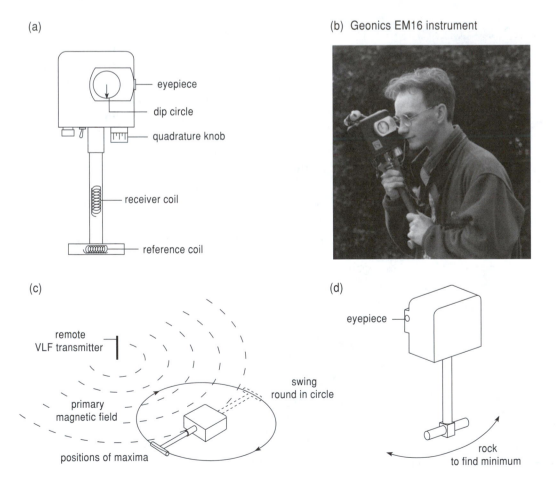

(a)

eyepiece

dip circle

quadrature knob

receiver coil

reference coil

(b) Geonics EM16 instrument

(c)

remote VLF transmitter

primary magnetic field

positions of maxima

swing round in circle

(d)

eyepiece

rock to find minimum

Figure 14.15
VLF instrument.

(a) tilt-angle measurement

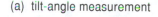
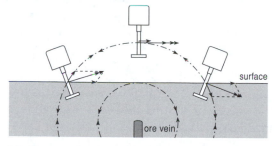

(b) tilt anomaly of a vertical sheet

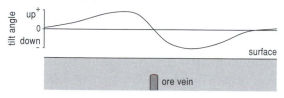

(c) tilt anomaly of a wide conductor

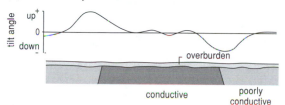

Figure 14.16 VLF profile over a vertical sheet.

If the sheet is thin and vertical it produces an anomaly as in Figure 14.16b, but if very thick, so that it has become a block of conductive rock between nonconductive ones, the anomaly is split apart into positive and negative parts, one near each boundary (Fig. 14.16c).

VLF surveying is best suited to mapping near-vertical sheets or contacts. When the target is extended but with no simple shape the length of the anomaly (i.e., where the tilt is not zero) is a measure of the size of the target, and an asymmetric anomaly reflects an asymmetrical target, such as a dipping sheet. The amplitude of the anomaly depends on the depth and conductivity of the target – loosely related to the ore concentration – and its depth.

Figure 14.17 shows an anomaly obtained for two, long, vertical sheetlike conductors. Two curves are shown, in phase and out of phase. The in-phase one corresponds to the tilt described above. The out-of-phase, or quadrature, reading is found using a dial connected to a second coil, the reference coil, in the crosspiece of the handle (Fig. 14.15a); the minimum signal found by tilting the instrument is not zero (except when there is no target), but it can

be reduced to zero using the quadrature dial (this is explained further in Box 14.2). The quadrature reading indicates how conducting the target is, and hence, for ores, how concentrated they are. Quadrature refers to the phase of the signal, an aspect of e-m surveying that has not yet been explained. As it is not confined to VLF instruments it will be explained separately in the next section.

14.6 Phase

Phase is a technical name for something quite familiar. To illustrate it, Figure 14.18 shows how the height of the sun – measured by its angle above the horizon – varies with time at three places around the world (on March 21, when the lengths of day and night are equal everywhere). The curves have a familiar shape, a sinusoid (i.e., the shape of a wave). All the curves have the same period of 24 hrs, but differ in their amplitudes and also – the point of this example – in being displaced with respect to one another: Midday in Calcutta (longitude 90° E) is 6 hrs before Greenwich (longitude 0°), which is 6 hrs before New Orleans (longitude 90° W).

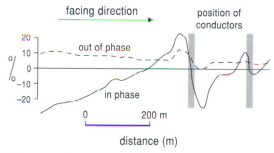

Figure 14.17 VLF profiles over two conductors.

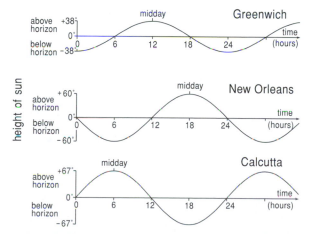

Figure 14.18 Height of the sun at three places.

BOX 14.2 Polarisation ellipse

As explained earlier, usually both primary and secondary fields induce currents in the receiver. These two fields differ in phase as well as in magnitude, and also generally have different directions. If they differed only in size and direction, but not in phase, they could be added vectorially, as has already been used for magnetic surveying (Fig. 11.15), for example; but the difference in phase makes the addition more complicated.

To start with, suppose that the two fields are equal in size, with the primary field *horizontal* and the secondary field *vertical* (Fig. 1a and b). Also suppose that the phase of the secondary field, due to the target, lags the transmitter by 90° (i.e., it is in quadrature). When the primary field is a maximum – point 1 on Figure 1a – the secondary field is zero –

point 1 on Figure 1b – so the combined field is therefore horizontal – point 1 on Figure 1c. An eighth of a cycle later the fields are equal (2), so the total field is midway between the horizontal and vertical, or 45° *from the horizontal* (this is an angle in space, not a phase angle). At 3 the secondary field is a maximum and the primary is zero, so the combined field is now vertical, and so on, until at 9 it is back where it started after one complete cycle. It turns out that the length of the combined field remains the same throughout and swings round like the hand of a clock, tracing out a circle on the diagram (Fig. 1c). This is called circular polarisation (this type of polarisation is quite different from the polarisation of IP but is related to the polarisation of sun spectacles and some camera filters).

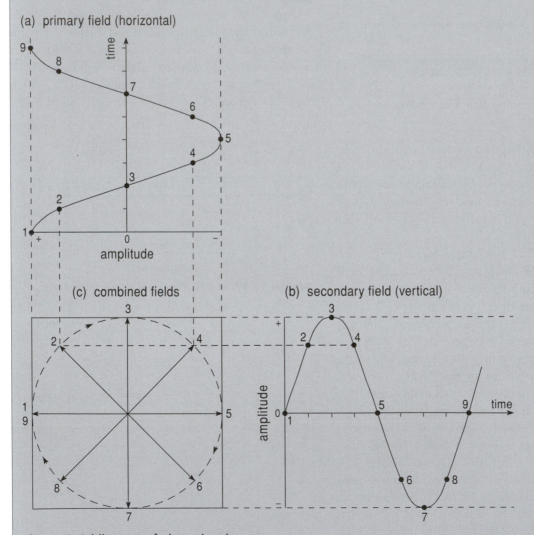

Figure 1 Adding out-of-phase signals.

BOX 14.2 Polarisation ellipse (continued)

The circle fits inside a square whose sides are twice as long as the maximum values of the fields (Fig. 2a). If the two fields are not equal and the primary field is the larger (as is usual), the enclosing square becomes a rectangle and the circle an ellipse (Fig. 2b). If the secondary field is no longer vertical, the enclosing rectangle becomes skewed (Fig. 2c); the two fields still combine to produce an ellipse but now one that is tilted. So far, the two fields have had a phase difference of 90°; changing the phase between them affects the width of the ellipse (the ratio of major to minor axes) and its angle of tilt within the skewed rectangle (Fig. 2d). (If the phase difference were zero – or 180° – the ellipse would 'collapse' to become a diagonal, which is what you would expect when there is no phase difference between the two and so normal vector addition can be used.)

In summary, for the usual situation, where the two signals are not equal in size, are not at right angles in space, and are not exactly 90° out of phase, they produce a polarisation ellipse whose major axis is tilted to the direction of the primary field.

It is the addition of the secondary field with a phase difference that produces an ellipse. If the secondary field is split into in- and out-of-phase components, as explained in Section 14.6, the tilt of the ellipse is found to depend largely on the in-phase component, while the ratio of minor to major axes depends largely on the out-of-phase (quadrature) component (assuming the secondary is considerably smaller than the primary field, which is usual). So the tilt and shape of the polarisation ellipse can be used to deduce the in-phase and quadrature parts of the signal from the target.

VLF. When taking readings (Fig. 14.15d) both receiver and reference coil are in the plane of the polarisation ellipse. When it is tilted to find a *minimum,* the receiver coil is aligned in the direction of the minor axis of the polarisation ellipse (Fig. 3). The signal is not zero because the minor axis is not zero, but it is then reduced to zero by cancelling it with a percentage of the signal in the reference coil, which is aligned along the major axis. Inside the instrument is circuitry, which includes the quadrature dial, which compares

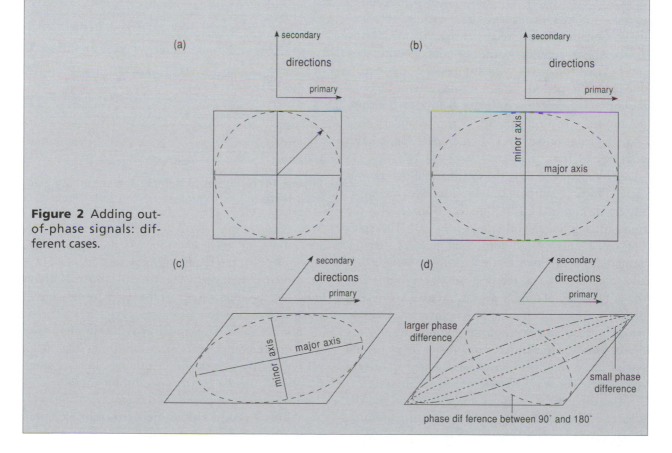

Figure 2 Adding out-of-phase signals: different cases.

(a) secondary directions primary

(b) secondary directions primary — minor axis / major axis

(c) secondary directions primary — minor axis / major axis

(d) secondary directions primary — larger phase difference / small phase difference / phase difference between 90° and 180°

BOX 14.2 Polarisation ellipse *(continued)*

the sizes of the signals in the two coils: When the sound is a minimum (ideally zero), the number on the quadrature dial gives the length of the minor axis as a percentage of the major. The tilt of the instrument gives an approximate but adequate value of the in-phase component as a percentage of the primary field – the dial is marked to give this rather than the angle – while the quadrature knob gives the quadrature component, also as a percentage.

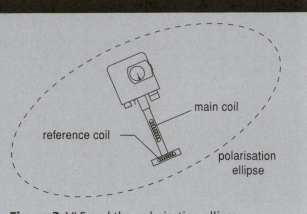

Figure 3 VLF and the polarisation ellipse.

This difference is the amount by which watches have to be adjusted when travelling by plane and is the cause of jet lag.

In describing the difference, or **phase** difference, we have to state which is being compared with which: We say that New Orleans **lags** Greenwich because midday there is later; equally, we could say Greenwich **leads** New Orleans. The *amount* of lead or lag depends on the difference of their longitudes, so we say New Orleans lags Greenwich by 90°. Similarly, Calcutta leads Greenwich by 90° (or lags by 270°). Angles are used instead of hours because they can be used for waves with any period: 360° equals one complete period. In e-m measurements phase is the difference from the primary field – or transmitter current – so in Figure 14.19 the secondary field lags (the transmitter) by 135° (compare Fig. 14.19b with Fig. 14.19a)

Another way to describe the phase of the receiver signal splits it mathematically into two waves, as shown in Figure 14.19c: One is in phase with the transmitter, while the other lags it by 90°. This splitting can be always be done, but requires that the amplitudes of the two waves be chosen so that they add together to equal the original wave; in this case the in-phase amplitude is negative (i.e., a lag or lead of 180°), which is why it is shown inverted. The two waves are called the **in-phase** component and the **out-of-phase** or **quadrature** component (in phase and out of phase are also known as real and imaginary respectively, for reasons irrelevant to geophysics).

The higher the conductivity of the target, the greater is the lag of the receiver signal, from little

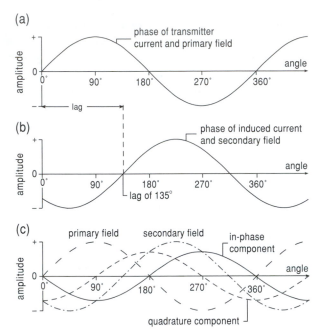

Figure 14.19 Phases of primary and secondary magnetic fields.

more than 90° to nearly 180°. As the lag increases towards 180°, the in-phase component grows as the quadrature dwindles (Fig. 14.20) and this effect indicates the conductivity of the target, a measure of the ore concentration.

Some e-m instruments display the in-phase and quadrature components as separate readings (e.g., VLF, Fig. 14.17), while in others, such as some Slingram instruments, the quadrature component is converted into apparent conductivity, using relationships beyond the scope of this book (such

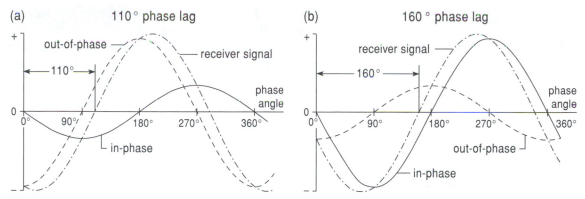

Figure 14.20 Receiver signal: In-phase and quadrature components for two phase lags.

instruments can give a negative conductivity reading for very-high-conductivity targets). In VLF instruments, the in-phase component is derived from the tilt of the instrument, while the quadrature dial gives the quadrature component (how the quadrature dial works is explained in Box 14.2).

In summary, the receiver signal usually differs in both magnitude and phase (with respect to the primary field). Phase offers information about the conductivity of the target, and is usually given as in-phase and out-of-phase or quadrature components, though conductivity may be displayed instead.

14.7 Magnetotelluric, MT, surveying: Looking into the deep crust and mantle

14.7.1 Basic concepts

It was explained in Section 14.4.2 that e-m waves are absorbed by the ground because it is conductive, but lower frequencies (longer wavelengths) penetrate deeper than higher frequencies. To investigate the deep crust or mantle requires very long – as well as powerful – waves, which cannot be generated by man-made equipment. However, such low-frequency waves are generated naturally by currents in the ionosphere, the electrically conducting upper atmosphere. Ultimately, both the existence of the ionosphere and the varying currents in it depend upon the Sun, in particular upon charged particles from it (when unusual activity on the Sun sends more particles, the fluctuations are much greater and can lead to magnetic storms and auroral displays that interfere with radio transmissions and magnetic surveys, Section 11.1.4). These variations have frequencies

from as little as about one cycle a day (about 0.00001 Hertz [Hz]), to thousands of Hertz; the longer period waves penetrate deep into the Earth – and are also responsible for the magnetic diurnal variation which needs to be allowed for in magnetic surveys (Section 11.1.4). These e-m waves induce currents in the Earth, called telluric currents ('telluric' means 'earth'), which flow in mainly horizontal whorls thousands of kilometres across; they are fixed in position relative to the direction of the Sun, on opposite sides of the Earth, so at most latitudes there are two maxima a day as the Earth rotates.

In any area, telluric currents tend to flow in the more conductive rocks. As an example, Figure 14.21 shows how current lines crowd through the gap over a ridge of less conductive rocks. This distortion of current flow paths, as in resistivity surveying, is the basis of MT surveying, explained later in this section. Because we have little initial knowledge of the 'transmitter' (i.e., the irregularly varying magnetic fields that induce the telluric currents), we have to modify the measurements, compared to the methods described previously.

We want to measure the resistivity of the subsurface. Ohm's Law (Section 12.1) relates the current, I, flowing through a resistance, R, and the potential difference, V, between the ends of the resistance: $V = I \times R$ (Fig. 14.22a). This relation can be used in different ways: For a given resistance, choosing the value V determines how large a current flows; conversely, if telluric current is flowing through a rock a potential difference must exist, proportional to the resistance of the rock. If the current is concentrated through a 'gap', as shown in Figure 14.21, the potential difference is therefore increased.

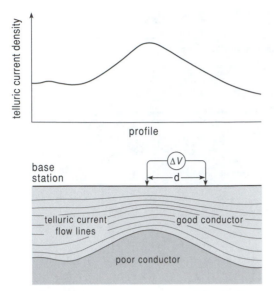

Figure 14.21 Telluric currents affected by a subsurface ridge.

Since the subsurface does not have definite boundaries within which the current flows, Ohm's Law has to be adapted. Suppose the subsurface is uniform, with resistivity ρ, and we imagine it made up of little cubes, each with sides 1 m long and positioned so that the current enters each cube only through one face, leaving through the opposite one, as shown in Figure 14.22b. The resistance of each cube (from Eq. 12.2a) is (ρ × length of cube)/(area of face of cube) = ρ (since the length and the area are both unity). If the current through a cube is i and the potential difference between opposite faces of the cube is E (this is the potential difference per metre, called the electric field, and is measured in volts/m) then, applying Ohm's Law to a cube,

$$E = i\rho \qquad\qquad \text{Eq. 14.2}$$

compared with:

$$V = IR \quad \text{Ohm's Law} \qquad \text{(Eq. 12.1)}$$

If two probes are pushed into the ground a distance d apart in the direction of the current flow (Fig. 14.22b), there will be a potential difference, ΔV, between them equal to $E \times d$, which can be measured, just as for ordinary resistivity surveys. To calculate ρ, we also need to know the current i. This cannot be measured directly, but it produces a magnetic field. Each row of cubes forms a current tube which produces a magnetic field at the surface

which is horizontal and at right angles to the current flow. The field at some point on the surface is the result of the fields of all the tubes. The calculation has to allow for the strength of the currents decreasing with depth – even when the subsurface is uniform, as was assumed above – because of the skin effect (Section 14.4.2). It gives

$$\rho_a = \frac{0.2 \times 10^{-6}}{f}\left(\frac{E}{H}\right)^2 \qquad \text{Eq. 14.3}$$

where f is the frequency, E is measured in volts/m, and H is in nT. Measuring this combination of magnetic field and potential difference is called **magnetotelluric, MT, surveying.**

Of course, as the subsurface usually does not have uniform resistivity, the value of ρ calculated is the apparent resistivity, ρ_a, as used in resistivity surveying (Section 12.3.3). As lower-frequency waves penetrate deeper than higher-frequency ones, the variation of resistivity with depth is found by measuring at a range of frequencies. By measuring at different places lateral variations can be investigated too.

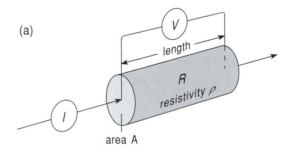

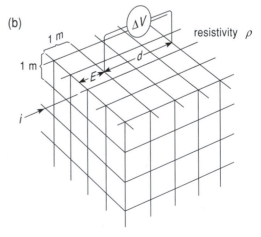

Figure 14.22 Ohm's Law for a resistance and the subsurface.

14.7.2 Carrying out an MT survey

The potential difference between the probes is measured in the same way as in resistivity or SP surveys, using two electrodes (non-polarising, as used in SP surveys [Section 13.2.2], are preferable). Potential gradients (electric fields) typically are only about 10 mV/km, so electrodes are placed several hundred metres apart to ensure that the p.d. between them is large enough to be measured accurately.

The magnetic field is very small – only a few thousandths of an nanoTesla – which requires a very sensitive instrument for measurement, generally a large coil with many turns of wire, coupled to an amplifier. Because we don't know the direction of the telluric currents and fields when setting up the equipment, two pairs of probes and two coils at right angles are used.

For relatively shallow and small-scale surveys, such as searching for mineral deposits, frequencies of tens or hundreds of Hertz are used, with stations a few tens of metres apart. For the deepest surveys the lowest frequencies are needed, so readings have to be taken over a period of hours, with perhaps only one station being occupied each day.

The amplitudes and phases of different frequencies are measured and converted into a vertical resistivity sounding. Repeated soundings along a traverse can be used to build up a resistivity section. An example is given in Figure 14.23. Soundings were made at 18 sites along a traverse perpendicular to the strike of the main structural features (Fig. 14.23a), using frequencies from about 100 Hz to 0.01 Hz. These were used to deduce vertical variation of resistivity assuming horizontal layering, and the values were then contoured to form the section shown (Fig. 14.23b). The presence of low-resistivity rocks has limited penetration, most severely at the northwest end, where the very low resistivity found below about a kilometre is thought to be due to an intensely fractured zone permeated by saline water. The high-resistivity rocks at about 40 km lie below granodiorite outcrops, so these are inferred to be plutons. The somewhat less resistive rocks at about 130 km are speculated to be the source that once fed the basalts that are found below sediments along much of the traverse. A further example, to detect intrusions at depth, is given in Section 21.7.

(a) geological map

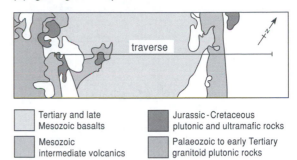

☐ Tertiary and late Mesozoic basalts

■ Jurassic-Cretaceous plutonic and ultramafic rocks

☐ Mesozoic intermediate volcanics

☐ Palaeozoic to early Tertiary granitoid plutonic rocks

(b) resistivity section

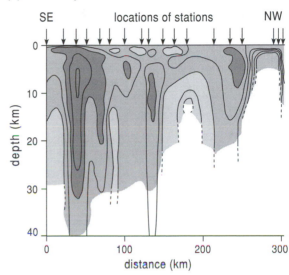

Figure 14.23 Magnetotelluric section of the southern Cordillera of Canada. Darker is more resistive.

14.8 Ground-penetrating radar, GPR

14.8.1 How ground-penetrating radar works

Ground-penetrating radar, GPR, has by far the highest frequency of any e-m method and displays the wave aspect most clearly. It depends on reflection of pulses of waves, and so resembles seismic reflection (Chapter 7). To follow this account *the reader needs to be familiar with the basic concepts of seismic reflection.* A pulse of e-m waves is transmitted downwards, reflects off interfaces, and is received back at the surface (Fig. 14.24) and the reflection time (two-way time, TWT) is used as a measure of the distance to the target. Before describing further the theory of GPR, we shall see how a survey is carried out, as with seismic-reflection surveying.

(a) time–distance graph

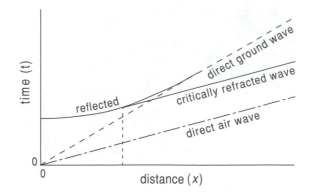

(b) ray paths

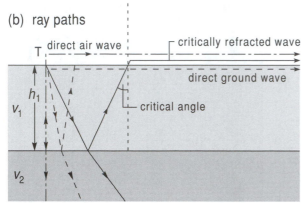

(c) Pulse EKKO IV ground-penetrating radar

Figure 14.24 GPR instrument and ray paths.

Carrying out a ground radar survey. The transmitter has an antenna (or aerial), producing an extremely short pulse of waves, lasting only a few nanoseconds (a nanosecond, ns, is a thousandth of a millionth, or 10^{-9}, seconds). These pulses contain frequencies in the range 25 to 1000 MHz, and the shorter the pulse the higher its frequency. Such short pulses are needed because e-m waves travel so fast that the TWTs, for interfaces that may be less than a metre deep, are extremely small.

The receiver may be the antenna used for transmitting, since it is needed for receiving only after the pulse has ended, or – as shown in Figure 14.24c – it may be a separate one. The transmitter and receiver, together with their necessary electronics, are often mounted on a trolley or sledge about the size of a lawn mower, with the antennae close to the ground. The trolley is pulled along a traverse, which needs to be fairly smooth because of the small ground clearance. In some surveys, separate transmitter and receiver antennae (Fig. 14.24c) are moved symmetrically away from a point for a common-midpoint (CMP) survey; this is useful for deducing how the wave velocity varies with depth in the ground, needed for accurately converting TWTs to depths. A cable supplies the trolley with power and also returns the signal to processors and recorders, which are often in the back of a van; data reduction and basic processing are carried out on site so that a 'picture' of the subsurface is usually available before leaving the site, though further processing may be needed.

Waves, or rays, can reach the receiver in several ways (Fig. 14.24b). As well as the desired reflections from interfaces, there are various unwanted rays: a direct ground ray, as with seismic waves, plus additional rays that arise because of a major difference of GPR from reflection seismology, that the velocity in air is always faster than the velocity in the ground (this occurs only rarely in seismology). This leads to the very important direct **air wave** or ray, which is the first arrival, followed by the direct ground wave (Fig. 14.24a); these produce the strong lines seen at the top of many sections (e.g., Figs 14.25 and 26.1). Another consequence of radar waves travelling faster in air is that rays reflected at a certain angle will critically refract at the surface and continue just above the surface. A further complication is that the air wave can reflect off surface objects including overhead metal objects, such as power lines, if the antenna is not shielded.

We next consider what sorts of targets can be imaged by GPR.

14.8.2 Velocity, reflection, penetration, and resolution

The velocity of radar waves depends on the material they are passing through; they travel at about 300,000 km/sec only in air or space. Their speed in the ground is slower by an amount that depends on its relative permittivity (also called the dielectric constant), ε_r (ε_r is an electrical property new to this subpart; it depends upon the extent to which different kinds of atoms, when placed in an electric field, develop positive and negative charges and become electric dipoles, similar to the way some atoms become magnetic dipoles, or tiny magnets, when in a magnetic field; relative permittivity is therefore analogous to magnetic susceptibility).

The value of ε_r for most rocks lies in the range 3 to 40 (Table 14.1); it depends to a large extent on the water content, for water has a high value of ε_r, and so the water table may be a reflector (see Section 26.4 for an example). The velocity decreases as the inverse of the square root of ε_r:

$$\text{speed}, v = \frac{c}{\sqrt{\varepsilon_r}} \qquad \text{Eq. 14.4}$$

where c is the velocity of light in air or space (this formula assumes that the magnetic susceptibility of the rock is small, which is usually the case). So a wave going from air ($\varepsilon_r = 1$) into, say, clay with $\varepsilon_r = 36$ slows to a sixth of its velocity.

The reflection coefficient – the ratio of the amplitudes of reflected to incident waves – depends on the velocities either side of the interface and hence on ε_r:

$$R = \frac{\sqrt{\varepsilon_2} - \sqrt{\varepsilon_1}}{\sqrt{\varepsilon_2} + \sqrt{\varepsilon_1}} = \frac{v_2 - v_1}{v_1 + v_2} \qquad \text{Eq. 14.5}$$

Penetration. Also important is the electrical conductivity of the rocks (Table 14.1) because, as explained in Section 14.4.2, this causes the waves to be absorbed, and is why, with its very high frequencies, GPR seldom penetrates rocks to more than a few tens of metres, and often much less. The depth is especially limited because the conductivity at the very high frequencies used in GPR is (for reasons beyond the scope of this book) several times the value found using any of the other methods in this division, which operate at much lower frequencies (i.e., the skin depth decreases with increase of conductivity and frequency in a complicated way). In practice, GPR can penetrate up to tens of metres in pure limestone, dry sand, or sand saturated with fresh water, but only centimetres into damp clay or salt water, so even a thin layer of either these materials usually prevents ground radar being used.

Table 14.1 GPR properties of some materials

Material	Relative permittivity, ϵ_r	Conductivity, σ (mS/m)	Velocity (m/ns*)	Wavelength (m)	
				at 50 Hz	at 1000 Hz
air	1†	0	0.30	6	0.3
water, fresh	81	0.5	0.033	0.66	0.033
water, sea	81	3000	0.01	0.2	0.01
ice, pure	3.2	0.01	0.16	3.2	0.16
clay, wet	25–40	50–100	0.5–0.6	10–12	0.5–0.6
granite	4–6	0.01	0.1–0.12	2–2.4	0.1–0.12
limestone	4–8	0.5–2	0.1–0.12	2–2.4	0.1–0.12
sand, dry	3–6	0.01	0.15	3	0.15
sand, wet	20–30	0.1–1	0.06	1.2	0.06
shale	5–15	1–100	0.09	1.8	0.09
silt	5–30	1–100	0.07	1.4	0.07

*Because radar waves take such a short time to reach the shallow interfaces, velocities are given in this form (1 ns = 10^{-9} sec).
†All values, which are from various sources, are approximate.

Resolution. When radar waves meet a sharp interface between rocks with different velocities (different values of ε_r) there are both reflection and transmission, as with seismic waves. As explained in Section 7.8.2, layers closer than about half a wavelength cannot be resolved apart. The wavelength in air ranges from a few tens of centimetres to a few metres, depending on the frequency used, but in rocks the wavelength is reduced in the same proportion as the velocity, since velocity = frequency × wavelength ($v = f \times \lambda$). Resolution can be improved by using higher frequencies, but only at the expense of more rapid absorption (reducing the wavelength to a quarter requires four times the frequency, which reduces the penetration to half, in a given rock). These effects are shown in Figure 14.25, where the same section has been surveyed at two frequencies. The survey at 200 MHz shows the positions of interfaces more clearly, but does not extend so deep.

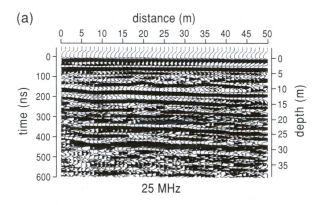

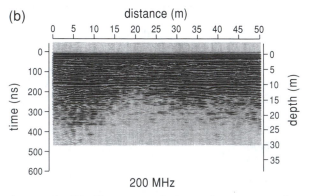

Figure 14.25 The same GPR section at two frequencies.

14.8.3 Data reduction

A survey measures TWTs to reflectors, and to convert them into depths needs velocities. Velocities can be found using moveout with increasing transmitter–receiver separation – as with seismic reflection – but this is seldom done, for usually the separation is fixed, in which case an average velocity is guessed. The data reduction and processing required are similar to those needed in seismic reflection (apart from the additional need to remove the strong air wave signal), with migration, removal of multiple reflections, statics and topographic correction, and so on being increasingly carried out, some resulting in 3D images in colour.

14.8.4 Uses of GPR surveys

GPR is increasingly used in situations where its high resolution is a great advantage and its limited penetration is sufficient. Uses include hydrogeology (particularly for finding the position of the water table – an example is given in Section 26.4), environmental geophysical surveying (polluted ground), site engineering (faulting, fractures, cavities, and buried metal structures), mineral exploration, and archaeology (an example is given in Section 28.5.2). It can also be used for mapping Quaternary soils, such as showing the structure of fluvial sediments (Fig. 14.25 shows onlap against a mound at the left), or to investigate the detailed structures of oil reservoir rocks, including its use in boreholes.

Summary

1. Electromagnetic (e-m) surveying (with the exception of GPR) responds, like resistivity surveying, to variations in the subsurface conductivity (inverse of resistivity). Most methods do not require electrical contact with the ground; instead currents are induced in the target by alternating magnetic fields produced by a transmitter; these induced currents produce their own magnetic fields, which are detected by a receiver.

2. E-m surveying – unlike resistivity – can be used for aerial surveying, or when the surface layer has very low conductivity. Conversely, very conductive overburden can prevent e-m being used.

3. Alternating magnetic fields can be regarded as e-m

waves; the wave aspects tend to be more important for higher frequencies, particularly GPR.

4. E-m waves are absorbed by conducting rocks, with shorter wavelengths/higher frequencies being absorbed more rapidly, so lower frequencies are often needed for deeper targets, though at the expense of resolution.

5. There are a variety of e-m methods, with variants, employing different instruments. One type is the moving source–moving receiver (Slingram), with two coils traversed together with a constant separation. Traverses can be repeated with different separations (and frequencies) to give different depths of penetration and hence allow pseudosections to be drawn. Another type is the fixed source–moving receiver method (e.g., Turam) which permits larger and more powerful sources to be used, needed for deeper or less conductive targets.

6. TEM (transient e-m) gets over the problem of detecting the target signal in the presence of the much larger transmitter field at the receiver by measuring the signal after the transmitter has been switched off. It is particularly useful for aerial surveys but is also widely used in ground ones.

7. The phase of the signal due to the target lags the transmitter current and primary field, by between 90° and 180°, being larger the higher the conductivity of the target, so that the phase lag is useful in interpretation. Phase information is usually expressed as in-phase (actually, 180° out-of-phase, or negative) and quadrature (90° out-of-phase) components.

8. The VLF method utilises distant radio transmitters and so the instrument is compact and light. For elongated targets, a transmitter is chosen that is approximately along strike. The tilt of the instrument is a measure of the in-phase signal, and is upwards (+ve) approaching a target, downwards on leaving it.

9. MT (magnetotelluric) surveying utilises widespread induced currents in the ground, which have a wide range of frequencies, some giving very deep penetration. Apparent resistivities are determined by measuring the ratio of potential differences, between electrodes, to the magnetic field, measured using a coil. A range of frequencies is used to investigate variation of conductivity with depth.

10. GPR (ground-penetrating radar) measures the times it takes pulses of very-high-frequency e-m waves to reflect from electrical discontinuities, to produce a section. Data is processed similarly to seismic reflection data. It offers high-resolution sections but is limited in penetration to a few metres and is almost useless in high-conductivity rocks such as clays.

11. You should be familiar with these terms: alternating field, magnetic flux, electromagnetic induction, induced current, conductivity, relative permittivity; transmitter, receiver, primary and secondary fields; Slingram, Turam, TEM, VLF, MT; GPR, air wave, skin depth; phase, lag, lead, in-phase, out-of-phase, quadrature.

Further reading

Accounts of most of the methods can be found in standard textbooks such as Kearey and Brooks (1991), Parasnis (1997), Reynolds (1997), Robinson and Coruh (1988), and Telford et al. (1990), with the last the most advanced, while guidance for field operations is given in Milsom (1996). Ground-penetrating radar has only recently entered general textbooks: Reynolds (1997) has a large chapter on it, and it is included in Parasnis (1997) and Milsom (1996), while Grasmueck (1996) shows how sophisticated the results can be.

Problems

1. Sketch the anomaly obtained using a Slingram system over a vertical sheet, showing how it relates to the position of the instrument. Explain the main features of the anomaly.

2. What are the advantages and disadvantages of e-m surveying compared to resistivity surveying for ground surveying?

3. A VLF survey is to be carried out to find whether a N–S–trending ore vein about 1 m wide continues beneath overburden. Describe how you would carry out a survey, including selecting a suitable transmitter and the direction of traverses. How could the presence of the vein be recognised during a traverse?

4. Curves A and B are identical sinusoids, except that A is a maximum when B is a minimum and vice versa.

(a) What is their phase relationship?

(b) What would it be if the amplitude of *A* were twice that of *B?*

5. A phase lag of 130° is the same as a lead of: (i) 30°, (ii) 40°, (iii) 50°, (iv) 70°, (v) 130°, (vi) 230°, (vii) 490°.

6. Describe in geophysical terms the type of target that can be found using GPR. Give two examples in geological terms.

7. How does MT differ, in application and method, from the other methods described in this chapter?

8. Describe how you would attempt to solve the following problems using e-m methods (there may be more than one suitable method):

(a) To locate a near-vertical fault zone in granite beneath 1 to 2 m of boulder clay.

(b) To measure the thickness of dry sand (up to 4 m) over shale.

(c) To locate any clay-filled, steeply dipping faults in an proposed extension to a limestone quarry, beneath a shallow layer of soil.

(d) To measure the depth to the water table (expected to be 30 to 80 m) in sandstone.

(e) To map the steeply dipping contact between a sandstone and a shale, beneath about 1 m of sandy soil.

(f) To investigate if there is saline water within an aquifer lying between 15 and 50 m, in chalk.

9. Which of the following systems are suitable for airborne surveying? (i) Slingram. (ii) Turam. (iii) MT. (iv) GPR. (v) TEM. (vi) VLF.

10. Explain why the amplitude of an e-m signal generally decreases with distance from the source.

11. A VLF transmitter is due east of an E–W–trending galena vein. Would it be better to take readings along north-to-south traverses or south-to-north ones?

12. What are the advantages of the TEM compared to the Slingram method?

13. What factors limit the vertical resolution of a GPR survey?

Radioactivity

*The atoms of isotopes of a number of elements spontaneously convert to other elements. This radioactive decay has two main applications in geophysics. The rate of decay provides a natural clock, for it is not affected by geological processes, and this is utilised in the many methods of **Radiometric dating,** the subject of the first of the two chapters within this subpart. **Radiometric surveying,** the subject of second chapter, is about measuring the concentrations of potassium, thorium, and uranium as an exploration and mapping aid, and of radioactivity for hazard monitoring, as well as for thorium and uranium ores.*

chapter 15

The Ages of Rocks and Minerals: Radiometric Dating

Radiometric dating provides a numerical date, in years, for a sample. This contrasts with geological dating, which provides a relative date, by assigning a sample to a position in a timescale, mainly using stratigraphic order and palaeontological correlations. Radiometric dating is chiefly applicable to igneous and metamorphic rocks; therefore it complements, rather than replaces, geological dating, which is mainly applicable to sedimentary rocks.

There are a number of radioactive elements that are found in rocks and hence a considerable number of radiometric dating methods. These have advantages and disadvantages, particularly in the types of rock and 'event' that they can date. Several dating methods used in

combination may provide significantly more information than any single method. The calculation of dates depends upon a number of assumptions, but these can be checked to some extent, depending on the method. A particular application of radiometric dating is to measure cooling histories of rocks, particularly following thermal metamorphism, and is termed thermochronometry.

15.1 The Atomic clock

Radiometric dating is possible because of **radioactive decay**, in which atoms of certain elements spontaneously change into a different element at a known rate. An atom consists of a small but comparatively massive nucleus surrounded by electrons with negligible mass. The nucleus is made up of positively charged protons and uncharged neutrons with the same mass. Which element an atom belongs to depends only upon its number of protons, called the **atomic number**, but some of the properties of the atom, such as decay, depend on the total number of protons plus neutrons, called the **mass number.** Atoms with the same atomic number but different mass number (different numbers of neutrons) are called **isotopes** of the element. For example, uranium has two important isotopes, $^{238}_{92}U$ and $^{238}_{92}U$; 92 is the atomic number, while 238 and 235 are the mass numbers. The atomic number will be omitted hereafter, for it is sufficient to specify the element, thus: ^{238}U and ^{235}U. Normally, the positive charge of the nucleus is balanced by an equal number of electrons, each of which has an negative charge, and so the atom as a whole is electrically neutral.

Radioactive decay follows an unlikely but simple form. Compare Figure 15.1a and b: The left-hand figure shows the fraction of a large number of people born in the same year that remains in successive years. The curve does not decrease at a constant rate because people are more likely to die at some ages than at others. After a sufficient time, we can be sure none will remain alive: There will be few after 100 years, none after 200. Radioactive atoms behave quite differently; their curve steadily decreases but never quite reaches zero because, remarkably, the probability of atom decaying in any second is quite unaffected by how long it has already existed. After some period of time, called the **half-life,** half will remain, after a further half-life only a quarter, and so on, halving in each successive half-life. The decays of all radioactive isotopes follow this form but differ in their half-lives.

Using half-lives is convenient provided the interval is an exact number of half-lives, but what fraction remains after 1½ half-lives, for instance? To answer this, we need to know the shape of the curve. The curve to the right of B (Fig. 15.1b) has the same shape as the curve to the right of A, though it is half as high; the curve to the right of C has the same shape as the curve to the right of B, and so on. In fact, the curve to the right of any point has the same *shape* as the curve to the right of any other point. This self-similarity is one property of an **exponential** curve; it is described mathematically by

$$P_t = P_i e^{-\lambda t}$$ Eq. 15.1

where P_i is the initial number of atoms and P_t the number remaining after a time t; the **decay constant,** λ, determines how fast decay occurs: If λ is small, decay is slow. λ has units of time^{-1}; for rocks, we use 'per year' or a or a^{-1} (a is annum, or year).

This equation is not enough to date a rock, for though we can measure the number of atoms remaining, we do not know how many were present initially, when the rock was formed. However, the isotope that is decaying away, called the **parent,** decays into another, the **daughter,** so the number of daughter atoms increases with time, as shown in Figure 15.2. The *ratio* of numbers of daughter to parent atoms at the present time (D_{now}/P_{now}), is larger the longer decay has been going on, and this ratio provides the clock. The ratio does not increase uniformly, and it is related to the age of the sample by

$$\left(\frac{D}{P}\right)_{now} = \left(e^{\lambda t} - 1\right)$$ Eq. 15.2

This is the **Basic Dating Equation,** which underlies many of the radiometric dating methods (it is derived in Box 15.1).

Relation between the half-life and the decay constant. These two quantities are clearly related, for the longer the half-life the slower the decay and therefore the smaller the decay constant. After one half-life, the number of parent atoms remaining and the number of daughter atoms formed are the same (Fig. 15.2), so putting the ratio 1 into the Basic Dating Equation gives

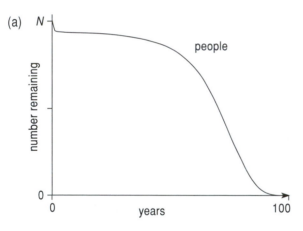

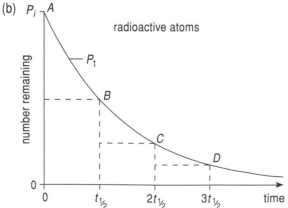

Figure 15.1 Survival of people and radioactive atoms.

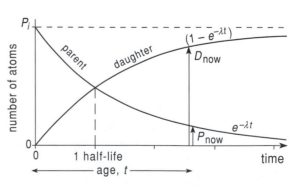

Figure 15.2 Parent and daughter curves.

BOX 15.1 The Basic Dating Equation

The probability that a particular radioactive atom will decay during, say, the next year is the same whether the atom was formed yesterday or a thousand million years ago, and all atoms of the same isotope have the same probability, which is λ, the decay constant. Though we cannot tell when any individual atom will decay, with millions we can give the fraction decaying per year quite precisely, just as, when tossing coins, we can only *guess* which side of a coin will show after a single toss (regardless of what it showed last time), but for a million tosses we can predict that heads will show close to half the time.

The number decaying in the first year, δP_1, is simply proportional to the number present initially, P_i:

$$\delta P_1 = -\lambda \times P_i$$

number decaying   decay × number
in one year   constant of atoms

Eq. 1

The minus sign shows that the number of atoms is decreasing.

After 1 year the number of atoms is slightly smaller, $(P_i - \delta P_1)$, so the number decaying in the next year, δP_2, is reduced in proportion, to $\lambda(P_i - \delta P_1)$, and so on for each subsequent year, leading to the shape of Figure 15.1b. The number, P_t, remaining after several years is found by subtracting $(\delta P_1 + \delta P_2 + \ldots)$ from the initial number P_i. This is done by putting Eq. 1 into calculus notation:

$$\frac{dP}{dt} = -\lambda P$$

Eq. 2

The left-hand side is the rate of decay, in decays per year. This integrates to give

$$\log_e P_t = \log_e P_i - \lambda t$$

Eq. 3a

or:

$$P_t = P_i e^{-\lambda t}$$

Eq. 3b

'$\log_e$' is often written 'ln'.

If each parent atom decays into one daughter atom, the number of daughter atoms formed in the period t, D_t, equals the number of parent atoms that have decayed, $(P_i - P_t)$. We do not know the value of P_i when the rock formed, but rearranging Eq. 3b gives

$$P_i = P_t e^{\lambda t}$$

Eq. 4

Therefore:

$$D_t = P_t e^{\lambda t} - P_t = P_t\left(e^{\lambda t} - 1\right)$$

Eq. 5

Hence:

$$\frac{D_t}{P_t} = \left(\frac{D}{P}\right)_t = \left(e^{\lambda t} - 1\right)$$

Eq. 6

This is the Basic Dating Equation.

$$\left(\frac{D}{P}\right)_{now} = 1 = \left(e^{\lambda t_{1/2}} - 1\right)$$

Eq. 15.3

So:

$$2 = e^{\lambda t_{1/2}}$$

Eq. 15.4

taking logs:

$$\log_e 2 = \lambda t_{1/2}$$

Eq. 15.5

hence:

$$\lambda = \frac{0.693}{t_{1/2}} \quad t_{1/2} = \frac{0.693}{\lambda}$$

Eq. 15.6

For example, to find the decay constant of an isotope with a half-life of 82 Ma (million years), first write the half-life as 8.2×10^7 a, then

$$\lambda = \frac{0.693}{8.2 \times 10^7} = 0.08451 \times 10^{-7}$$
$$= 8.451 \times 10^{-9} \text{ year}^{-1} \text{ or } a^{-1}$$

Our first use of the Basic Dating Equation is with the uranium–lead dating method.

15.2 The uranium–lead (U–Pb) method

Uranium has two radioactive isotopes that each decay (via many intermediate isotopes, see Section 15.12.1) to different lead isotopes:

$$^{238}U \rightarrow {}^{206}Pb; \lambda_{238}, 1.551 \times 10^{-10} \text{ a}^{-1};$$
$$\text{half-life, 4468 Ma}$$

$$^{235}U \rightarrow {}^{207}Pb; \lambda_{235}, 9.849 \times 10^{-10} \text{ a}^{-1};$$
$$\text{half-life, 704 Ma}$$

Though both parents belong to the same element, as do both daughters, their decays are quite independent, and each daughter/parent ratio can be used in the Basic Dating Equation, with the appropriate decay constant.

Lead is widely distributed in small amounts in rocks, whereas uranium is rather rare, but a few minerals form with significant amounts of uranium in them but negligible amounts of lead. Most important is zircon – zirconium silicate, $ZrSiO_4$ – found as an accessory mineral in many igneous rocks. To obtain a date, zircons are extracted from the rock – a single zircon may be sufficient – dissolved in acids, and, after chemical manipulations, the ratios $^{206}Pb/^{238}U$ and $^{207}Pb/^{235}U$ are measured using a mass spectrometer (Box 15.2) and the values are substituted in the Basic Dating Equation.

As an example, suppose the $^{206}Pb/^{238}U$ ratio is 1.562×10^{-2}.

$$0.01562 = \left(e^{\lambda t} - 1 \right)$$

so:

$$1.01562 = e^{\lambda t}$$

Taking logs:

$$\ln\left(1.01562\right) = \lambda t$$
$$0.015499 = \lambda t$$

so:

$$t = \frac{0.015499}{\lambda} = \frac{0.015499}{1.55 \times 10^{-10}}$$

therefore:

$$t = 10^8 \text{ years} = 100 \text{ Ma}$$

This could also be done with the $^{207}Pb/^{235}U$ ratio, and we would expect the dates to be **concordant** – that is, the same – which provides a check on the validity of the dates. The U–Pb method can be applied to a number of other minerals, including sphene, uraninite, monazite, and apatite.

15.3 Assumptions of the Basic Dating Equation

The Basic Dating Equation makes several assumptions, which should be understood:

   i. *The system has remained closed since the rock or mineral formed;* that is, there must have been no loss or gain of either parent or daughter atoms in or out of the sample, so their ratio changes only because of decay of parent into daughter. For example, escape of daughter atoms at any time before we measure the ratio would give a spuriously low (young) age.

   ii. *One parent decays into one daughter atom.* This is implicit in Figure 15.2.

   iii. *There were no daughter atoms present at the time of closure.* If, for instance, there had been lead atoms in the zircon crystal of the previous section when it formed, it would have given a spuriously high (old) age.

There are additional assumptions for some of the other dating methods, which will be pointed out when we come to them.

In the case of the U–Pb method, assumption (ii) is satisfied because of the decay scheme, (iii) is satisfied by choosing zircon or another suitable mineral, and (i) is also generally satisfied by the choice of mineral. But we shall find that these assumptions are not always satisfied, either for the U–Pb method or for the other methods described in the following sections. Any uncorrected violation of these assumptions can lead to erroneous dates, which will not be indicated by the quoted error, for this reflects only errors of measurement.

15.4 The potassium–argon (K–Ar) dating method

This, probably the most used single method, has two versions, the 'conventional' potassium–argon ($^{40}Ar/^{40}K$) and the argon–argon ($^{40}Ar/^{39}Ar$) methods, with the term 'potassium–argon method' referring to either method.

(a)

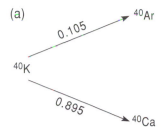

(b)

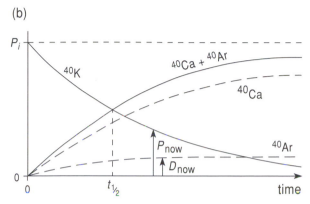

Figure 15.3 Decay scheme and decay curves for ^{40}K.

15.4.1 The conventional K–Ar method

Potassium is the commonest and most widely distributed of the 'primeval' radioactive isotopes (i.e., those that have been present since the Earth formed). It has three isotopes, of which only ^{39}K and ^{40}K concern us; only ^{40}K is radioactive, and it forms 0.01167% (a little over a hundredth of 1%) of potassium. It decays into two daughter elements, argon, ^{40}Ar, and calcium, ^{40}Ca (Fig. 15.3a):

$$^{40}K \rightarrow {}^{40}Ar\ (+{}^{40}Ca);\ \lambda,\ 5.543 \times 10^{-10}\ a^{-1};$$
$$\text{half-life, 1250 Ma}$$

This double decay violates assumption (ii), for we measure only the ^{40}Ar, not all the daughters, but as the fraction of decays that form argon is constant at 0.105 (Fig. 15.3a), it is only necessary to insert this figure in the Basic Dating Equation to give the equation for the growth curve of ^{40}Ar (Fig. 15.3b):

$$\frac{^{40}Ar}{^{40}K} = 0.105\left(e^{\lambda t} - 1\right)$$

Eq. 15.7

Suppose we wish to date an igneous rock, such as a lava or granite intrusion. Inside the Earth, ^{40}K

has been decaying into ^{40}Ar, so the magma rising to the surface contains argon. However, before the magma solidifies, the argon is usually lost with the other volatiles, for it is an inert gas and so has no chemical bonds holding it within the rock, unlike its parent potassium. Once the magma has solidified, the argon formed by decay is physically trapped within the rock, particularly within the component crystals, and so begins to accumulate: The radiometric clock has started.

To date the rock, we crush a fresh sample and extract suitable minerals, such as hornblende, micas, and some feldspars, or take small pieces (whole-rock sample). One portion (aliquot) is weighed and chemically processed, and then its total potassium content is measured (usually by flame photometry, which measures the colour produced when a solution is sprayed into a flame). The amount of ^{40}K is calculated by multiplying the total amount by 0.01167%, the fraction of potassium that is ^{40}K, and used to calculated the amount per gram of the sample.

Another portion is used for argon determination. The amount of argon is very small: For example, a rock 100 Ma old and containing 1% of potassium will contain only about 0.004 mm^3 of argon per gram. To measure such a small amount, a weighed portion is melted within a vacuum system. After impurities such as water vapour have been removed, the argon is measured and the amount per gram calculated. The amounts per gram of both ^{40}K and ^{40}Ar are converted into numbers of atoms per gram and put into Eq. 15.7 to yield the age of the lava.

How young a rock can be dated by the K–Ar method? For any dating method, the younger the sample the smaller the amount of daughter that has been produced, so there is some age so young that the amount cannot be measured precisely. For the K–Ar method this limit depends partly on the presence of atmospheric argon.

About 1% of the atmosphere is argon, and mostly ^{40}Ar, identical to the radiogenic argon that has formed by decay inside our samples, for it has mostly been produced by decay of potassium over the history of the Earth. As the amount of radiogenic argon in a sample is very small, even a tiny

potassium content of the sample: For example, when it reaches 50%, argon corresponding to half the potassium has been released. Not all age spectra are as simple as that shown, and much may be learned from the different shapes of age spectra or how they vary even within different parts of a grain, but this will not be explained further.

Age spectra have been obtained for samples of less than a milligram using a laser to the heat the sample, which allows tiny crystals to be selected from a sample or to detect argon loss by diffusion, by dating the edges and centres of crystals. (Examples of Ar–Ar dating, with age spectra, are given in Fig. 25.6a and b.)

BOX 15.2 Mass spectrometers

Isotopic ratios are most often measured using a mass spectrometer. This is an instrument that first sorts atoms according to their masses and then measures their amounts. It has three parts (Fig. 1): a source, which produces a beam of charged atoms, or ions; an analyser, which sorts them; and the collector, which measures their amounts. The first step is to convert uncharged atoms into charged ones, or ions, so that they can be manipulated electrically. Positive ions are formed by removing an electron from an atom, either by bombarding a gas, such as argon, with electrons (gas source mass spectrometer) or by heating a solid compound of the element concerned (solid source). The positively charged ions are gently repelled through slit S_1 by a small positive voltage on the repeller plate, and then are rapidly accelerated by the high voltage, V_{acc}, of a second plate with a slit, S_2, in it. This creates a beam of ions, with the heavier ions having a slower speed.

The analyser is a wedge-shaped magnet, which bends the ions in a circle, but the paths of the heavier ions are less curved. This causes the various ion masses to diverge, just as white light passing through a triangular prism splits into the colours of

the rainbow – hence the name spectrometer. The beam leaving slit S_2 is spreading, but the magnet, as well as separating the ions by mass, brings them to a focus as shown, because its apex is at A in line with the source and collector slits.

The entrance to the collector is another slit, S_3. Only ions of one mass will pass through the slit, to be stopped by an electrode. Since the ions are charged, they produce an electrical current, though an extremely small one, and this is amplified and measured. By adjusting either the strength of the magnetic field of the magnet or the accelerating voltage, V_{acc}, each beam in turn can be made to pass through S_3, and the ratio of the currents gives the ratio of the numbers of the different ion masses and hence of the original atoms. Alternatively, a row of collector slits can be placed in the correct positions to measure all the required masses simultaneously. The mass spectrometer operates at a very high vacuum to prevent the ions being diverted by collisions with gas molecules. There are other designs of mass spectrometer, such as a double or tandem instrument used for carbon-14 dating, but they have the same function of sorting and counting atoms.

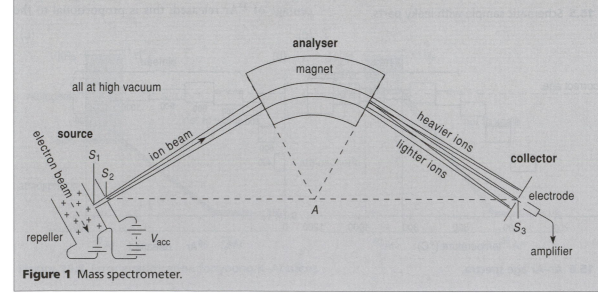

Figure 1 Mass spectrometer.

To determine the proportion, p, of ^{39}K converted to ^{39}Ar, standard samples of known age (called flux monitors) are irradiated with the samples and the value of p that gives the known age is used for the samples being dated. The age of the standard is measured by the K–Ar method, so the Ar–Ar method is not a truly independent method.

15.5 The rubidium–strontium (Rb–Sr) dating method

There are several isotopes of rubidium and strontium, but the ones relevant to dating are

$$^{87}\text{Rb} \rightarrow {}^{87}\text{Sr}; \lambda_{87}, 1.42 \times 10^{-11} \text{ a}^{-1};$$
half-life, 48,800 Ma

^{86}Sr (reference isotope, not formed by decay)

Rocks or minerals that contain rubidium also contain strontium, so we have the problem of daughter atoms being present initially. To allow for it we also measure the amount of ^{86}Sr, because it is not formed by decay and so can be used for comparison. It is called a **reference isotope**.

Suppose we wish to date the **crystallisation age**, the time when a rock crystallised from magma. We assume the magma was well mixed so that the ratios ^{87}Sr/^{86}Sr and Rb/Sr were everywhere the same. But once the minerals crystallise Rb/Sr is no longer uniform, because minerals differ in their chemical compositions. However, the ^{87}Sr/^{86}Sr ratio is not affected, for the two isotopes are chemically the same. Therefore, just after crystallisation, different types of mineral, A, B, C, . . . have the same ^{87}Sr/^{86}Sr but different Rb/Sr ratios, as shown by points A_0, B_0, . . . in Figure 15.7a. Note that the Rb/Sr ratio is actually measured as ^{87}Rb/^{86}Sr, the parent and reference isotopes.

Within each closed mineral grain ^{87}Rb decays to ^{87}Sr, causing ^{87}Sr/^{86}Sr to increase and ^{87}Rb/^{86}Sr to decrease. As one ^{87}Rb atoms decays to one ^{87}Sr atom, the isotopic composition of each grain evolves along a line at 45°, $A_0A_1A_2$, . . . For mineral B, with twice as large an ^{87}Rb/^{86}Sr ratio as A, the rate of increase in the strontium isotopic ratio will be twice as fast. Therefore, the isotopic compositions of the different grains at some instant lie on a straight line, for example, $A_1B_1C_1$ . . . that passes through I; the line is called an **isochron**, meaning at the same time. (We shall meet other isochrons in this chapter.) As

time passes, this line rotates counterclockwise about I. I corresponds to a grain – were it to exist – that contains Sr but no Rb.

To calculate this time, we need an equation relating the slope of the line to the age of the rock. The Basic Dating Equation is modified by adding ^{87}Sr$_i$, the number of daughter atoms present initially, to those formed by decay:

$$^{87}\text{Sr}_{\text{now}} = {}^{87}\text{Sr}_i + {}^{87}\text{Rb}_{\text{now}}\left(e^{\lambda t} - 1\right) \qquad \text{Eq. 15.9}$$

The value of ^{87}Sr$_i$, varies from one mineral to another, so we divide by ^{86}Sr to provide a ratio that was the same everywhere when the rock first crystallised:

$$\frac{^{87}\text{Sr}_{\text{now}}}{^{86}\text{Sr}} = \frac{^{87}\text{Sr}_i}{^{86}\text{Sr}} + \frac{^{87}\text{Rb}_{\text{now}}}{^{86}\text{Sr}}\left(e^{\lambda t} - 1\right) \qquad \text{Eq. 15.10}$$

(a)

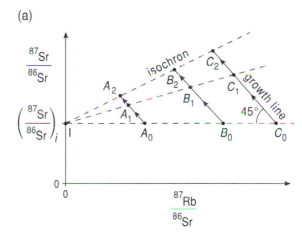

(b)

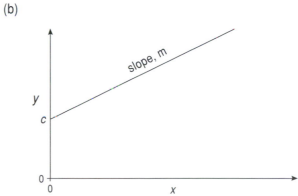

Figure 15.7 Rubidium–strontium isochron diagram.

As the amount of the reference isotope ^{86}Sr does not change, we can regard it as either ^{86}Sr$_i$ or ^{86}Sr$_{now}$ as it suits us. The equation can be rewritten as

$$\left(\frac{^{87}\text{Sr}}{^{86}\text{Sr}}\right)_{now} = \left(\frac{^{87}\text{Sr}}{^{86}\text{Sr}}\right)_i + \left(\frac{^{87}\text{Rb}}{^{86}\text{Sr}}\right)_{now}\left(e^{\lambda t}-1\right)$$

$$y \quad = \quad c \quad + \quad x \qquad m \qquad \text{Eq. 15.11}$$

This is the equation of a straight line, which has the form $y = mx + c$ (Fig. 15.7b); m is the slope of the line and c is the value of y when x is zero, that is,

$(^{87}\text{Sr}/^{86}\text{Sr})_i$. It shows that the *increase* in y, the present strontium isotopic ratio, above its **initial isotopic ratio** $(^{87}\text{Sr}/^{86}\text{Sr})_i$, is in proportion to x, the present rubidium/strontium ratio. The proportionality, m, is the slope of the line, which is $(e^{\lambda t} - 1)$ – just the right-hand side of the Basic Dating Equation – and so we can calculate the date.

In summary, the Rb–Sr method assumes that the magma everywhere had the same ^{87}Sr/^{86}Sr ratio, but at crystallisation – the event to be dated – it is divided into separate volumes or subsystems with different Rb/Sr ratios, so that thereafter the ^{87}Sr/^{86}Sr ratios diverge (Fig. 15.8). Dating a rock by this method means deducing how long ago the present-day samples, with different Rb/Sr and ^{87}Sr/^{86}Sr ratios, had the same ^{87}Sr/^{86}Sr ratio.

To date a rock, it is crushed and various types of minerals, such as micas and feldspars, are extracted. A small amount of each mineral type is dissolved in acids, it is chemically manipulated, and its isotopic ratios are measured and plotted on an isochron plot. Suppose the results are as in Figure 15.9a. The slope is 0.00574, so

$$\text{slope} = 0.00574 = \left(e^{\lambda t}-1\right)$$

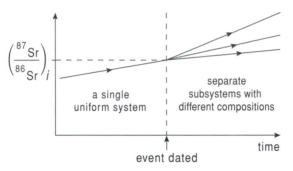

Figure 15.8 Divergence of strontium isotopic ratios.

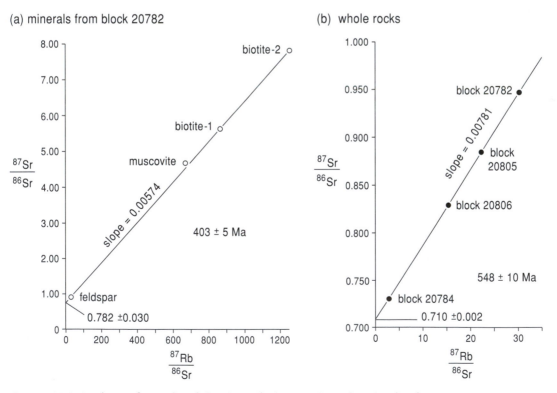

Figure 15.9 Isochrons for rocks of the Carn Chuinneag Complex, Scotland.

Therefore

$$1.00574 = e^{\lambda t}$$

taking logs:

$$\log_e(1.00574) = 0.005724 = \lambda t$$

so

$$t = \frac{0.005724}{1.42 \times 10^{-11}} = 403 \text{ Ma}$$

If the points do not lie on a straight line – within their errors – it is evidence that some of the underlying assumptions have been violated; thus the isochron method has an intrinsic check on the validity of its assumptions.

It is not necessary when dating a rock to find the value of $(^{87}Sr/^{86}Sr)_i$, but this ratio is useful for geochemical purposes, for it can characterise the source of the magma. It is often used in conjunction with the initial ratios of other isochron methods, particularly the samarium–neodymium method, described shortly.

Whole-rock isochron. Using different minerals is not the only way to obtain an Rb–Sr isochron. The requirement is simply that, at the time to be dated, there was a range of samples with the same initial isotopic composition but different Rb/Sr ratios, and this condition can often be met by selecting blocks from different parts of an outcrop (rather than different types of minerals from a single block) separated by tens of metres. These are ground up to homogenise their minerals, and then a representative sample is taken from each and analysed. The resulting isochron is called a whole-rock isochron (in contrast to the mineral isochron described above). One is shown in Figure 15.9b (the reason why it gives a higher date than the minerals is explained in Section 15.9.4).

A whole-rock isochron depends upon the original magma having a range of Rb/Sr ratios over distances of several metres but still requires it to be homogeneous in $^{87}Sr/^{86}Sr$. Generally, the range of values of $^{87}Rb/^{86}Sr$ between blocks is much smaller than between different minerals, as Figure 15.9 illustrates.

15.6 The samarium–neodymium (Sm–Nd) dating method

This is similar to the Rb–Sr method, for there are several isotopes of both elements and the daughter, Nd, is present at closure and so an isochron is needed. The relevant isotopes are

$$^{147}Sm \rightarrow {}^{143}Nd; \lambda_{147}, 6.54 \times 10^{-12} \text{ a}^{-1};$$
$$\text{half-life, } 1.06 \times 10^5 \text{ Ma}$$
$$^{144}Nd \text{ (reference isotope)}$$

Comparison with the Rb–Sr method shows that ^{147}Sm corresponds to ^{87}Rb, ^{143}Nd to ^{87}Sr, and ^{144}Nd to ^{86}Sr. Substituting these values in Eq. 15.11 gives

$$\left(\frac{^{143}Nd}{^{144}Nd}\right)_{now} = \left(\frac{^{143}Nd}{^{144}Nd}\right)_i + \left(\frac{^{147}Sm}{^{144}Nd}\right)_{now}\left(e^{\lambda t} - 1\right)$$
$$\text{Eq. 15.12}$$

The method is more applicable to basic and ultrabasic rocks than to acidic ones and so complements the Rb–Sr method, which is best applied to acidic rocks. Sm and Nd are less affected than K, Rb, and Sr by regional metamorphism and alteration, making the Sm–Nd method useful for dating rocks in which these processes have occurred. Because of the extremely long half-life of ^{147}Sm, the $^{143}Nd/^{144}Nd$ ratio evolves slowly, so the method is mainly applied to old rocks, often thousands of millions of years old, including some meteorites.

15.7 The lead–lead (Pb–Pb) dating method

15.7.1 Theory of the method

$$^{238}U \rightarrow {}^{206}Pb; \lambda_{238}; 1.551 \times 10^{-10} \text{ a}^{-1};$$
$$\text{half-life, } 4468 \text{ Ma}$$
$$^{235}U \rightarrow {}^{207}Pb; \lambda_{235}, 9.849 \times 10^{-10} \text{ a}^{-1};$$
$$\text{half-life, } 704 \text{ Ma}$$
$$^{204}Pb \text{ (reference isotope)}$$

Most rocks and minerals that contain uranium also have lead present initially (zircon and the other minerals mentioned in Section 15.2 are exceptions), so we use isochrons analogous to that for the Rb–Sr method (Eq. 15.11):

$$\left(\frac{^{207}Pb}{^{204}Pb}\right)_{now} = \left(\frac{^{207}Pb}{^{204}Pb}\right)_i + \left(\frac{^{235}U}{^{204}Pb}\right)_{now}\left(e^{\lambda_{235} t} - 1\right)_{\text{Eq. 15.13a}}$$

$$\left(\frac{^{206}\mathrm{Pb}}{^{204}\mathrm{Pb}}\right)_{\mathrm{now}} = \left(\frac{^{206}\mathrm{Pb}}{^{204}\mathrm{Pb}}\right)_i + \left(\frac{^{238}\mathrm{U}}{^{204}\mathrm{Pb}}\right)_{\mathrm{now}} \left(e^{\lambda_{238}t} - 1\right) \qquad \text{Eq. 15.13b}$$

These could be used to deduce independent dates, which should be concordant, as with $^{206}\mathrm{Pb}/^{238}\mathrm{U}$ and $^{207}\mathrm{Pb}/^{235}\mathrm{U}$ dating. However, if we forgo this check we can derive some extra information by recasting the equations. First, since $^{238}\mathrm{U}/^{235}\mathrm{U}$ has everywhere the same value of 137.88 we can replace $^{235}\mathrm{U}$ by $^{238}\mathrm{U}/137.88$. Next, dividing the first by the second equation, $^{238}\mathrm{U}$ cancels out, giving an equation without either parent isotope! After some rearranging, we have

$$\underbrace{\left[\left(\frac{^{207}\mathrm{Pb}}{^{204}\mathrm{Pb}}\right)_{\mathrm{now}} - \left(\frac{^{207}\mathrm{Pb}}{^{204}\mathrm{Pb}}\right)_i\right]}_{y_r} = \underbrace{\left[\left(\frac{^{206}\mathrm{Pb}}{^{204}\mathrm{Pb}}\right)_{\mathrm{now}} - \left(\frac{^{206}\mathrm{Pb}}{^{204}\mathrm{Pb}}\right)_i\right]}_{x_r} \underbrace{\left[\frac{1}{137.88}\frac{\left(e^{\lambda_{235}t} - 1\right)}{\left(e^{\lambda_{238}t} - 1\right)}\right]}_{m} \qquad \text{Eq. 15.14}$$

Though the actual quantities that vary, y and x, are $(^{207}\mathrm{Pb}/^{204}\mathrm{Pb})_{\mathrm{now}}$ and $(^{206}\mathrm{Pb}/^{204}\mathrm{Pb})_{\mathrm{now}}$, the equation is simpler if we regard the whole brackets y_r and x_r, as the variables, for then it is simply the equation of a straight line with slope m. The variables y_r and x_r are the *increases* in the ratios since closure; in effect, we have shifted the origin (0, 0) of the plot to the point defined by the initial lead isotopic ratios, I (Fig. 15.10).

The expression for m is complicated, but the only quantity in it that is not constant for a suite of samples from the same rock is t. Thus, the slope yields the age of the sample (as with the Rb–Sr

isochron, though the expressions for the slope are not of the same form).

The underlying model is as for the Rb–Sr method: An originally homogeneous system subdivides at some time into separate, closed subsystems, as illustrated in Figure 15.8, each with different U/Pb ratios, but all with the same lead isotopic compositions; all plot at the same point, I. As time elapses, the isotopic compositions of these subsystems evolve separately, along individual growth curves, and at different rates, but Equation 15.14 tells us that at any instant they lie on a straight line passing through I. This isochron rotates clockwise with time, unlike the Rb–Sr and Sm–Nd isochrons (this difference merely results from choosing to plot $^{207}\mathrm{Pb}/^{204}\mathrm{Pb}$ on the y, rather the x, axis).

The Pb–Pb method is not widely used, but its especial advantage is that, because the parent isotopes have been eliminated from the isochron equation, it can be applied to samples that have experienced chemical processes that have changed the U/Pb ratio, even removing all the uranium. Examples are lead ores and samples that have been weathered, and it has been applied to sedimentary limestones. Another, very important, application follows.

15.7.2 The 'age of the Earth'

Before we can discuss the dating details, we need to ask what the age of the Earth means – what event is being dated? Or rather, what event could start the radiometric clock? This leads to asking about the Earth's formation, and to answer this we need to understand the formation of the Solar System, of which the Earth is a part.

Formation of the Solar System. Some of the details are unclear, but the consensus is that the Solar System formed out of a rotating disc of gas and tiny solid particles ('dust'), termed the Solar Nebula. The bulk of the matter ended up as the Sun, but some of the dust, composed of metals and silicates, aggregated into progressively larger masses, called planetesimals, orbiting about the Sun. These in turn

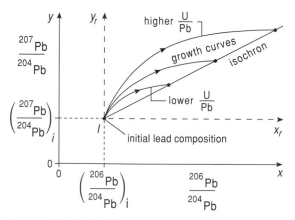

Figure 15.10 Pb–Pb plot.

aggregated into larger masses, and in most cases the process ended with the formation of planets.

However, some of the planetesimals didn't progress the whole way to planets and some remain as the numerous asteroids, up to a few hundred kilometres in diameter, mostly orbiting between Mars and Jupiter. Collisions in this asteroid belt produce debris, some of which occasionally hits the Earth as meteorites. These are important as being, with lunar samples, the only extraterrestrial material we can analyse in our laboratories. The majority of meteorites are chondrites, remarkable assemblages often with rounded bodies a few millimetres across called chondrules, and literally like nothing on Earth. Their origin is not fully understood, but they are believed to have formed very early in the development of the Solar System. The other types of meteorite have all been formed from melts, and some – achondrites – are similar to terrestrial igneous rocks, such as basalts, while others – iron meteorites – are made mostly of iron–nickel alloy. These are believed to have come from asteroids that had melted and formed a mantle and core rather like the Earth, and later were broken apart by collisions. These, too, are believed to have formed early in the development of the Solar System, though not quite so early as the chondrites.

Remarkably, isotopic compositions of many *different* chondrites lie close to a single Rb–Sr isochron (Fig. 15.11a). This shows both that they derived from a single system and that they have remained closed systems ever since. More precise dates on

individual meteorites show them to have been formed about 4560 Ma ago – a U–Pb zircon age of 4563 ± 15 Ma on one meteorite illustrates the precision that can be obtained – and confirms the early formation of meteorites though (because of the high closure temperature of zircon, Section 15.9.2), not the lack of heating since.

Leads from different chondrites also lie close to an isochron (Fig. 15.11b). Iron meteorites do not contain significant amounts of uranium, potassium, or rubidium, and so cannot be dated directly. However, some contain inclusions of iron sulphide – called troilite – and these contain significant amounts of lead but negligible uranium. Troilites from different iron meteorites have closely the same lead isotopic composition, and this common composition plots on the chondrite isochron, but only after it has been extended down to the left (Fig. 15.11b). This is evidence that the iron meteorites derived from the same homogeneous system as the chondrites, and so they are assigned the same age. Because they contain negligible uranium, their lead isotopic composition has not changed, and they are believed to have the initial lead isotopic composition – *I* on Figure 15.10 – at the time the Solar System formed, called primeval lead. Achondrite leads also lie close to the line.

So far the Earth has not been mentioned. We know of no terrestrial rocks that have remained undisturbed since the Earth formed; indeed, it is unlikely any could have survived the violent events of the first few hundred million years of the Earth's existence, when there were huge impacts and proba-

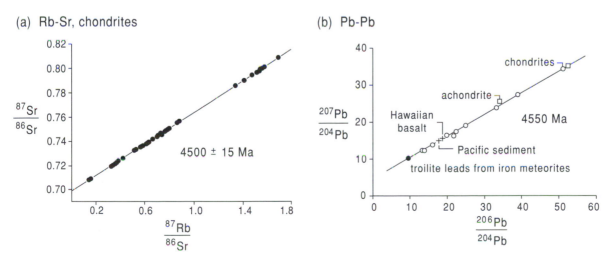

Figure 15.11 Rb–Sr and Pb–Pb isochrons for meteorites and the Earth.

bly much volcanism. The oldest *rocks* known are the Acasta Gneisses of north-central Canada, dated at 3900–4000 Ma ago, though detrital zircons as old as 4270 Ma have been reported from Australia. However, the mantle probably formed very early in the Earth's history and as a whole has remained an approximately closed system. There is no need actually to sample the mantle directly, and several types of sample can be used instead. Oceanic basalts derive from it and their lead isotopic compositions will be the same as the mantle from which they derive. Another is lead from oceanic sediments, which is an average of many fairly recent sources, while a third is lead ores derived recently from the mantle. (Any evolution of isotopic ratios since extraction from the mantle will be small, but can be corrected for, knowing their present U/Pb ratios, and their age, found using another dating method.) These terrestrial compositions also fall on the chondrite isochron (Fig. 15.11b).

What does it all mean? The fact that leads from various chondrites, iron meteorites, and the Earth lie close to a single isochron implies the existence of an originally homogeneous system which divided into separate, closed subsystems with different U/Pb ratios. We identify the original system with the Solar Nebula, and the subsystems with the Earth and the various meteorite sources. The date of c. 4550 Ma age is then the time the Solar Nebula divided into separate subsystems. This need not mean the time the various bodies became much as we now see them but only when they had acquired different U/Pb ratios and become separate from one another, sometime during the aggregation process. Therefore it is not so much the age of the Earth (whatever that might be taken to mean), as the time at which the Solar Nebula became the Solar System, which we believe took only a few million years (too small to show up on Fig. 15.11). The subsequent formation of crust, mantle, and core was a complex process that can be partly deduced by carefully considering dating and other isotopic evidence, but will not be considered further here.

Finally, why was it necessary to combine the two equations, Eqs. 15.13 a and b, and sacrifice a check on the validity of the date? There are two reasons. Without eliminating uranium it would have been impossible to have included the iron meteorites, which contain no uranium. And it would have been

hard to get a representative terrestrial U–Pb ratio, for this ratio is affected by the partial melting process that produces rocks from the mantle. So the sacrifice was worth making.

15.8 Fission-track (FT) dating

Most atoms of ^{238}U, when they decay, ultimately end up as ^{206}Pb atoms (Fig. 15.20), but about a millionth decay in a different way, spontaneously splitting into two roughly equal parts (plus some neutrons), a process called **fission**. The two main parts recoil with sufficient energy to force a way through the host crystal for many lattice spacings, leaving a **fission track** of displaced atoms about 15 microns (millionths of a metre) long. The number of tracks increases with the age of the sample and this replaces the number of daughter atoms for calculating the age.

The tracks can be seen using an electron microscope, but it is more convenient to enlarge them and use an optical microscope. To date a sample, it is sliced and given a polished surface, which is then etched with chemicals that preferentially dissolve any tracks that intersect the surface (Fig. 15.12a). The shape and length of a track depends on the angle at which it cuts the surface and how much was removed in preparing the surface (Fig. 15.12b). The track density – the number of tracks per square centimetre – is counted. The density of parent atoms, ^{238}U, has also to be measured, as with the other dating methods. This is done by exposing the sample to neutrons in a nuclear pile, which induces fission in ^{235}U atoms, an isotope that is easily fissioned in this way, unlike ^{238}U. Some of the fission particles recoil out of the surface and produce tracks in a film of special material pressed against the surface; the film is removed after irradiation and the density of induced tracks is counted, and the density of ^{238}U is calculated by multiplying by 137.88, the ^{238}U/^{235}U ratio. Alternatively, the natural tracks are destroyed (by heating) after they have been counted and then the sample is irradiated to produce artificial tracks, which are counted once a new polished surface has been made. By irradiating samples with known uranium content at the same time and comparing the densities of tracks, the ^{238}U density of the sample can be deduced.

(a) track enlargement by etching

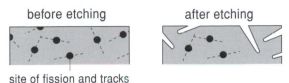

(b) photomicrograph of tracks

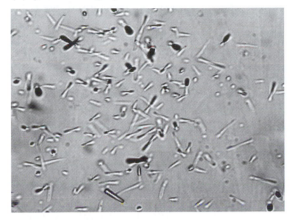

Figure 15.12 Fission tracks.

There are two main sources of error. Decays of individual ^{238}U atoms are quite random in time – as with all types of radioactive decay – and it is only when there are many decays that we can make precise predictions. Therefore, there must be sufficient tracks, which depends upon the age of the sample and its uranium content. At the other extreme, if there are too many tracks some will overlap and not be counted. The requirement of 'not too many, not too few' tracks imposes a dateable range on each mineral used. The other main source of error is disappearance of the tracks by annealing, in which some of the displaced atoms move back to regular lattice positions, and this can lead to spuriously low dates, just as loss of argon does in the K–Ar method. The rate of annealing increases with temperature, but also depends on the mineral used, such as zircon, sphene, micas, and apatite.

15.9 What event is being dated?

A radiometric 'clock' starts when a rock or mineral closes and daughter atoms begin to accumulate (tracks, in the case of fission-track dating). We have assumed so far that closure occurred when the rock or mineral formed, but this is not always the case, because closure requires the temperature to fall to a sufficiently low value. For K–Ar dating of a lava, the time between the extrusion of the magma and its cooling to near ambient temperature is so short – only a few years – that the precise moment when argon began to be retained is not important. Other examples are not so simple: Large intrusions can take millions of years to cool, rocks can be heated by later intrusions, and regional metamorphism can maintain high temperatures for tens or hundreds of millions of years. To understand how high temperatures affect the date we need to consider diffusion.

15.9.1 Diffusion

To illustrate **diffusion**, think of a single mineral grain in a cooling igneous rock that has just crystallised and that is later to be dated using the K–Ar method. Further, suppose the grain is spherical; though an implausible shape for a crystal, it approximates to a roughly equidimensional grain.

Argon atoms are not held chemically in the crystal lattice, for they are inert; instead, being large atoms, they occupy spaces between lattice atoms (Fig. 15.13). All atoms vibrate, the more vigorously the higher the temperature, and at a sufficiently high temperature some argon atoms will occasionally have sufficient thermal energy to jump from one space to another. (This is similar to the mechanism underlying solid-state creep, Section 9.2, and is related to thermal demagnetisation, Section 10.3.3). Jumping is a random process, so the direction of each successive jump is unrelated to that of the previous one – for instance, the second jump might bring it back to its initial position – but on average an atom will wander through the lattice.

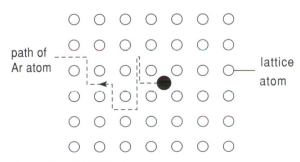

Figure 15.13 Diffusion of an argon atom.

Suppose, to further simplify the explanation, that an atom can move only left, L, or right, R, that is, 1D diffusion (Fig. 15.14a). After 2 jumps, say, the possible outcomes are LL, LR, RL, and RR. As an L and an R together cancel, the net number of jumps travelled are 2L, 0, 0, and 2R, so the atom has 2 chances in four of arriving back where it started and one chance each of being 2 jumps to the left or right. Figure 15.14b shows the probabilities of travelling different distances after 4, 8, and 12 jumps. This shows, as you would expect, that there is a very small chance of all jumps being in the same direction; however, while the *single* most likely outcome is to end up at the starting point, there are so many other possibilities that this is actually quite unlikely, so the most likely outcome is that the atom has moved some distance, and the average distance moved increases with the number of jumps. These probabilities are the same as for tossing a coin, if we equate heads with left, say, and tails with right. The theory of coin tossing is well understood (except by gamblers), and the shapes are known as binomial distributions. With large numbers of tosses, or jumps, these become smooth bell-shapes. In 2D, atoms follow an irregular path, as shown in Figure 15.13, but – as with 1D diffusion – most end up away from their starting point, and similarly for 3D diffusion.

Though diffusion is an inefficient way of travelling through a lattice, at high temperatures jumps occur so frequently that atoms soon reach the crystal surface, where they escape. Therefore, while the crystal is hot, argon atoms escape soon after they are formed and so do not accumulate. As the temperature falls, diffusion slows, so some argon accumulates, just as water will rise to some level when poured into a bucket with only a small hole in it. At sufficiently low temperatures diffusion is so slow that even over geological periods of time few atoms escape, and the crystal is effectively closed. So as the crystal cools, it passes from being an open system, through an transitional state, to being closed (equivalent to the hole in the bucket becoming smaller and smaller). If the transitional state lasts a significant time – because cooling is slow – we need to consider what date will be measured.

15.9.2 Closure temperature

Figure 15.15 shows, above, temperature slowly cooling. The corresponding accumulation of argon atoms is shown below: At high temperatures there is no accumulation, but as the temperature falls, there is a transitional period with incomplete but increasing accumulation, until – at D – retention becomes 100%. When we use the Basic Dating Equation we are calculating the time at which the number of daughter atoms was zero, by following the daughter curve of Figure 15.2 back until it reaches the axis; so if we apply it when cooling has been slow, this is equivalent to extrapolating the line ED of Figure 15.15 back in time to C. Going vertically up to the cooling curve above shows that at this time the temperature was T_c, called the **closure temperature**. Therefore, the date measured was the time when the sample cooled through its closure temperature.

Diffusion and the closure temperature concept are not limited to argon. For example, lead atoms are incompatible with the zircon lattice and so diffuse out at sufficiently high temperatures. Even compatible atoms in a crystal, such as strontium in a mica, will move around and so exchange with strontium atoms from nearby crystals, preventing the crystals from evolving the different isotopic ratios that produce an isochron, so if a sample is reheated above its closure temperature isotopic ratios will tend to rehomogenise and so destroy the dating information. Full rehomogenisation **resets** the atomic clock, and the date measured is then the time when the sample subsequently cooled below its closure temperature.

(a) one-dimensional diffusion

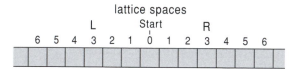

(b) binomial distribution

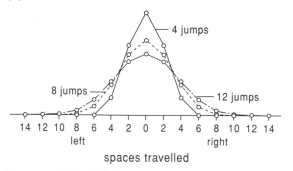

Figure 15.14 Diffusion in 1D.

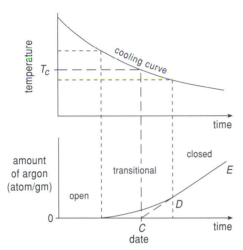

Figure 15.15 Slow cooling and the closure temperature.

The value of T_c depends mainly on the types of atoms and of crystal, and so will have a different value for each dating method/mineral combination. However, it also depends to some extent upon the cooling rate, for if the sample cools very slowly the time it spends at even a comparatively low temperature is sufficient for significant diffusion to occur, even though it is slow. It also depends on grain size, for atoms will reach the edges and escape more quickly from small crystals than large ones. Therefore, the closure temperature values given in Table 15.1 are only approximate, and ideally should be adjusted for grain size and cooling rate.

Table 15.1 Closure temperatures

Dating method	Mineral	Closure temperature T_c (°C)
U–Pb	zircon	800
Sm–Nd	whole-rock	650–800
Rb–Sr	whole-rock	650–800
K–Ar	hornblende	530
U–Pb	sphene	500
Rb–Sr	muscovite	400
K–Ar	muscovite	350
Rb–Sr	biotite	320
K–Ar	biotite	280
Rb–Sr	feldspar	350
K–Ar	glauconite	230
K–Ar	microcline	130
fission track	sphene	300
fission track	zircon	250
fission track	apatite	90

Most values are from Harland et al. (1990); cooling rates are mostly a few °C/Ma.

15.9.3 Cooling histories

Because closure temperatures differ (Table 15.1), a rock or region that has cooled slowly can yield a range of values if dated using different method/mineral combinations, and these can reveal its cooling history, a branch of radiometric dating known as **thermochronometry**. The results for the Glen Dessary Syenite, Scotland (Fig. 15.16a) show that it took roughly 50 Ma to cool from 1000 to 200°C, and much longer to cool below 100°C. The Separation Point Batholith, New Zealand, which was intruded about 114 Ma ago, cooled quickly at first (Fig. 15.16b), but the results suggest that the cooling rate slowed about 90 Ma ago. Both these examples show the importance of fission track dates for measuring the cooling at low temperatures. Temperature increases with depth in the Earth (Chapter 17), and at a depth of a few kilometres often exceeds 200°C.

(a) Glen Dessary Syenite

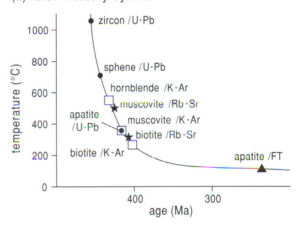

(b) Separation Point Batholith

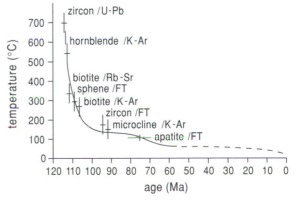

Figure 15.16 Cooling curves.

Cooling below this temperature often has to await for erosion to bring the rocks slowly closer to the cool surface; in fact, the method has been used to deduce erosion rates.

Closure temperatures depend on cooling rates, as explained above. Once an initial cooling curve has been found, it can be used to give approximate cooling rates at different temperatures by drawing the tangent to the cooling curve, and this can be used to refine the closure temperatures and modify the cooling curve; a single modification is sufficient.

15.9.4 Two dates from a single rock, using the Rb–Sr method

Table 15.1 shows that Rb–Sr whole-rock dates have a considerably higher closure temperature than do commonly used minerals such as micas and feldspars. This sometimes allows both the formation (crystallisation) and reheating times of a rock to be measured. If, long after its crystallisation, a rock is heated to, say, 500°C, strontium (and other) atoms will be able to diffuse between minerals. If this mixing of ^{87}Sr and ^{86}Sr atoms continues long enough, it will cause the minerals to have the same $^{87}Sr/^{86}Sr$ ratio, resetting the *mineral* isochron. The mineral isochron therefore gives the time when the minerals cooled again; if the heating was brief, the date is effectively the time of reheating. Because diffusing atoms take such erratic paths, it would take a million times as long to homogenise isotopic ratios over a distance of 10 m as 10 mm, so – depending on the duration of heating – minerals within a block can be homogenised but not blocks separated by tens of meters. The consequence is that a whole-rock isochron using well-separated blocks can give the age of crystallisation, but minerals extracted from a single block give the age of reheating. Figure 15.9 shows an example. The dates of crystallisation and reheating can also sometimes be measured using the U–Pb method, as described next.

15.9.5 Two dates from a single rock, using the U–Pb discordia method

Section 15.2 explained that when minerals such as zircon form closed systems, their $^{206}Pb/^{238}U$ and $^{207}Pb/^{235}U$ ratios, though different, give the same date. A pair of such concordant ratios defines a point such as R on Figure 15.17, and as the ratios

increase with time they trace out a curve, called the concordia, as shown by the ages marked on it. However, sometime ratios are not concordant due to severe heating in the past.

Suppose the isotopic ratios have evolved – in closed crystals – along the concordia curve to H, at which time the temperature briefly rises high enough to cause lead atoms to diffuse out of the lattice; as ^{206}Pb and ^{207}Pb atoms diffuse at (almost) the same rate they are lost in the proportions in which they occur – if a grain loses a quarter of one it will lose a quarter of the other – so lead loss causes the isotopic composition to move down the line HO towards the origin. If different grains lose different amounts of lead – perhaps because they are of different sizes – then immediately after the heating their lead isotopic compositions will be distributed along HO.

Following cooling, their isotopic ratios evolve from these points. A grain that had lost all its lead and so had been completely reset starts from O and so moves up the concordia, reaching, say, R at the present time. Any grain that had lost no lead would simply continue to evolve along the concordia, reaching C. It turns out that the grains that had lost some lead all lie on the line RC. Therefore, if grains are extracted from the rock and individually analysed, their compositions define the **discordia** line RC; the intercepts of the discordia line with the concordia curve, at C and R, give the times of crystallisation and reheating, in this example 2500 and 500 Ma ago. An example is given in Figure 25.6c in Section 25.3.3.

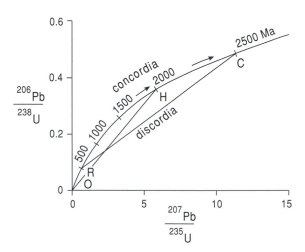

Figure 15.17 Concordia curve and a discordia line.

15.9.6 Dating palaeomagnetism of slowly cooled regions

A cooling rock acquires its remanent magnetism when it cools through its blocking temperatures (Section 10.3.3). Determining the time at which this occurred can be uncertain if cooling is slow but sometimes can be measured by choosing a dating method/mineral combination with a closure temperature that matches the blocking temperature at which the magnetism was acquired.

Blocking temperatures, T_b, in palaeomagnetism and closure temperatures, T_c, in radiometric dating (these are the correct uses of the terms, though they are sometimes used interchangeably) have similarities and differences. Both depend upon thermal fluctuations, in one case to alter domains and in the other to allow diffusion, but differ because, whereas a domain can be unblocked deep in a sample as soon as the temperature rises high enough, diffusion frees an atom only locally, and it has to diffuse right out of a crystal for the crystal to be reset. Thus closure and blocking both depend upon the temperature and the time spent at it, but closure also depends upon grain size.

To date when palaeomagnetism was acquired, the palaeomagnetic direction of the sample is measured and then its blocking temperature is found using thermal demagnetisation (Section 10.3.3). Then another sample of the rock is dated using a combination of dating method and mineral with closure temperature as close as possible to the blocking temperature (ideally, after allowance for cooling rate and grain size). Alternatively, a cooling curve could be deduced as described in Section 15.9.3 and the time when it cooled through the blocking temperature read off. An example of measuring the age of magnetisation of a gabbro is given in Armstrong et al. (1985).

15.10 Dating sedimentary rocks

Sediments are difficult to date by the above methods. Most dateable minerals found in sediments are detrital, deriving from the source rocks; dating them, therefore, does not date the sediments, though it can help to identify their source. What is needed is a mineral formed at the same time as the sediments (i.e., authigenic) that contains a suitable parent isotope and forms a closed system. There are few such minerals; probably the most commonly used is glaucony, a general term for green granules of poorly defined composition, which includes glauconite, a clay mineral that contains several percent of potassium and significant amounts of rubidium. These are often dated using the K–Ar method, rather than the Ar–Ar one, because argon is lost even at low temperatures; their suitability for Rb–Sr dating is not well understood. It is probable that a glaucony grain does not close until up to some millions of years after the sediment in which it is found was laid down. For these and other reasons, consistency checks are needed, such as that several glaucony dates are concordant, or consistent with the stratigraphic order of the layers they are in. Carefully selected and checked glauconies can give reliable K–Ar dates for Cenozoic samples, but often give low ages for older rocks.

Sediments are therefore most often dated indirectly, using igneous rocks that are close in age. Most useful are ash falls, because they are commonly deposited in waterborne sediments over wide areas. The chief causes of error are alteration of the minerals and contamination by older material, which can be from earlier eruptions carried in by rivers or entrained in the eruption as it rises through older rocks. For these reasons, minerals have to be carefully selected and the Ar–Ar method on single grains is often used. Lavas and intrusions are also sometimes used, if they can be closely related in time to the sedimentary succession. These various approaches have been used to establish the geological timescale.

15.11 The geological timescale

The geological timescale – devised long before radiometric dating – was established using two main tools. The first is the Principle of Superposition, that sediments in a column are normally younger than those below them, is used to establish the relative ages of rocks whose depositional order can be observed (Fig. 15.18). The second method is palaeontological dating, the recognition that certain life forms existed only for a limited period, so that even widely separated rocks containing a given species are contemporaneous. The relative ages of fossils is established using the Principle of Superposition, and together the two methods have been used to construct the Phanerozoic timescale, in which the divisions reflect major and minor changes in the range of species in existence. For instance, the transition from Mesozoic to Tertiary periods is marked by the extinction of the dinosaurs and many other species.

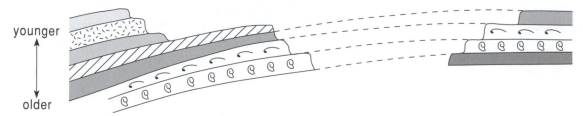

Figure 15.18 Relative dating by stratigraphic order and fossils.

This geological timescale is based on sedimentary rocks, whereas radiometric dating is applicable mostly to igneous and metamorphic rocks.

The geological timescale is given numerical values by dating whatever suitable samples can be found: glauconies or igneous rocks whose position in the timescale is clear and preferably close to a boundary. The majority of igneous samples used have been of volcanic rocks, for these cool quickly and their stratigraphic position – particularly that of volcanic ashes – is often better defined than for intrusive rocks. The great majority of dates are by the K–Ar method (including Ar–Ar dates), but a few are by the Rb–Sr method, and the U–Pb method on zircon is increasingly important.

The timescale is revised from time to time as dates that are more precise or closer to geological boundaries are obtained. Figure 15.19 shows a recent version of the Phanerozoic, plus the much less-detailed earlier time divisions.

15.12 Dating young rocks

The youngest sample that can be dated by a given method partly depends upon the decay constant of the parent isotope, for the slower the decay, the longer it takes before the daughter has accumulated to the point where it can be measured precisely enough. The decay methods described so far depend on 'primeval' isotopes, isotopes that were present when the Earth formed about 4550 Ma ago, and these necessarily have long half-lives or they would have decayed to negligible amounts. A number of dating methods that are particularly useful for dating young rocks – that is, as a rough guide, less than a million years old – use short-lived parents; these are being produced at the present time, in a variety of ways.

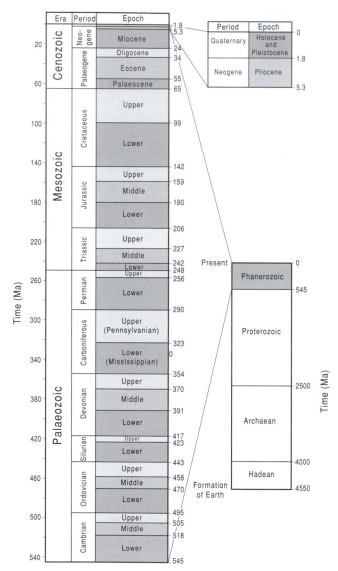

Figure 15.19 The Phanerozoic timescale of Gradstein and Ogg (1996).

15.12.1 Uranium-series disequilibrium methods

The primeval isotopes, ^{238}U and ^{235}U – and also ^{232}Th (thorium) – all decay to isotopes of lead, but each passes through a decay series of many intermediate radioactive isotopes before reaching a stable lead one (Fig. 15.20). Atoms progress through a series by either (i) emitting an **α-particle,** which contains two protons and two neutrons, so reducing the atomic number by two and the mass number by four, or (ii) emitting a **β-particle,** which is simply an electron and so *increases* the atomic number by one, leaving the mass number unchanged. (Many radioactive atoms also emit γ rays, but as these have no electric charge they change neither the atomic number nor mass number; however, they are important for radiometric surveying, the topic of the next chapter.) Provided none of the intermediate isotopes

escapes from the sample, uranium decays as a closed system into lead, and the U–Pb method (Section 15.2) may be used. But if a rock weathers, the intermediate isotopes may escape and later be incorporated into a sediment and become parent isotopes.

Because the daughter, as well as the parent, isotope decays, the Basic Dating Equation is not applicable. Instead, dating relies on measuring **disequilibrium:** If an isotope of a series remains in a closed system for many half-lives, its daughter reaches a **secular equilibrium** in which it decays into the next member of the series at the same rate it is being produced (it is not a true equilibrium, for the amount slowly decreases with time as the whole series decays, but often this is too slow to concern us).

The number of atoms decaying per second is called the **activity** and equals the number of atoms present multiplied by the decay constant, $n\lambda$, so when a chain is in equilibrium,

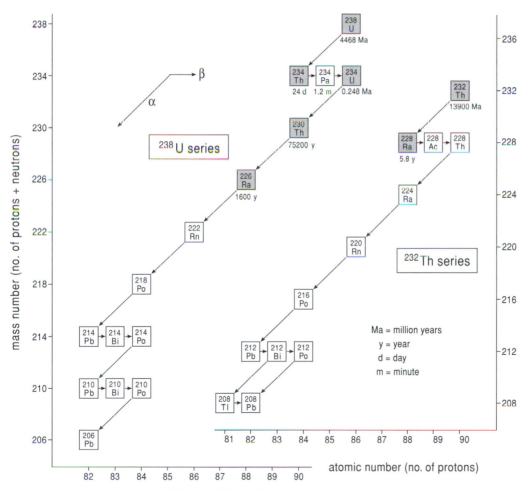

Figure 15.20 Decay series of ^{238}U and ^{232}Th (shaded isotopes are mentioned in the text).

$$n_1\lambda_1 = n_2\lambda_2 = n_3\lambda_3 = \cdots \qquad \text{Eq. 15.15}$$

where the subscripts 1, 2, 3, . . . indicate the successive members of the series. But extracted isotopes incorporated into a new rock do not have the proportions of equilibrium, because they have been selected by their different chemical properties. For example, if only uranium is incorporated into a carbonate there will be ^{238}U and ^{234}U, but no ^{234}Th or ^{230}Th (Fig. 15.20). However, the missing members of the series start to be formed once the carbonate becomes closed and gradually equilibrium is restored. ^{234}Th forms by decay of ^{238}U and, because it and the next isotope have short half-lives, soon decays at an equal rate into ^{234}U. But it takes much longer for ^{234}U to reach equilibrium because of its longer half-life. It practice, it takes about five half-lives for any isotope effectively to reach equilibrium. Dating relies on detecting disequilibrium, for once equilibrium has been reached it will continue indefinitely, and then all that can be deduced is that at least five half-lives have elapsed.

Disequilibrium means that the activities of successive members of the series are not equal. Isotopes are usually measured as their activities (by counting the α-particles emitted) rather than as ratios of numbers of atoms as hitherto. However, mass spectrometric measurement of ratios of atoms is increasingly being used, its greater precision and sensitivity sometimes justifying its extra cost, but equations will be given only in terms of activities. Activity ratios are often indicated by being enclosed in parentheses ( ) but to avoid confusion with other sorts of ratio the ratios in the rest of this section which are of activities are shown in square brackets [ ].

There are a number of these **uranium-series** dating methods, covering a range of ages from a year to a million years, but only a few will be described. All are susceptible to violation of their assumptions, to varying degrees.

The $^{234}U/^{238}U$ method

^{238}U: λ_{238}, 1.55×10^{-10} a^{-1};
half-life, 4468 Ma

^{234}U: λ_{234}, 2.794×10^{-6} a^{-1};
half-life, 248,000 years

This method is used for marine carbonates, for these incorporate uranium (but not thorium) from seawater.

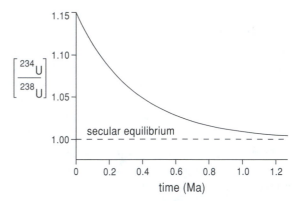

Figure 15.21 Return to equilibrium of [$^{234}U/^{238}U$]].

This uranium has a near-constant activity ratio of [$^{234}U/^{238}U$] of 1.15 and so is in excess of equilibrium. Once incorporated in a carbonate, ^{234}U decreases by decay faster than it is formed from ^{238}U and the ratio of their activities tends towards the equilibrium value of one (Fig. 15.21).

The curve of Figure 15.21 can be regarded a decay curve decreasing exponentially towards a value of one (rather than zero as is the case for numbers of atoms (Fig. 15.1b). Therefore, if we subtract 1 from the value of [$^{234}U/^{238}U$], the remainder, or excess above equilibrium, decays as $e^{-\lambda_{234}t}$. Hence,

$$\left[\frac{^{234}U}{^{238}U}\right]_{now} = 0.15 e^{-\lambda_{234}t} + 1$$

$$\text{Eq. 15.16}$$

The method can be used to date carbonates up to about a million years old. Corals and some other rocks are often suitable, though sometimes uranium leaches out, violating the closure assumption. Detrital thorium may also be a problem (see the next method). Corals have been used to measure climatic changes in the recent past (with temperatures determined using their oxygen isotope ratios, a method not described in this book).

The $^{230}Th/^{232}Th$ method

^{230}Th: λ_{230}, 9.217×10^{-6} a^{-1};
half-life, 75,200 years

^{232}Th: λ_{232}, 4.986×10^{-11} a^{-1};
half-life, 13,900 Ma

In contrast to the previous method, this depends on the sediment incorporating thorium but not uranium, which is possible because of the different behaviours of uranium and thorium in surface waters. In this oxidising environment uranium stays in solution (except when incorporated into carbonates), whereas thorium is soon chemically precipitated on to detrital grains. Once such a sediment layer has been buried and become a closed system, the ^{232}Th remains virtually unchanged because of its huge half-life, but the ^{230}Th decays away, so that

$$\left[\frac{^{230}\text{Th}}{^{232}\text{Th}}\right]_{now} = \left[\frac{^{230}\text{Th}}{^{232}\text{Th}}\right]_{i} e^{-\lambda_{230}t} \qquad \text{Eq. 15.17}$$

This method is suitable for measuring rates of sedimentation for periods from thousands to hundreds of thousands of years. However, it assumes that no uranium is also deposited (which can occur if carbonate is present) and that the detrital grains did not already contain thorium.

15.12.2 Carbon-14 (^{14}C) and other dating methods using cosmogenic isotopes

^{14}C emits β-particles; λ_{14}, 1.209×10^{-4} a^{-1};
half-life, 5730 years

Carbon-14 is formed in the upper atmosphere from ^{14}N, the common isotope of nitrogen in the atmosphere, by the action of cosmic rays, which are energetic particles that come from space. It is soon distributed throughout the atmosphere, and some, together with the common and stable isotope ^{12}C, is taken into living organisms by photosynthesis or in food. When the plant or animal dies its ^{14}C is no longer exchanged and its amount decreases by decay, so the ^{14}C/^{12}C ratio halves with the passing of each half-life; the ratio yields the age.

There are two ways of measuring this ratio. The original and commoner method is to measure the activity, the number of decays per second in a known amount of carbon. To count the β-particles, carbon from the sample is converted into carbon dioxide gas, which is purified and put into a counting tube. By knowing the amount of carbon dioxide in the tube, the ratio ^{14}C/^{12}C can be calculated. (A gas is used and put into the tube because β-particles are easily absorbed by solids and liquids.) Precautions are taken to reduce spurious counts due to other sources of electrons, such as radioactive decay of other elements and cosmic rays.

The more advanced – but much more expensive

– way is to use a special kind of mass spectrometer, called a tandem accelerator mass spectrometer, to measure the ^{14}C/^{12}C atomic ratio. It is much more precise than counting decays, and needs only milligram samples; this was the method that showed that the Turin shroud was made in the 13th century, using only a few threads of the cloth.

As the half-life is only 5730 years, the maximum age that can be dated, even using a tandem accelerator, is about 40,000 years. This makes it useful mainly for dating objects of archaeological interest, but it has some geological and geomorphological uses, such as constraining the time of a fault by dating carbon in sediments offset by the fault, dating a fossil shoreline using shells, or dating a lava from engulfed trees.

An assumption of the ^{14}C, radiocarbon, dating method, additional to those stated in Section 15.3, is that the rate of ^{14}C production in the atmosphere is constant, so that the atmospheric ^{14}C/^{12}C ratio had the value in the past it has now. But it is known to be only roughly constant, and a correction is routinely applied based on dendrochronology – established by dating annual growth rings of trees counted back from the present – or, for older times, U-series dating. Figure 15.22a shows how the radiocarbon date compares with the true age for the past 2000 years; dates are given as years B.P., which stands for 'before present', though it is actually before 1950 A.D., for otherwise dates would need constant revision. For samples about 1500 years old the radiocarbon date is roughly 100 years too old (for 20,000-year-old samples it is about 2000 too young). The corrections are difficult to work out, and small adjustments to the correction curve are made every few years.

Though analytical (measurement) errors in radiocarbon dates are usually less than 10 years for young samples, increasing to several tens of years for old ones, after correction these can result in quite different errors, depending on the shape of the correction curve for that time. At some times there can be three possible calendar dates corresponding to a single radiocarbon date; for example, Figure 15.22b shows that a radiocarbon date of 950 ± 10 B.P. would convert to possible actual ages of 800^{+8}_{-3}, 840^{+14}_{-7}, or 907^{+5}_{-5} years. Unless there is additional information to choose between them, such as a succession of dates spanning the period, this can introduce considerable uncertainty.

(a) past 2000 years

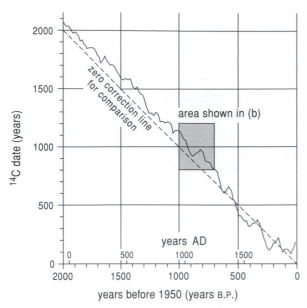

(b) detail of (a)

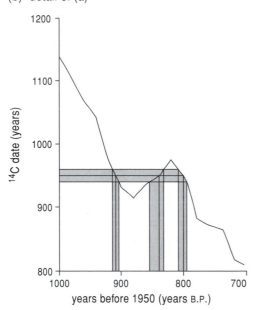

Figure 15.22 Conversion curve for carbon-14 dates.

Cosmic rays penetrate surface rocks and soils and produce radioactive isotopes, mostly within the topmost metre. The most important are ^{10}Be (beryllium) and ^{26}Al (aluminium), with half-lives of 1.51 and 0.7 Ma, respectively. When a rock is newly exposed, the concentrations of these isotopes at first increase but tend to equilibrium as the number decaying also builds up. As long as the concentration is less than

the equilibrium value, the **exposure age** can be calculated but is limited to a few million years by the half-lives. The presence of ^{10}Be in igneous rocks above subduction zones shows that surface material can be incorporated in magma at depth and return to the surface in less than 10 Ma (Section 20.4.2).

15.13 Why so many radiometric dating methods?

The methods described above – plus others – are all in use because no single one is universally applicable. Factors to be considered when choosing a method include these:

i. What radioactive parents are likely to be present, and are there any suitable minerals? The answer to this basic question depends on the sample; for instance, a granite may contain micas (potassium and rubidium) and zircons (uranium).

ii. What is the expected age of the sample? Methods have age limits, depending upon the half-life of the parent and on other factors.

iii. What event is to be dated? Method/mineral combinations with low closure temperatures are susceptible to resetting and so are unsuitable to measure the crystallisation age if slow cooling or subsequent heating is likely; conversely, they can be used to check if resetting has occurred or to date it.

The methods described in this chapter are now considered in the light of these and other points.

Potassium–argon method. This is widely used, because potassium is abundant and widely distributed. It can often date rocks significantly younger than 1 Ma, and occasionally down to a few thousand years old, and up to all ages. Most minerals are moderately susceptible to thermal resetting, but hornblende has a high closure temperature. Argon may leak out, giving spuriously old dates; this and the absence of an internal checks of the validity of a date are the chief drawbacks of the conventional K–Ar method, but the Ar–Ar step-heating method gets over these.

Uranium–lead. Though this method is limited to the few minerals that do not contain lead at formation, these occur in a range of igneous rocks. Zircon is moderately common and has the highest closure temperature, and so is particularly suitable for dating crystallisation (formation) ages of rocks likely to have been reheated, or to date the earliest stages of cooling from high temperatures. Occasionally, the very high closure temperature of zircon, coupled with its resistance to erosion, allows it to be inherited from older rocks, and it can even recommence growing, resulting in spurious ages unless recognised. Single zircons thousands of millions of years old can be dated to 1 Ma or better, but precision becomes poorer for dates less than 100 Ma. A discordia plot sometimes yields both crystallisation and resetting ages.

Rubidium–strontium. Acidic rocks (60% or more silica) are needed. Many of the minerals used for K–Ar dating are suitable (but not hornblende) and closure temperatures are similar (Table 15.1), but whole-rock samples are more resistant and may yield the crystallisation age when minerals give the reheating age. Because of rubidium's long half-life and the presence of initial strontium, the youngest rocks that can be dated are usually at least tens of millions of years old. A well-defined isochron is internal evidence that the dating assumptions have been satisfied.

Samarium–neodymium. This is resistant to thermal resetting, which makes it suitable for dating old rocks; in any case, because of its very long half-life, it is not used for rocks less than several hundred million years old. The most suitable rocks are basic and ultrabasic, so it tends to complement the Rb–Sr method.

Lead–lead. This is mainly used when subsequent processes have changed the U/Pb ratio. A particularly important application is dating the formation of the Solar System, or 'age of the Earth'.

Fission track. Its chief advantage for the geologist is its low closure temperatures, useful for dating the late stages of cooling or erosion rates. Tracks are susceptible to fading at low to moderate temperatures. It can be applied to a range of minerals and also to glasses and some archaeological objects, and it is mostly used at the younger end of the dating range, from less than a million to several hundred Ma, depending on the mineral used.

U-series methods. There are many methods applicable to sediments of various types, for ages up to a million years, depending on the particular method. The assumptions on which they depend are often violated.

Cosmogenic isotope methods. The most important of these is carbon-14, which uses carbon of organic origin and is limited to the past 40,000 years. Because the rate of production of ^{14}C has varied, dates have to be corrected, sometimes resulting in considerable errors. Other cosmogenic isotopes are ^{10}Be and ^{26}A, useful for deducing exposure ages and recycling times.

Summary

1. Radiometric dating utilises radioactive decay, in which atoms of a parent isotope decay at a rate unaffected by geological processes.
2. The Basic Dating Equation relates the present ratio of the numbers of daughter to parent atoms to the age of the sample. It assumes (i) a closed system, (ii) that one parent atom decays into one daughter atom, and (iii) that there were no daughter atoms present when the system closed. The equation is modified when (ii) or (iii) are not satisfied.
3. Closure occurs when the temperature has dropped through the closure temperature, which is when diffusion effectively ceases. Different sample/dating method combinations have different closure temperatures (which also depend upon grain size and cooling rate). Thermochronometry utilises this to date, variously, the time of crystallisation of the sample or of its subsequent cooling, or deduce a cooling curve.
4. The most common methods using primeval parent isotopes are K–Ar, U–Pb, Rb–Sr, Sm–Nd, Pb–Pb, and fission track. The Ar–Ar method is a version of the K–Ar method; it permits step heating, and if a plateau results this is usually a valid date. The Rb–Sr and U–Pb methods can each sometimes date both crystallisation and reheating times.
5. Fission-track dating measures the number of decays that have occurred by the number of tracks formed, rather than the number of daughter atoms.

were radiogenic and quantities could be measured to 1% precision? Think of a situation in which the result could be useful despite such large errors.

24. The following dates were obtained from a variety of rocks in a small area:

Zircon/U–Pb: 550 Ma.

Hornblende/K–Ar: 550 Ma.

Biotite/K–Ar: 340 Ma.

K-feldspar/K–Ar: 340 Ma.

Which of the following could account for these results?

(i) The rocks formed 550 Ma ago, but argon has leaked out over a long time.

(ii) The rocks formed 550 Ma ago but were heated to about 400°C 340 Ma ago.

(iii) The biotite and feldspar came from a younger rock.

(iv) There has been slow cooling of the area.

25. Which of the following isotopes are primeval?

(i) ^{40}K, (ii) ^{234}U, (iii) ^{235}U, (iv) ^{10}Be, (v) ^{14}C, (vi) ^{86}Sr.

What is the significance for dating whether an isotope is primeval?

chapter 16

Radioactivity Surveying

Radioactivity surveying measures the natural radioactivity due to potassium, thorium, and uranium in near-surface rocks, which has applications in geological and geochemical mapping, and is used to find ores of uranium and thorium or other types of ore that have associated radioactivity. It also has environmental applications, mapping radon, a hazard to health, in surface rocks and waters.

The most common surveying method detects γ rays, which can be used to identify the source element as well as detect the presence of radioactivity, and can be employed in ground or airborne surveys, but radon measurement often requires sampling below the surface.

16.1 Radioactive radiations

The previous chapter explained how radioactivity could be used to date rocks because isotopes decay from one element to another. This chapter is mainly concerned with the 'radiations' that accompany the decays, as a way of detecting and identifying the source elements. There are three principal types of radiation, all of which originate from the nuclei of radioactive atoms (Section 15.12.1). α-**particles** consist of two protons and two neutrons, and so have a positive electrical charge (ultimately, they each combine with two electrons to form helium atoms, and this is the origin of the helium used to inflate balloons). β-particles are electrons – produced when a proton converts to a neutron plus a electron – and so have a negative charge. Because of their electric charges, α- and β-particles cannot travel far through matter, no more than a few centimetres in air, or a few millimetres of rock, and so are little used in surveying. γ **rays**, the third type of radiation, are different: They have no charge and are like X-rays, but even more penetrating. They have no definite range – unlike α- and β-particles – but their numbers are progressively reduced as they travel through matter,

the more rapidly the denser it is, but also depending on their energy. In practice, few travel more than 100 m in air, or half a meter in rock. The energy of a γ ray depends on the particular isotope that produces it and so can be used to determine the element producing the radioactivity. How the energy is measured is described in the next section.

16.2 γ ray surveys

16.2.1 Measurement: the γ ray spectrometer

γ rays can be detected in a number of ways, but the most useful instrument is the γ **ray spectrometer** (Fig. 16.1). The detecting part consists of a large, transparent crystal, usually of specially treated sodium iodide (Fig. 16.1a). If a ray is absorbed within the crystal, it gives rise to a tiny pulse of light, or scintillation, proportional to the energy of the γ ray. Each light pulse is measured by a special light amplifier, called a photomultiplier, and pulses of different sizes are counted separately. Because it distinguishes γ rays with different energies and does not simply count the total number, the instrument is called a spectrometer (or sometimes a spectral scintillometer) to distinguish it from a total-count **scintillometer**.

Natural radioactivity is due almost entirely to the three most abundant radioactive elements: uranium (^{238}U and ^{235}U) and thorium (^{232}Th), plus their decay series products (Section 15.12.1), and potassium (^{40}K). The objective is to measure the concentrations of K, U, and Th, but in practice, the γ rays used to measure uranium and thorium come from some of their decay products, which emit much more energetic γ rays and so are also more penetrating and easier to detect (measuring decay products, rather than the parents, has consequences when weathering has occurred recently, as will be explained in the following section). The spectrometer is required to count only pulses due to the particular γ rays selected for K, U, and Th, distinguishing them from pulses due to other decays and processes; to achieve this it has to be carefully tuned and should be checked every few months by measuring an artificial sample with known γ rays.

The instrument gives the number of measured decays of U, Th, and K occurring per second or minute, plus their total, or – as explained in the next section – the concentration of these elements assuming a standard geometry. If the total exceeds the counts of U, Th, and K in sum, there must be other radioactive isotopes present, almost certainly produced artificially. This is found in some of the estuaries in northwest England, due to radioactivity from the Sellafield nuclear site in Cumbria. To identify what isotopes they are requires a spectrometer that can discriminate other isotopes, but this will not be discussed further.

16.2.2 Carrying out a γ ray survey

Because γ rays can travel many metres through air, they can be measured by either ground or airborne surveys.

(a) schematic diagram

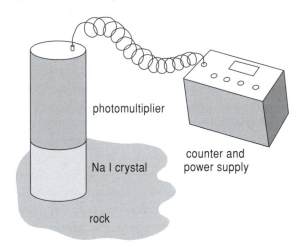

photomultiplier

Na I crystal

counter and power supply

rock

(b) Scintrex GAD4 spectrometer

Figure 16.1 γ ray spectrometer.

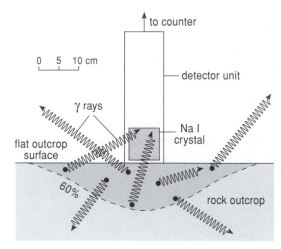

Figure 16.2 Source volume being measured.

Ground surveying. If a γ ray spectrometer is placed on a flat rock surface (Fig. 16.1b), most of the rays it detects come from a saucer-shaped volume about 2 m across, and 60% originate in the darker-shaded volume of Figure 16.2, for few produced further away will penetrate through the rock to the spectrometer. The volume is smaller with denser rocks, and Figure 16.2 is for a granite, one of the most radioactive common rocks. The spectrometer does not count all rays produced even within this volume, for some do not travel towards the spectrometer, some are absorbed before reaching it, and some pass through the crystal. However, this does not matter if the instrument is used to compare how the counting rates of the K, U, and Th vary between stations. Alternatively, the instrument can be calibrated to convert counts into concentrations of the elements by taking readings over an area containing known concentrations of these elements.

The count rate also depends upon the solid angle of rock about the spectrometer. If the spectrometer were placed on an edge of a step or cliff (Fig. 16.3a), the count would be reduced to about half because of the absence of radiation from the 'missing' rock, while on a corner it would be reduced to a quarter; conversely, at the base of a cliff (Fig. 16.3b) it would give a reading 50% larger, and double inside a cave; these geometrical effects do not affect the ratios of elements. A soil covering usually has less radioactivity than the rock below and so reduces the count, though the contact between rocks of widely differing radioactivities –

such as shale and sandstone – can sometimes be mapped if the covering is only a few centimetres thick. Weathered rock may also give erroneous readings because, as explained above, U and Th are determined using γ rays emitted from some of their decay products, and these can be removed by weathering. For these various reasons a reading is best taken where there is flat, fresh, exposed surfaces, rather than keeping to a regular spacing of stations. Errors can also occur if too few rays are counted at each station, for whether a particular atom decays while the spectrometer is in place is purely chance; errors from this cause equal the square root of the number of readings, so for 100 counts the error is about 10, or 10%, while for 1000 it is about 32, or 3%. The user can usually choose the duration of counting, which should be long enough for the errors to be low enough for the purpose of the survey, though for weakly radioactive rocks holding the spectrometer against a surface can become tedious. Radioactivity usually varies from exposure to exposure of a rock, particularly when there are mineralised veins or pegmatites, and anomalies are usually considered significant only when they are at least three times the average.

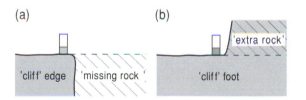

(a)   (b)

'cliff' edge   'missing rock'   'cliff' foot   'extra rock'

Figure 16.3 Counts depend on the geometry of the outcrop.

Airborne surveys. If a spectrometer is raised above the ground the source volume increases, particularly its area (Fig. 16.4), but despite the larger source volume, the counting rate needs to be increased because the moving spectrometer spends little time over any particular part of the surface; this is done by using much larger (and more expensive) detector crystals than those employed in ground-based surveys, to intercept more γ rays. Even so, surveys will be limited to where the rocks are sufficiently radioactive to give significant readings, such as uranium ores (and provided they are close to the surface). As the spatial resolution is poor because of the large source area, a detailed ground survey is often carried out if a significant anomaly is found. Airborne radiometric surveys are often combined with e-m and magnetic surveys, as the total cost is increased only slightly by adding additional instruments.

Figure 16.5 shows the results of an airborne survey of the Ranger orebodies in northern Australia. It was flown with 80 m of terrain clearance (i.e., above the ground surface) along E–W lines 180 m apart, using a 50-litre crystal. Anomalies 1 and 3 were proved to be orebodies with 50,000 and 30,000 tonnes respectively of uranium oxide; anomalies 2 and 4 appear to be due only to secondary mineralisation, while anomaly 5 was not investigated further as it was in a national park.

Seabed surveys. The spectrometer can be adapted for surveys at sea or in lakes, rivers, and so on. Because water is much more absorbing than air, the instrument is towed along the seabed, to minimise absorption of γ rays by the water.

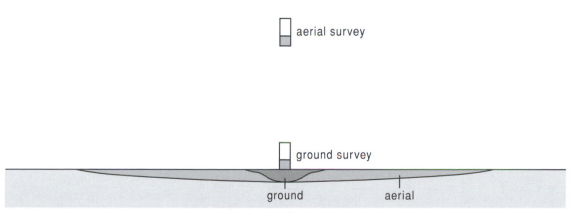

aerial survey

ground survey

ground   aerial

Figure 16.4 Source volumes for ground and airborne surveying.

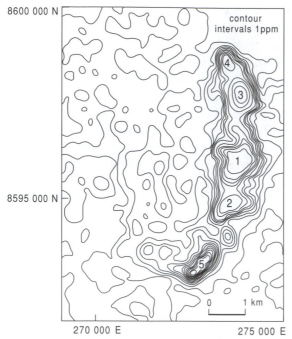

Figure 16.5 Airborne survey of uranium orebodies, Ranger, Australia.

16.2.3 Geological mapping

Uranium ores are an obvious surveying target, but with the end of the Cold War and with few nuclear power stations being built, the demand for uranium currently is not large. But γ ray surveys can also be used for mapping, and sometimes it is the ratios of the radioactive elements, rather than their amounts, that is most revealing. Table 16.1 gives values of U, Th, and K for some rocks.

Igneous rocks. Uranium and thorium are both rare in the Earth as a whole but tend to be concentrated into magmas, particularly those with higher silica content, so that, for example, granite usually has more of these elements than does basalt (Table 16.1). Within acid rocks, the U and Th – unlike K – are concentrated into accessory minerals such as apatite, monazite, sphene, and zircon (which is one reason why these minerals are used for U–Pb and fission-track dating, Sections 15.2 and 15.8). Uranium and thorium tend to be equally concentrated into magmas, for they are chemically similar in the reducing conditions of melts, so many rocks have a similar U/Th ratio, about 0.25. However, in oxidising conditions U, but not Th, forms compounds that are soluble in water (exploited in the ^{230}Th/^{232}Th method, Section 15.12.1), and this can remove U from a rock; therefore, measurement of the U/Th ratio can indicate whether rocks have been affected by hydrothermal alteration.

Two Caledonian granites from southwest Scotland exemplify these behaviours of U and Th. The Loch Doon Granite, dated as 408 ± 2 Ma by the Rb–Sr method, is zoned in composition (Fig. 16.6), with the silica content increasing inwards, from about 50% (diorite), to 72% (granite). This zoning is thought to be due to primary differentiation of the magma, as there is little evidence of alteration that could indicate secondary concentrating processes. This is supported by the results of a γ ray spectrometer survey, which showed that though the concentrations of U, Th, and K all increase greatly towards the centre of the intrusion – consistent with their tendency to concentrate into more acidic magmas –

Table 16.1 Concentrations and ratios of U, Th, and K in some rocks and minerals

Rock or mineral	Concentration of element, ppm (parts per million)			Ratio of U/Th
	Uranium	Thorium	Potassium	
Sediments				
limestone	2	2	3,000	1.6
sandstone	2	11	27,000	0.35
shale	4	12	25,000	0.27
Igneous and metamorphic				
andesite	2	6	25,000	0.3
basalt	1	3	10,000	0.28
gabbro	0.05	0.15	800	0.33
granite	4	25	40,000	0.25
pegmatite	10–100			
schist	3	11	27,000	0.35
ultramafic	0.001	0.004	30	0.26

The data, which are derived from various sources, are intended only as a guide to values.

the U/Th ratio has a much smaller range, and with the values expected for a magma.

A different result is found in the nearby Cairnsmore of Fleet Granite (dated by the U–Pb method as 390 ± 6 Ma); this too is zoned (Fig. 16.7) but with a much narrower silica range, from about 71% at the rim to 75% in the centre. A further difference from the Loch Doon Granite is the extensive mineralisation, mainly in the aureole, thought to be due to hydrothermal fluids associated with crystallisation of the magma. The distribution of Th reflects the zoning, but U (and K) are more irregularly distributed, and the U/Th ratio has a much greater range than is found in the Loch Doon Granite, probably due to remobilisation of U and K by the hydrothermal solutions (high concentrations of U are found in the mineralised aureole, not shown in Fig. 16.7). The γ ray surveying provided a quick way of mapping zoning and hydrothermal mobilisation of elements.

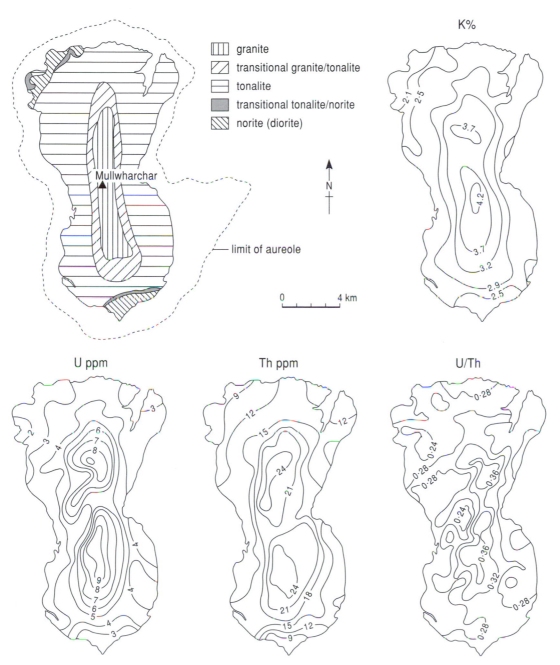

Figure 16.6 Radiometric survey of the Loch Doon Granite, Scotland.

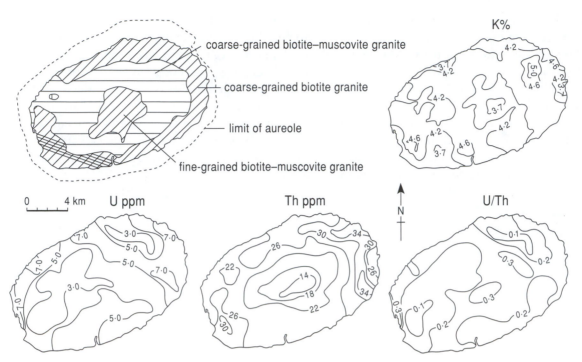

Figure 16.7 Radiometric survey of the Cairnsmore of Fleet Granite, Scotland.

Sedimentary rocks. Clastic sediments often have the U/Th ratio of igneous rocks, around 0.25, but when chemical precipitation has occurred other ratios are often found. Uranium carried in solution in oxygenated groundwaters is precipitated if it meets reducing conditions, such as rocks containing sulphides, organic (carbon-rich) material, or hydrocarbons; most of the commercial deposits of uranium in the United States are in sandstones and were formed in this way. Shales, which are deposited in reducing conditions and may contain carbon, often have a relatively high concentration of uranium. As was pointed out in Section 15.12.1, thorium precipitates on the insoluble particles of sediments, giving a low U/Th ratio, while uranium is precipitated with sedimentary carbonates, resulting in a high U/Th ratio of about 1.3. The variation of radioactivity with lithology is exploited in wireline logging, particularly in hydrocarbon prospecting (Section 18.5.4).

16.3 Radon monitoring

Radon is a radioactive gas, and as it is chemically inert (like argon and helium), it is not immobilised by forming chemical compounds, so it leaks through pores and cracks in the ground or is carried by water.

Radon is formed in each of the three uranium and thorium decay series (Section 15.12.1), so there are three isotopes of radon (the one deriving from thorium is often called thoron). However, only ^{222}Rn (which derives from ^{238}U, the more abundant of the two natural uranium isotopes) has a long enough half-life – 3.82 days – to allow it to travel a significant distance from its source, up to several hundred meters, before it decays. Radon within the soil can be measured in a variety of ways. One of the most reliable is to bury a thin-walled container of paper or plastic in the ground; radon diffuses into it and thence is sucked through a tube to a counter at the surface where some atoms decay. Decays are usually detected by the α-particles they emit, often by a simple scintillometer, which detects the flashes of light produced when α-particles hit certain materials, such as zinc sulphide. Such a scintillometer is smaller and cheaper than a γ ray spectrometer, for only a thin layer is needed to absorb an α-particle; the flashes are relatively large and do not have to be discriminated by their size. Radon in air can be measured in much the same way, but to monitor buildings special photographic films are often left for months to record the α-particles resulting from radon decay. Radon concentration is measured in terms of its activity, given as becquerels per cubic metre (Bq/m^3) – a becquerel is one decay a second – or as curies per litre (Ci/l), which is a unit 3.7×10^{13} times larger.

Radon dissolved in water usually moves by being carried along in a flow, for it can diffuse only slowly in a liquid. Its concentration is measured by extracting the radon from a sample of the water.

The concentration of radon broadly reflects the concentration of uranium in the underlying rocks, so it can be used as a prospecting tool. Locally, however, the highest concentrations will be where radon can travel most easily from its source to the surface, and so tend to reflect the presence of fractures or groundwater flow, and radon can be used to detect these. Its concentration should be measured at a depth of about a metre, for rain tends to seal the surface, resulting in a buildup of radon there, while wind tends to remove it. Because of the number of factors involved, radon monitoring as a mapping or prospecting tool has limited use; however, it is increasingly measured because it is a possible environmental cause of cancer; a concentration in air of 100 Bq/m^3 is regarded as a significant risk. A first step to reducing the risk is to map areas of potentially high radon production to identify where its concentration is dangerously high. In England, they are mainly associated with the granites of the southwest and the mineralisation of Derbyshire and Northamptonshire, but occur locally elsewhere. Another application is to detect hydrocarbon pollution, such as leakage of petroleum products, for uranium in groundwater is deposited when it meets reducing conditions, as explained above. In turn, this will produce radon, which is carried away by groundwater and can be detected. Radon concentration has sometimes changed before an earthquake, presumably due to changes in groundwater flow, but this is only one of many possible earthquake precursors (Section 5.10.2).

Summary

1. Radioactive decay is accompanied by α- and β-particles and γ rays; only the last penetrates far enough through rock and air to be detectable by surface or airborne surveys.

2. γ rays are usually detected using a γ ray spectrometer, which counts the number of decays due to each of K, U, and Th (for the last two, the γ rays counted are actually due to decay products).

3. For surface surveys the γ ray spectrometer is normally placed on a flat surface of fresh rock about 2 m in diameter (correction can be made for other surface geometries). Counting should continue long enough to reduce errors to an acceptable level. Airborne surveys are carried out at a height of 50–100 m (often in conjunction with magnetic and e-m surveys). They can survey much more rapidly than ground surveys but are less sensitive and have poorer resolution of how activity is distributed.

4. Count rates can be converted into concentrations, but often relative concentrations are sufficient; a significant anomaly has a counting rate three or more times the local average. The U/Th ratio may be important.

5. In igneous rocks, the highest concentrations of radioactive elements are found in the most differentiated ones (highest silica content) and particularly in pegmatites. Uranium and thorium mostly reside in accessory minerals such as zircon and monazite. The U/Th ratio is usually about 0.25. Uranium but not thorium is relatively easily leached by oxygenated waters; thus leaching (e.g., by hydrothermal alteration) is revealed by a lower U/Th ratio. Uranium is deposited from solution in reducing conditions and hence high concentrations are found in rocks such as shales, and also in sedimentary carbonates.

6. Radon in soil or groundwater is measured by extraction and counted by an α-particle detector; its concentration in air is measured either by passing the air through a scintillation counter or by long exposure of special photographic film. Its concentration depends upon the uranium content of the underlying rocks, the presence of fissures, and groundwater flow; near-surface concentrations are also affected by the weather.

7. You should understand these terms: γ ray; γ-ray spectrometer, scintillometer, radon.

Further reading

Durrance (1986) covers the physics of radioactivity and its detection, the geochemistry of uranium and thorium, and all geological aspects of radioactivity. Shorter accounts are given in Parasnis (1997) and

Telford et al. (1990), the latter going further into the physical principles involved. Milsom (1996) gives a short account with emphasis on field procedures. Gates and Gunderson (1992) describe a number of radon investigations in the United States.

Problems

1. Measurement of naturally occurring radioactivity is confined to the elements uranium, thorium, and potassium. This is because of which of the following?

    (i) Their compounds are insoluble and so do not move easily.

    (ii) Their half-lives are much longer than the age of the Earth.

    (iii) Their half-lives are comparable with the age of the Earth.

    (iv) They are the most abundant naturally occuring radioactive elements.

    (e) They are volatile.

2. Why are ground surveys to measure concentrations of U and Th preferably made on rock exposures that are (a) flat and (b) fresh?

3. How would the count rate of a γ-ray spectrometer, placed in the following situations, compare with that on a wide flat surface of the same rock?

    (a) In a tunnel.

    (b) At the foot of a cliff.

    (c) Halfway up a cliff.

    (d) On a corner of a large rectangular block.

4. What determines the counting time to be used in a surface γ ray survey?

5. Summarise the chemical properties of U and Th in oxidising and reducing conditions, and explain their significance when surveying igneous and sedimentary rocks.

6. The U/Th ratio over an exposure of granite varies from 0.08 to 0.35. What might be the reason for the variation?

7. In a micaceous sandstones/shale sequence, the total γ-ray count is higher in the sandstone than in the shale. A γ-ray spectrometer shows the percentage of thorium to be much higher in the shale than in the sandstone. What do these observations mean?

8. What are the advantages and disadvantages of airborne γ-ray surveying compared to ground surveying?

9. Why is radon more of a health hazard than most elements formed in decay series?

Geothermics

chapter 17

Geothermics: Heat and Temperature in the Earth

Temperature generally increases with depth in the Earth, and this is responsible for the maturation of hydrocarbons, many forms of mineralisation, thermal metamorphism and, of course, volcanism and other igneous activity. At sufficient depth, rocks are too hot to fracture and deform ductilely, while at even greater depths they are hot enough to flow like a very viscous liquid. It is this ability to flow that allows the Earth to be an active planet, with mountain building, earthquakes, and plate tectonics; without its hot interior the Earth would be a dead planet, like the Moon.

To understand these processes we need to know the temperatures at different depths within the Earth. Unfortunately, there is no direct way of measuring temperature below the few kilometres that boreholes reach, and temperatures at greater depths are mainly inferred from measurements of the temperatures and thermal properties of rocks near the surface.

17.1. Basic ideas in geothermics

17.1.1 Introduction

Temperature increases with depth in the Earth; this is why, for instance, deep mines are hot and deeply buried rocks are thermally metamorphosed. The increase is partly because heat is coming out of the hot interior, just as heat leaks out of a hot oven, and partly because heat is being produced within the rocks by radioactive heating. This chapter is mainly about how temperature varies with depth in different geological situations, but as we cannot measure temperature directly (except at shallow depths) we are forced to deduce its value indirectly, using a combination of what we can measure and an understanding of how heat is produced and transported.

One important measurement is heat arriving at the surface, called the **heat flow,** though, in practice, to measure this requires measuring temperatures below the surface. This is an example of how, in geothermics, heat and temperature are intimately related, and so it is important to be clear about these and related terms.

17.1.2 Temperature and heat

We are familiar with **temperature:** A summer's day at a temperature of 25°C is hotter than a winter's day at 5°C; boiling water at a temperature of 100°C is much hotter than freezing water at 0°C; magma is liquid because it is far hotter than the rock it will become when it cools and solidifies.

Objects become hotter if heat is added to them. **Heat** is a form of energy: To heat 2 litres of water from, say, 10 to 20°C requires twice as much energy as heating 1 litre and so consumes twice as much electricity in a kettle or gas on a stove. Heating one litre from 10 to 30°C takes more heat – approximately twice – than heating it from 10 to 20°C. Thus the amount of heat depends on the increase in temperature and the amount of material heated, but it also depends on the type of material. Heat is measured, like other forms of energy, in joules; for instance, heating a litre of water from freezing to boiling needs 420,000 joules. (An older unit, the calorie, is sometimes seen, particularly on the sides of food packets; it equals 4.18 joules.)

Heat always tends to move from higher- to lower-temperature places, which is why coffee cools and ice cream warms up; insulation slows but does not stop the movement of heat. The tendency of heat to move from higher to lower temperatures is similar to the tendency of water to move from higher to lower pressures or electric current from higher to lower potential (voltage). So if we find a temperature difference between two places in the Earth we know heat is moving between them, and if heat is moving there is a temperature difference (just as, in resistivity, a potential difference between two places causes a current to flow while, if a current is flowing, there must be a p.d.), and this offers a way of measuring the heat transported (Section 17.2.1).

Temperature increases with depth, as mentioned above, and the rate of increase, the **temperature gradient,** measured in boreholes on land is typically 25 to 30°C per kilometre (°C/km). However, this value cannot persist to great depths, for if it did, the temperature 100 km down would be over 2500°C, well above the melting point of all known rocks, whereas we know from seismology that the Earth at that depth is solid (Section 4.5). Therefore, the temperature gradient must decrease with depth, partly because heat travels more easily at depth, so we shall next examine the ways that heat travels.

17.1.3 How heat travels: Conduction and convection

In **conduction,** heat moves, or flows, *through* a material from a hotter to a cooler part (Fig. 17.1a). If one side of a piece of material is heated its temperature rises, and this in turn heats the adjacent material, and so on until heat arrives at the other side. This is how the outside of a cup becomes hot soon after coffee has been poured into it.

In **convection,** heat is transported by *movement* of hot material; an example is circulating hot water carrying heat around a central heating system. Convection often occurs spontaneously when a fluid is heated at its base, for a fluid usually expands as it heats up and the resulting reduced density tends to cause it to rise (Fig. 17.1b); at the cooler top surface the fluid loses heat, contracts, and sinks.

This natural **thermal convection** is due to the combination of hotter base and cooler top, and the heat, as with conduction, moves from hotter to cooler places. This contrasts to forced convection, as in a pumped central heating system (sometimes 'advection' is used for the forced convection, and 'convection' for thermal convection). In the Earth, heat may be transported locally by forced convection where groundwater flows downhill, but much more important is natural convection of groundwater in the vicinity of hot rocks and – as we shall see in the next section – flow of the deep mantle. In summary, conducted heat travels *through* a material; convected heat is *carried by* moving material.

Heat can also travel by thermal radiation (Fig. 17.1c), which is related to light, radio waves, and other types of electromagnetic radiation (mentioned in Section 14.4.1). Such radiation can travel through space, which is how the Sun warms us; it can also travel through a few materials (just as light can travel through glass though not most solids), but this is not important inside the Earth.

Convection is only possible in a fluid, whereas conduction occurs in both fluids and solids. However, in fluids that are poor conductors heat often travels so much more rapidly by convection that conduction may be neglected. When water is heated in a saucepan, the heat is conducted up through the solid metal base – a good conductor – but the water is a poor conductor and so thermal convection usually develops naturally within it and transports heat from the hot base to the cool top, circulating the water. The effectiveness of convection is shown by soups or porridge that are too thick to convect easily: Without stirring, they tend to burn at the base, or sputter, because the heat is not transported away, and in a plate they cool slowly because convection does not operate there either. In fact, it was through the slowness of thick liquids to cool that convection was first recognised.

17.1.4 Convection and conduction within the Earth

We know that nearly all of the Earth's crust and mantle is solid, for seismic S-waves can travel through them (Section 4.5), so it might seem that convection could not take place except locally where there is groundwater or magma. However, it was

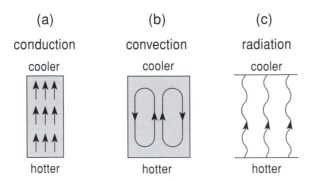

Figure 17.1 Ways that heat travels.

explained in Section 9.2 that rocks can flow slowly by solid-state creep at temperatures that, though high, are below the melting point, and as a result the Earth can be divided into a cool, fairly rigid layer – the lithosphere – over a hot asthenosphere that 'flows' by creep. Though the asthenosphere is extremely viscous, heat is transported within it mainly by convection, but the lithosphere is too rigid to convect, so heat travels through it by conduction. Thus heat originating in the deep interior rises up by convection until it reaches the lithosphere, where it continues to the surface by conduction (Fig. 17.2a).

When a volume of the mantle rises in convection, negligible heat conducts in or out of it, for it is a poor conductor and its mass is huge, as it will be at least tens of kilometres across. However, though it does not lose heat, it cools slightly because, as it rises, the pressure on it – due to the weight of overlying layers – decreases, and so it expands slightly; in turn, this leads to cooling, just as a spray-can cools when its contents expand in spraying.

(a) heat transport

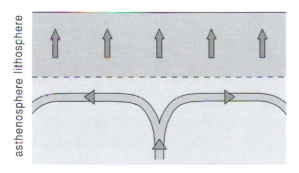

(b) temperature versus depth

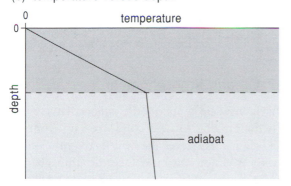

Figure 17.2 Heat transport within the Earth (schematic).

The resulting temperature gradient is called the adiabatic temperature gradient, or **adiabat** (Fig. 17.2b), and its gradient is only about ½°C/km, much less than in the lithosphere above. (The adiabatic cooling of air when it rises is much greater because air is much more compressible, and this accounts for the decrease in temperature up mountains.) We need to take account of this if we wish to know, for instance, whether ascending mantle will produce magma when it nears the surface; it is not sufficient to know whether its temperature is greater than the temperature of partial melting near the surface, because of adiabatic cooling as it ascends. One way to allow for this is to calculate the **potential temperature,** the temperature it would have at the surface after allowing for adiabatic cooling. Then the potential temperature can be compared with the temperature of melting. However, the adiabat is so small compared to the gradient in the lithosphere that sometimes we may treat it as negligible.

Thus the division between lithosphere and asthenosphere is important in geothermics as well as in isostasy (the lithosphere is defined in various ways, depending on the geophysical property of interest, and its thickness depends on the definition, but in general, the lithosphere is cooler and less yielding than the mantle below). Within the lithosphere there is a transition from brittle to more ductile behaviour, though not to flowing, at about 450°C (depending on the rock type). This is important in structural geology because faults will tend to detach at the transition.

Convection occurs locally in parts of the lithosphere. This can be by magma rising, because it is hot and less dense, in diapirs and plumes (Section 20.9.2), or forced convection by the movement of tectonic plates (Section 20.9.4). Thermal convection occurs in groundwater when it is heated by a hot intrusion, resulting in hydrothermal circulation, which can produce mineralisation (Section 23.2). Such circulation is particularly important near oceanic spreading ridges, where ascending magma rises close to the surface (Section 20.8). In addition, forced convection occurs when groundwater flows downslope, but this is significant only locally. The temperature profile through the Earth (e.g., Fig. 17.2b) is called the **geotherm,** and it varies with location or time, as we shall see later.

17.2 Heat flow and temperature

17.2.1 Measurement of heat flux

The *rate,* or power, at which heat moves up towards the Earth's surface is the heat flow, Q, and is measured in watts (joules/sec). The flow up through a single square metre is the **heat flux, q,** and is measured in watts/m² (it is also commonly called 'heat flow', which can be confusing, or sometimes 'heat flow density'). There is a heat flux at all depths, but unless stated otherwise 'heat flux' refers to the value at or close to the surface, where it is measured. Its measurement is used to help make deductions about temperatures and heat production at depth.

Heat flux could be measured, in theory, by putting a tank of water on the ground and measuring how fast it heats up, but this is impracticable because it would takes months for the temperature to rise even a few degrees and meanwhile most of the heat would escape. Instead, heat flux is calculated indirectly.

The amount of heat, Q, conducting up through a block of rock (Fig. 17.3) is directly proportional to the temperature difference between its base and top, and to the area of cross-section, but inversely to its length (so doubling the length halves Q). This is summed up by the heat conduction equation:

Q also depends on the **thermal conductivity, K,** of the rock, which depends on the ease with which heat conducts through it (K has units of watts/metre-degree, W/m-°C).

Conductive heat flow is similar to electrical current flow – compare Figures 12.3 and 17.3, with thermal conductivity corresponding to electrical conductivity. As the heat flux, q, is the amount of heat flowing up through a block with an area of 1 m²,

$$q = \frac{\Delta T}{L} \times K$$

$$\begin{array}{ccc} \text{heat} \\ \text{flux} \end{array} = \begin{array}{c} \text{temperature} \\ \text{gradient} \end{array} \times \begin{array}{c} \text{thermal} \\ \text{conductivity} \end{array} \qquad \text{Eq. 17.2}$$

Therefore, to find the heat flux, the temperature difference, ΔT, between two depths L metres apart is measured using sensitive thermometers, and the conductivity of the rock is measured separately; more detail is given in Box 17.1. Because heat flux is *calculated* using heat conduction equations, it only includes *conducted* heat and ignores any heat *convected* to the surface, which can sometimes be significant (Section 17.4.1).

As the values of heat flux are small, they are given in mW/m², and are mostly in the range 40 to 200 mW/m². Heat flux varies in value between places mainly because of differing thickness of the lithosphere and differing concentrations of radioactive elements in it, and these are discussed in the following

$$Q = A \times \frac{\Delta T}{L} \times K$$

$$\begin{array}{c} \text{heat} \\ \text{conducted} \end{array} = \begin{array}{c} \text{area of} \\ \text{cross-section} \end{array} \times \frac{\text{temperature difference}}{\text{length}} \times \begin{array}{c} \text{thermal} \\ \text{conductivity} \end{array} \qquad \text{Eq. 17.1}$$

sections. Heat flux is also affected by orogenies, intrusions, erosion, deposition, and deformation, some of which are discussed in Section 17.3.

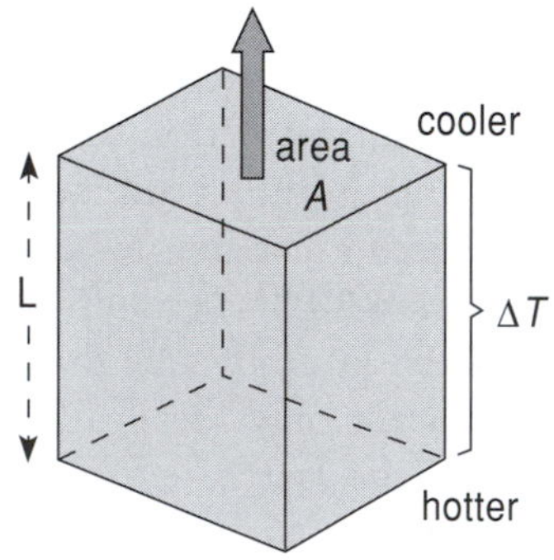

17.2.2 Oceanic lithosphere

Oceanic lithosphere contains little radioactivity, so heat flow is mainly due to heat conducted up from the mantle below it. However, the heat flux varies because its thickness is not constant. Oceanic lithosphere is composed mainly of ultrabasic mantle rocks below a few kilometres of basaltic crust (Section 20.8), both rock types that contain little radioactivity (Table 16.1).

Figure 17.3 Heat conduction through a block.

BOX 17.1 How heat flux is measured on land and in the oceans

As explained in Section 17.2.1, heat flux is calculated by multiplying the values of the vertical temperature gradient and thermal conductivity.

Measuring heat flux on land. This usually needs a borehole for measuring the vertical temperature gradient. Drilling the hole disturbs the temperature gradient, due mainly to the fluid (called drilling mud) pumped down to cool the bit and to remove chippings (Section 18.2), so either the hole has to be left to return to its original temperature or allowance is made for the disturbance by measuring how fast it changes following cessation of drilling; this disturbance is least at the bottom of the hole, where mud has circulated for the least period.

A further possible problem is groundwater flowing up or down the hole, which disturbs the gradient, for such forced convection is an effective transporter of heat; to prevent this, the hole should be cased. In addition, the temperature gradient near the surface is affected by past changes of surface temperature (see Section 17.5), so to get a true measure of the heat flux it should be measured at a depth of several hundred metres. A cased borehole of this depth which can be left to reach equilibrium, and from which some cores have been taken, is expensive, and so comparatively few boreholes have been drilled primarily for heat flux measurements, though less-reliable measurements are often made in holes drilled for other purposes, such as for oil or water. The temperature is measured using electrical thermometers that can measure accurately to about 1/100°C. It is best measured at intervals of a few metres over a wide range of depths, to provide a more accurate measurement of the temperature gradient and to help identify if any of the above disturbances have been occurring.

Thermal conductivity can be measured on a slice of core (Fig. 1) from the part of the hole where the temperature gradient is measured. The sample is sandwiched between two metal bars, and one end of the apparatus is heated while the other is cooled by piped water. Thermometers measure the temperature gradient, shown to the side. The same amount of heat flows through the sample as in the bars (the apparatus is surrounded by insulation), and Eq. 17.1 shows that the temperature gradients in rock and bar are in inverse proportion to their conductivities, thus allowing the conductivity of the rock to be compared with that of the bar, which is

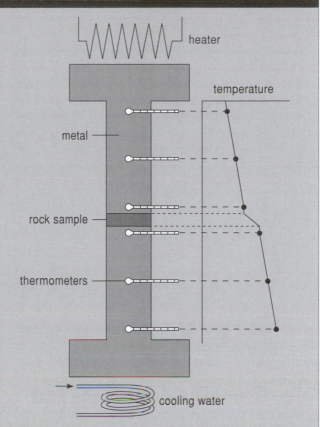

Figure 1 Measurement of thermal conductivity in the laboratory.

known. Alternatively – following the measurement of temperature in the borehole – a long heater within a cylindrical probe is switched on and the rate at which its temperature rises is measured. The more conductive the surrounding rock, the more rapidly heat is conducted away from the heater, and so the slower the rise of its temperature, allowing the value of the thermal conductivity to be calculated.

Measuring heat flux through the ocean floors. Surprisingly, this is easier than on land, because most of the ocean floor is covered with many metres of soft sediment and a special probe can be forced into it by a heavy weight (Fig. 2) without the need to drill a hole. The probe is only a few metres long so temperatures have to be made very precisely, to about 1/1000°C. The frictional heating due to forcing the probe into the sediments is then significant, so the electronic thermometers are mounted away from the main tube.

BOX 17.1 How heat flux is measured on land and in the oceans (continued)

Figure 2 shows one design, with several thermometers in a much thinner tube a few centimetres to the side of the main tube. Temperatures are measured before frictional heating resulting from pushing the probe into the sediments can conduct out to them; several thermometers are used to give a more accurate value of the temperature gradient.

Thermal conductivity can be measured on a core taken from within the probe (removed after the probe has been pulled up to the surface), or the probe can contain a heater, as described above for land measurements; the latter saves having to pull the probe to the surface to remove the sample.

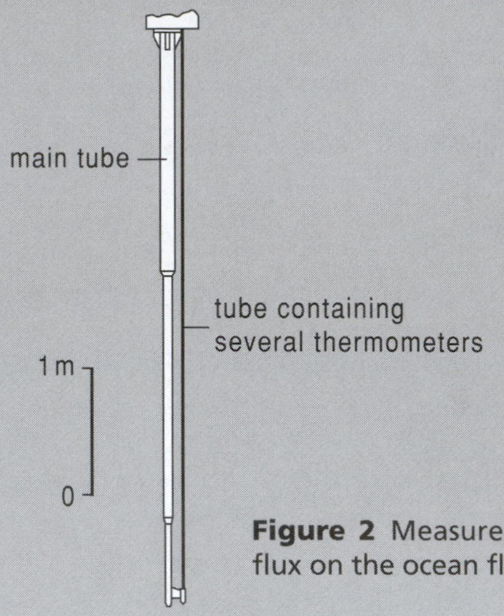

main tube

tube containing
several thermometers

1 m

0

Figure 2 Measurement of heat flux on the ocean floor.

Therefore, the heat flux down any column through the lithosphere is almost constant, so the temperature gradient is nearly uniform, assuming the thermal conductivity also is uniform, which is approximately true. However, the heights of columns below different parts of the ocean differ, because oceanic lithosphere is formed at an oceanic spreading ridge and thickens as it moves away, as will be explained further in Section 20.2.2. This is shown schematically by (a) and (b) of the upper part of Figure 17.4.

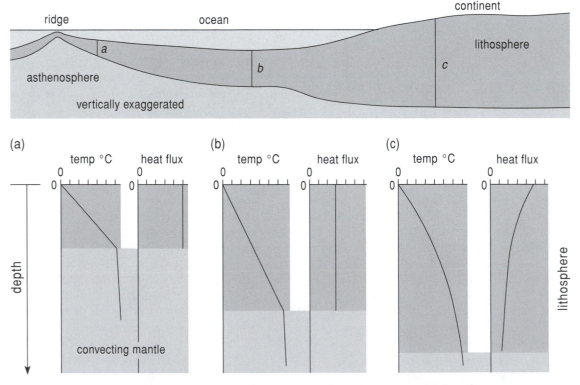

Figure 17.4 Temperatures and heat fluxes through oceanic and continental lithosphere.

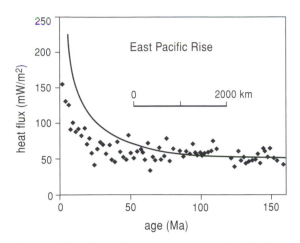

Figure 17.5 Heat flux near the East Pacific Rise (distance scale is approximate).

The base of oceanic lithosphere is where the temperature is high enough for convection to be significant (Section 17.1.4), and is roughly the same everywhere, about 1300°C. As the ocean bottom temperature is close to 0°C, the temperature difference between base and surface of oceanic lithosphere is approximately constant, so both the temperature gradient and the heat flux decrease as the thickness increases (Fig. 17.4a and b). Figure 17.5 shows heat flux results for ocean floor of different age in the Pacific, which is proportional to the distance from the East Pacific Rise, the spreading ridge in the Pacific Ocean (Section 20.6). The dots show the values, which, on average, decrease more and more slowly with age, because the lithosphere thickens more slowly (Section 20.2.2). The curve is the estimated total heat flux when heat transported by hydrothermal convection – greatest near the ridge – is included.

17.2.3 Continental lithosphere and radioactivity

Continental lithosphere differs from oceanic lithosphere in containing a considerable amount of the radioactive elements uranium, thorium, and potassium. These generate heat, and the amount of heat produced each second within a cubic metre of rock is called the **heat productivity.** The heat productivity of surface rocks can be found by measuring the concentrations of these elements but we also need to know how it varies with depth. We know it must decrease, for if all crustal rocks contained the same concentration as surface ones the total heat produced would exceed the entire heat flow out of the Earth. We can directly measure heat flux at depth only where there are suitable bore holes, and even these extend only a few kilometres. However, we can deduce its variation below heat flow provinces, which are plutons where the heat flux over the surface is simply proportional to the heat productivity of the *surface* rock, as shown in Figure 17.6.

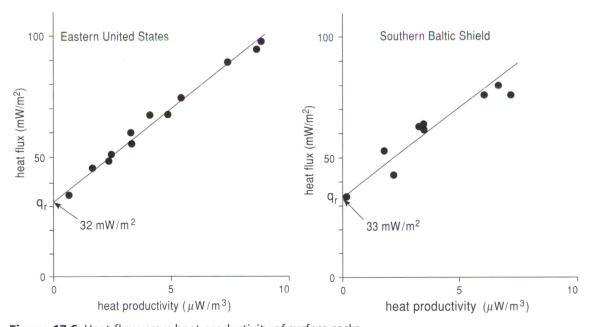

Figure 17.6 Heat flux versus heat productivity of surface rocks.

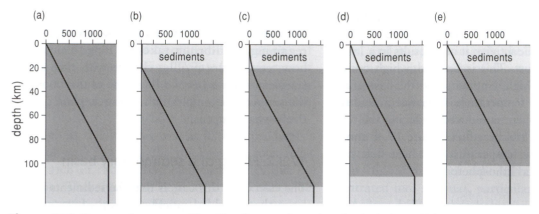

Figure 17.9 Temperature reequilibration for a sedimentary basin (schematic).

17.3.3 Overthrusting and underthrusting

The thrusting of a layer of rocks over the surface brings hot rocks from depth into contact with the cold original surface. We shall assume that the rocks are continental, with a curved geotherm due to concentration of radioactivity in the crust. Successive geotherms are shown overlapped in Figure 17.10, with the equilibrium geotherm shown dashed for comparison. Immediately after overthrusting, the temperature profile has a sawtooth shape, which produces a large inverse temperature gradient; as a result, heat at first flows *down* into the underthrust rocks, raising their temperature, but this inverse gradient has disappeared after a few million years.

The longer-term evolution depends on two factors: the extra heat production due to there being two layers enriched in radioactive elements, and ero-

sion, which slowly removes the upper layer and so moves hot rock upwards (forced convection); after 60 Ma these two factors have caused the geotherm to exceed the equilibrium value, which is not attained for well over a hundred million years, depending upon the erosion rate. The overthrusting rate need only exceed a few centimetres a year for the thermal consequences to be essentially the same as for an instantaneous overthrusting.

17.3.4 Crustal thickening and orogenies

Orogenies are the result of lateral compression of continental lithosphere (Section 20.4.3). This causes complex overthrusting, which can continue for tens of millions of years, as is happening in the Alps and Himalayas, and the thickness of the crust may be doubled. The resulting great thicknesses of rocks with relatively high concentrations of radioactivity

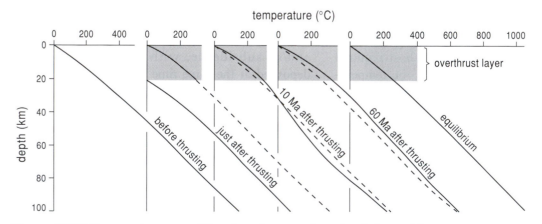

Figure 17.10 Temperature reequilibration after overthrusting (schematic).

produce high grades of metamorphism, while subsequent cooling – by a combination of erosion and conduction of heat – can take hundreds of millions of years, as is demonstrated using radiometric dating methods with different closure temperatures (e.g., Section 15.9.3). For these and other reasons that are not fully understood, continental heat flow tends – like oceanic heat flow – to decrease with time, though much more slowly (Fig. 17.11).

17.4 Global heat flow and geothermal energy

17.4.1 Global heat flow

The total amount of heat arriving at the Earth's surface is estimated by adding the heat flux values over its surface. There are many areas with few measurements, but values can be estimated from our understanding of how heat flux varies, particularly how it decreases with age (Figs. 17.5 and 17.11); an addi-

tion is made for hydrothermal convection near oceanic ridges, estimated to be about a third of the oceanic heat flow. The total is about 4.2×10^{13} watts. This is 50 to 100 times larger than the power required to cause earthquakes and raise mountains, both of which ultimately derive their energy from the Earth's internal heat through the movements of plates, which in turn are moved by thermal convection (Section 20.9).

Though the global heat flow is large, it is small compared to the heat the Earth receives from the Sun, which is roughly 10,000 times larger; so the temperature at, and above, the *surface* depends hardly at all on internal heat flow. Solar heat is responsible for most erosion, caused by the downhill movement of water (including ice) raised by evaporation of surface water, or by particles moved by winds, which are also powered by the Sun. Thus rocks are raised by the Earth's internal heat but worn down by the Sun's heat.

17.4.2 Sources of the Earth's heat

How was the heat flowing out of the Earth's interior produced, and where within the Earth did it originate? We saw in Section 17.2.3 that radioactivity in the continental lithosphere, particularly the crust, generates a considerable amount of heat, whereas nearly all oceanic heat flow comes from beneath the lithosphere. Heat productivities due to radioactive decay can be estimated for all parts of the Earth, based on likely compositions, though these estimates are less accurate the deeper we go (Table 17.1). The continental crust generates a considerable amount of heat, despite its small volume, because of its relatively high concentrations of radioactive elements, in rocks such as granite, but though the mantle has small concentrations of these elements, it generates the most heat because of its huge volume.

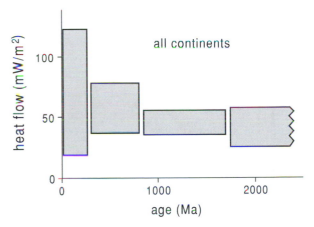

Figure 17.11 Decrease of continental heat flux with age.

Table 17.1 Radioactive heat generation

Part of Earth	% of Earth's volume	Rock types	Thermal productivity ($\mu W/m^3$)	% of Earth's total heat outflow
continental crust	0.7	granites, etc.	0.55	c. 10%
oceanic crust	0.2	gabbro	0.03	c. 0.15%
mantle	84	peridotite	0.014	c. 30%
core	16	mainly iron	negligible	negligible
Total				c. 40%

Based on Table 8.1 of Brown and Mussett (1993).

The total heat production is probably less than half that arriving at the surface; as no other significant sources are known, the conclusion is therefore that the Earth is cooling. It may seem surprising that the Earth can still be cooling after 4550 Ma, but it is so large and the lithosphere is such a good insulator – being very thick – that the interior temperature is estimated to be cooling at only about 0.2°C in a million years. This rate of cooling is so slow that lithospheric temperatures and heat flow remain close to equilibrium, though it probably powers the geodynamo and also means that tectonic processes were probably more vigorous in the remote past.

17.4.3 Geothermal energy

The total global heat flow of about 4.2×10^{13} watts (Section 17.4.1) is large – greater than the power of 10,000 large power stations – but is only a few times the total power consumed in the burning of oil, coal, and gas, plus production by nuclear and hydroelectric power stations. Though this suggests that geothermal power could supply much of our energy needs, the greater part is released beneath the oceans. A bigger problem is that heat flow is very difficult to utilise because it is mostly spread so thinly over the Earth's surface. Average continental heat flow is about 50 mW/m², so a saucepan resting on the ground would heat up about a million times more slowly than one on a cooker. However, in a few regions of the world geothermal energy can make a useful contribution; for example, the Philippines produces a fifth of its electricity this way.

Natural steam. The most useful areas are where steam naturally reaches close to the surface, usually in areas of current or recent volcanism, such as parts of California, Iceland, Italy, Japan, the Philippines, New Zealand, and Kenya (Section 21.6). Because most power stations use their fuel to produce steam, which drives the machinery that produces electricity, it is relatively simple to substitute steam from the ground. Potential geothermal areas can sometimes be mapped using resistivity surveying, for high temperatures or the presence of water, often in hydrothermally altered rocks, may produce very low resistivities.

Hot water. Less useful is where there is a supply of hot water, perhaps as hot springs or hot groundwa-ter within porous rocks at depth (thermal aquifers). This is usually not hot enough to produce steam to generate electricity, but it can be used directly where lower temperatures are sufficient, such as heating houses or greenhouses; this is a more efficient use of its heat, for heating buildings by burning fuel, either directly or via electricity, wastes much of the heat in flue gases, and so on. An example is Southampton, England, where water at over 70°C is pumped from sandstones at a depth of about 1.7 km and used to heat buildings in the city centre within a radius of 2 km, providing several megawatts of heat. There is a similar scheme in the Paris Basin.

Hot dry rocks. There is far more heat in 'dry' rocks than in thermal aquifers or as steam, for they are far more common, so potentially this offers a much more widespread energy source, but the problem is to extract it. One way is to pump water down a borehole, which returns to the surface via a second hole after being heated by the intervening rock (Fig. 17.12). The water needs to reach a considerable depth, for rocks are hotter there; if possible, an area with a high temperature gradient would be chosen.

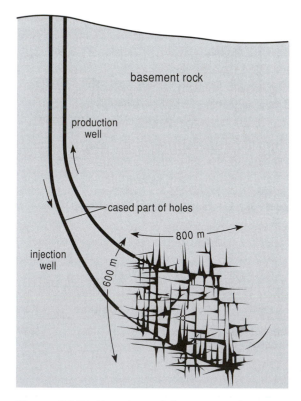

Figure 17.12 Hot, dry rock heat-extraction system.

Experiments have been carried out in various places, particularly at Rosemanowes Quarry in southwest England, where a granite rich in radioactive elements has a high – for Britain – temperature gradient of 30 to 35°C/km. Holes have been drilled to 2.6 km, where the temperature is about 100°C (a commercial system would need 200°C or more). Much effort has been invested to find ways of opening fractures in the rock so that heat can be extracted from as large a volume of rock as possible. So far, none of the experiments have succeeded in extracting heat at an economical cost, but there is some promise for areas with a much higher temperature gradient.

The total production of geothermal power worldwide is about 40,000 MW, most as electricity and nearly all from natural steam areas. This is less than ½% of the total global power production.

None of the above ways of using geothermal power is truly renewable, for heat is removed much faster than it can be replaced by radioactive decay in the rocks or heat flow from below. In most areas of natural steam, such as New Zealand and the Geysers, California, pressure has decreased in the years since steam has been extracted. The depletion has sometimes been monitored seismically, for as steam is extracted some of the remaining hot water vaporises, changing the ratio of v_P/v_S through the reservoir.

17.5 The effect of surface temperature changes: A record of past climates

Measuring temperatures down a borehole can reveal how the climate has changed in the past several hundred years, long before reliable meteorological records began about 150 years ago. If the surface temperature rises, the subsurface temperature profile also begins to rise, starting at the surface and extending progressively downwards. If the surface temperature then falls, a second adjustment propagates downwards behind the first. Suppose the temperature varies sinusoidally (a fair approximation for daily and annual variations) as shown at the top of Figure 17.13; this causes a succession of peaks and troughs of temperature to propagate downwards from the surface. Because individual peaks and troughs broaden as they propagate, they begin to overlap and so partially cancel, so that the amplitude decreases with depth. As the decrease is exponential

and so never reaches zero, the **skin depth** – the depth in which the amplitude decreases to $1/e$ or about one third – is often used to describe how rapidly the amplitude decreases with depth. This decrease has similarities with the reduction of e-m waves within conductors (Section 14.4.2), and as with e-m waves, the skin depth is proportional to the period, so longer period changes penetrate deeper. The skin depth for daily variation is only a few tens of centimetres, and for annual variation only a few metres, which is why caves have an almost constant temperature, but a glacial period, lasting roughly 100,000 years, affects down to a kilometre.

Figure 17.14 shows results of measurements in Canada. The climatic effects are superimposed on the usual geothermal gradient, which gives a straight line if the thermal conductivity does not vary with depth (Fig. 17.14a). This long-term steady temperature gradient is subtracted to give Figure 17.14b, the climatic effect alone. These figures have two curves because temperatures were measured twice, 22 years apart. They differ because about 8 years before the first measurement the area around the borehole had been deforested, raising the surface temperature by about 4°C. It was calculated that by the time the second measurement was made the increase would have propagated down to about 170 m, with a maximum difference between the two profiles at about 80 m; this is confirmed by the record and gives confidence in the measurements.

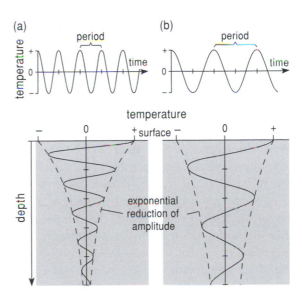

Figure 17.13 Propagation of alternating temperature changes at two frequencies.

(a) borehole temperature (°C)

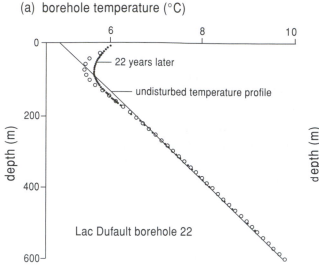

(b) effect of climate change (°C)

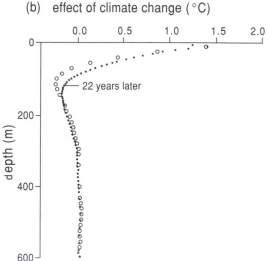

(c) variation of surface temperature

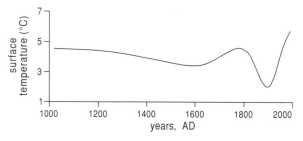

Figure 17.14 Measurement of past surface temperatures, Lac Dufault, eastern Canada.

The profiles converge towards the surface because the near-surface rocks have returned to equilibrium. Below 170 m, the curves are identical and contain information about the surface temperatures before the deforestation. This has been modelled, taking into account measured thermal conductivities of borehole rocks, to deduce the history of the surface temperature (Fig. 17.14c). It shows that the temperature has been rising since about 1900 A.D.; there was an earlier, longer minimum around 1600 A.D. known from historical records and called the Little Ice Age. Though past temperatures can be measured in other ways, such as oxygen isotope ratios in ice cores (not explained in this book), these can be affected by very short-term changes, such as a particularly cold spell one spring; the temperature profile method provides a more accurate average temperature.

Borehole measurements help extend our knowledge of climatic changes a thousand or more years back, needed to measure any global warming (probably occurring because of the known increase in atmospheric carbon dioxide resulting from the burning of fossil fuels) and to discover what natural variations have occurred.

Summary

1. The temperature within the Earth generally increases with depth, due to radioactive decay and heat retained from its formation.

2. Heat is transported within the Earth by conduction through rigid parts and convection in fluid parts. Convection may be thermal or forced, but thermal convection is more widespread. In particular, convection occurs in the asthenosphere and locally within the lithosphere in magmas, hydrothermal systems, and groundwater flow, but in the lithosphere transport is mainly by conduction.

3. Heat reaches the Earth's surface mainly by conduction, but also by hydrothermal convection, particularly near oceanic spreading ridges.

4. Heat flux (sometimes called heat flow or heat

flow density) is the amount of heat passing up through 1 m^2 in 1 sec. Its value is calculated from paired measurements of temperature gradient and thermal conductivity, and does not include convective heat transport.

5. Heat flux is mostly in the range 40 to 200 mW/m^2. In the ocean floors it decreases as the lithosphere ages, halving in a few tens of millions of years. On continents it tends to decrease with age since the last orogeny, but much slower than for oceans.

6. Radioactive heat production is significant within continental but not oceanic crust. Within continents, radioactivity is concentrated towards the surface. In heat flow provinces its concentration decreases exponentially with depth, resulting in heat flow being proportional to the thermal productivity of the surface rocks.

7. Temperature normally increases with depth. In the oceanic lithosphere the temperature gradient is almost constant with depth, its value decreasing with the age of the lithosphere. In continents it increases towards the surface, and 25 to 30°C/km is a typical near-surface continental temperature gradient. Below the lithosphere, in the convecting asthenosphere the gradient – the adiabat – is only about ½°C/km.

8. Heat flux and temperatures are affected by deposition and erosion, overthrusting/underthrusting, lithospheric stretching, and igneous activity. It can take many millions of years before temperatures return close to equilibrium values. Past temperature gradients were important for causing thermal metamorphism and the maturation of oil and gas.

9. The near-surface temperature gradient is affected by climatic changes and so can be used to deduce average temperatures during the past several hundred years.

10. Average heat flow is far too small to be commercially useful as an energy source, but in some areas – mainly those with natural steam – geothermal energy provides a useful contribution, though small in global terms. The huge amount of heat in hot, dry rocks has not yet been successfully exploited. Geothermal power is not strictly a renewable energy source, for heat is extracted far faster than it is replaced.

11. Important terms: heat, heat productivity, heat flow, heat flux; temperature, temperature gradient, geotherm, adiabat, potential temperature; thermal conductivity, capacity, specific capacity; skin depth; conduction, convection, thermal convection; steady and transient states.

Further reading

Heat flow and temperature measurements are discussed in few geophysics textbooks but are covered in appropriate physics books. However, their role in the lithosphere is described (in a fairly mathematical way) in a chapter of Fowler (1990), while Brown and Mussett (1993) discuss them in connection with mantle convection and related topics. Verhoogen (1980) explains how the Earth's internal heat powers orogenies, earthquakes, and plate motions.

Durrance (1986) discusses heat production from radioactivity. Geothermal power is discussed in Economides and Ungemach (1987). An article on the Larderello geothermal field, one of the first to produce power, is given by Minissale (1991). The deduction of past climatic temperatures is described by Pollack and Chapman (1993) and by Beltrami and Chapman (1994).

Problems

1. Distinguish between conducted and convected heat transport, and between thermal and forced convection. Give examples where each occurs within the Earth.

2. Why are two quantities measured to deduce the heat flux. Explain why this measurement does not include any convected heat.

3. Temperature corresponds to which of the following electrical quantities?
(i) Current. (ii) Voltage. (iii) Potential. (iv) Resistivity. (v) Conductivity. (vi) Charge.

4. Sketch geotherms through (a) oceanic and (b) continental lithosphere. Why do they differ?

5. Show by means of sketches the effect upon the temperature profile of (a) a rapid deposition of sediments, (b) a rapid overthrusting.

6. Though the global total of heat reaching the Earth's surface from its interior exceeds the global energy consumption, explain why it is unlikely to provide a high proportion of our energy.

7. Explain why geothermal energy is not truly a renewable energy source.

8. The internal heat flow of roughly what area of average continental surface (in km^2) would be needed to match a 2000 MW power station? (i) 5, (ii) 50, (iii) 500, (iv) 5000, (v) 50,000.

9. Explain why caves seem cool on hot days but warm on cold ones.

10. Explain, with sketches, how temperature varies with depth below ground due to daily and annual temperature variations.

11. Why does it sometimes take millions of years for the temperature profile in the Earth to return to equilibrium following a disturbance to it?

12. A thick layer of clay overlies granite, and the thermal conductivity of the clay is half that of the granite. If the temperature gradient in the granite is 16°C/km, the gradient in the clay, assuming thermal equilibrium, would be (in °C/km): (i) 4°, (ii) 8°, (iii), 12°, (iv) 16°, (v) 20°, (vi) 24°, (vii) 25°, (viii) 32°.

Subsurface Geophysics

chapter 18

Well Logging and Other Subsurface Geophysics

The geophysical methods described so far in this book have nearly all investigated the subsurface using measurements made at the surface. This chapter describes measurements made underground, mainly in boreholes but also in mines and tunnels. The main advantages compared to surface measurements are much increased detail and close correlation with geological observations at precisely known depths; the disadvantages are the cost of boreholes and often the limited volume surveyed.

The most important application is in the exploration, evaluation, and production of oil and gas, by providing information on porosity, permeability, fluid content, and saturation of the formations penetrated by a borehole. Other applications are in mineral exploration and evaluation, and in hydrogeology. Subsurface measurements between holes or between holes and the surface may be combined to deduce the intervening geology.

Most of the geophysical principles used are the same as those used to 'look down' from the surface, described in the preceding chapters, but instruments and measurements have to be adapted to 'look sideways or upwards' in the special conditions of the subsurface, particularly the confined space of a borehole, where they also have to overcome the alterations produced in the surrounding formation by the drilling.

Well logging differs from most other geophysical methods, not only in being carried out down a borehole, but in relating the physical quantities measured more specifically to geological properties of interest, such as lithology, porosity, and fluid composition.

18.1 Introduction

Thousands of holes are drilled every year throughout the world, ranging from a few metres deep for engineering purposes, to over 14 km in the Kola Peninsula of northeast Russia, drilled to investigate the crust at depth. Most important economically are those for hydrocarbon exploration and extraction, which commonly reach depths of a few kilometres.

Their primary purpose is to provide samples of the subsurface or to extract fluids (oil, gas, water), but boreholes are expensive and the interpretation of data from them can be misleading: For example, a hole may be very close to a fault. Therefore, their value is considerably enhanced by also carrying out subsurface geophysical surveys in them to determine the properties of the rocks surrounding the borehole.

Compared to surface geophysical surveys, borehole surveys may offer deeper penetration, but their main advantage is greater detail, for measurements are made within formations of interest and can be compared with the geological information obtained from the same hole. Just as surface surveys 'look down' at the subsurface, so surveys in near-vertical boreholes through gently dipping strata can 'look sideways' and so can investigate beyond the immediate zone surrounding the borehole that, as we shall see shortly, has been affected by the drilling process, so providing some types of information that are not available even from cores taken from the hole.

18.2 Drilling and its effects on the formations

The physical properties of the geological formations in the vicinity of a borehole are often affected by the drilling process, so this will be described briefly. Drilling is usually done using a rotating bit that grinds away the formations to produce cuttings from shales, sand from unconsolidated sandstones, or chips up to a few millimetres in size from solid rock. The bit is fitted to the bottom of the drill stem, which typically consists of 10-m lengths of steel pipe screwed together. The bit may be turned by rotating the whole drill string of 10-m pipes from its top, but more commonly nowadays the bit is turned by the mud being pumped through a turbine just behind the bit. Turbine drilling allows much greater control of drilling direction. Down through the drill stem is pumped drilling fluid, usually termed **mud**; this has several functions: To cool and lubricate the bit, to carry the chips up to the surface, to stabilise the wall of the borehole, and – by its weight – to prevent any high-pressure gas encountered from blasting out of the hole. Therefore, it is not common or garden mud, but a mixture of ingredients formulated to produce the required rheological properties such as density and viscosity. Usually these solids are made into a thick liquid using freshwater, but sometimes salty water or oil are used. The nature of this base affects the physical properties of the mud (principally electrical conductivity), which have to be measured for they affect the geophysical measurements, both because the measuring instruments are immersed in it and because the mud penetrates the surrounding rocks and modifies their properties.

Because a permeable formation such as sandstone acts as a fine filter, a layer made up of the solids in the mud is deposited on the sides of the hole as **mudcake** (Fig. 18.1). The formation is invaded by the liquid of the mud, the **mud filtrate**, to a distance that depends on the porosity of the formation and characteristics of the mud. Nearest to the hole is the **flushed zone**, in which the fluids are entirely mud filtrate (apart from some residual fluid

that cannot be removed); next is the **transition zone,** with decreasing replacement until at a sufficient distance is the **uninvaded zone,** quite unaffected by the drilling. In very permeable and porous formations the mudcake builds up quickly and forms a barrier to further invasion. In less-permeable, low-porosity formations the buildup is slow and invasion can be very deep.

These radial zones and the fluids in them are identified by subscripts, for example ρ_m, ρ_{mc}, ρ_{mf} ρ_{xo}, ρ_i, ρ_t, ρ_g, ρ_w, and ρ_s refer to resistivities of mud, mudcake, mud-filtrate, flushed zone, invaded zone, uninvaded zone, gas, water, and adjacent bed respectively. (An oddity is that o is usually used for oil except in resistivity, where ρ_o is the resistivity of formation 100% saturated with water.)

Before hydrocarbons are extracted, holes are cased with steel piping, which is cemented to the wall of the hole, and as this provides an additional obstacle to measuring the properties of the surrounding rock, most measurements are taken in the **open hole** before casing. However, measurements are sometimes taken in a **cased hole,** either because the walls are unstable and the topmost part of the hole (where the pressure of the mud is low) has had to be cased, or because they are needed after extraction has begun. The diameter of the casing is reduced downwards in steps, with that at the bottom of a hole used for producing hydrocarbons commonly 120 mm, though a deep hole can start as a metre across at its top.

Later, the casing of a hole that is to produce is perforated where it passes through formations containing hydrocarbons, to allow their extraction. To facilitate precise location of the perforating gun down the hole, the collars where the lengths of drill pipe screw together are sometimes 'marked' by radioactive 'bullets', which can be located later by the γ ray logging tool (Section 18.5.4).

18.3 Sources of information from a borehole: Logs

The variation of a property down a borehole is recorded against depth as a **log** (e.g., Fig. 18.2). The most obvious log is a description of extracted cores, to provide information of lithology, microfossils, permeability, porosity, and fluids. These logs are used for geological correlation between different

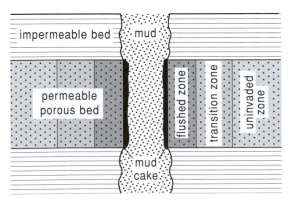

Figure 18.1 Effect of drilling mud.

wells across an oil field and for the quantitative interpretation of the geophysical logs. The ideal, of course, would be to obtain continuous core, but this is very slow and expensive, for it requires the drill string to be withdrawn for each 10-m length of core; therefore, it is done only for formations of special interest, usually potential oil-producing formations when they can be identified in advance. After drilling, samples may be obtained by **sidewall sampling.** Samples of fluids are also taken with a **formation tester,** for identification and measurement of their properties, to aid interpretation of some of the geophysical measurements. As well as being very expensive, cores do not provide all the required information, which is obtained in other ways.

The first log to be obtained is the **drilling time log,** which records the rate of drilling progress, because this depends upon the nature of the formation. The chips carried to the surface are filtered from the mud and examined to provide the **mud log.** This gives an imperfect record of the formations penetrated by the drill below, because chips travel to the surface at rates that depend on their size and density, so chips of different formations may become mixed, while soft formations that wash out or soluble ones like salt may not yield any chips at all.

Geophysical measurements are made by sophisticated instruments suspended by a wire, as they are pulled to the surface by a winch fitted with a depth counter. The instruments are called **sondes** or **tools,** and the measurements they produce are known variously as **wire-line logs, geophysical well logs,** or simply **well logs.** Figure 18.2 illustrates a typical arrangement, and also shows logs for SP (self-potential) and resistivity, with scales increasing to the left and right respectively. Well logging is sometimes carried out during drilling – which has therefore to be interrupted for a few hours – as well as at its completion, and also before and after casing, and – in the case of a production well – at subsequent intervals.

18.4 Geophysical well logging in the oil industry

In Section 7.10.1 it was explained how a hydrocarbon trap often consists of a porous reservoir rock containing the oil or gas, which is prevented from leaking to the surface by an impervious cap rock. Oil accumulations occur in such reservoirs, and well

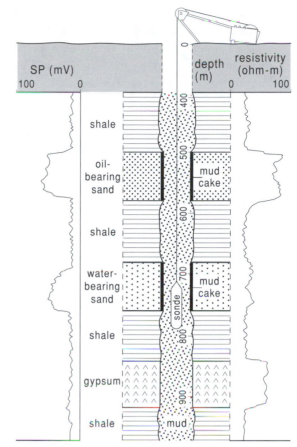

Figure 18.2 A typical well-logging arrangement and two logs.

logs have particular importance for their identification and evaluation, but the most important properties of the rocks of a reservoir – such as porosity, hydrocarbon saturation, and permeability – are not simply related to the quantities that can be measured by geophysical tools, such as electrical resistivity or self-potential. Consequently, many different measurements are usually made and combined to estimate the quantities of interest, as well as making allowance for the changes produced by drilling. This section describes the main reservoir properties with an outline of how they relate to the physical properties that are measured. To simplify the discussion, the porous reservoir rock will usually be taken to be a sandstone and the impervious cap rock as shale.

(i) The **porosity,** ϕ, is the fraction of the rock that is occupied by pore space. It is usually the primary or matrix porosity of sandstones due to the spaces between the grains, but its value will be decreased by any shale content (shaliness) of the sands, for the shale will tend to block the pores. Its value may be increased by secondary porosity created by fractures or – in the case of limestones – by solution.

We saw in Section 12.2.1 that most rocks are electrically conducting only because the pore spaces contain water (except at shallow depths). Thus porosity and electrical resistivity (or conductivity) are related; however, conductivity also depends on the fraction of the porosity that contains water – rather than hydrocarbons, which are insulating – the salt content of the water, and the temperature. Porosity affects other physical properties of the rocks, notably density and seismic velocity, so it can be measured indirectly, while the fluid that fills the pores can also be detected because – whether water, oil, or gas – it is rich in hydrogen. Density, seismic velocity, and hydrogen content can be measured in boreholes, although they too partly depend on other rock characteristics, such as lithology, what the fluid is, and the shale content or shaliness. However, by combining the results of various logs porosity can be estimated.

(ii) The **hydrocarbon saturation, S_{hc},** the fraction of the pore volume occupied by hydrocarbons. The greater its value the less the **water saturation, S_w** – the fraction occupied by water – as $S_w(1 - S_{hc})$, and so the higher the resistivity of the formation, for hydrocarbons are electrically insulating.

(iii) The **permeability, k,** is a measure of the ease with which fluids can pass through a bed. In general, it depends strongly on porosity but cannot be reliably estimated from it. Fine-grained rocks of high porosity usually have low permeability, while some rocks such as limestone of low porosity may have high permeabilities due to fractures.

Permeability can be measured directly in the laboratory on samples of rock but not as such in boreholes; instead, it is estimated from the porosity after comparing the values of porosity and permeability of a number of samples from the same formation. The practical unit

of permeability is the millidarcy, abbreviated to md, and high-permeability rocks are in the range 10 to 100 md. (A material has a permeability of 1 darcy when a fluid of one centipoise viscosity moves at 1 cm/sec under a pressure gradient of 1 atmosphere/cm.)

(iv) The thickness of the permeable formations, from which any hydrocarbons will be extracted, can be estimated from the logs, for most show a change at the junction with impermeable beds, as illustrated in Figure 18.2.

These are the most important quantities that need to be known about a reservoir. Ultimately, what needs to be deduced is the amount of hydrocarbons present and the ease with which they can be extracted. The amount depends on the volume of the reservoir times the fraction of it that is hydrocarbons. In turn, the volume is found from the average of its thickness measured at a number of places over the extent of the field times the area of the field, while the fraction is the product of the porosity, ϕ, and hydrocarbon saturation, S_{hc}.

(v) But not all of the hydrocarbon in a formation, particularly oil, can be extracted, for some adheres to the grains. This **residual hydrocarbon saturation** can be measured before the hydrocarbons are extracted, for it is the amount remaining in the flushed zone, which can be estimated from its resistivity, ρ_{xo}.

The above quantities are determined from measurements of resistivity, SP, and so on. Other quantities, such as temperature, are also needed because they affect resistivity, while further logs are needed when holes have had to be cased, to improve the estimates. Thus many different types of logs are in use, with a selection employed in any given hole, and their results are combined.

The sections that follow describe the most common geophysical logs developed for use by the hydrocarbon industry, as well as some physical ones like the calliper log. How they are combined to give quantities needed for hydrocarbon exploration and extraction will be explained where appropriate. Some of these logs are useful outside the oil industry, so some other applications will be pointed out in Sections 18.6 and 18.7.

18.5 The most commonly used logs

18.5.1 The measurement of strata dip, borehole inclination, and diameter

Geophysical measurements in boreholes are used to estimate properties of reservoir rocks away from the borehole, and as some of these depend on the dip of the strata, the inclination of the borehole, and its diameter at the point of measurement, these quantities need to be measured.

Though the regional dip of a particular interface can sometimes be obtained from its positions in three or more boreholes not in a straight line, in practice there are seldom enough suitable boreholes to allow for variations in dip, and there may also be doubt whether major discontinuities occur between them. Instead, a **dipmeter** is used to determine the dip of strata within a single borehole by recording the value of the SP (spontaneous potential) value at three electrodes 120° apart (Fig. 18.3), each relative to an electrode at the surface.

The method relies on formations having different potentials (how these are produced is explained in the next section) so that if the formation boundary is not perpendicular to the hole the electrodes record changes at different depths of the tool in the hole.

The orientation of the tool and the deviation of the hole from the vertical must also both be known, and these are measured by a **photoclinometer** (Fig. 18.3a), which is run at the same time as the dipmeter. This has guides to keep its axis parallel to that of the hole. A steel ball rolling in a concave graduated glass dish shows the inclination by the distance of the ball from the centre. Below the dish is a compass to measure orientation. A camera with a flash, operated from the surface, takes photographs at intervals to show the compass needle, the ball, and the three dipmeter electrodes (Fig. 18.3b).

The diameter of the hole varies with the lithology. For example, in a sand–shale sequence, mudcake builds up on the permeable sands, reducing the diameter, as shown in Figure 18.2; in contrast, the diameter in the shale is increased because the vigorous flow of drill mud washes it away, so the presence of mudcake is a good indicator of high permeability. The thickness of the mudcake also needs to be known for interpreting the results of several of the geophysical logs to be described, for measurements of the properties of the surrounding rock have to be made through it. The diameter of the hole is also needed for calculating the amount of cement needed to fix the casing.

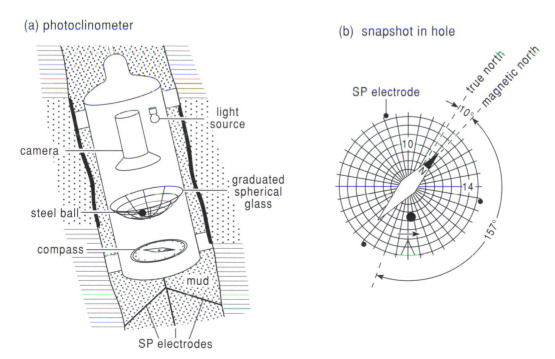

Figure 18.3 The photoclinometer with dipmeter electrodes.

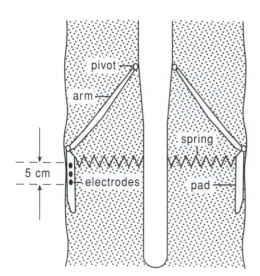

Figure 18.4 The calliper log with microlog electrodes.

The **calliper log** measures the diameter using a calliper sprung lightly against the sides of the hole (Fig. 18.4), and their extent is converted by an electromechanical device into an electrical signal suitable for recording. Due to subsurface stresses, holes are seldom circular in section, so the tool usually has more than one calliper. Sometimes an arm of the calliper log includes electrodes for the microlog described in Section 18.5.3.

18.5.2 The self-potential, SP, log

The **SP log** has two important uses, in addition to revealing the dip. One is that it indicates the positions of permeable formations, as already shown in Figure 18.2. The second use is to measure ρ_w, the resistivity of the formation fluid, needed to help calculate S_{hc}, the fraction of the pore space occupied by hydrocarbons.

Self-potential, as described in Section 13.2.1, was mainly of interest in connection with minerals that are electronic conductors, such as massive sulphides and graphite. Self-potential in oil wells arises in quite a different way, where there are adjacent formations with different concentrations of electrolyte such as a shale–sand contact (Fig. 18.5). This is because shale has a sheetlike structure with negative oxygen ions at the edges, which repel the negative ions of the salts dissolved in the water, so allowing positive but not negative ions to pass through.

Because the formation water in the uninvaded sands has a much higher concentration of salts than the water normally used to make up the drilling fluid, ions tend to flow between them to equalise the concentrations, but because negative ions are unable to pass through the shale, the current flows in elliptical paths as shown. The associated potential difference (p.d.) is less than 100 mV, considerably smaller than the p.d.'s that can be generated by sulphides.

The reading of the SP log does not change abruptly at the contact but varies either side over a distance that depends on the ratio of the resistivity of the uninvaded formation to that of the mud, ρ_t/ρ_m, as comparison of columns (i) and (ii) of Figure 18.6 reveals. The position of the contact is given by the point of inflection shown by the dashed lines. The SP only reaches a constant value in sufficiently thick beds, and this value is called the static SP, or **SSP**. The cleaner the sand the greater the SSP, so the actual value is important, for the presence of shale within the sand reduces its porosity and its permeability, both of which adversely affect its value as a reservoir. The percentage of shale within the sand is roughly equal to the value of the SSP of the formation expressed as a percentage of the SSP of clean sand. How the value of the SSP is involved in the calculation of the water saturation, S_w, is explained in Box 18.1.

The second use of SSP is to determine ρ_w, the resistivity of the formation fluid, using the formula

$$\text{static self-potential,} \quad SSP = -K \log \frac{\rho_{mf}}{\rho_w} \qquad \text{Eq. 18.1}$$

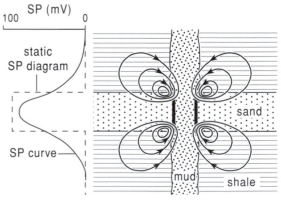

Figure 18.5 Origin of SPs in an oil well.

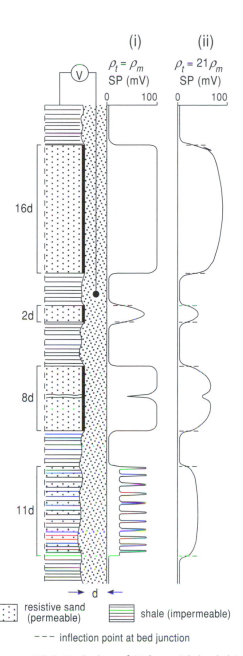

resistive sand (permeable)

shale (impermeable)

--- inflection point at bed junction

Figure 18.6 Variation of SP log with bed thickness.

where ρ_{mf} is the resistivity of the mud filtrate determined from a sample in a laboratory and K is a coefficient that depends on temperature, an average value being 71 at 25°C. Here ρ_w is not important in itself but is needed to determine the saturation, S_w, the proportion of the pore space occupied by water, using the equation

$$\text{saturation,} \quad S_w = \sqrt{\frac{F\rho_w}{\rho_t}} \qquad \text{Eq. 18.2}$$

where F is the **formation resistivity factor,** defined as the ratio ρ_o/ρ_w; ρ_o is the resistivity when the formation is 100% saturated with water, and ρ_t is the true or uninvaded **formation resistivity,** measured using one of the deep resistivity tools described in the next section.

18.5.3 Resistivity logs

The **resistivity log** is used to determine the fraction of the pore space that is occupied by water, called the water saturation, S_w, (in turn used to calculate S_{hc}, the fraction that is hydrocarbons, using $S_{hc} = 1 - S_w$). This is possible because the resistivity of sediments, ρ_t, depends on the amount of water in their pores. However, resistivity also depends on the resistivity, ρ_w, of the water in the pores (which depends on the salt content) and what volume of the rock is pore space (i.e., its porosity, ϕ), and these have to be taken into account.

A further complication is that some rocks, particularly shale, have quite a low resistivity though they contain little water. However, using the SP log in addition to the resistivity one can distinguish such rocks (see Fig. 18.2) from porous ones containing water. Thus shales and water-bearing sands both have low resistivity, but whereas the sands show a high SP deflection due to their high permeability, shales show a low SP. Oil-bearing sands differ from both these rocks by having high SP and high resistivity. Other rocks that have low porosity, such as gypsum, have high resistivity and low SP. This simple example illustrates how the value of logs is enhanced by combining their results.

In Chapter 12 we saw how electrical resistivity variations beneath the ground may be measured using a four-electrode array on the surface, and in Chapter 14 how its inverse, conductivity, may be obtained by inductive methods. Both can be adapted to measure variation of resistivity down a borehole. In surface resistivity methods, electrodes are put into the ground to make contact; in boreholes, contact is through the mud, which therefore must be conducting. An example is the lateral sonde shown in Figure 18.7, which includes two potential electrodes, M and N, and a current electrode, A, the second current electrode, B, being on the surface.

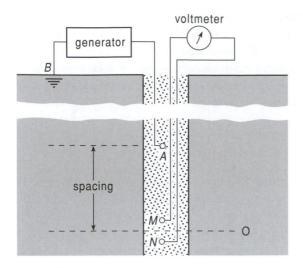

Figure 18.7 The lateral sonde.

Determining the important quantity S_w (water saturation) requires measuring the true formation resistivity, ρ_t (i.e., the resistivity in the uninvaded zone), but the electrodes in the hole are separated from it by, successively, mud, mudcake, the flushed zone, and the transition zone, which have differing resistivities. In principle, this could be solved by using a 'sideways' version of vertical electric sounding, VES, but this would be valid only provided the electrodes were not near a change of lithology and

would be time-consuming to carry out. The requirement is for a tool with a large radius of investigation and fine vertical resolution. This cannot be met by a single tool. Resistivity and conductivity tools have been developed with different radii of investigation in a variety of geological situations, and combination tools have been developed in which more than one kind of log is obtained at the same time.

A related drawback for this purpose of the arrays described in Chapter 12 is that, if the mud is much less resistive than the formation, little current would penetrate beyond the mud unless the electrodes were far apart, which, as was just pointed out, would probably place them within a different lithology. This situation could occur in offshore drilling, where salty muds are used. To overcome this problem, extra electrodes are used to concentrate or 'focus' the current sideways through the mud into the formation. Such sondes are known as **laterologs**, the simplest being the laterolog 3, which employs three electrodes (Fig. 18.8a). The current electrode, A_0, is at the centre of the sonde, with symmetrically to either side electrodes A_1 and A_1', which are maintained at the same potential as the central electrode A_0 to prevent the current flowing towards them (current can only flow from higher to lower potential), thus driving it into the formation. (A current and a potential electrode at the surface make up the usual four electrodes.)

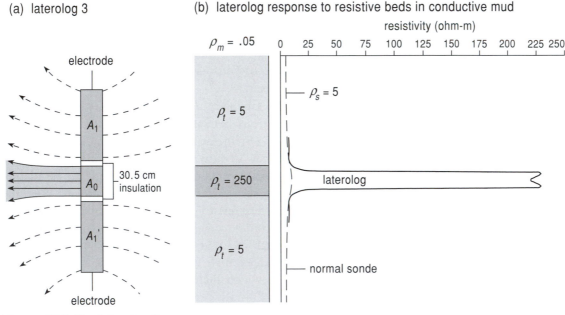

Figure 18.8 The laterolog 3.

To maintain the three electrodes at the same potential requires that the current be varied depending on the apparent resistivity of the formation the sonde is in, and so the current can be calibrated in units of resistivity, ohm-m.

The great advantage of the laterolog compared to an unfocussed sonde in high-resistivity formations is illustrated in Figure 18.8b. The formation resistivity is 250 ohm-m, whereas that of the mud is only 0.05 ohm-m. The laterolog gives over 90% of the correct value, whereas the unfocussed sonde gives less than 10%. Modern laterologs include different electrodes to give deep and shallow penetration into the walls of the borehole, and so allow ρ_t and ρ_i to be determined.

To measure the resistivities of the various zones between the sonde and the uninvaded zone, needed for correction, requires microresistivity sondes with electrode separations considerably smaller than those of the sondes for deep penetration described so far. An example is the **microlog**, in which the electrodes are mounted close together and make contact with the mudcake through button-shaped electrodes mounted flush with the surface of a rubber pad; this is pressed against the side of the hole by springs and so is unaffected by the mud. Figure 18.4 shows a version in which the microlog electrodes are built into one of the pads of the calliper logging tool. It contains 3 electrodes about 2.5 cm apart in a vertical line. They can be connected in two combinations, together with an electrode at the surface, to give logs from which ρ_{mc} and ρ_{xo} may be estimated. If ρ_{mf} is known from mud samples and the residual oil saturation is known or assumed, then the formation factor, F, and hence the porosity, ϕ, may be estimated for a nonshaly sand.

An important use of the microlog is to determine the exact location of formation boundaries, especially with thin beds or in conducting muds when the SP log is not very precise (Fig. 18.8). Its response in mudcake is affected by fluids that leak out of the formation next to the hole, with oil giving a higher resistivity than water. The log can therefore reveal the boundary between oil and water in a permeable formation.

The microlog is run with the springs extended on the way up but is also run on the way down with the springs collapsed to give a mud log that shows the variation of mud resistivity with depth. This can

be used to identify levels at which fluids – such as oil or water – are entering the hole, for they will have resistivities different from that of the mud. This is useful for the identification of fluid-bearing permeable formations and oil-water interfaces.

Resistivity sondes require that the hole contains a conducting fluid to allow electrical contact with the formation. The **induction log** was introduced for those holes that do not meet this requirement (i.e., if they were drilled with air or an oil-based mud, or if they are dry or – as some old wells – lined with concrete or bakelite). The induction logging sonde has transmitting and receiving coils like the Slingram system used for surface surveys (Section 14.2.1), but these are arranged with their axes along the borehole (Fig. 18.9). The strength of the currents they induce – 'eddy currents' – depends on the resistivity of the surrounding rocks. In practice, the sonde has additional coils to produce focussing analogous to the extra electrodes of the laterolog so that different depths of penetration can be selected. The induction sonde has the greatest penetration of any resistivity tool, up to 5 m or more from the borehole. It is therefore best suited for the determination of ρ_t and so is routinely used in water-filled holes as well as the special holes mentioned above.

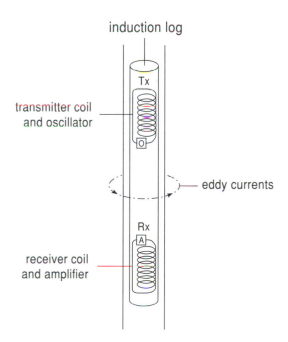

induction log

Figure 18.9 The induction log.

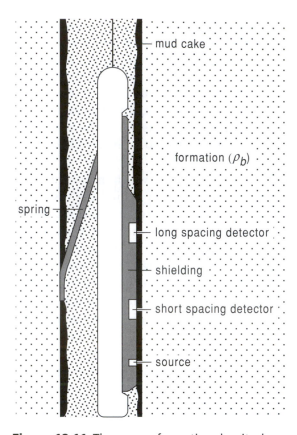

Figure 18.11 The γ–γ, or formation density, logger.

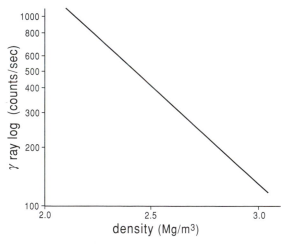

Figure 18.12 Relation between density logger count and density.

Some γ rays are also absorbed or scattered by the mud and mudcake between the tool and the formation, affecting the reading. To reduce this, the tool is held against the side of the hole by a strong spring to scrape through the mudcake as it is pulled up the hole. However, this may not be totally effective, so the thickness of any remaining mudcake is measured by having two detectors placed at different distances from the source; these are affected differently by the mudcake, so comparing their readings allows a correction to be made.

The bulk density of a porous rock, ρ_b, depends on the density of the matrix, ρ_{ma}, the porosity, ϕ, and the density of the fluid in the pores, ρ_f:

$$\phi = \frac{\left(\rho_{ma} - \rho_b\right)}{\left(\rho_{ma} - \rho_f\right)} \qquad \text{Eq. 18.3}$$

The values of density obtained from the log are also used in the interpretation of surface gravity and seismic data, as both of these depend on density (Chapter 8 and Box 4.1). It is also useful, sometimes combined with other logs, for recognising fractures, in compaction studies, and – as demonstrated below – in the identification of lithology.

The second type, the **neutron** or **porosity logging tool,** is similar to the gamma–gamma tool in having two scintillation counters held against the side of the hole but differs in having a radioactive source (e.g., plutonium–beryllium) that bombards the formation with fast neutrons rather than γ rays. Neutrons travelling through the formation only slow down significantly when they collide with atoms of a similar mass, that is, hydrogen atoms. Once they have been slowed by repeated collisions, they are absorbed into the nucleii of the heavier atoms present and cause them to emit γ rays, some of which are recorded by the counters. The more rapidly the neutrons slow down, the nearer to the counters the γ rays are produced, resulting in a stronger signal. As hydrogen is an important component of both water and oil, which fill the pore spaces, the response increases with porosity, which is why the log is also known as the porosity log. But the tool is calibrated to give the true porosity only in clean limestones filled with water; allowances have to made for other lithologies. This can be done by combining the neutron log with the density or the sonic log, described in the next section, as their responses also depend on porosity and lithology. The tool also responds to gas, which has many fewer hydrogen atoms per unit volume (referred to as having a 'lower **hydrogen index**'), but shows a porosity that is too low. Oil and water have almost the same hydrogen index.

As the two tools, γ ray and neutron, respond to permeability and fluids in a similar way to SP and resistivity respectively, they are used to replace them in cased holes.

18.5.5 The sonic log

The **sonic log** is a record of the seismic velocity, v_p, and is mainly used in the interpretation of seismic reflection sections. Values are combined with those of the density log to calculate the variation of acoustic impedance (Section 7.8.1) down the borehole. This is then used to calculate a synthetic seismogram, which can be compared with the observed one (Section 7.8.3). This allows seismic sections for parts of the hydrocarbon field that lack boreholes to be interpreted with more accuracy. Other uses are to measure fracture porosity and help identify lithologies.

The log operates in the same way as that of a surface seismic refraction survey of a dipping interface (Section 6.4), the dip occurring if the tool is oblique to the borehole (Fig. 18.13a), with the wall rock behind the mudcake being the higher velocity dipping layer. As with a surface survey, pulses are produced at each end (alternately), and for each transmitter there are two receivers (R_1 and R_3 for the lower transmitter in Fig. 18.13a) positioned far

enough along the tool that the first arrivals at both are rays refracted from the wall of the borehole (Section 6.2). Therefore, their separation divided by their difference in arrival times gives the apparent velocity in the wall rock. The upward and downward velocities (corresponding to forward and reverse in surface surveys) will differ, but as the deviation of the tool from the borehole axis is small, it is sufficient to take their average for the velocity of the wall rock. The results are not presented as a velocity but its reciprocal, the transit time, Δt, usually measured in μs/foot, for imperial units are still widely used in the oil industry (1 μs/foot = 3.281 μs/m), but shown in μs/m in Figure 18.13b. Just to the right of the depth scale is a line with ticks or pips; these show the total travel time down from the surface, so that to deduce the travel time between any two depths it is only necessary to count the number of pips between them. This can be used to calculate interval velocities (Section 7.2), which are needed for converting two-way travel-times of surface seismic waves to depths; their values are obtained more directly than those calculated from moveout; and because they operate at frequencies of 20 to 40 kHz, compared with 5 to 50 Hz for surface reflection surveys, sonic logs resolve the boundaries more precisely – 50 cm compared with 50 m (Section 7.8.2).

(a) sonic logging tool

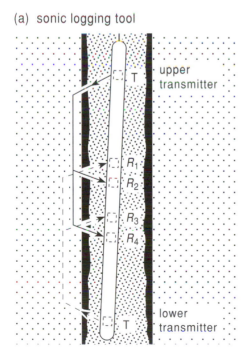

(b) sonic log

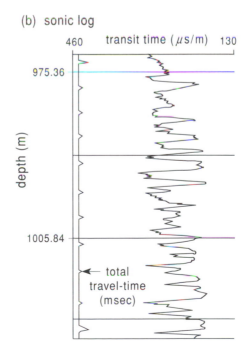

Figure 18.13 The sonic log.

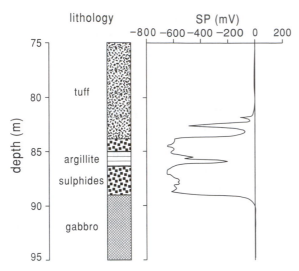

Figure 18.15 Self-potential log through a massive sulphide deposit, New Brunswick.

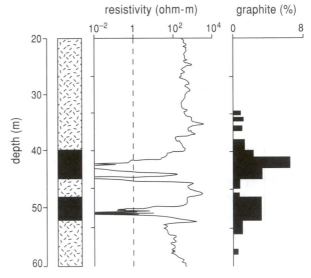

Figure 18.16 Resistivity through a graphite deposit, Canada.

industry, and some of their uses are described in the sections that follow, but others have been developed for use in mineral exploration and these are described later. The physical properties of interest are magnetic susceptibility, density, conductivity, and the IP effect of the rocks themselves, rather than resistivity, and so on related to the fluids in the pore spaces as in the hydrocarbon industry.

Figure 18.15 shows the SP log obtained through a volcanogenic sulphide deposit containing pyrite (FeS_2), galena (PbS), sphalerite (ZnS), and chalcopyrite ($CuFeS_2$). It was drilled to evaluate a surface anomaly, and continuous SP measurements were made between a nonpolarising electrode at the surface and an inert lead electrode in the hole. Large SP anomalies up to about 600 mV were observed in the mineralised zones intersected by the borehole, but smaller anomalies where there was no borehole mineralisation indicate other zones at some distance.

Figure 18.16 shows a resistivity log through a high-grade graphite deposit. The resistivity correlates strongly with the graphite content, increasing by several orders of magnitude as the content decreases. This correlation reduces the need for detailed sampling and chemical analysis of borehole samples, so illustrating the potential of logs in evaluating economic deposits.

Another example is given by Figure 18.17, which

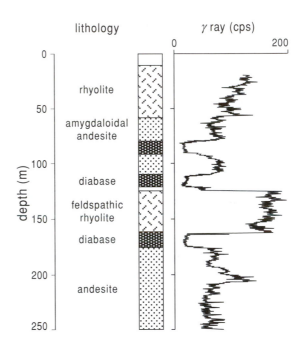

Figure 18.17 γ ray log, Buchan's Mine, Newfoundland.

shows the γ ray response through a sequence of volcanic rocks. The rhyolitic rocks, which contain the most potassic feldspars, show higher responses than the diabase, which contains the lowest, and the andesites, with intermediate content.

We turn next to logs that are not adaptations of those used in the hydrocarbon industry but have been developed for mineral exploration and assessment.

18.6.2 Magnetic logs

Two kinds of magnetic measurement are made in boreholes, magnetic field and magnetic susceptibility (Section 10.6). We saw in Chapter 11 how magnetometers can be used on the ground, from the air, or from ships. Boreholes can be used to extend measurements downwards. The magnetometers used in a **magnetic field log** are fluxgates (Box 11.1) because they give continuous readings; by measuring three perpendicular components the total field can be calculated in direction as well as strength. Figure 18.18 shows the position of a borehole, BH1, that had been drilled in a surface magnetic anomaly but failed to intersect any ore. Magnetic measurements in it showed large magnetic anomalies that could not be attributed to the rocks intersected by the borehole. It was therefore deduced that all the significant sources of magnetisation lie between the borehole and the surface; a second hole, BH2, positioned as a result of this information, successfully intersected ore.

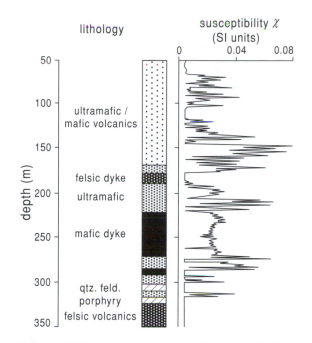

Figure 18.19 Borehole magnetic susceptibility log, nickel deposit.

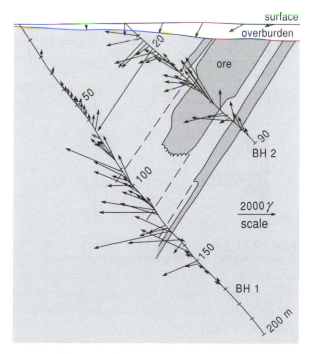

Figure 18.18 Magnetic field log near sulphide bodies in Sweden.

Magnetic susceptibility logs can also be made continuously down a hole, using a coil system similar that in the induction logging tool (see Fig. 18.20). In fact, the two measurements can be carried out by a single instrument, the magnetic susceptibility and electrical conductivity being proportional to the in-phase and out-of-phase components respectively (Section 14.6). They can be used to identify mineralised zones through changes in susceptibility resulting from alteration of the magnetic minerals or to detect the presence of ore-bearing magnetic intrusions. Figure 18.19 shows the susceptibility logged through a nickel deposit, with high susceptibilities only in the mafic and ultramafic rocks and low in the felsics. The log may be used to quantify the nickel content more easily than by collecting samples for chemical analysis.

Another, fairly obvious, application is the estimation of the grade of iron ore deposits, for Figure 18.20 shows that there can be a strong relationship between susceptibility (measured on samples in the laboratory) and the ore grade. The figure also shows the ability of the γ ray and resistivity logs to discriminate between the shaly rocks and magnetic cherts.

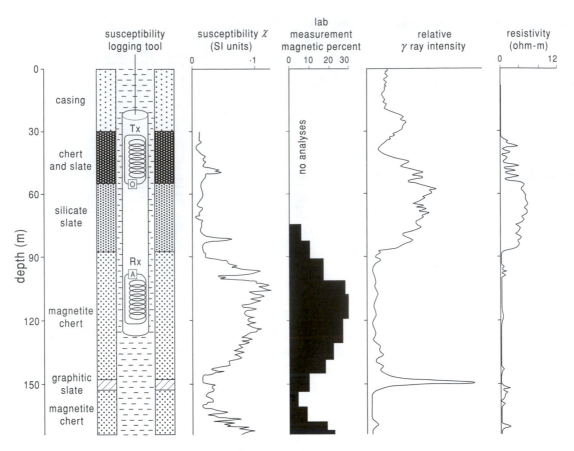

Figure 18.20 Susceptibility and other values versus iron content.

18.6.3 The IP–resistivity log

The **IP (induced-polarisation) log** is particularly use-ful for detecting disseminated sulphide ores, as explained in Section 13.1. Time-domain equipment can be used with little adaptation in water-filled boreholes, using four electrodes made of lead, attached at intervals to the lowering wire to form an array. As with surface measurements, resistivity measurements are also taken, since they need no extra apparatus. Measurements are usually made with the array stationary, and as they take a few sec-onds are made at intervals. However, a slow-speed system has been devised to allow continuous mea-surements at a slow rate of ascent. Figure 18.21 shows the results of such a log in a carbonate-hosted zinc deposit. Although sphalerite itself is noncon-ducting, there is an IP anomaly due to the presence of pyrite, whose concentration correlates with the sphalerite, whereas the resistivity log does not reveal it so clearly. Similarly, IP logging can be used for

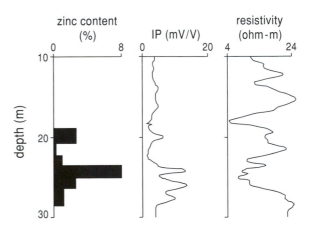

Figure 18.21 Induced-polarisation resistivity log in a carbonate-hosted zinc deposit in Newfoundland.

gold prospecting because of the common association of pyrite with gold.

The above are the main logs used in connection with metallic ores. There are also nonmetallic ores sufficiently valuable to drill for, such as sulphur, evaporite, and coal. Various combinations of the tools described can be used but will not be described.

18.7 Other well-logging applications

Borehole logging is also used for a variety of other purposes. One is in water location and extraction, where – as with hydrocarbons – porosity and permeability are important. Another hydrogeological application is to measure saline contamination by exploiting the large effect it has on conductivity. The selection or monitoring of a waste disposal site provides yet another use, for the presence of fractures that could provide pathways for leachates needs to be known, and fractures can be located from their effect on a number of tools, including sonic, neutron, resistivity, and calliper ones. In assessing the potential of the subsurface for geothermal energy, temperature is obviously important, but so also is the presence of water and permeability, whether through pores or fractures. The borehole logs have been a very useful complement to the information obtained from recovered core in the ocean and continental deep-drilling programs that investigate the Earth's deep crustal structure. However, in a book of this nature it is not possible to go further into these applications.

18.8 Other subsurface geophysics

Once boreholes, mines, and tunnels become available, they can be used to make geophysical measurements between the surface and subsurface or entirely in the subsurface and thus overcome some of the limitations of making measurements confined to the surface. For example, we saw in Chapter 7 how seismic surveys with sources and detectors on the surface are extensively used to provide models of the subsurface velocity distribution and structure. However, the ability of such surveys to resolve small features is limited because observations have to be made through a highly variable weathered overburden whose effects cannot be allowed for exactly. The problem can be overcome by carrying out a **crosshole** tomographic (Section 4.6) survey if two boreholes are available – a seismic source can be placed in one hole and detectors placed in the other and below the surface layers. By measuring the travel times and amplitudes of direct P and S waves from sources at various depths in one hole to detectors in another, it is possible to use computer-assisted tomographic (CAT) imaging techniques, similar to those used in medicine, to make a detailed section of the velocity variation between the holes. Special high-frequency sparker sources and detectors which can be clamped in the hole have been developed for such studies.

Electrical tomographic surveys can also be carried out in a similar manner using electromagnetic sources to map the conductivity distribution between boreholes in the exploration for minerals, oil, and water, as well as in environmental and crustal studies. Another subsurface seismic application is the use of reflection surveys in mines to map steeply dipping orebodies by directing the seismic energy sideways rather than vertically as in the surface surveys. The mise-à-la-masse method, in which a conducting orebody encountered in a drill hole is energised by means of a current electrode and potentials are measured on the surface or in other boreholes (Section 12.4.2), is used to determined the shape and extent of the body.

Summary

1. The most used form of subsurface geophysical surveying is well logging in boreholes. Other types of measurement are made from borehole to surface, between boreholes, and in mines and tunnels.

2. The records of the variation of various quantities down boreholes are called logs, and are of both geophysical and nongeophysical quantities.

3. Geophysical logs are made by instruments in sondes, or tools, suspended on a wire, with readings usually taken as they are pulled to the surface. Sondes often contain several instruments that can operate without mutual interference.

4. Measurements of geophysical quantities in boreholes differ from those in surface surveys in three main ways:
   (i) Measurements are made 'sideways' into the formations around the borehole.
   (ii) Instruments have to be adapted for the dimensions of a borehole and usually to operate submerged in drilling mud.

(iii) Measurements have to take account of the changes resulting from the drilling.

5. Geophysical logging is not only cheaper than continuous coring but it provides information that cannot be obtained on cores, partly because of the alterations produced by the drilling, and reduces the need for sampling. Geophysical logs are often more valuable when combined than when used singly.

6. By far the largest and most sophisticated application of well logging is in the hydrocarbon industry, where it is used to assess the amount of hydrocarbon in a reservoir and also the ease with which it can be extracted. This requires estimating the volume of reservoir, the hydrocarbon concentration within it, and its permeability. These are calculated from the lateral and vertical extent of the reservoir formation, and from its porosity and hydrocarbon saturation. The last two cannot be *measured* directly by geophysical well logs but are *deduced* by combining the results of several geophysical and nongeophysical logs. Permeability is estimated from the porosity.

7. In hydrocarbon exploration and assessment, the most important geophysical logs from open holes with conducting mud are SP and resistivity, used in conjunction with dip, inclination, and calliper logs.

8. Self-potentials in oil wells usually result from there being different concentrations of salts in permeable sandstones, impermeable shales, and the mud. The SP logs reveal shale boundaries; as shale is a common cap rock, they therefore often reveal the boundary of the reservoir rock. The SP value attained in a sufficiently thick porous formation is called the SSP (static SP); it is used in the estimation of ρ_w, which is needed to calculate saturation.

9. The induction log also measures resistivity (or conductivity), but – unlike the resistivity log – it also functions in a dry hole or one filled with nonconducting mud.

10. Radioactivity logs belong to two groups. In one, the natural γ ray activity is measured and used to identify lithologies. In the other, γ ray activity induced by γ ray (density log) or neutron (neutron or porosity log) sources carried in the sonde are recorded. The γ ray and neutron logs

are recorded together as a pair and can be used instead of the SP and resistivity logs in a cased hole.

11. The sonic log measures the seismic velocity and is used to improve interpretation of seismic reflection sections. It can also be used to estimate porosity and to improve identification of lithologies when used in a cross plot with the neutron log.

12. The temperature log is mainly needed to correct for the effects of temperature on resistivity. It is also useful to locate formation boundaries and high-pressure gas zones, and in monitoring the progress of cementing operations.

13. The most commonly used logs in mineral exploration are SP and resistivity, for they respond to most massive sulphides and graphite, but magnetic logs (field and susceptibility) are useful for magnetic ores, while IP is valuable with disseminated ores.

14. In hydrogeology, electric logs are useful in delineating aquifers, determining their porosity, and estimating the quality of the water in them.

15. Geophysical sources and/or sensors can be used between boreholes, mines, and tunnels as well as at the surface for tomographic studies or to extend the depth of exploration.

16. You should understand these terms: drilling mud, mudcake, mud filtrate; flushed, transition and uninvaded zones; open hole, cased hole, sidewall sampling, formation tester; log, wireline log, geophysical well log, sonde, tool; reservoir rock, cap rock; porosity, permeability, hydrocarbon saturation, residual hydrocarbon saturation, hydrogen index, water saturation; SP, SSP, formation resistivity, formation resistivity factor; cross plot, cross-hole tomography.

You should know what the following logs are and what they measure: drilling time, mud, calliper, dipmeter, clinometer, SP, IP, resistivity, laterolog, microlog, induction, natural γ ray, spectral γ ray, gamma–gamma, formation density, neutron, porosity, sonic, temperature, magnetic field, magnetic susceptibility.

Further Reading

Rider (1996) gives an excellent account of the principles and practice of well logging and their geological interpretation in hydrocarbon exploration. Baltosser and Lawrence (1970) is a good early introduction to the use of well logging outside the oil industry. Chapellier (1992) is a very readable account of the principles and uses of well logs written for engineers and hydrogeologists.

Problems

1. Give two general reasons why geophysical logging would still be needed in hydrocarbon exploration even if complete cores were available.

2. In what circumstances are radioactivity logs preferred to electrical ones?

3. What are two advantages of an induction tool over a resistivity one?

4. What are the main factors that determine the electrical resistance of a formation?

5. Explain the uses of SP logs in (a) hydrocarbon exploration and (b) mineral exploration.

6. What does the natural gamma ray log respond to, and what is it used to determine? Why is the spectral γ ray log an improvement?

7. If you suspected a lithology was dolomite, how would you confirm this and measure its composition using logs?

8. How would the signal of a neutron log in dry sand compare with that in water-saturated sand and in sand filled with gas? Explain why.

9. Why can the sonic log provide more precise estimates of seismic velocities and more precise locations of interfaces than surface seismic reflection surveys?

10. (a) The transit times for the matrix and fluid in a sandstone are known from laboratory measurements to be 180 μs/m and 656 μs/m respectively. What is the porosity of a bed for which the sonic log gives a value of 213 μs/m?

    (b) The neutron log opposite the same sandstone gave a porosity of 9.5%. What is the most likely cause of the difference?

11. Name three logs that would probably reveal the presence of shale.

12. In a sedimentary sequence of shales and sandstones, the γ–γ log gives values of 60 and 20 API units opposite a shale and a clean sandstone respectively. What is the percentage of shale by volume in a shaly sandstone for which the reading is 25 API units?

13. How are water saturation and hydrocarbon saturation related?

14. Sketch the SP and resistivity logs for the following downward succession of formations: thick sand with the water table halfway down it, thin shale, thick water-saturated sand, thick gypsum, thin water-saturated sand, thick shale, thick oil-saturated sand, shale.

PART II

EXAMPLES
OF APPLICATIONS

Part I described a range of geophysical methods, with a number of examples provided for illustration. In reality, of course, the problem normally comes first, and then the question arises, Which methods – geophysical and others – will help to solve it? In Part II we give examples of such problems. They have been chosen to illustrate the range of problems that geophysics has helped solve, and also to involve a range of geophysical – plus other – methods. But the range of problems is not comprehensive, nor does geophysics always plays so large a role or so successfully; these are simply a variety of problems where different geophysical methods have been able to make a significant contribution.

Part II begins with a short chapter describing how appropriate geophysical methods are chosen.

chapter 19

Which Geophysical Methods to Use?

19.1 Introduction

In Part I, deciding which method to use in any of the examples given was not a problem, for they were chosen to illustrate the particular method being described, but when a geological problem is first encountered it is necessary to decide which – if any – geophysical methods to use and how best to employ them. Choosing the most suitable one or combination needs experience and perhaps some luck, but considering the following questions should narrow the choice.

19.2 Does the problem have geophysical expression?

Geophysical surveys do not respond to geological features as such, but to differences in physical properties, so the first requirement is that the geological

situation has geophysical expression; that is, there must be some related subsurface body or structure that can be detected geophysically. For example, a granite pluton, which rose into place because of its low density, gives rise to a negative gravity anomaly (Fig. 8.16), and this may be used to locate it and estimate its size. In this example, the geophysical expression – the negative anomaly – is *directly* due to the body to be detected because its density is an intrinsic property of the granite, but sometimes geophysical expression is *indirect*. For example, a fault may be detectable by a seismic reflection survey if it has produced a vertical offset in subhorizontal layers (Fig. 7.10) but not if there are no layers or they are not offset vertically; or a concealed shaft may be directly detected by its negative gravity anomaly, but indirectly, for example, by a magnetic survey if it happens to contain ferrous objects (Section 27.2.3). The geophysical expression may be only *associated*; for example, when gold is present it is in such low concentrations that it produces no detectable change in the physical properties of the host rock, but in some areas it is associated with banded iron formations, which are magnetic, or with disseminated sulphide ores that can be detected using IP. With the wide range of geophysical techniques it is often possible, with some ingenuity, to find an indirect or associated geophysical expression, if there is no direct one.

So deciding which physical properties are likely to vary spatially as a result of the geological situation is the first step to choosing a method: A resistivity contrast suggests a resistivity or e-m survey, a density difference suggests a gravity one, and so on.

19.3 Is the variation lateral or vertical?

Though, for example, a density difference suggests that a gravity survey could be used, a gravity anomaly is produced only by a *lateral* variation of density, for a uniform horizontal sheet simply produces a constant increase or decrease in *g* measured at the surface; only where the sheet ends is there an anomaly (Section 8.4). This is also true for magnetic surveying.

Methods that utilise waves – seismics and GPR (ground-penetrating radar) – mostly require subhorizontal interfaces, which usually means a layered subsurface. If the waves are being reflected the interface must form a *discontinuity*, which requires that the thickness of the interface must be smaller than about a quarter-wavelength (Section 7.8.2). In some survey techniques subhorizontal interfaces may be more an *assumption* of the interpretation than a necessary requirement to get results; for example, modelling of VES (vertical electrical sounding) results usually assumes that the subsurface consists of electrically uniform horizontal layers, though these may not actually exist.

The value of tomographic methods is that they can be used whether variations are vertical or lateral or when there are no discontinuities, though resolution is usually poor and often the methods are complex to carry out.

19.4 Is the signal detectable?

Even if the above requirements have been met, the geophysical 'signal' may not be large enough to be measured with useful precision. This might be because the target is too deep: In seismic reflection, the reflections from progressively deeper boundaries become weaker as the downgoing pulse loses energy by reflections from higher layers, by absorption, and by simply spreading out. The amplitude of a gravity anomaly decreases with the depth of the causative body – as well as with decreasing density contrast with its surroundings – so there is some depth below which its anomaly is too small to be detectable.

In practice, what limits detectability is usually not that the signal has become too small to be measured but that it is submerged in the noise. In the gravity example above, the anomaly due to the granite may be within the capability of the gravimeter to measure it, but the varying thickness and density of overburden may obscure it; that is, the signal-to-noise ratio (Section 2.3) is too low. Various things can be done to improve the signal-to-noise ratio. In passive methods, such as gravity and magnetics, where the signal is generated entirely by the target, the signal itself cannot be increased, but readings can be averaged and corrections can be made with greater care. With active methods, where a signal is sent into the ground, the geophysicist has more control. In seismic reflection, for example, the size and duration of the pulse can be chosen to improve depth of penetration or resolution; in resistivity surveys the type of electrode array, the electrode separation, and the current through them can

be selected for the same purpose. Even so, there is usually a trade-off between depth of penetration and resolution, and there may be no pulse or electrode setting that allows a small target at depth to be detected. In seismic reflection, two reflecting interfaces cannot be resolved if they are less than about a quarter wavelength apart, but a pulse with sufficiently high frequency may be too attenuated before it reaches the required depth to give a detectable reflection. Other ways of improving the signal-to-noise ratio are stacking and filtering. Signal-to-noise ratio needs to be considered when designing a survey, to allow the above factors to be taken into account.

Knowing the limit of detectability is particularly important if the object is to check that some feature is *not* present, so as to be sure that when no signal has been found this is because no causative body larger than some size is present. For example, suppose that for a proposed building site it is necessary to take remedial action if there are cavities in the underlying limestone more than say 5 m across at a depth less than 10 m below; the precision of measurement and correction need to be sufficient to detect the corresponding anomaly. To find whether a boundary is present, a seismic refraction line needs to be long enough for the refracted rays to be first arrivals over a significant distance. The station spacing is also important, for this must be less than the width of the anomaly, while the traverse must be long enough to extend beyond the anomaly.

19.5 Will the result be clear enough to be useful?

All geophysical methods suffer from some degree of ambiguity of interpretation, which limits what may be expected from a survey. Sometimes the ambiguity is intrinsic to the technique, as with the non-uniqueness of gravity and magnetic modelling, where, even in theory, many different bodies (with varying degrees of plausibility) could be producing the observed anomaly. Other examples are the principle of equivalence in VES, where different combinations of layer thickness and resistivity can give the same readings, and hidden and low-velocity layers in seismic refraction. It has also been pointed out that geophysical interfaces are not necessarily geological boundaries and vice versa. Ambiguity also arises, of course, from limited resolution and from errors on readings.

19.6 Is a survey practicable?

Because surveying costs time and money, measurements need to be concentrated where they will be most useful. Extensive geophysical surveys, which are likely to be expensive, may start with a reconnaissance survey (after, of course, considering the available geological and other evidence) to see if any likely targets are present, before employing more detailed surveys. As illustration, in an area where mineral veins are possibly present, the first survey could be to measure profiles across the likely strike of the veins, perhaps using an airborne e-m technique such as TEM, and then, if any significant anomalies are found, to map them in more detail using ground surveys, perhaps on a grid. Follow-up surveys need not use the same technique as the reconnaissance survey; for example, in another situation magnetic and gravity surveys with widely spaced stations may indicate the presence of a sedimentary basin and hence the possibility of hydrocarbon fields; but high-resolution seismic reflection would be used where hydrocarbon traps were likely (Section 22.4). Using more than one technique often provides complementary information, which together may reveal far more than separately, as was explained, for instance, for wire-line logging in the previous chapter.

The full consideration of all relevant factors needs great experience, far beyond the scope of this book to impart, but the chapters that follow describe a number of case studies that illustrate some of the points.

Problems

1. Basement rock is overlain by twenty to thirty metres of sand, which is beneath several metres of clay. To measure the depths to the top and base of the sand you could use which of the following methods?
   (i) Seismic refraction.
   (ii) Gravity.
   (iii) Magnetics.
   (iv) Resistivity.
   (v) GPR.

   (vi) Radioactivity surveying.
2. Strata seen in a sea cliff can also be seen on the beach below in a few places where it is not covered by a thin layer of sand. Which of the following methods should be considered for investigating (during low tides) the geometry of the strata below the beach?
   (i) Seismic refraction. (ii) Seismic reflection. (iii) Resistivity. (iv) Slingram. (v) GPR. (vi) Magnetic.
3. You wish to survey for normal faulting of limestone beneath an overburden of glacial deposits. Discuss whether and how each of the following techniques might be useful:
   (a) Microgravity.
   (b) Seismic refraction.
   (c) Resistivity.
   (d) γ ray survey.
   (e) Magnetometer survey.
   (f) Magnetic gradiometer.
   (g) GPR.
4. A new road is to be laid across an area underlain by Carboniferous limestone that is known to contain sinkholes filled with clay. What quick and cheap site investigations would you recommend?
5. Two boreholes in a Permian sandstone produce freshwater, but a third is quite saline. Overlying sands and gravels plus faulting are thought to control the distribution of saline water, which probably has a source in evaporite deposits. How would you employ geophysical surveys to determine the extent of the saline water and the factors determining its distribution?
6. A gas pipeline is to cross an area of sedimentary rocks that is intersected by both igneous dykes and valleys eroded in the sediments. The dykes will cause difficulty when excavating the trench needed to hold the pipeline, while the valleys may contain saline water at depth that could corrode the pipe. What geophysical surveys could help reveal such features?
7. You need to locate a plastic pipe beneath a road surface. It was buried in a trench cut about a metre into granite, then infilled with limestone chippings and surfaced. Discuss which of the following it would be sensible to try:
   (a) Microgravity.
   (b) Seismic refraction.

(c) Resistivity.

(d) γ ray.

(e) Magnetometer.

(f) Magnetic gradiometer.

(g) GPR.

(h) Slingram e-m.

8. As the previous question except the pipe is steel.

9. What is the problem of nonuniqueness? How may it be surmounted, at least partly?

10. Which geophysical methods can be employed in aeroplanes? What are the advantages and disadvantages of doing so?

chapter 20

Global Tectonics

In the 1960s the theories of continental drift and sea floor spreading (hitherto largely regarded with scepticism) fused to give birth to plate tectonics, the idea that the surface of the Earth consists of huge rigid pieces that move independently, with most tectonic and igneous activity taking place at their margins as a consequence of their relative movements. Plate tectonics provides a framework for much of geology, being relevant to topics as diverse as continent formation, orogenesis, earthquakes, volcanoes, past climates, and palaeontology. It has been particularly successful when applied to oceans and their margins, but less so at explaining tectonic processes within continents, where deformation extends far from the plate margins.

The success of plate tectonics posed further questions: How deep do plates extend, and what moves them? How does intracontinental tectonics relate to plate collisions? What causes the volcanism – sometimes very extensive – found far from plate margins? This has led enquiry deeper within the earth, particularly to convective flows within the mantle, and this larger framework can be termed global tectonics.

This chapter is mainly concerned with the basic concepts of plate tectonics, which were established largely by geophysical evidence, and geophysics, with its ability to investigate the deep Earth, continues to play a major part in extending our understanding of its processes.

Geophysical techniques employed: Many geophysical techniques have played a part, but seismology, seismicity, magnetics, palaeomagnetism, gravity, radiometric dating, and heat flow have had the major roles.

20.1 The basic concept of plate tectonics

Most geologists are familiar with the concepts of plate tectonics, but fewer know the evidence supporting the theory, in which geophysics played the major role, or how geophysics continues to help work out the details. Until the 1950s, most earth scientists thought the continents were fixed in position, with areas that have been largely constant except for flooding by shallow seas, while the oceans – largely from lack of data – were conceived rather as submerged continents. Mountain building, thrusting, and basin formation were recognised, of course, but there was no adequate theory of why they occurred.

Evidence for continental splitting provided by matching of features across oceans had been dismissed by most earth scientists, but the advent of palaeomagnetism provided new evidence that continents could move slowly or 'drift' over the globe. However, widespread acceptance did not come until a better understanding was gained of the ocean floors, and, as these were largely inaccessible to conventional geology, this was mainly acquired using geophysics. A crucial discovery – also involving magnetism – was that ocean floors are splitting apart at oceanic ridges, with continuous formation of new ocean floor. The combining of the ideas of **sea floor spreading** and **continental drift** led to the theory of **plate tectonics,** according to which the surface of the Earth is divided into a number of pieces or **plates** (Plate 2c and Fig. 20.25). The plates are more or less rigid, and relative movement between them leads, at their margins or boundaries, to most of the observed tectonic activity, including orogenies, volcanism, and seismicity (Plate 2b). Plate margins are often marked by topographic features, as Plate 2a shows; margins may be edges of continents, but not all continental edges are plate margins, for a plate may comprise both continental and adjacent oceanic areas.

What processes occur at a plate margin depend on the relative motion of the plates. There are three types of margins. Where plates move apart new plate surface is created, and these are called **divergent** or **constructive margins;** where they move together and plate is destroyed are called **convergent** or **destructive margins.** In addition, in some places plates slide past one another without construction or destruction of plate, and these are called **conservative margins.** We shall examine the evidence for the different processes that occur at these different types of margins.

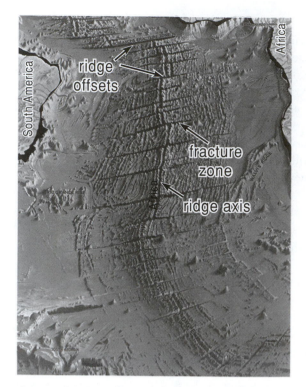

Figure 20.1 Southern part of the Mid-Atlantic Ridge.

20.2 Divergent, or constructive, margins

Divergent margins are marked by an ocean ridge, where material wells up (Fig. 20.1 and Plate 3a). They are also called **spreading ridges** because the newly formed plate moves away from the ridge axis as further material upwells (there are also ridges that are not spreading; these lack earthquakes and are called aseismic ridges; an example is the N–S ridge in the Indian Ocean, Plate 2). Magnetic, gravitational, seismic, and other types of evidence have allowed us to deduce what is happening.

20.2.1 Ocean-floor magnetic anomalies

A magnetometer towed behind a ship crossing a ridge shows strong anomalies, often exceeding 500 nT. Adjacent profiles are similar (Fig. 20.2a), so when a succession of parallel profiles are contoured the positive (shaded black) and negative (unshaded) anomalies form strips, often called sea floor **magnetic stripes.** These are parallel to the ridge axis and are fairly symmetrical about it (Fig. 20.2c), quite different from anomalies found on land.

They are formed as follows: As the magma

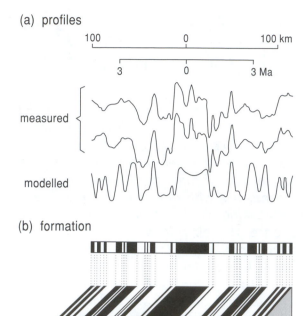

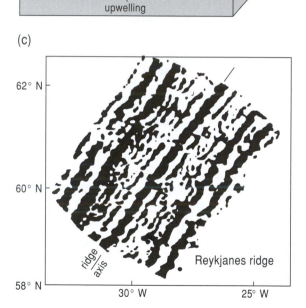

Figure 20.2 Ocean-floor magnetic anomalies.

upwelling at the ridge axis cools to form new plate, it becomes magnetised in the magnetic field at the time and then moves away from the axis, but because the Earth's magnetic field reverses repeatedly (Sections 10.1.3 and 10.5.1), new plate is successively magnetised in the present direction of the field (normal magnetisation) or the opposite (reverse) (Fig. 20.2b). The shape of the resulting anomalies depends on the strike and latitude of the ridges (Section 11.2.5): For an approximately N–S ridge at high latitudes, the normally and reversely magnetised strips of ocean floor produce positive and negative anomalies respectively. The anomalies are symmetrical about the ridge axis because plate usually spreads equally on either side. If the spreading rate is constant the widths of the anomalies will be in proportion to the N and R intervals of Figure 10.21. Calculated values are shown in the modelled profile (Fig. 20.2a), which is similar to the measured profiles but not identical, because new plate does not form as uniformly as the model assumes.

We can use the anomalies to deduce the spreading rate. For instance, suppose somewhere along a spreading ridge the width of the innermost (positive) anomaly is 15 km *to either side of the axis*; as the last reversal occurred 0.71 Ma ago (Fig. 10.21), each plate has grown at 15/0.71, or 21 km in a million years, which equals 2.1 cm a year. This is the *half*-spreading rate, the rate at which each plate moves away from the ridge; the plates are separating at twice this rate, which is therefore the rate at which the ocean is widening.

We can also use the magnetic anomalies to date the ocean floor. We count the number of anomalies from a ridge axis to some point on the ocean floor; then we count down in Figure 10.21 the same number of anomalies to find the age of the point. Plate 3b shows the ages of the ocean floors deduced in this way (the individual anomalies are not shown as there are too many). The oldest ocean floor is in the western Pacific and is about 180 Ma old (Jurassic). This is far less than the oldest continental rocks, some of which exceed 3000 Ma in age, because ocean plates are destroyed at convergent margins. (As explained in Section 10.5.1, the polarity timescale was first established, back to about 4 Ma in age, by radiometric dating of separate continental lavas of known polarity, and this was used to calibrate the ocean-floor anomalies, which in turn were used to extend the polarity timescale further back in time, initially assuming a constant spreading rate. These extended ages have been checked and slightly adjusted using radiometric dating of the rocks or by the palaeontological ages of sediments resting on the topmost igneous layer of the ocean floor; however, such dates are far too few to map by themselves the ages of the ocean floors in the detail given by using the polarity timescale.)

20.2.2 The shape of spreading ridges

Spreading ridges are wider (being thousands of kilometres across, Plate 2a and Fig. 20.3a) and more regular in shape than continental mountains, and slope down in a smooth curve (apart from local irregularities) to merge into ocean basins, as illustrated by Figure 20.3a. What produces this shape? The first piece of evidence is from gravity. The free-air anomaly (Section 8.7) is almost flat overall (Fig. 20.3b, solid line), and far less than would be expected if the material beneath the ridges had uniform density (dotted line). This shows that the ridges are close to isostatic equilibrium (apart from the small anomalies, which correspond to the peaks and troughs on the ridge, which are too small to be compensated). This means that the highest part of the ridge is underlain by the least dense rocks, according to the Pratt mechanism of isostasy (Section 9.1.4). As ridges are spreading, these least dense rocks are continually being carried away from the ridge axis and so cannot be due to laterally different compositions (unlike the 'roots' of continental mountains). This leaves temperature as the most likely cause of the density difference, for heating rocks causes them to expand and lowers their density.

This suggests the model of Figure 20.4: Mantle so hot that it behaves as a fluid upwells under the ridge crest, where it cools and becomes rigid and so part of the plate. As material moves away from the crest, following along the flow lines, the plate continues to cool from the top down by conduction, so that cooler temperatures extend deeper and deeper, as shown by the isotherms (lines of equal temperature); therefore, the plate thickens. Because the heat has to conduct up through a greater thickness, the temperature gradient lessens, so reducing the heat flux; in turn, the plate thickens more slowly, as mentioned in Section 17.2.2. As the material of the plate cools it contracts, which is why the plate surface lowers as it moves away from the ridge axis. This model therefore predicts three changes as the plate moves away from the ridge axis: It thickens, the heat flux decreases, and the plate subsides, with the changes becoming progressively slower with time.

When worked out mathematically (Box 20.1), this model predicts that the depth of the ocean floor *below the ridge crest* should increase as the square root of its age (which is proportional to the distance

(a)

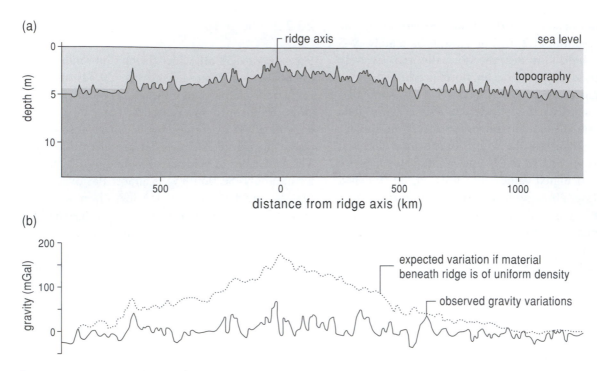

Figure 20.3 Topography and free-air anomaly of the Mid-Atlantic Ridge.

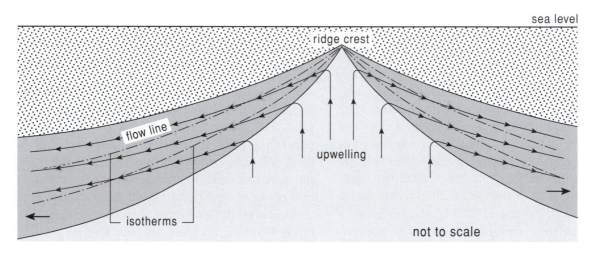

Figure 20.4 Flow lines and temperatures beneath a spreading ridge.

from the ridge axis), while the heat flux decreases as the square root of its age (e.g., at four times the age the depth is doubled and the heat flux is halved). Figure 20.5 shows that depth to the ocean floor generally follows prediction (shown by the line), though there is increasing deviation from it beyond about 80 Ma of age, showing that some other process is operating, though what this is is not clear.

This model of oceanic plate is a simplification. For instance, it does not allow for the crust having a different composition and so density from the rest of the plate (Section 20.8); however, the effect is small because its density is not much less than that of mantle, and its thickness is small. Also, ridges differ in their depths below the ocean surface, probably because the underlying mantle varies in temperature. Nevertheless, cooling is the main reason for the shape of ridges. Heat flux values are less simply compared with the model because some of the heat reaching the ocean floor is transported by hydrothermal convection. Because of the way heat flux is measured this is not included (Section 17.2.1), so the value recorded is less than the true

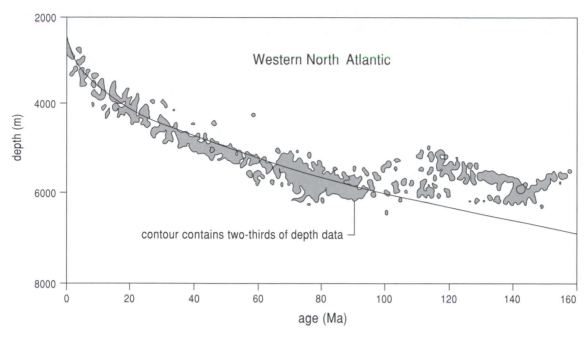

Figure 20.5 Shape of part of Mid-Atlantic Ridge.

value, in particular close to the ridge axis. However, the results do show a decrease with age.

20.3 Conservative margins

Oceanic spreading ridges have offsets, usually at right angles to the ridge axis; some can be seen in Figure 20.1. It might appear that the ridge crests once were continuous but have been displaced by faulting at the offsets. However, there is a second explanation, and the two explanations are compared in Figure 20.6. In Figure 20.6a, the offset is due to a transcurrent fault, with block A moving *left* with respect to block B (i.e., they are separated by a *sinistral* strike-slip fault). The fault extends along the fracture zone (Fig. 20.1) beyond the ridge crests, but

---

BOX 20.1 Spreading ridges: Heat flux and shape

Figure 1 shows an oceanic plate, its base defined by the T_b°C isotherm, the temperature below which the material is sufficiently rigid to move with the plate (this is about 1300°C). Below the plate the mantle is convecting, so the temperature increases only slowly with depth because the temperature gradient of the mantle – the adiabat – is so small (Section 17.1.4); to simplify the calculation we shall treat it as zero.

Because of the temperature difference between base and top of the plate, heat conducts up through it (Fig. 1c), causing mantle material to 'freeze' to its lower surface, just as water freezes to the base of a sheet of ice on a pond during cold weather. This thickening of the plate reduces the temperature gradient, as shown in Figure 1b, in turn reducing the heat flow.

The heat conducted up through the plate – which can be measured as the heat flux when it reaches the plate surface – is mainly supplied by cooling of the plate, for the radioactivity within oceanic plate is very small (Section 17.2.2) and the mantle beneath remains at about the same temperature. We could calculate the change of heat at every depth in the plate and add them together, but it is simpler to calculate the total amount of heat in the plate and then subtract the amount in it a short interval δt later, when it has thickened by a small amount δh_p. The thermal capacity of a m² column height, h, through the plate is ch, where c is the specific thermal capacity, so the total quantity of heat in the column after it has thickened is

$$Q_{after} = \left(h_p + \delta h_p\right)cT_{av} \qquad \text{Eq. 1}$$

We use T_{av}, the average temperature, as the temperature in the plate varies from 0 at the surface to T_b at its base; T_{av} obviously equals $T_b/2$.

BOX 20.1 Spreading ridges: Heat flux and shape *(Continued)*

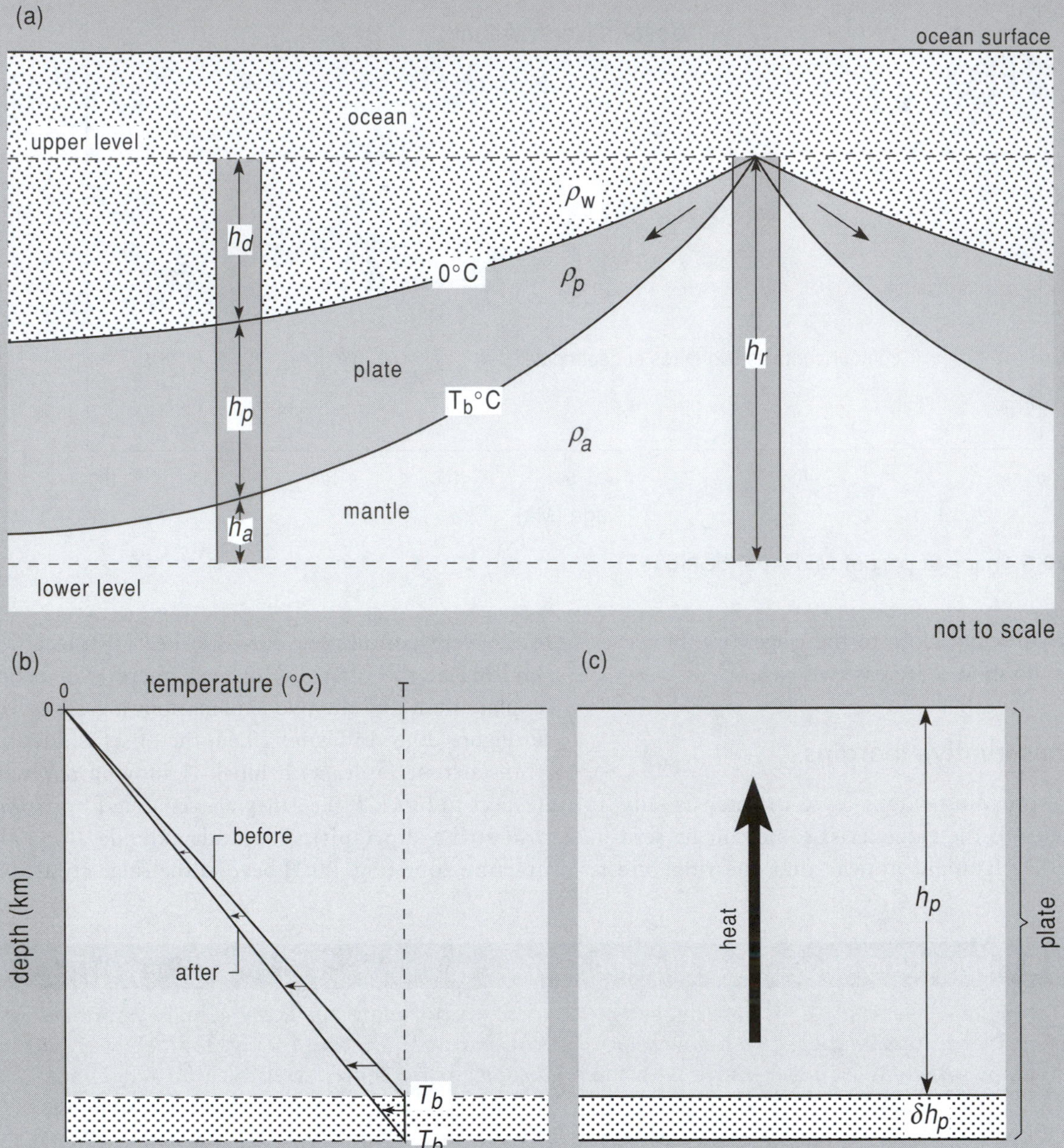

(a)

(b)

(c)

Figure 1 Thickening of oceanic plate with age.

The quantity of heat before thickening is

$$Q_{before} = h_p c T_{av} + \delta h_p c T_b \qquad \text{Eq. 2}$$

where the second term is the amount of heat in the layer δh_p about to be added to the plate.

Subtracting Q_{after} from Q_{before} and replacing T_{av} by $T_b/2$, gives the change of heat, δQ:

$$\delta Q = \delta h_p c\left(T_b - \frac{T_b}{2}\right) = \delta h_p c \frac{T_b}{2} \qquad \text{Eq. 3}$$

where δQ arriving at the surface is the heat flux, q, multiplied by the interval δt, which (assuming the plate thickens is so slowly that it remains close to thermal equilibrium) is also given (using Eq. 17.1) by

BOX 20.1 Spreading ridges: Heat flux and shape (Continued)

$$\delta Q = q\delta t = K\frac{\Delta T}{h_p}\delta t \qquad \text{Eq. 4}$$

where ΔT is the temperature difference between base and top of the plate and equals T_b. Equating these two expressions for δQ gives

$$\frac{KT_b}{h_p}\delta t = \frac{cT_b}{2}\delta h_p \qquad \text{Eq. 5}$$

Rearranging:

$$h_p\delta h_p = \frac{2K}{c}\delta t \qquad \text{Eq. 6}$$

Taking the small quantities to the calculus limit and integrating:

$$h_p = \sqrt{\left(\frac{4K}{c}\right)}\sqrt{t} \qquad \text{Eq. 7}$$

Therefore, a plate thickens in proportion to the square root of its age. As the heat flux is proportional to $1/h_p$ (Eq. 4) and h_p is proportional to the square root of the plate's age (Eq. 7), heat flux *decreases* as the square root of the age of the plate.

The shape of a ridge. To deduce this we note that ridges are close to isostasy (Section 20.2.2), so columns through the ridge will have the same weight. We employ Eqs. 9.1 and 9.2. For the upper level we choose the height of the ridge crest, for the weight of ocean water above this level is the same everywhere; for the lower level we need only ensure that it is below the plate. To simplify the equations, we also choose one column to be

through the crest, so that it consists only of asthenosphere (Fig. 1a). Equating the weights of the two columns shown gives

$$h_d\rho_w + h_p\rho_p + h_a\rho_a = h_r\rho_a \qquad \text{Eq. 8}$$

Equating the heights of the columns:

$$h_d + h_p + h_a = h_r \quad \text{or} \quad h_a = \left(h_r - h_d - h_p\right) \qquad \text{Eq. 9}$$

Substituting for h_a from Eq. 9 into Eq. 8:

$$h_d\rho_w + h_p\rho_p + \left(h_r - h_d - h_p\right)\rho_a = h_r\rho_a \qquad \text{Eq. 10}$$

Collecting terms:

$$h_d\left(\rho_w - \rho_a\right) = h_p\left(\rho_a - \rho_p\right) \qquad \text{Eq. 11}$$

Rearranging:

$$h_d = h_p\left(\frac{\rho_p - \rho_a}{\rho_a - \rho_w}\right) \qquad \text{Eq. 12}$$

Note that though the density of the plate varies down the column (because the temperature increases), the average density will be the same in any column (this ignores the somewhat lighter density of the crust, but the error is small); also, the densities of the oceanic water and the asthenosphere beneath the plate are each approximately constant. Therefore, the expression in brackets in the last equation is the closely the same for any column, and so the depth, h_d, *below the ridge crest*, is proportional to the thickness of the plate, h_p. Since plate thickness increases in proportion to the square root of its age (Eq. 7), so does h_d.

its displacement decreases with distance until it dies away. The blocks in Figure 20.6b are different, with their common edge following the ridge axis and offset; as block *A* moves away from block *B* new plate is formed along the ridge crests, so the sides of the offset slide past each other as a *dextral* strike-slip fault. The fault, which does not extend beyond the ridge crests, is called a **transform fault**. Critical differences between the two explanations are given in Table 20.1.

These predictions can be tested in various ways.

One is by fault-plane solutions. The lower two solutions shown in Figure 20.7 are for earthquakes that occurred along offsets. There are always two possible faults that satisfy the first arrival data (Section 5.3.1); here they are a dextral strike-slip fault striking roughly E–W, or a sinistral N–S one. The second solution is implausible as it is inconsistent with both explanations for the offset and, taking into account nearby similar earthquakes, would require many faults perpendicular to the offset; so the evidence supports dextral faulting. The uppermost solution is for an earthquake on the ridge axis and it has mainly normal displacement with a small strike-slip component. The solutions are for a fault striking oblique to the ridge (azimuth 353°), which is implausible, or one striking parallel to the ridge, consistent with plates pulling apart. Magnetic anomalies also support transform faulting for they show that offsets do not decrease with distance from the ridge, but are consistent with new crust being created at ridge crests.

The transform fault explanation does not at first seem to account for why offsets often appear to extend outside the ridge crests, as a gash in the ocean floor called a **fracture zone** (Fig. 20.1). However, these occur because the parts of the plates on either side of a point on a fracture zone, such as *P* in Figure 20.8, are different distances from the ridge crests where they formed and so have different ages; they are therefore subsiding at different rates, producing the fracture zones and the occasional earthquake that occurs along them.

The offset of a transform fault does not increase with time – as it does with transcurrent

faulting – but is established soon after a plate splits in two.

20.4 Convergent, or destructive, margins

About half the ocean floor has been formed in the past 65 Ma, a rate sufficient to replace the whole of the Earth's surface in only 5% of its age. As there is no convincing evidence that the Earth is expanding at the required rate, plate has to be consumed, and this happens at converging margins. Constructive margins are always between two oceanic plates – because oceanic plate is what is 'constructed' – but converging margins can be between two continental or two oceanic plates, or one of each. What happens in a collision depends upon which types of plate are colliding, for whereas oceanic plate is denser than the mantle, continental plate is less dense, having been formed from lighter elements extracted from the mantle (Section 17.2.3).

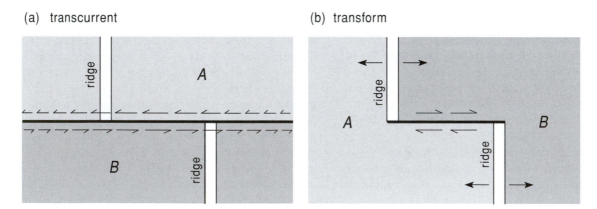

Figure 20.6 Two possible explanations for ridge offsets.

Table 20.1 Alternative predictions for ridge offsets

	Non-plate tectonic explanation – transcurrent fault	Plate tectonic explanation – transform fault
movement on fracture zone *between* ridges	strike-slip to *left*	strike-slip to *right*
movement on fracture zone *beyond* ridges	strike-slip to left, decreasing with distance	none
movement at ridges	none	tensional – normal faults

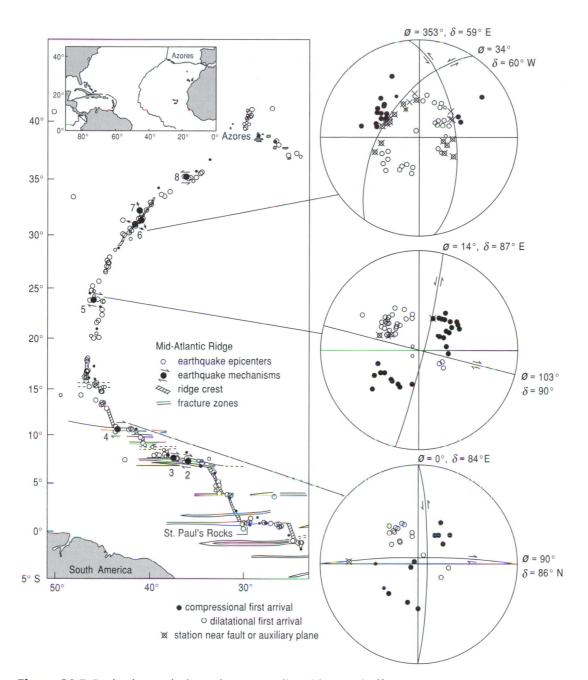

Figure 20.7 Fault-plane solutions along spreading ridges and offsets.

20.4.1 Ocean–ocean convergent margins and subduction zones

These are marked by **island arcs,** curved lines of volcanic islands such as are found in the western Pacific, for example. The *epicentres* of the many earthquakes form wider bands than those along ridges (Plate 2b); however, the *hypocentres* are found to lie in a fairly thin dipping zone (sometimes on two parallel planes), called a Benioff–Wadati zone, (Fig. 20.9; also Fig. 20.10 and Plate 1). This provides major evidence that a plate is being **subducted,** diving into the mantle and forming a **subduction zone.** The shape of a Benioff–Wadati zone can be quite complex, as Figure 20.10 shows for the northern end of the Tonga–Kermadoc–New Zealand subduction zone, which lies between the Pacific and Australian Plates (see Fig. 20.25).

Figure 20.8 Formation of fracture zones.

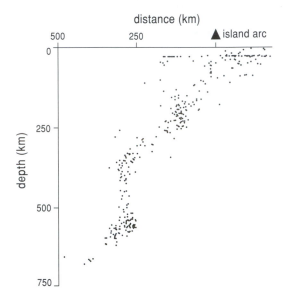

Figure 20.9 Earthquake hypocentres on E–W section through Tonga at 19° S.

A subduction zone has several characteristic features (Fig. 20.11). There is an **oceanic trench** 50 to 100 km wide where the subducting plate dives below the surface one, and these form the deepest parts of oceans. Nearer to the arc is an accretionary wedge or prism, shown by seismic reflection surveys to be imbricated slices of sediments scraped off the subducting plate. This is followed by a forearc basin, and then the arc of volcanic islands. The arc is believed to result from partial melting initiated at depth by release of water that is carried down in sediments that escape the accretionary wedge, and from hydrated rocks of the topmost igneous layers of the subducting plate. On the side of the arc remote from the trench there is often a **backarc spreading ridge**, whose origin is not well understood.

Stresses in the subducting plate can be deduced from fault plane solutions. Figure 20.12 shows some for earthquakes occurring near part of the Aleutians island arc. At the edge of the trench remote from the arc they show that normal faulting is occurring, attributed to bending of the plate (*A* in Fig. 20.12b). Most of the earthquakes beneath the island arc are due to thrusting of the subducting plate beneath the surface plate (*B* in Fig. 20.12b), at a shallow angle, with relative motion as shown by the arrows on the beach balls. The underthrusting is oblique to the trench, a feature commonly found in subduction zones. (The alternative solution would require each earthquake to be due to a steep fault trending

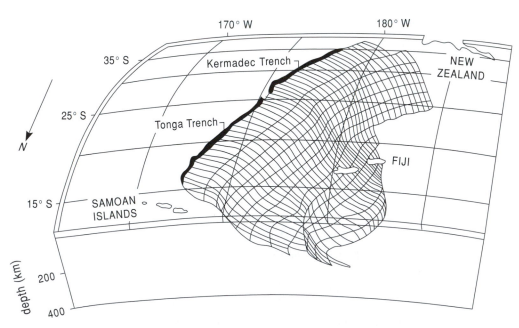

Figure 20.10 Benioff–Wadati zone beneath the Tonga–Kermadoc Trench.

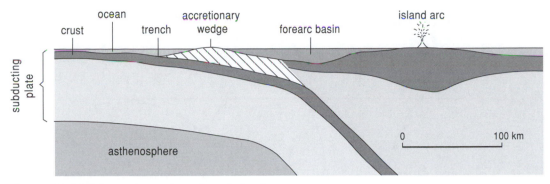

Figure 20.11 Schematic section of a subduction zone.

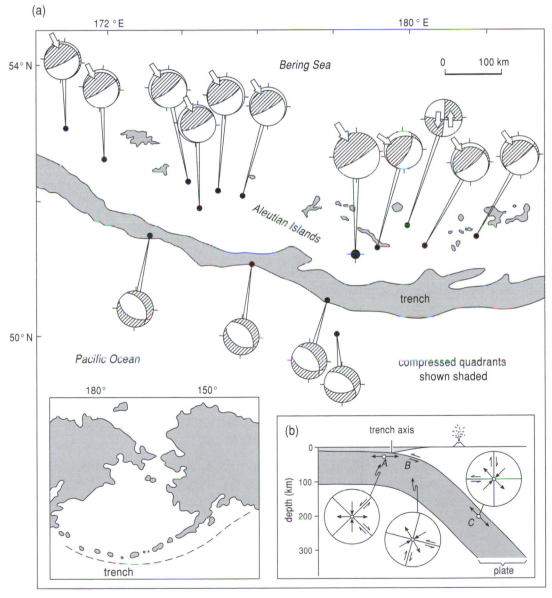

Figure 20.12 Fault-plane solutions for part of the Aleutian island arc.

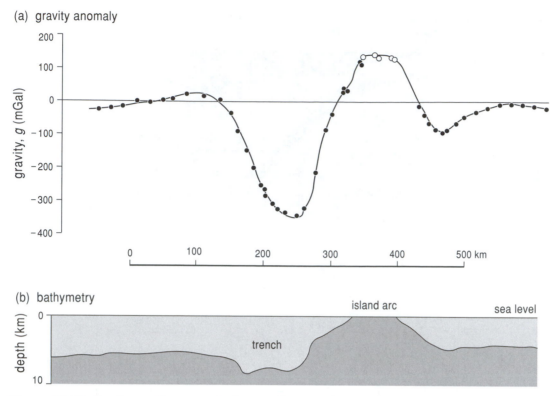

(a) gravity anomaly

(b) bathymetry

Figure 20.13 Gravity profile across the Puerto Rico Trench.

roughly NE–SW; where faulting is seen at the surface it confirms the first interpretation.)

Earthquakes occurring deeper than the base of the surface plate are not interpreted as faulting between plate and mantle because the mantle is believed to be too weak to be able to sustain large elastic strains. Instead, they are explained in terms of stresses within the plate (Section 5.3.2). Often the stresses show that the plate is in tension (C in Fig. 20.12b). The tension is due to the negative buoyancy of the plate, its weight pulling it down, thought to be the major force moving plates (Section 20.9). (More correctly, there is no actual tension but the compressive force is less than that expected because of lithostatic compression due to the weight of overlying rock; conversely, we talk of a compressive force when it is greater. The difference from the lithostatic compression is called the deviatoric force.)

The tension accounts for the very large negative gravity anomalies found at trenches (Fig. 20.13 and Plate 3a), which reach –200 mGal, and are the largest known anywhere. The immediate explana-tion for the anomaly is that the trench is filled with water and sediments, which have low densities. However, the anomalies are far too large and extensive to be supported by the strength of the lithosphere, so one would expect trenches to disappear by isostatic adjustment; as they do not, some downward force must be preventing it, identified as the negative buoyancy of the subducting plate. There is also a large positive anomaly over the arc, supported by the compressional forces resulting from the collision of the two plates.

20.4.2 Ocean–continent convergent margins

Plates may have both oceanic and continental parts. If an oceanic part continues to subduct, any attached continent will be brought to the subduction zone but will not subduct because of its great buoyancy; the subducting plate in such a collision is always the oceanic plate. The subduction zone is similar to that of an ocean–ocean one, but with the island arc replaced by a chain of mountains, which

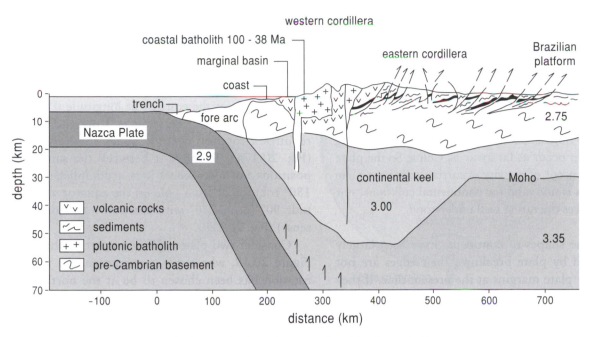

Figure 20.14 Section through the Andes (numbers are densities in Mg/m³).

are both folded – due to the compression – and volcanic – due to the magma ascending from the subducting plate. An example is the Andes of South America (Fig. 20.14), where the half-arrows indicate magma ascending from the subducting plate to form the coastal batholith of the western cordillera, while the eastern cordillera is primarily formed by compression. The mountains are supported by a root, the 'keel'.

The volcanic rocks found above subducting plate in both ocean–ocean and ocean–continent collisions contain minute but measurable amounts of ¹⁰Be, which is formed only at the surface, as explained in Section 15.12.2. This shows both that the ascending magmas derive some of their material from the subducting plate, and that this returns to the surface in less than 10 Ma, for otherwise the ¹⁰Be would have decayed to negligible amounts.

20.4.3. Continent–continent convergent margins

An ocean–continent convergent margin will further evolve into a continent–continent one when a continent joined to the subducting oceanic plate reaches the trench. Since the collision is now between plates both too buoyant to subduct, further subduction

largely ceases and instead the two continental plates compress each other, resulting in major orogenic mountains with folding and overthrusting; examples are the Alps, resulting from Africa moving northwards against Europe, and the Himalayas, due to India moving northwards into Asia (Fig. 20.15).

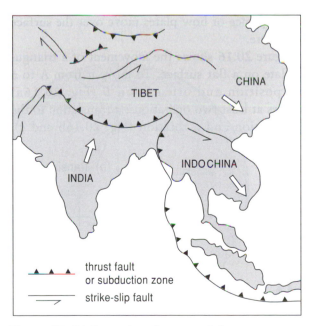

Figure 20.15 Tectonics of eastern Asia.

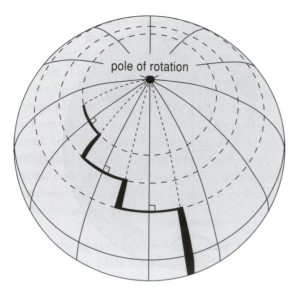

Figure 20.19 Finding the pole of rotation.

Converging margins also rotate about poles, so the rate of subduction increases with distance from the pole. This helps explains why the maximum depth of earthquakes varies along the Tonga–Kermadoc–New Zealand subduction zone (Fig. 20.20). As plates subduct they heat up until they become too hot to be able to sustain the elastic strains that produce earthquakes, but the faster they subduct the deeper they get before this temperature is reached; as the pole of rotation lies far to the south of New Zealand the deepest earthquakes are at the Tonga end, where subduction is fastest. The lines show calculated isotherms within the subducting plate.

20.5.2 Triple junctions and plate evolution

Where three plates meet is a **triple junction.** As the three margins involved can be divergent, convergent, or conservative, there are many possible combinations, 16 in total, and they evolve differently with time. Firstly, we shall consider how the relative velocities between the plates are related.

The simplest triple junction is three spreading ridges, an example being where the three Indian Ocean ridges meet (Fig. 20.25). Figure 20.21a shows a hypothetical triple junction with ridges spreading at different rates. The shaded strips are the amounts of new plate formed in each of the past four years. A point on Plate B is being carried away from one on Plate A at a velocity – the full spreading rate – shown by the arrow $_AV_B$ (i.e., velocity of plate B relative to plate A), while plate C is being carried away from B at velocity $_BV_C$. (As we are dealing only with a small area there is no need to use poles of rotation and rotation angles.) Added together, these two velocities must equal the velocity of C with respect to A, $_AV_C$, as shown in Figure 20.21b. Perhaps easier to remember is that on going completely around the triple junction the relative velocities of successive pairs of plates, $_AV_B$, $_BV_C$, and $_CV_A$, add to zero and so form a complete triangle (Fig. 20.21c).

This method can be extended to triple junctions with various types of margin and allows us to deduce the relative velocity between two plates on either side of a margin if those of the other two margins are known; this reveals what kind of margin it is. Figure 20.22a shows a triple junction, where a

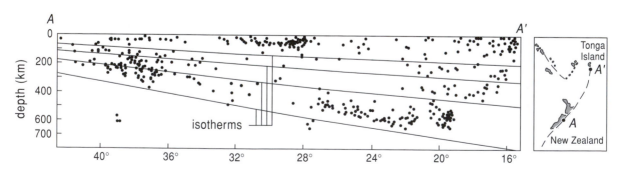

Figure 20.20 Depths of earthquakes along the Tonga–Kermadoc–New Zealand subduction zone.

(a)

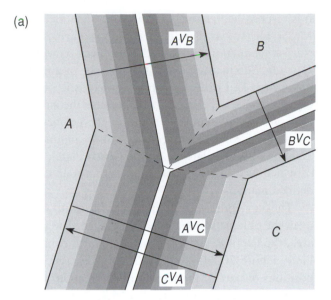

(b)

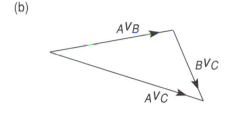

(c)

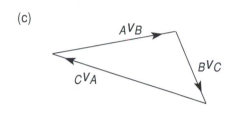

Figure 20.21 Triple junction of three spreading ridges.

ridge, a transform fault, and an unknown margin meet, with the spreading and strike-slip rates known. Figure 20.22b shows the two known velocities added end to end, and the triangle is completed to give $_cV_A$. As this shows that plates A and C are moving towards one another, the margin is a trench, though it does not show where the margin is nor which of the two plates is subducting.

This type of calculation was used to deduce the existence of an unknown plate, called the Juan de Fuca Plate, between the Pacific and North American Plates (Fig. 20.23). The various transform faults on the ridge apparently separating the Pacific and North American Plates are not parallel, so relative movement of these plates would lead either to gaps or to highly distorted plates. An additional margin was postulated, as shown by the long dashed line, and when the triple junctions at each end were solved, its relative velocity showed it to be a trench, subducting at about 2.6 cm/year, with a ridge–transform–trench triple junction at its northern end (like Fig. 20.22) and a transform–transform–trench one at its southern end. Investigation has confirmed the existence of a subduction zone beneath the western United States.

(a)

(b)

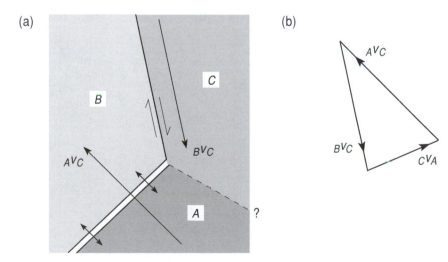

Figure 20.22 Triple junction of ridge, transform fault, and unknown margin.

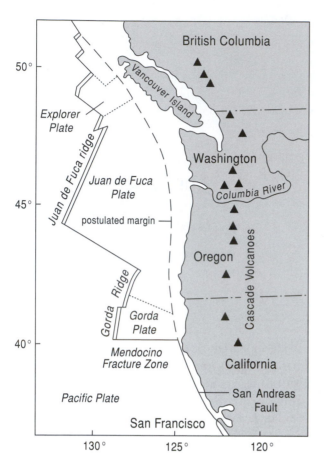

Figure 20.23 Discovery of the Juan de Fuca Plate.

It is responsible for the Cascade volcanoes, including Mt. St. Helens, with andesitic composition characteristic of subduction zones, and it has been named the Cascadia Subduction Zone. The absence of an obvious trench and other features characteristic of subduction zones is believed to be because the subduction rate is slow. Understanding of the geometry of the area has since been refined, with recognition of additional, small plates (e.g., the Gorda Plate), some of which have deformation zones rather than clearly defined margins.

The San Andreas Fault System to the south results from the oblique subduction of an earlier plate in this area, the Farallon Plate, beneath the North American Plate. Figure 20.24 shows, schematically, successive positions of part of the Farallon Plate margin, which consists of a length of spreading ridge between two transform faults with opposite senses; positions are shown superposed in Figure 20.24a and separately in Figure 20.24b to e. Once the spreading ridge has begun to subduct (Fig. 20.24c) the Farallon Plate has been separated into two parts; between them, the Pacific Plate is in contact with the North America Plate, and because it is not moving with the same velocity as the Farallon Plate there is a triple junction at each end, *T* and *T*. Calculation of velocities shows that the Pacific–North American plate margin is a transform fault, which takes up the oblique motion relative to the North American Plate. The margin between the Pacific and North American plates lengthens until the full length of the spreading ridge

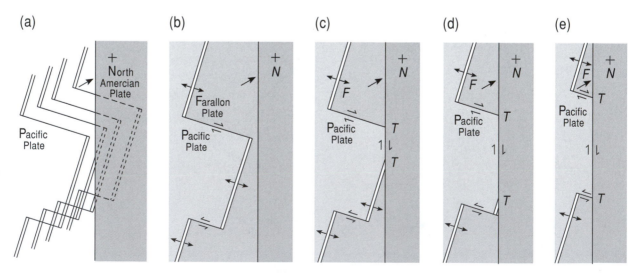

Figure 20.24 Formation of the San Andreas Fault, shown schematically.

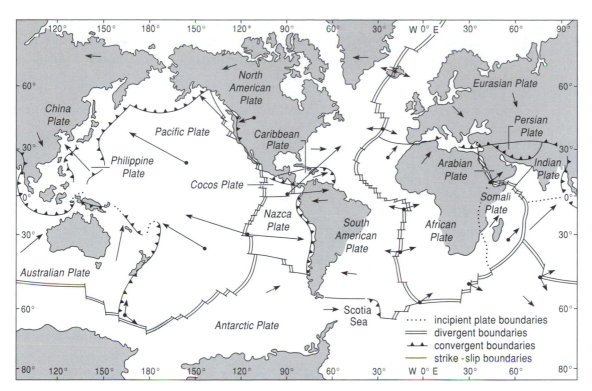

Figure 20.25 Plate tectonic map showing relative plate motions.

has been subducted, and then the southern triple junction becomes a transform–transform–trench junction and moves northwards until the next section of ridge reaches the margin. The transform margin between the Pacific and North American Plates has produced the San Andreas Fault, which formed about 20 Ma ago, while the remnants of the Farallon Plate form the Juan de Fuca and Cocos Plates (Fig. 20.25; subducted fragments of the plate have been imaged under North America by tomography, as high velocity zones extending down through the upper mantle).

20.6 The globe according to plate tectonics

Figure 20.25 shows the tectonic plates and their relative motions, together with spreading ridges, subduction zones, and conservative margins. Plates 2 and 3 provide other details. The longest feature is the interconnecting spreading ridge system of divergent margins, found in every ocean, including crossing the Arctic Ocean and surrounding Antarctica, with a length totalling about 75,000 km. Convergent margins (subduction zones) are prominent around most of the Pacific Ocean, so this ocean is contracting, despite the fast spreading rate of the

East Pacific Rise, in contrast to the Atlantic, Indian, and Antarctic Oceans which are expanding. The Alpine–Himalaya mountain chain results from the continent–continent collision of Africa, Arabia, and India with Euroasia.

As the East Pacific Rise spreading ridge nears the North American coast, the sections of spreading ridge are becoming shorter and the offsets or transform faults longer, until it consists almost entirely of the latter. It continues through the Gulf of California as a dextral strike-slip fault and continues into the United States as the San Andreas Fault system, the margin between the Pacific and North American plates. It is one of the most studied faults in the world because of the huge earthquake damage caused in the past and anticipated for the future, as well as for its plate tectonic interest. As mentioned in the previous section, the San Andreas reemerges into the Pacific to the north of San Francisco before it ends at the Mendocino triple junction. In the southwest Pacific, the Chile Rise is being subducted, together with parts of the Nazca and Antarctic Plates, beneath South America, the only current example of the subduction of a spreading ridge.

Recently formed margins are found in several places. India and Australia have probably been on different plates for about the past 8 Ma, with Australia rotating counterclockwise with respect to India, about a pole roughly on the equator to the south of India. At present, the relative motion is being absorbed by buckling and stretching of ocean floor without either spreading or subduction, though these are expected to develop as the deformation increases. A later stage of ocean development is found in the ocean floor at the western tip of New Guinea, where magnetic anomalies and other evidence reveal the complicated initial stages of ocean formation (Taylor et al., 1995).

The Red Sea is at a later stage of oceanic development, being an extension of the Central Indian Ridge connecting with the Alpine–Himalayan compression belt via a fault up the Dead Sea Rift. It formed when the Afro-Arabic Plate split 20 to 30 Ma ago, and the presence of magnetic stripes along the centre of the Red Sea shows that ocean floor is being formed (described further in Section 21.1). The angle between the Red Sea and Gulf of Aden spreading ridges probably marks a triple junction, with an third, incipient spreading limb extending southwest through Ethiopia into eastern Africa. This suggests that the East African Rift may become a new ocean, as discussed further in Chapter 21.

20.7 Continental positions in the past

Plates are moving, so their positions must have been different in the past. The easiest way to reconstruct past positions is to 'delete' the most recently formed ocean floor. For example, cutting out those parts of the Atlantic Ocean that formed in the past 20 Ma, say (Plate 3b), and closing the gap shows where the Americas were with respect to Europe and Africa 20 Ma ago. However, this approach cannot be used to restore ocean floor that has been subducted, which is a severe limitation to reconstructing the Pacific Ocean in the past, nor can it be used for reconstructions earlier than the oldest ocean floor, which is about 180 Ma old. However, for older reconstructions we can call upon palaeomagnetism (Section 10.2.3), though this has the limitation that, as we cannot determine palaeolongitudes from the magnetic data alone, easterly or westerly movements may not be recognised, and

also errors may be quite large. These can be reduced by a third approach, which is by matching geological formations and structures that have been separated by continental splitting, or recognising sutures formed when continents have collided, marked by old mountain chains such as the Urals.

Figure 20.26 shows the results of combining these approaches, for a succession of times in the past. About 450 Ma ago, in the mid-Ordovician, the globe looked very different from today, with many small land masses. As time went on, these progressively amalgamated to form a single huge supercontinent, named Pangaea, which existed for a period, then broke up to form the continents we have today. Some of the stages that shaped our present globe are described briefly.

The Iapetus Ocean – which was roughly where the Atlantic now is but with Scotland on the 'American' side, and with Newfoundland in two parts – closed, to form the suture of which the Caledonian and Appalachian Mountains form part. Asia assembled from several pieces and then fused with Europe about 250 to 300 Ma ago, producing the Urals. By the late Triassic nearly all the continents had combined to form Pangaea, though Asia was separated from Afro-Arabia by the Tethys Ocean. When Pangaea broke up about 200 Ma ago, the Atlantic, Indian, and Antarctic Oceans opened, but Tethys closed, with the Mediterranean its relic. The southern continents – South America, Africa, Antarctica, Australia, and India, collectively known as Gondwanaland – broke up in stages, with India moved rapidly north during the Cretaceous, to collide with Asia in the early Tertiary. Australia and New Guinea separated from Antarctica only in the Tertiary and since have been moving north, as did South America, which connected with North and Central America only late in the Tertiary.

There were probably earlier supercontinents, in the Precambrian. The sequence of breakup of one supercontinent to the formation of a subsequent one, as the fragments recombine in a new combination, is known as the **Wilsonian cycle**. At present, we are probably in mid-cycle, with the Americas moving away from Europe and Africa as the Atlantic opens, but towards Asia and Australasia as the Pacific Ocean contracts.

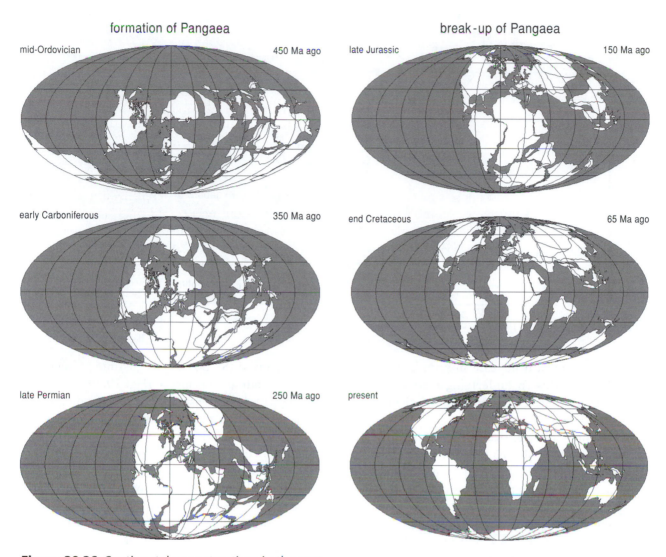

formation of Pangaea

mid-Ordovician   450 Ma ago

early Carboniferous   350 Ma ago

late Permian   250 Ma ago

break-up of Pangaea

late Jurassic   150 Ma ago

end Cretaceous   65 Ma ago

present

Figure 20.26 Continental reconstructions in the past.

20.8 Crust formation at ridges

In Section 20.2 spreading ridges were merely described as where new plate forms from upwelling mantle, without explaining, for instance, how it acquires a crust; this section examines in more detail what happens below a ridge. Seismic studies show that there are several layers beneath the ocean floor, with increasing seismic velocities. The topmost ones have been sampled by drilling, while the full section can be compared to the layers seen in ophiolites, portions of ocean floor that have been **obducted** (i.e., thrust on to land). Figure 20.27 shows the layers: Layer 1 is sediments, which have been deposited since the plate formed and thicken

with distance from the ridge axis; Layer 2 at its top is pillow lavas (formed when magma is extruded into sea water), but these progressively give way downwards to dykes, which once fed magma to the lavas, until there are only dykes, forming a sheeted complex; Layer 3 is gabbro. The lavas, dykes, and gabbros all have basaltic composition, but their seismic velocities differ because the upper layers are less compacted and have been hydrothermally altered by sea water while still hot. Ocean water penetrates 2 to 3 km down cracks, where it heats up and alters or dissolves minerals before returning to the surface in a convective system; the cooling resulting from mixing with cold ocean often precipitates sulphides as 'black smokers' and sulphide deposits.

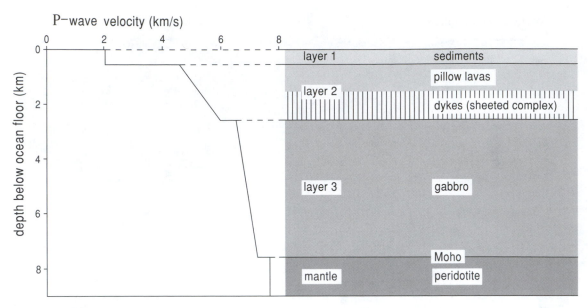

Figure 20.27 Oceanic plate layers.

The next downward increase in seismic velocity is from the gabbros to underlying ultrabasic peridotites and dunites, a compositional change that forms the Moho.

How does the basaltic crust derive from the ultrabasic mantle? Magma forms when the temperature of the upwelling material exceeds the **solidus** (the lowest temperature at which partial melt exists); the solidus temperature depends upon pressure and hence depth, as shown in Figure 20.28. The temperature of upwelling mantle follows an adiabat (Section 17.1.4), and when upwelling mantle crosses the solidus (M), melt begins to form, with basaltic composition. Being both liquid and less dense than the solid from which it derived, it percolates up towards the surface and combines to form a magma. However, geochemical evidence shows that this magma does not directly produce the basaltic crustal rocks, for there is a further stage of differentiation, which suggests a crustal magma chamber. Seismology confirms this and reveals its extent.

Seismic reflection surveys have shown (Fig. 20.29a) that below the East Pacific Rise there is a narrow, near-horizontal reflector, 1 to 2 km below the ridge axis and 1 to 2 km wide. The strong reflection indicates a high percentage of liquid and hence a magma. No base is seen, which might be either because it grades into a mush of crystals plus liquid or because it is too thin for the lower surface

to be resolved (Section 7.8.2). Seismic tomography provides additional evidence (Fig. 20.29b). There is a zone with lower seismic velocity extending down from the magma chamber, widening towards the base of the crust, and with its lowest velocities in a roughly circular zone at the top. The lowered velocities are probably due to the presence of liquid, but this cannot be more than a few percent, for the P-wave velocity reduction is no more than about 1 km/sec compared to normal mantle away from the ridge axis at this depth (i.e., Layer 3 gabbro).

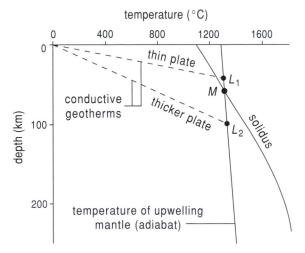

Figure 20.28 Solidus and mantle temperatures below a ridge.

(a) seismic reflection

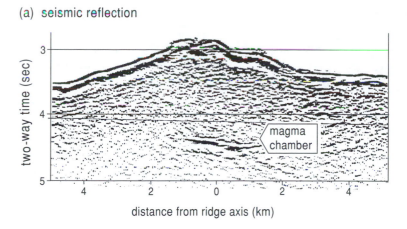

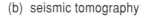

(b) seismic tomography

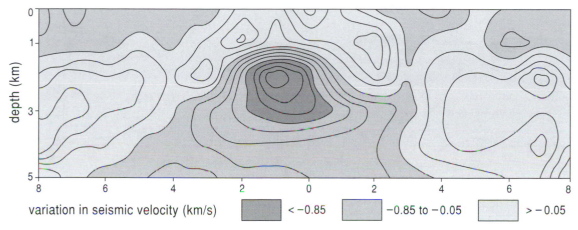

Figure 20.29 Seismic tomographic and reflection results for the East Pacific Rise.

Its boundaries are probably gradational. Thus the seismic tomography and reflection results together indicate that the magma chamber is quite small but lies above a crystal mush zone with the percentage of liquid decreasing downwards and sideways, until it becomes solid gabbro (Fig. 20.30a and b).

Once magma has solidified to form plate it no longer transports heat upwards by convection; instead, heat *conducts* up through it, which results in much steeper temperature profiles, or geotherms (Section 17.1.4), shown by the dashed lines of Figure 20.28; the point where such a profile meets the mantle adiabat (*L*) marks the depth of the base of the plate or lithosphere; it is not a sharp boundary but gradational. As the thickness increases *L* moves downwards until it is below the solidus and partial melting is not possible, which is why magma production is limited to near the ridge axis (except

where exceptionally hot mantle rises, as in plumes, Section 20.9.2).

The seismic evidence given above was for the East Pacific Rise, which spreads rapidly, exceeding 10 cm/year (half-spreading rate) in places. **Slow-spreading ridges,** such as the Mid-Atlantic Ridge (half-spreading rate about 2 cm/year), have a rift valley along their crests and differ from **fast-spreading ridges** in other ways. No strong reflector is seen in reflection surveys of the Mid-Atlantic Ridge; together with evidence that small earthquakes occur down to 8 km, this indicates that there is no semi-permanent magma chamber (Fig. 20.30c). Instead, eruptions are likely to be fed more or less directly from the mantle, with magma accumulating temporarily in discrete centres below the ridge crest, as indicated by positive gravity anomalies of tens of milliGals (Lin et al., 1990).

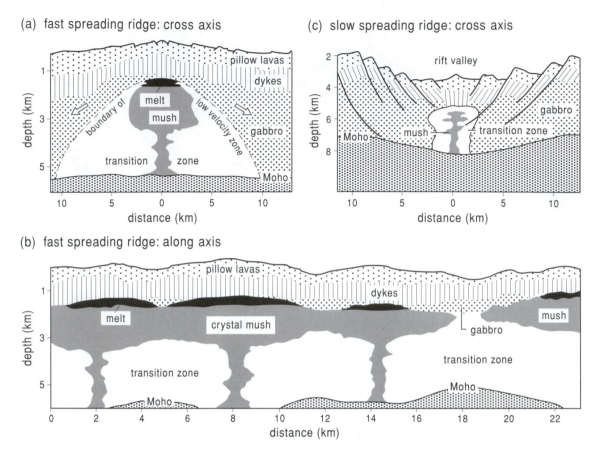

Figure 20.30 Schematic sections of fast- and slow-spreading ridges.

The actual processes at ridges undoubtedly are more complex than described above, as shown by the complexity of the topography of ridges, variations in composition of rocks along the ridge crests, and evidence that the spreading axis can move or have overlaps. There is also evidence that lavas are sometimes erupted up to 4 km from the ridge axis, revealed by uranium-series dates (Section 15.12.1) that show them to be significantly younger than the date calculated from their off-axis distance and the spreading rate.

20.9 What moves the plates?

Plate tectonic theory is largely kinematic, for it is mainly about how plates move and the tectonic consequences, but, of course, we also want to know what moves them.

20.9.1 Forces on plates

There is little doubt that plates are moved by some form of convection: High temperatures deep in the interior of the Earth result in hotter and so less dense material rising, while elsewhere cooler, denser material sinks to complete the circulation. This suggests that spreading ridges are over upwellings, with plates being carried away from the ridges by the horizontal flow below them, like logs on a stream (Fig. 20.31a). But there are good reasons for thinking plates are probably not moved by such **basal drive.**

Two other possible drive forces have been suggested. **Slab pull** (Fig. 20.31b) is the idea that plates subduct because they are denser than the mantle, as mentioned in Section 20.4.1; the subducting plate is often called a **slab.** When ocean plate forms at a ridge axis it has about 7 km of basaltic crust, which

(a) basal drive

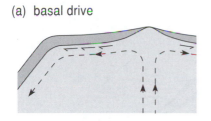

(b) slab pull

(c) ridge push

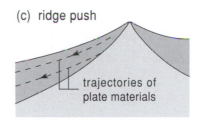

trajectories of
plate materials

Figure 20.31 Possible drive forces.

is significantly lighter than the mantle beneath, but the plate thickens by the addition of cooled mantle to its base and the added material is denser than the mantle beneath it because it is cooler. By the time the plate is about 10 Ma old its average density equals that of the mantle, and thereafter it becomes denser. So old oceanic plate is gravitationally unstable (unlike continental plate) and so liable to sink. A similar situation is seen in lava lakes within volcanoes where a solid skin forms and then founders, reproducing many of the features of plate tectonics (Duffield, 1972). Once a plate dives into the mantle, forming a subduction zone, it will be slow to heat up by conduction of heat into it from the hotter surrounding mantle, and so the temperature difference below its surroundings will increase, increasing its downward pull. (Its negative buoyancy is further increased when it sinks deep enough for a phase change to a denser form occurs; this will be at a shallower depth than in the surrounding mantle because it is cooler.) Evidence that plates are being pulled down by their weight comes from fault-plane solutions and gravity, as described in Section 20.4.1.

The other possible drive force is **ridge push**, which derives from the shape of ridges (Fig. 20.31c). Just as lava tends to flow downhill, so does the plate itself: Any point within the plate follows a descending path as it moves away from the ridge axis. The force does not arise at the ridge between the two plates, forcing them apart, but on the slopes of the ridge.

As well as these three possible drive forces (shown on Figure 20.32 as F_{BD}, F_{SP}, and F_{RP}) there are possible retarding forces – for instance, where plates slides past one other at transform faults (F_{TP}) and converging margins (F_C), demonstrated by the occurrence of earthquakes. In addition, a subducting plate has to overcome the resistance of sliding through the viscous mantle, to which may be added extra resistance if it reaches the 660-km discontinuity (F_{SR}). Further, if

plates are driven by slab pull or ridge push, rather than basal drive, there will be a **basal drag** force as they move over the underlying mantle (i.e., the plate will tend to drag the mantle with it, rather than vice versa). These forces are shown in Figure 20.32, where F_{BD} stands for basal drag or drive as the case may be.

How can we tell which of these many forces are important? Suppose that slab pull is the major drive force. Then a plate with a large amount of subducting edge would go faster than one with little. This idea can be extended to all the forces. To simplify the analysis, the forces are classified as either edge or areal forces: Apart from the basal-drive or drag force, all the forces act predominantly at or near the edges of plates. This simplification is justified as an approximation because, though ridge push, for example, acts for hundreds of kilometres before a plate has descended most of the way from a ridge crest, and a descending slab can be equally long, these distances are still considerably smaller that the widths of most plates, which are thousands of kilometres across. To estimate the size of a force we add up the percentage of edge that is subducting, spreading, or whatever, or its area in the case of basal forces. However, we need to make a correction if an edge force acts on opposite edges of a plate. For example, the Africa Plate has ridges on both its east and west sides, so any ridge-push forces these may exert can be expected to cancel. After allowing for such cancellation we have the corrected percentage of edge force. Thus Antarctica, which is almost surrounded by ridges (Fig. 20.25), has only a small corrected percentage of ridge edge.

To carry out the analysis we also need the velocities of the plates. This is not their velocities relative to one another, but their 'absolute' velocities over the mantle, for this is what determines the size of the mantle force on the base of a plate. Finding these velocities is a problem in its own right.

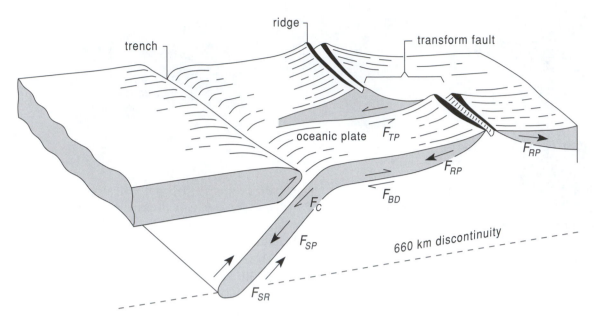

Figure 20.32 Forces acting on plates.

20.9.2 The hot-spot frame of reference: Plate velocities

Hot spots are localised volcanic areas not related to plate boundaries (though they may occur at plate boundaries). The best-known example is the Hawaiian Islands, far from any plate margin (Fig. 20.33). These islands are at the end of a long chain of mostly submerged volcanic seamounts, known as the Hawaiian–Emperor seamount chain, and potassium–argon dating has shown that the age increases progressively along the chain away from the currently active volcanoes, reaching 78 Ma where it is being subducted under the Aleutians. The source of volcanic activity is believed to be a **plume,** a column of hot material rising from deep below the plate, possibly originating at the core–mantle boundary, and the chain marks the passage of the Pacific Plate over it.

There are dozens of other hot spots (such as Yellowstone in the United States, the Canaries in the Atlantic, and Iceland on the Mid-Atlantic Ridge) and though their relative velocities are not zero – hardly to be expected in a convecting mantle – their velocities of 2 to 5 mm/year are much smaller than those between plates, and so they approximate to a frame fixed in the mantle, known as the **hot-spot frame of reference.**

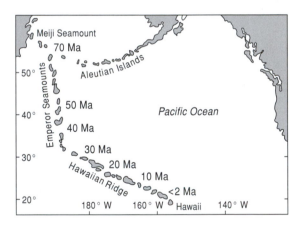

Figure 20.33 Hawaiian–Emperor seamount chain.

Starting with, say, the Pacific Plate, its velocity relative to the hot-spot frame is 9 cm/year along the direction of the Hawaian–Emperor chain towards the southeast. Then the velocity of, say, the Nazca Plate is found by adding to this the *full* spreading velocity of the East Pacific Rise, and so on for all the plates. In fact, this is done using rotation rates about poles of rotation, as explained in Section 20.5.1. Figure 20.34 shows the resulting 'absolute' velocities; the arrows within a single plate are not parallel because plates are wrapped over the surface of the Earth and motion is about a pole of rotation.

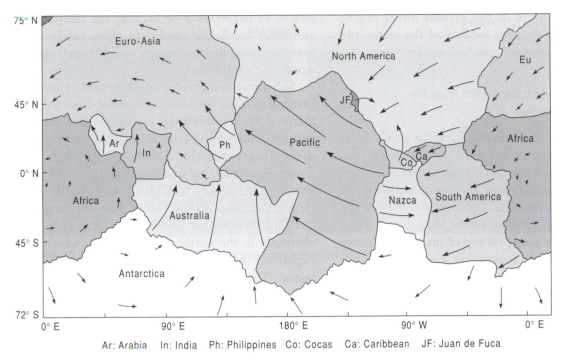

Figure 20.34 "Absolute" velocities of the plates. Speeds are proportional to lengths of arrows.

Ar: Arabia   In: India   Ph: Philippines   Co: Cocas   Ca: Caribbean   JF: Juan de Fuca

20.9.3 Deducing the dominant drive forces

We are now ready to determine which are the dominant forces. In Figure 20.35, plate velocities relative to the hot-spot framework are shown along the base of each of the diagrams; they have a wide range. In Figure 20.35a, this is compared with the areas of the plates, which also range widely, from the tiny Nazca and Cocos plates to the giant Pacific, yet these three plates have nearly the same velocity, so we conclude that basal drive is not important (nor basal drag either), for if it were the Pacific would be much the fastest. However, Figure 20.35b

shows that, in general, the larger the percentage of trench edge after correction the faster the velocity. This supports slab pull as being the dominant drive force. Velocity also increases with percentage of ridge edge (Fig. 20.35c), though the correlation is not so strong, so ridge push is a less important drive force. The existence of a significant ridge-push force explains why some pairs of plates, such as the South American and African ones, are moving apart despite neither having a significant amount of subducting edge to move them. None of the retarding forces gives a significant correlation with velocity.

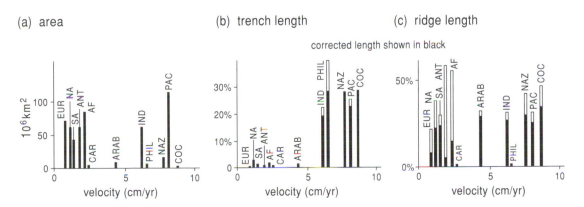

Figure 20.35 Plate velocities versus possible drive forces.

the slab is in tension unless it penetrates to the base of the upper mantle. Above a slab subducting beneath oceanic plate there are successively, starting close to the hinge line of the bend, an oceanic trench, an accretionary wedge/prism, a volcanic arc, and then sometimes a back-arc spreading ridge. If subduction is beneath continental plate, the succession is similar except that the island arc is replaced by complex mountains, formed by both the emplacement of magma and by compression; there is no back-arc ridge, of course.

Continent–continent collision results principally in mountains formed by compression, with folding and overthrusting.

5. Conservative margins do not affect plate area. They occur between offsets of a spreading ridge, as a strike-slip fault. Their main manifestations are fracture zones (on the ocean floor), and movement on the fault, which results in earthquakes.

6. The motion of a plate can be fully described by a pole of rotation and the rate of rotation about it. Poles often remain fixed for millions of years, but if they move successive instantaneous poles are used to define their path. Poles of rotation can be deduced from the orientation of ridge offsets, or by the tapering of sea floor magnetic stripes. Relative plate motions are often given as velocities, measured in cm/year, though this strictly applies only to one place along a margin, as the rate varies with distance from the pole.

7. Triple junctions are where three plates meet. If the relative motions of two of the pairs of plates are known, the relative motion of the third pair can be calculated. This has been used to deduce the presence of a hitherto unknown plate margin.

8. A continental margin may or may not be a plate margin. One that is a plate margin is described as an active margin, because tectonic processes occur there, while one that is not is a passive margin.

9. Continents follow the Wilsonian cycle, coming together to form supercontinents, which in turn break apart with new oceans forming in the gaps, whose growth leads to new aggregations of continents. The present distribution of conti-

nents results from the breakup of Pangaea. The present topography of the earth is due to both present and past tectonic processes.

10. The depth to which the base of a plate extends is not well defined. Roughly, it corresponds to the thickness of the lithosphere (itself defined in various ways) and represents the crust and upper mantle sufficiently cool to be approximately rigid. Oceanic plate thickens progressively away from the spreading ridge in proportion to the square root of its age.

11. Plumes are narrow columns of material actively ascending from deep in the mantle, because of their high potential temperature. Relative velocities between plumes are much smaller than those between plates, and so plumes provide the hot spot frame of reference, which is used to deduce absolute plate velocities.

12. Plates are moved by convection. They form an intrinsic part of the convection cycle and provide most of the driving force, primarily through the negative buoyancy of subducting slabs (indicated by the large negative gravity anomalies over trenches), but aided by ridge push at spreading ridges. Upwelling beneath ridges is passive, resulting from the separation of plates, not causing it.

13. The concept of plates as being rigid with tectonic activity confined to their margins is only an approximation to what actually happens. It is least validly applied to continental collisions, where deformation can occur in zones that extend for thousands of kilometres from the boundary. Nor does it include the effects of plumes.

14. The main geophysical methods used for investigating plate tectonics are seismic – particularly the distribution of earthquakes and fault-plane solutions but including seismic refraction, reflection, and tomography – gravity, and magnetics, both palaeomagnetism and surveying. Heat flow and radiometric dating have also made important contributions. Conventional geology has contributed through observation of ocean floor that has been obducted and of mountain chains, where these have been dissected, as well as being the main source of knowledge about plates and their processes in the past. Geochemistry is essential to help

understand igneous processes associated with plate margins. All three approaches have been applied to drilled holes and their cores, mostly in the ocean floors, thereby yielding much detailed information.

15. You should understand these terms: plate, plate tectonics; divergent, convergent, conservative, constructive, destructive, active and passive margins; sea floor spreading, spreading ridge, backarc spreading ridge, fast- and slow-spreading ridges, (half-) spreading rate, magnetic stripe; obduction, subduction, subduction zone, slab, Benioff–Wadati zone; transform fault; Wilsonian cycle; island arc, oceanic trench; fracture zone; pole of rotation, Euler pole, instantaneous pole; triple junction; slab pull, ridge push, and basal-drive/drag forces; plume, hot spot, hot-spot frame of reference; solidus.

Further reading

Kearey and Vine (1996) give a thorough account of global tectonics. Cox and Hart (1986) explain the basic ideas, and concentrate on geometrical concepts, with lots of numerical examples. Fowler (1990) covers the basic theory at a more advanced level and relates plate tectonics to many features of the lithosphere, while Park (1988) integrates structural geology with plate tectonics. Cox (1973) contains most of the papers that established plate tectonics and how it functions. Sinton and Detrick (1992) give a review of processes below ridges, while MacDonald and Fox (1990) provide an nontechnical account of how ridge topography evolves. A detailed investigation of a fast spreading ridge is given in *The MELT Seismic Team* (1998). Duffield (1972) describes the behaviour of lava lakes, which show many of the features of plate tectonics.

Problems

1. Distinguish among constructive, conservative, convergent, destructive, divergent, passive, and active margins.
2. A tectonic region that has earthquakes with mostly shallow hypocentres and tensional focal mechanism is most likely to be
   (i) A continent/continent collision zone.
   (ii) A large continental shear zone.
   (iii) A transform fault.
   (iv) A mid-ocean ridge.
   (v) A subduction zone.
3. Japan suffers earthquakes because it is associated with a
   (i) Spreading ridge.
   (ii) Ridge offset.
   (iii) Subduction zone.
   (iv) Transform fault.
   (v) Transcurrent fault.
   (vi) Hot spot.
4. The fault movement associated with conservative plate boundaries is which one of the following?
   (i) Thrust faulting.
   (ii) Strike-slip faulting.
   (iii) Normal faulting.
5. The east coast of South America is which of the following types of margin?
   (i) Conservative.
   (ii) Constructive.
   (iii) Destructive.
   (iv) Passive.
   (v) Active.
6. Which of the following types of margin are applicable to the east coast of Japan?
   (i) Conservative.
   (ii) Constructive.
   (iii) Convergent.
   (iv) Destructive.
   (v) Divergent.
   (vi) Passive.
   (vii) Active.
7. Which one of the following is *not* associated with an oceanic spreading ridge?
   (i) High heat flow.
   (ii) Earthquakes in the top 20 km.
   (iii) Earthquakes below 100 km.
   (iv) Pillow lavas.
   (v) Linear magnetic anomalies.
8. Active volcanoes are associated with which of the following tectonic settings?
   (i) Divergent plate boundaries.
   (ii) Convergent plate boundaries.
   (iii) Conservative plate boundaries.
   (iv) Transform faults.
   (v) Hot spots.
9. Which one of the following is likely to have

earthquakes only between the depths of 0 and 15 km?

(i) A conservative margin.

(ii) A zone of continental collision.

(iii) A passive continental margin.

(iv) A hot spot.

(v) An ocean–ocean subduction zone.

10. The deepest earthquakes are found in which one of the following tectonic locations?

(i) Mid-ocean ridges.

(ii) Subduction zones.

(iii) Himalayan-type mountains.

(iv) Sedimentary basins.

(v) Deep-ocean trenches.

11. What are the major forces that move plates, and what is the evidence that they do so? What are the forces that retard the motions of plates?

12. The simplest way to move a rigid body from one position to any other on the surface of a sphere is

(i) A translation.

(ii) A rotation.

(iii) A translation followed by a rotation.

(iv) A rotation followed by a rotation.

(v) Two translations.

(vi) Two rotations.

13. What is a pole of rotation? Why is it useful in plate tectonics?

14. If the spreading rate measured at angular distance of 90° from the pole of rotation is 10 cm/year, the spreading rate at a distance of 180° is:

(i) 0.0, (ii) 4.00, (iii) 5.00, (iv) 7.01, (v) 8.66.

15. What is the approximate location of the pole of rotation for the relative movement of North America with respect to Europe?

16. Relative plate motions *must* be which of the following?

(i) Perpendicular to a ridge.

(ii) Parallel to a ridge.

(iii) Perpendicular to a trench.

(iv) Oblique to a trench.

(v) Parallel to a transform fault.

17. Explain why we believe that though there is a large volume of hot material below the axis of a spreading ridge, the volume of magma is small.

18. Explain how oceanic magnetic anomalies form. What is their importance for plate tectonics?

19. The age of the oldest ocean crust found in situ is (in Ma):

(i) 25, (ii) 180, (iii) 580, (iv) 2100, (v) 3900.

20. The Moho is deepest under which one of the following?

(i) Fennoscandia.

(ii) The Mid-Atlantic Ridge.

(iii) Hawaii.

(iv) The Himalayas.

(v) The East African Rift Valley.

(vi) Australia.

21. Explain how ocean ridge offsets might be explained in terms of either transcurrent or transform faulting. Describe how the evidence of earthquake locations, fault-plane solutions, magnetic anomalies, and ocean-floor topography can be used to discriminate between the alternatives.

22. Explain briefly why oceanic ridges are higher than ocean basins.

23. Plates A, B, and C meet at a triple junction, with the A–C boundary extending northwards with plate C on the east side, the B–C boundary southwards, and the A–B boundary westwards. The boundaries between A and B and between B and C both have dextral strike-slip motion of 6.0 cm/year. What is the velocity of A with respect to B, and what kind of boundary is it?

24. Plates A, B, and C meet at a triple junction, with the A–B boundary extending northwards with plate B on its east side and the B–C boundary southwards, while the A–C boundary has an azimuth of 240°. The boundary between A and C is a spreading ridge with half-spreading rate of 3 cm/year, and its azimuth is 240°, while plate C is subducting beneath B at 3 cm/year, with both spreading and subduction perpendicular to the margin. What is the relative velocity of A with respect to B, and what kind of boundary is it?

25. Distinguish between relative and absolute plate velocities. Explain how the relative velocities of the South American and African plates can be measured. Explain how the absolute velocity of the Pacific plate can be measured.

26. Give one piece of evidence that suggests the 660 km discontinuity is a barrier to convection, and one that convection flow crosses it.

chapter 21

Is the Kenya Rift a New Plate Margin? A Regional Geophysical Study

Tectonic plates go through the Wilsonian cycle, supercontinents splitting into continents with the formation of new oceans between; elsewhere established oceans contract due to subduction and ultimately bring continents together to form a new supercontinent. The breakup of continents is poorly understood, but the East African Rift System is probably an example of the process in its early stages.

In this chapter we examine this idea by presenting investigations into the deep structure below part of the Rift using a variety of geophysical techniques, as an example of regional geophysical surveying.

Geophysical methods employed: Mainly gravity, seismicity, and seismic refraction, but also e-m (TEM) and geothermics

21.1 Introduction: The East African Rift System

In Section 20.6 it was suggested that there may be a triple junction where the axes of the Red Sea and the Gulf of Aden meet at an angle, with the third arm extending southwestwards through Ethiopia and then through East Africa to the Zambezi and perhaps beyond. If this is indeed the case, the third arm has not yet developed to the point where oceanic crust is forming, and it therefore offers the best location on earth to investigate what happens in the early stages of continental breakup.

The Gulf of Aden is an extension of the Central Indian Ocean Ridge, and it shows the pattern of symmetrical linear magnetic anomalies found near active spreading ridges (Fig. 21.1). The anomalies extend into the Red Sea but taper noticeably, showing that they are nearing the pole of rotation (Section 20.5.1) (the pole is in the eastern Mediterranean). The oldest anomaly is number 5, which formed about 10 Ma ago (Section 10.5.1), showing that oceanic crust began to form at about this time. Independent evidence of rotation about a pole in the eastern Mediterranean is provided by palaeomagnetism, for 10-Ma-old rocks in Arabia have palaeomagnetic directions about 7° west of the present field (Fig. 21.2), but 'closing' the Gulf restores the direction to north.

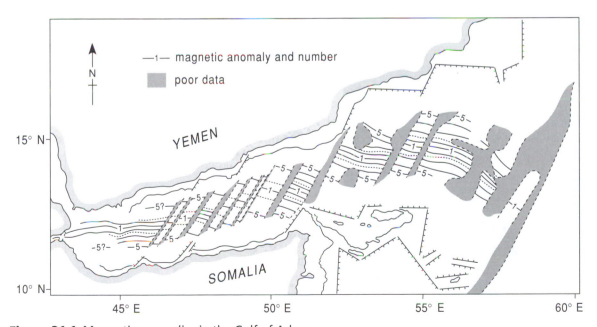

Figure 21.1 Magnetic anomalies in the Gulf of Aden.

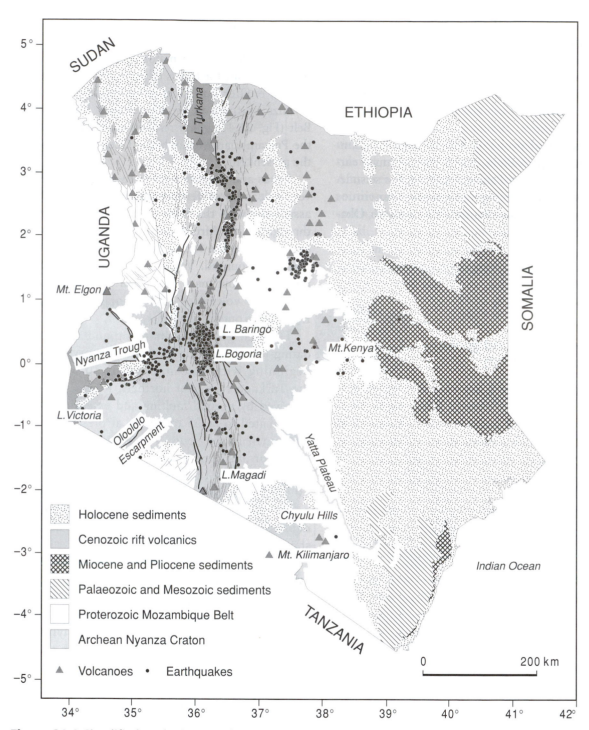

Figure 21.4 Simplified geologic map of Kenya with earthquake epicentres and volcanoes.

21.3 Gravity studies

From a general understanding of rifts we expect that gravity anomalies could result from a number of causes. The infilling on the rift floor by low density lavas and sediments will produce a negative anomaly. However, the shallowing in depth to the Moho resulting from crustal thinning, which brings the denser mantle closer to the surface, will produce a positive anomaly, but this positive anomaly is likely to be reduced by the mantle beneath the rift being hotter and so having a lower density than normal mantle to either side. There may also be lateral variations of density within the crust due to juxtaposition of rocks of different density. Resolving all these above causes is not possible by gravity alone, but because of the effort and cost of obtaining seismic and other geophysical data on a regional scale, many of the early ideas about the deep structure of the rift were based upon gravity data, and we shall describe some of the early models.

The first study was carried out in the early 1930s (using pendulum measurements rather than modern gravimeters, Section 8.5). It was a large-scale survey and established that the uplifted plateau on the east side of the rift was isostatically compensated but over the Rift gravity is negative. Later, more detailed surveys showed that gravity in the rift is much more complicated than the simple broad negative anomaly found in the early studies, as the gravity profiles of Figure 21.5 demonstrate. For example, a gravity profile across the rift in the vicinity of the volcano Suswa (Fig. 21.6a) showed a positive anomaly within the broad negative (Fig. 21.5a). This was interpreted by Searle (1970) in terms of a dense mantle-derived intrusion 20 km wide extending from a depth of 20 km to within 2 or 3 km of the rift floor. Independently, in 1971 Baker and Wohlenberg interpreted the positive anomaly observed along a profile further north as also being due to high-density material intruded into the crust. But in addition, they attributed the broad negative anomaly to anomalous mantle material with lower than average density due to higher temperature beneath the rift.

These two models stimulated a lot of interest in the rift because they implied, by their wide intrusions along the axis of the rift, that crustal separation had already taken place by ten or more kilometres (i.e., that the continent had already split). Clearly, the gravity models proposed are speculative, largely due to the nonuniqueness of the method (Section 8.8.1), as they lack independent information on densities or depths of the postulated bodies. This could be provided from boreholes or seismic surveys, and we shall next describe what has been found using various kinds of seismic survey, but later we shall combine seismic and gravity data to give a more refined model.

(a) Searle model at Suswa

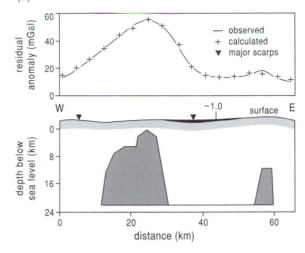

(b) Baker and Wohlenberg model at Menengai

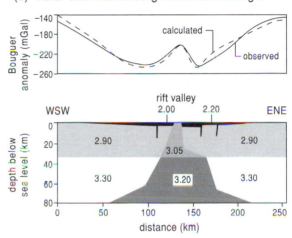

Figure 21.5 Gravity profile models across the Kenya Rift (densities in Mg/m³).

21.4 Seismic surveys

21.4.1 The seismicity of Kenya

We have already noted the continuous belt of earthquakes extending from the Carlsberg Ridge into the Gulf of Aden, the Red Sea, and the East African Rift System (Fig. 21.3b). These maps were derived from events recorded by a global network of 125 seismographs known as the World Wide Standard Seismograph Network (WWSSN). This was set up by the U.S. authorities in the 1960s, primarily to detect and identify underground nuclear explosions, but it also resulted in better location maps for earthquakes with Richter magnitudes greater than 5 occurring anywhere in the world, and these were crucial to the development of plate tectonics by delineating the plate boundaries (Plate 2b). Though the seismicity of Figure 21.3b is clearly associated with the Rift, fewer earthquakes were being recorded in the Kenya Rift than the Western Rift even though the greater volcanism in the Eastern Rift suggests there would be greater activity there. However, additional earthquakes might be occurring in the Eastern Rift but were too small to be recorded by the global network, which had only a single station in Kenya. To check this, temporary local networks with up to a few tens of stations were set up in the Kenya Rift in 1992. They showed that tens of small earthquakes did occur every day, with magnitudes down to −3; the ones big enough to be located are included in Figure 21.4.

Most of the seismic activity occurs as swarms – periodic bursts of activity – within the rift, at depths of 5 to 10 km; surprisingly few occur on the main rift faults. Some of this activity has been attributed to movement of geothermal fluids or the propagation of dykes in the upper crust, which is consistent with petrological data mentioned in Section 21.2 and also with seismic data on velocity and structure described in the following sections that suggest there is thermal upwelling and partial melting at a shallow depth in the crust.

In addition to the location of earthquakes, seismic recordings can reveal their fault planes and sense of motion, or the tensional and compressional axes associated with them (Section 5.3.2). Data recorded by the WWSSN and by the local networks in Kenya show that the stress directions are those expected for crustal extension across the rifts.

21.4.2 Teleseismic studies

The seismic velocities of rocks depend on temperature – decreasing as the temperature increases – as well as on the composition of the rocks, especially towards the melting points of the constituent minerals. This can be used for comparing temperatures in the upper mantle beneath different parts of a region, for at this depth the composition is likely to be fairly uniform laterally.

Ideally, velocities would be best found using refraction surveys, but these would need to be

extremely large because of the depths of interest, and hence expensive because of the size of the explosions and numbers of recording stations required. A less costly, though less accurate, alternative is to utilise distant earthquakes as seismic sources. For teleseismic sources, which are at least 2000 km (18° epicentral angle) distant, the rays travel up to the surface at a fairly steep angle and so spend minimal time in the variable crustal rocks (see Fig. 4.20). If some parts of the subsurface are hotter, the seismic velocity will be lower and so rays that pass through them will be delayed relative to those travelling through cooler rocks. By recording arrivals from earthquakes in different directions, the hotter region can be located and its form roughly estimated tomographically. To locate a hot region requires several steps. The origin time and location of the earthquakes are determined from the WWSSN or a similar network. Then the expected travel-time to the recording station is found from a published travel-time table such as for model iasp91 (Section 4.4.3); the difference between the expected and actual travel-times gives the delay due to a low-velocity region somewhere along the ray path. However, this does not reveal where along the path the region is, and there is also the possibility that the delay may be at least partly due to errors in timing or locating the earthquake. These uncertainties can be eliminated by comparing the delay with that of a station that is nearby but above normal mantle. In Kenya, the WWSSN station NAI near Nairobi (Fig. 21.7), which is close to the Rift, recorded arrivals about 2 sec late relative to Bulawayo, which is on normal continental crust, indicating that the rocks at depth near the Rift are hotter.

To provide more detail, a temporary network of 85 stations was set up for several months in 1990, as shown in Figure 21.6a, forming part of the Kenya Rift International Seismic Project (KRISP). It recorded 250 earthquakes from different parts of the world, and the 'residuals' – with positive denoting a late arrival – are shown in Figure 21.6b. After allowance for local crustal variation found from separate small-scale refraction surveys, they were used to deduce the velocity perturbations at depth across the Rift, as shown in Figure 21.6c, where velocities are given as percent slower than the standard model. This shows slower material, presumed to be hotter mantle, in the lower part of the lithosphere and mantle below.

(a) station locations

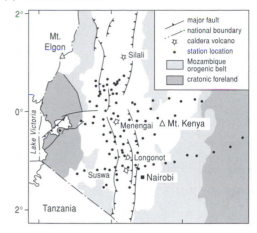

(b) P-wave residuals

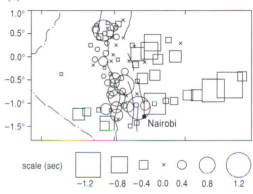

(c) P-wave velocity perturbations

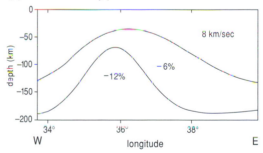

Figure 21.6 Teleseismic delay-times experiment.

21.4.3 Seismic refraction and wide-angle reflection surveys

Teleseismic studies are useful and economical, as they use natural rather than expensive controlled sources. However, they lack detail, for their resolution is limited by station spacings, and uncertainties in the origin time, location, and source characteristics of the earthquakes recorded. For higher resolution, controlled sources and closely spaced stations are needed, though

the depth of investigation is limited to about 50 km for practical reasons. In the KRISP series of experiments the controlled sources were underwater explosions where lakes with adequate depths of water existed at suitable positions, or more expensive ones in boreholes where they did not. The shot depths and amounts of explosives were carefully chosen to ensure sufficient signal at the frequencies recorded; with ranges up to 600 km, as much as two tonnes of explosive were needed. The signal-to-noise ratio was improved by shooting during the night, when noise due to wind and human activity was least.

As the objective was to compare the crustal structure beneath the Rift with that of surrounding areas, five profiles were shot (Fig. 21.7). For the required resolution an average station spacing of less than 2 km was needed and over 200 portable recorders were used. Even with this number it was not possible to record simultaneously along the longest profile, so this was broken up into three shorter lines, giving seven lines in all, A–G. One profile was along the Rift axis (A to C), two across the Rift (D and G), and two outside the Rift (E and F). Figure 21.9 displays a sample of the data, with each trace showing the vertical component of a seismic recorder. The traces are in order of distance of the recorder from the shot point and run from bottom to top, but they have been shifted in the time direction to give a reduced plot. This is because on a conventional time–distance plot, as in Figure 6.6, or Figure 21.8a, stations distant from the shot point do not receive any signal for some time, so the lower right part of the plot is without arrivals. To utilise this 'wasted' space, each trace is displaced 'downwards' by an amount proportional to the distance of its station from the source to give a reduced plot, Figure 21.8b.

Since the line has been drawn on a t–x plot, its slope is equivalent to a velocity, and this arbitrary velocity is called the reduction velocity, v_r; its exact value is not important, and it is usually chosen to be a whole number, close to the velocity of the rocks at the level of interest. In crustal studies, values of 6 km/sec or 8 km/sec, close to the average for the crust and mantle, are usually chosen, though in Figure 21.8b it is 1.7 km/sec. The reduced travel-time, t_r, relates to the actual travel-time, t:

$$t_r = t - \frac{x}{v_r}$$

Eq. 21.1

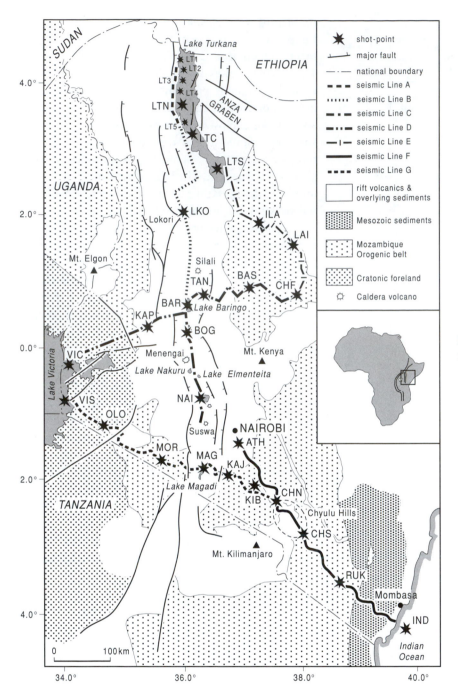

Figure 21.7 Locations of seismic refraction lines.

If the average velocity of travel to a station equals the reduction velocity, it will plot on the horizontal axis of the reduced plot (Fig. 21.8b); if slower above, and if faster below. In Figure 21.9 the reduction velocity was 6 km/s, so the phase labelled P_g, which plots on the line, has a velocity of 6 km/sec, while phases P_1 and P_n, which are head waves from the lower crust and mantle, are faster as they plot below it. The amplitudes have also been 'trace normalised' so that the maximum

amplitude of each trace is made the same, to facilitate correlation. In addition, records were filtered to reject frequencies outside the interval 2 to 20 Hz, to reduce noise.

The interpretation of the record section depends on identifying the different phases. As well as the refracted phases P_g, P_1 and P_n, there are some later phases, $P_{i_1}P$, $P_{i_2}P$, $P_{i_3}P$, and P_mP, which are due to wide angle reflections from the tops of intracrustal boundaries and the Moho. In the insert figure the

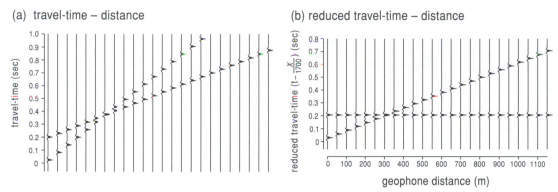

Figure 21.8 Normal and reduced travel-times versus distance plots.

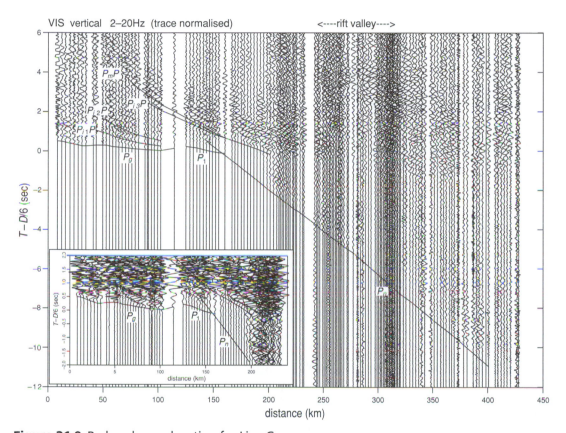

Figure 21.9 Reduced record section for Line G.

amplitudes have been increased to show the mantle arrivals more clearly.

Figure 21.10 shows the model deduced to account for the arrival times of the phases identified on Figure 21.9. First, a simple model was constructed from the slope and intercept times as described in Chapter 6. Then the positions of the interfaces and the velocities were adjusted until the travel times calculated for the rays shown on the model (Fig. 21.10a) matched the

ones recorded. The travel-times calculated from the model and those shown on Figure 21.9 are given on Figure 21.10b. This was also done for other shots into the same array of recorders, and Figure 21.10c is the final model, taking into account all the arrivals from all shots into the line. Other profiles were treated in the same way, and the resulting model sections are displayed together in a fence diagram shown in Figure 21.11.

(a) ray paths from shot VIS to stations on Line G

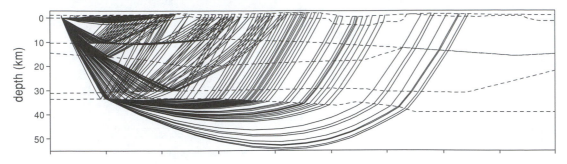

(b) observed and calculated travel-times for phases in Figure 21.9

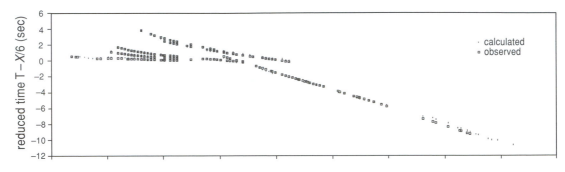

(c) crustal structure beneath Line G

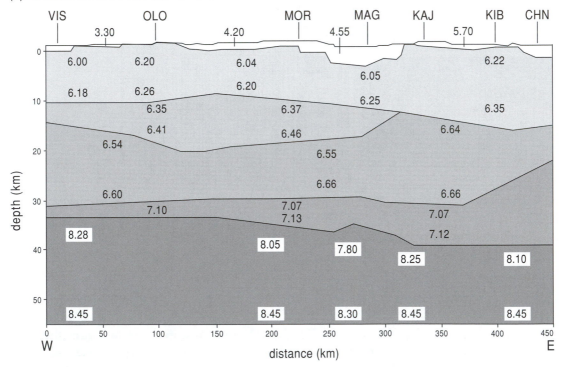

Figure 21.10 Seismic model for Line G.

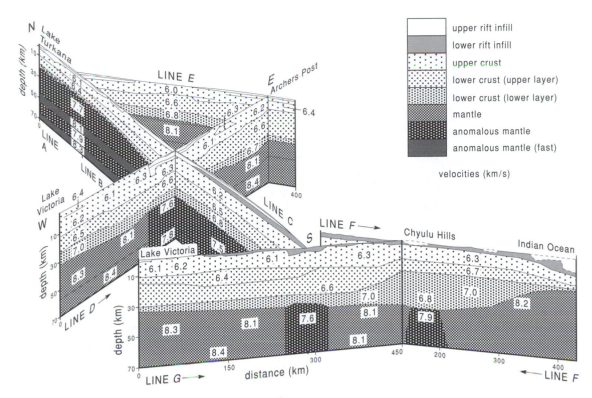

Figure 21.11 Crustal structure of the Kenya Rift.

There are some clear differences in crustal structure below the Rift compared to outside it. Away from the Rift, the crust is layered, with thickness varying between 33 and 44 km, and overlies mantle with the typical velocity of 8.0 to 8.2 km/sec. Beneath the rift the mantle velocity is anomalously low, being between 7.5 and 7.7 km/sec. The crustal thickness also differs, being thinner, and varies from about 35 km north of Nairobi and close to the crest of the Kenya dome, where the Rift is about 60 km wide, to only 20 km in Turkana, where the rifted zone is about 180 km wide, indicating that extension has been much greater in the latter area. Comparison of the sections shows that the thinning is greatest in the lowermost layers. A thinned crust is expected from the combination of elevation and high Bouguer anomaly in the region.

If we look in more detail at the E–W section of Line *D* where it crosses the Rift near the top of the Kenya dome (Fig. 21.11), we see at the top an infilled basin filled with sediments and volcanics, which have low seismic velocities. The basin is asymmetric, being about 8.7 km thick against the bounding fault on the west but thinning eastwards. Velocities increase from 3.8 km/sec at the top of the infill to about 5.8 km/sec at its base. Below the basin, in

the upper crust, velocities range from 6.1 to 6.4 km/sec, the highest values being in the Archaean rocks at the western side. The velocity increases downwards through the middle and lower crust to 7.1 km/sec. The section also shows that crustal thinning of about 5 km, overlying anomalous mantle, is confined to the Rift zone itself.

The structure beneath Line *G* across the southern flank of the domal uplift shows similar but more subdued features, compared with Line *D*. The Rift infill is thinner, and the crustal thinning is less. A major difference is in the asymmetry of the crustal structure across the rift beneath Lines *G* and *F* – indeed the thickest crust observed anywhere in Kenya is beneath the Chyulu–Kilimanjaro region, where it is 44 km thick. This overlies a zone where the mantle velocity decreases to about 7.9 km/sec, but the transition is less sharp compared with that towards the Rift itself, where the velocity is lower, only 7.5 to 7.8 km/sec. The differences between the properties of the crust east and west of the Rift support the idea that the rifting is developing along the old suture zone between the Archaean Nyanza craton to the west and the Proterozoic Mozambique Belt to the east, as illustrated in the schematic cross-section (Fig. 21.12).

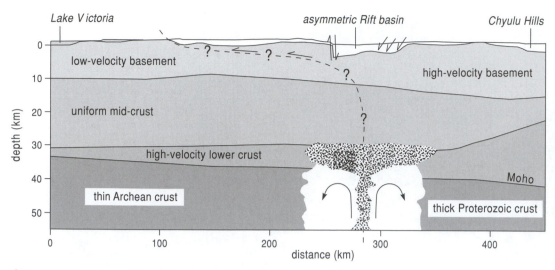

Figure 21.12 Schematic section across the Rift along Line G.

21.5 Combined seismic and gravity models

Seismic and gravity data used together provide better models than either separately, not just because they depend on different physical variables – seismic velocities and density – but also because these are related to each other, allowing a seismic model to be tested against the gravity results as well as the seismic ones. This depends on being able to assign densities to the seismic velocities in the seismic model. During the course of the gravity surveys described in Section 21.3, many density measurements were made, so values are available for the near-surface rocks. For the deeper layers, densities were estimated from the seismic velocities using the empirical Nafe–Drake curve, which is named after two geophysicists who plotted all the data for rocks for which both density and seismic velocity were known. The density model corresponding to the seismic model of Figure 21.10c is shown in Figure 21.13d, while Figure 21.13a compares the calculated and observed Bouguer anomalies. The main features of the observed profile are reproduced in the calculated one except for the general rise from west to east shown in Figure 21.13b. Removing this regional effect gives Figure 21.13c, which agrees well with the observed values; the small misfits are attributed to local near-surface variations in density.

The regional increase in gravity from west to east is to be expected from the observation noted in Section 21.3, that the elevated topography is isostatically compensated (Section 9.1.6). The compensation

mechanism must have its origin below the base of the model, which is 60 km. A model of the lithosphere extending to over 100 km that accounts for this compensation is shown in Figure 21.14. It shows a low-density region beneath the plateau, which could be due to the higher temperatures that produced the velocity perturbation described in Section 21.4.2.

21.6 Heat flow studies

One would expect high heat flux values in the rift, not only because the widespread occurrences of volcanics obviously require heat sources, but also because continental splitting would be accompanied by upwelling of hot mantle, as shown in the models described in the previous section. In fact, upwelling may not simply be a consequence of the splitting but actually initiate it, by the force of buoyancy, particularly if it occurs as a plume (Section 20.9.2). Measuring heat flux offers a way to understand more about thermal processes at depth, though so far data are limited.

Heat flux is best measured in boreholes drilled for the purpose, but as these are expensive to drill, measurements are often made in holes drilled for other purposes. In Kenya, measurements have been made in 69 holes drilled for water (which are less deep than preferred for heat flow measurements), supplemented by others drilled in exploration for oil. The results are summarised in Figure 21.15. The values are somewhat scattered but in general are higher in the Rift than outside it, as expected. A

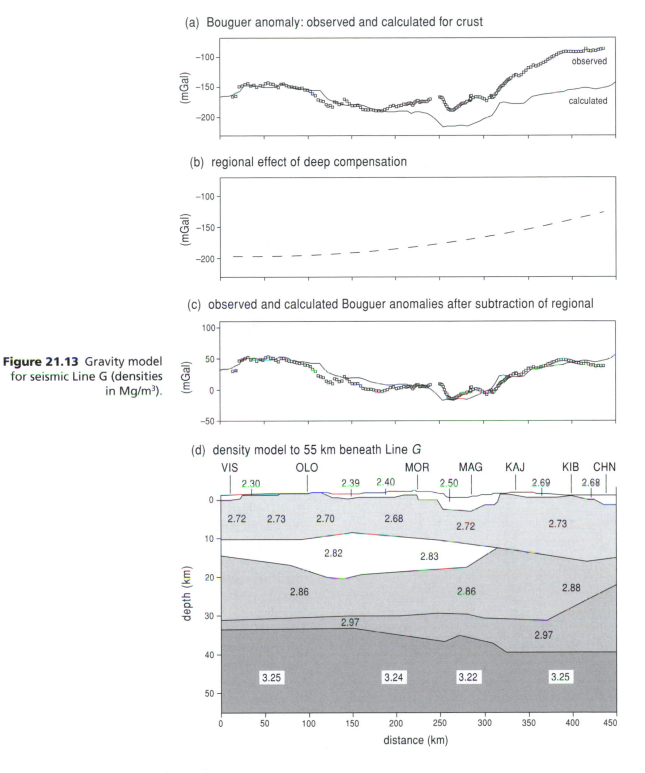

(a) Bouguer anomaly: observed and calculated for crust

observed

calculated

(b) regional effect of deep compensation

(c) observed and calculated Bouguer anomalies after subtraction of regional

Figure 21.13 Gravity model for seismic Line G (densities in Mg/m³).

(d) density model to 55 km beneath Line G

range of values in any area is not surprising, for the measured heat flux near the surface depends not only upon the temperature at depth in the crust and mantle, but also upon the rocks the heat travels through to reach near to the surface where the heat flux is measured (Section 17.2). It is therefore affected by the variability in conductivity of near-surface rocks and by groundwater flow, which itself depends upon the variable permeability through the pores and fractures of the rocks.

(a) observed and calculated Bouguer anomaly for crust and upper mantle

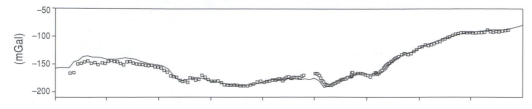

(b) density model to 110 km beneath Line *G*

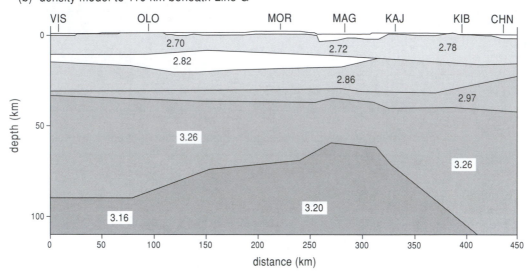

Figure 21.14 Extended gravity model for seismic Line G (densities in Mg/m³).

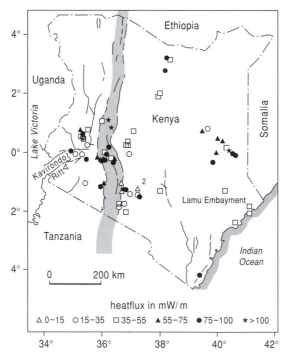

Figure 21.15 Heat flux data for Kenya.

The high heat flux values in the Kenya Rift could be due in part to heat being conducted up from an anomalously hot mantle, but it is thought more likely they are due to magmatic activity and hydrothermal circulation quite high in the crust, because there has not been sufficient time for heat to conduct up from the Moho, about 30 km deep, to the surface before the volcanoes formed.

A second objective of heat flow studies is to locate places where geothermal power might be extracted. This is of some importance to Kenya, for it has no fossil fuels and few hydroelectric sites remain unexploited, but the conditions within the Rift are favourable for geothermal power production. Sources of heat are evidently present in the shallow magma chambers that have fed the volcanics, but for a productive geothermal field, water is also needed to provide the steam (Section 17.4.3). Permeable water reservoirs occur in the brecciated and fractured volcanics, while fault zones offer pathways for surface water to charge them.

Hydrothermal alteration of the volcanic rocks near the surface leads to the formation of an impermeable layer, which has a confining role similar to that of the cap rocks in oil and gas fields, providing a seal for containing the superheated water until it is penetrated by drilling and gives rise to high-pressure steam, which is used to drive turbines. Kenya already produces a significant amount of electricity (about 100 MW) in this way, but though most of the 15 large central volcanoes in the Rift are potential sources, only one has been exploited so far. In addition, there may be geothermal fields associated with dyke injection that has no surface expression but has been inferred from seismicity (Section 21.4.1). The geophysical work described in this chapter is being used in the exploration and assessment of the country's geothermal resources.

21.7 Electrical conductivity

Another way of investigating temperatures at depth is through electrical conductivity studies. In Section 12.2.1 it was explained that conductivity often depends on the amount of water the rock contains and the amount of dissolved salts in the water. However, high conductivities are often found in the middle and lower crust associated with high temperatures due to the presence of magma or fluids released from them. Such depths are usually too deep for investigation using resistivity arrays because these would need extremely large arrays with many tens of kilometres of cable. However, in Section 14.7 it was explained that conductivities at depth can be found using magnetotelluric (MT) surveys, which measure variations in the magnetic and electric fields. A limitation of the method is that it shows little detail, only indicating large volumes of higher or lower conductivity.

MT surveys have been made in Kenya to investigate the conductivities of the deep crust and uppermost mantle. They indicate a region of high conductivity about 20 km beneath the Rift axis, which has already been associated with positive gravity anomalies and several very prominent volcanoes (Fig. 21.5) and is probably due to the presence of magma. There is another high-conductivity zone 50 km deep near Mount Kenya, which could also be due to magma. This greater depth is consistent with the petrologic evidence that volcanics outside the Rift originate at greater depth than those inside.

A later survey was designed to measure conductivity below seismic lines *F* and *G* of Figure 21.7, and preliminary results are shown in Figure 21.16. As expected, there is higher conductivity in the crust and mantle below the Rift. The Rift infill is particularly conducting, probably due to the porous volcanics and sediments containing saline water. The infill has higher conductivity to the west than the east, consistent with the seismic section of Figure 21.11, which shows the infill to be much thicker to the west. There is a sharper contrast between the Rift and western flank than for the eastern one. Beneath the western flank the upper mantle has low conductivity but is underlain by conductive mantle.

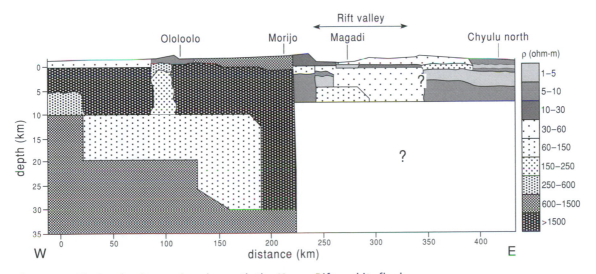

Figure 21.16 Conducting regions beneath the Kenya Rift and its flanks.

Away from the Rift to the east are the Chyulu Hills, a young volcanic field, and there are highly conducting zones within 1 km of the surface. The high conductivities in the lower crust and upper mantle support the models, derived from the gravity and seismic data, which are consistent with partially molten zones.

21.8 Summary

Topographic, geological, palaeomagnetic, magnetic, gravity, and seismic data indicate that the East African Rift System is part of the global system of divergent plate boundaries. It has developed over a domal uplift which has been superimposed on the eastern edge of the East African plateau. Seismic and gravity results show that the structure of the crust and upper mantle below the Rift differ considerably from those to either side: layers are thinner, with the Moho in particular at a shallow depth. Heat flow and electrical conductivity values are higher and indicate that the Rift (and volcanic areas outside it) are underlain by hotter, perhaps partially molten material.

The accepted explanation of these results is that the Rift is indeed a constructive plate boundary and lithospheric extension is occurring, accompanied – or caused by – upwelling mantle material due to one or more plumes. However, there has been significant extension for no more than about 10 Ma, and crustal thinning is some way short of allowing oceanic lithosphere to form.

Further Reading

Olsen (1995) gives a general account of rifts. Two special issues of *Tectonophysics* (Prodehl et al., 1994; Fuchs et al., 1997) describe recent KRISP surveys and what has been learned about the Kenya Rift from them. Other recent work has been indicated by references in the text.

chapter 22

Hydrocarbon Exploration

This chapter is primarily about the role of geophysics in finding and extracting petroleum, for geophysics has a crucial role to play in the successive stages of hydrocarbon exploration and then its extraction. After looking at energy sources and demands in general, and the particular importance of hydrocarbons, we outline how petroleum originates in sedimentary basins and the conditions needed for it to collect into accumulations that are worth extracting. Gravity, aeromagnetic, seismic refraction, and reflection surveys are used to help locate basins. More detailed surveys, particularly using seismic reflection, are used to identify structures within them, such as anticlines, faults, or salt domes, that are likely to contain hydrocarbon traps, and to determine whether these contain pools of oil. This is followed by exploratory drilling, with well logging used to measure their vertical extent and the permeability, porosity, and hydrocarbon content of the reservoir rocks. Drilling is very expensive, so siting the wells has to be done with great care and relies primarily on detailed seismic reflection surveys. Later, surveys may be carried out to monitor extraction. The methodology of exploration and extraction of hydrocarbons is illustrated with two case studies from the North Sea, the first of gas, the second of oil.

Geophysical methods employed: Mainly seismic reflection and well logging, with some gravity and magnetics.

22.1 Introduction: Energy sources and the demand for hydrocarbons

For most of the thousands of years that civilisation has existed, humans have had to rely mainly on muscle power to make things and to move them – either their own muscles or those of the few suitable animals that have been domesticated. Results could be remarkable: The ancient Egyptians were able to quarry and move blocks of stone weighing over 1000 tonnes, and they raised the pyramids (the largest 146 m high and with a mass of 6 Mt) by human power. The only mechanical sources of power until recent times were simple waterwheels and windmills, mainly used to save the chore of grinding grain by hand, and sails to propel ships. The main fuel used globally was wood, which is still important in many underdeveloped parts of the world.

Change came with the onset of the industrial revolution. The invention of semiautomatic machines – particularly in the textile industry – required larger sources of power and resulted in the workers leaving the spinning wheels and looms in their cottages for concentrations of these machines in the first factories, which were located where streams could provide the power to turn waterwheels. The increasing numbers of people in towns needed new sources of fuel, and this began to be met by coal, far more compact and hence more easily carried and stored than wood. Carrying it by packhorse or wagon on the poor roads of the time was slow and expensive, and so canals were built: One horse towing a boat could transport 30 tonnes, instead of the fraction of a tonne before. Canals were also useful for transporting manufactured goods to markets at home or abroad, particularly such fragile ones as the products of the potteries. The growing demand for iron (and later steel) required that iron ore and coal be brought together and the resulting products taken to where they were needed, leading to further transport demands. One of the engineering surveyors involved in canal building projects in Somerset in the late 18th century was William Smith, who realised that the continuation of geological strata observed in cuttings could be predicted, and he produced a geological map of England and Wales to facilitate canal planning, a remarkably high-quality product which is still useable today.

The late 17th century saw the appearance of the steam engine, first developed to replace horses, which struggled to pump water from mines that were being extended to greater depths. At first, the primitive engines merely provided the reciprocating motion needed for pumping, but when they had been developed to turn cogs they could be used to drive machines, allowing factories to be sited away from the streams, which were becoming inadequate for the ever-growing power demands. Later, the engines became compact enough to be placed on wheels and the locomotive was born. Railways quickly spread across the world, replacing canals as the major means of transport over longer distances. On the oceans, steam ships were faster and more reliable than sailing ships.

Coal suited the technologies of the time, for, being solid, it could be transported in simple containers, stored in bunkers, cellars, or simply in heaps on the ground, and shovelled into fires or furnaces. The first coal was dug where seams outcropped at the surface; when this was used up people dug deeper, and mining technology developed to allow this. There was little need initially for geological understanding, for seams were simply followed into the ground. Now, of course, understanding of faults and folding is used to forecast how seams will continue ahead of the coal face, and the existence of coal below areas where there are no outcrops has been successfully predicted. Geophysics has a modest role in the coal industry, mainly reflection seismology to map the subsurface at high resolution for extraction by machines, and also specialist forms of seismology that aid the mining engineer by 'looking ahead' along seams or by detecting the microseismicity that warns of incipient collapse.

In turn, the heavy and cumbersome steam engine gave way to the internal combustion engine, particularly for transport, and this created a demand for liquid fuel, which is mainly derived from petroleum; without it our roads would be nothing like so congested, and successful aircraft could not have been invented. Liquid and solid petroleum products have been used since the beginning of civilisation for building, ornamental, medicinal, shipbuilding, and other purposes. In warfare, for instance, oil-soaked tow was used in incendiary arrows by the Persians in the siege of Athens around 500 B.C., while 'Greek Fire' – incendiary bombs with paraffin obtained by distilling oil, and ignited by gunpowder and fuses – was used in military campaigns over 2000 years ago. Gradually, a demand for petroleum for use as a fuel and for illumination developed as a replacement for whale oil.

Baku, in the Caucasus, has long been known for its oil seeps, for these gave rise to fires, which, seemingly eternal, were regarded as holy. By 1725, the traffic of extracted petroleum up the Volga was sufficient to require regulation. (The area is still producing, and exploration is being carried out in the expectation of finding yet more oil.) However, the amounts transported were small enough to be carried in barrels, and though use of such containers has given way to tankers, the unit, the barrel (35 imperial gallons), survives; production from a single modern well is typically a few thousand barrels a day but occasionally exceeds 100,000. In global terms, however, it is more usual to measure production of oil in millions of tonnes (1 tonne = 7.36 barrels). Gas, which is often measured in billions of cubic metres (bcm), is compared by being converted to millions of tonnes of oil equivalent (MTOE): 1 bcm = 0.9 MTOE.

Though petroleum is found at the surface in limited quantities, in oil seeps, the great bulk of it necessarily is underground, for once it reaches the surface and comes into contact with air, it is destroyed by oxidation. Petroleum can exist as solid, liquid, or gas, liquid being the preferred form at present for its ease of transport . The greatest amount exists in solid form as oil shales and tar sands, and though small amounts of fluid have been extracted by distillation, no one has yet found an acceptable and economic way of extracting them in quantity. Gas, which often occurs together with oil, was for a long time burnt off because it could not easily be transported or stored, but the building of pipelines or shipment in special vessels after being compressed and cooled to a liquid has made it a valuable fuel, especially as it burns cleanly and produces less carbon dioxide – of concern as a 'greenhouse gas' – than coal or oil for the same amount of heat produced.

Petroleum products have to a large extent replaced coal, not just for transportation but also for heating and electricity production. Their dominance may be gauged from the fact that whereas coal (including lignite) comprised 94% of fossil fuel consumption at the beginning of the 20th century, by 1965 it had shrunk to less than 40%, the rest being mostly oil and gas. Since then the proportion has fallen further, though the amount has increased (Fig. 22.1).

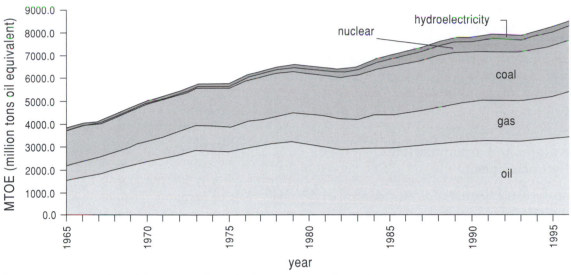

Figure 22.1 Annual world consumption of primary energy since 1965.

There are other sources of energy apart from fossil fuels, such as hydroelectricity and nuclear energy, which are also shown on Figure 22.1. Hydroelectricity contributes about 2.7% of global energy, and new dams are being built, though often in the face of environmental objections. However, the scope for further development is limited, for most of the suitable sites have already been used. The main role of the geophysicist here is to help investigate the foundation on which a dam is to be built and to assess the seismicity of the area to ensure that structures are built to withstand the maximum magnitude of earthquakes likely to occur during their lifetime.

Nuclear power was once forecast to take over from fossil fuels, and its use rose rapidly (Fig. 22.1) to account for 7% of the world's energy consumption, most of this being in Europe. However, it has failed to live up to its early promise, when it was even overoptimistically claimed that it would be so cheap that it would be given away. It has fallen so far from favour – partly due to the Chernobyl catastrophe – that few if any new nuclear powerstations are being built. There is only a limited demand for new uranium deposits even to continue to fuel the existing reactors, for disarmament of nuclear weapons has resulted in a glut. The major challenge of the nuclear industry to the earth scientist is to address the so far unsolved problem of how to dispose of the high-level radioactive waste products, by finding an underground formation in which they can be buried safely for about 250,000 years.

Although the Earth's internal heat is vast, most is in the inaccessible deep interior. Geothermal energy can be utilised only when it is concentrated within a few kilometres of the surface, as explained in Section 17.4.3. At present, it accounts for only about 0.1% of global energy production and so is too small to appear on Figure 22.1. Most of this production utilises naturally occurring high-pressure steam in a few favoured locations, and unless there is a major technological advance permitting the cheap extraction of energy from hot, dry rocks, as discussed in Section 17.4.3, its contribution is not likely to increase significantly; indeed, in many areas of current production, output is declining.

Of the renewable sources of energy, wind power is currently the most popular, being already competitive in price with nuclear- and coal-powered electricity in favourable locations. Wave and tidal power are only of importance locally, and although the direct generation of electricity using solar cells is becoming steadily cheaper, it is still considerably more expensive than using fossil fuels, except for special, small-scale applications and in areas remote from mains electricity. The contribution that geophysics can make to the development of renewable energy sources is small.

The petroleum industry is therefore likely to remain the dominant energy source for the forseeable future – especially for transportation. It is also needed to supply the petrochemical industry, whose products include asphalt, explosives, fertilisers, fibres, medicines, paints, plastics, solvents, and synthetic rubber. As the petroleum industry is by far the largest user of geophysics, this chapter is devoted to it.

22.2 The origin and accumulation of hydrocarbons

Before discussing the contribution that geophysics can make to the exploration and extraction of hydrocarbons, we shall outline how hydrocarbons originate and then accumulate in reservoirs, for this determines the targets that have to be discovered and exploited. Hydrocarbons have an organic origin, as explained briefly in Section 7.10. They derive from microorganisms – algae, plankton, and bacteria – buried in sediments under anaerobic conditions so that they escape destruction by oxidation. As the most common of these source rocks is shale, much of our discussion is based on this rock. Within the source rock, the organic remains are first converted to a semisolid called kerogen that contains a range of hydrocarbons, including methane, which can accumulate in commercially useful quantities.

As the sediment is buried progressively deeper by subsequent deposits, the temperature rises because of the geothermal gradient, which first kills off the bacteria and then, at about 65°C, converts the kerogen into heavy oil, which is thick and has constituents with high boiling points. Lighter oils are produced as the temperature rises, until at about 175°C only gas, largely methane, is present. The process of conversion of kerogen into fluid hydrocarbons that can be extracted is called maturation, and it depends on time to some extent as well as temperature; as it occurs between depths of 2 and 5 km, hydrocarbons mostly originate in deep sedimentary basins. Gas can also derive from coal, and this source rock is important in the southern North Sea (Section 22.5).

The amount of oil formed clearly depends on the original organic content, and a source rock should have at least 0.5% by weight, but 5% is more typical of oil fields, while for oil shales it may reach 50% or more. To be extractable, the hydrocarbons must have been squeezed out of the shale, which has very low permeability, and concentrated by primary migration into a reservoir, composed of porous rock (Fig. 22.2), commonly a sandstone or limestone including – as in parts of the North Sea – chalk, where the hydrocarbons enter the spaces between grains or in fractures. As the reservoir rock usually also contains water, the hydrocarbons, being less dense, move upwards by secondary migration. They will reach the surface (and then be lost) unless they are prevented by an impermeable caprock, commonly shale or evaporite. The final stage of concentration is the movement of the hydrocarbons from the large area underlain by the reservoir into more localised traps, which may be structural or stratigraphic, as explained in Section 7.10.1; an anticline is a common example of a structural trap. The vertical distance between the highest point of a structure and the lowest is referred to as the closure, as shown in Figure 22.2.

The accumulation of hydrocarbons within part of a trap forms a pool, with any gas stratified above the oil. All the pools within a single geological structure together form a field, and there may be several fields in a geological or geographic province. Fields have areas ranging from 5 to 3000 km², 25 being a common size, and contain upwards of a million barrels (136,000 MTOE). Individual pools are much smaller, ranging from less than 0.5 km² to a few hundred km², with a vertical extent of a few tens to a few hundreds of metres, usually at depths of 0.5 to 3 km, but occasionally to more than 6 km.

22.3 Where sedimentary basins form

The first requirement for the formation of hydrocarbons is a sedimentary basin. Most of the world's oil fields are found in basins formed at passive plate margins (Section 20.4.3) soon after continental splitting has occurred. Marine life is abundant, and the dead organisms are deposited in silty sediments derived from the erosion of the continental margins and deposited in the deeper and anoxic central parts of the young, narrow ocean; the Red Sea (Section 21.1) is a present-day example.

Some of the other requirements for extractable hydrocarbons to occur are also at least partly met about the same time. Potential reservoir rocks are often provided by sands, which, being coarser, are

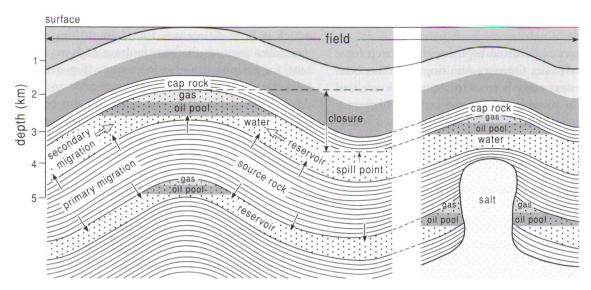

Figure 22.2 Migration of hydrocarbons into traps.

deposited nearer to the shore or, in tropical seas, by shallow-water limestone reefs. Potential cap rocks are provided by shales deposited on the sands or by evaporites, which are layers of salts precipitated when sea water is evaporated in nearly closed seas; evaporites formed during the Tertiary beneath the Red Sea and parts of the Mediterranean, and during the Permian and Triassic periods in the seas that covered much of western Europe, including parts of what is now the North Sea.

There are several types of trap. Some are structural traps formed by later folding or faulting of the sediments (Fig. 7.25). Others result from salt deposits, for salt is an unusual rock, being both of low density (2.2 Mg/m^3) and plastic under pressure; consequently, when buried it tends to flow upwards in places to form diapiric structures. It may bulge the overlying sediments into an anticline known as a salt dome, or it may penetrate them as a salt plug with a roughly circular cross-section, or as a salt wall when elongated. As salt is impermeable to hydrocarbons, traps may occur in the overlying dome or the bulged sediments against the flanks of the salt, as shown in Figure 22.2 (also Fig. 7.27). Stratigraphic traps (Fig. 7.26) are becoming increasingly important, because although the accumulations are often small, they can be economic, especially if they are close to an existing infrastructure, as illustrated in Figure 22.15.

Not all basins are formed at passive margins.

Some occur in the late stages of the closure of an ocean near a subduction zone, where basins are particularly deep. The associated compressional forces can form very large anticlinal traps, such as those of the Middle East, which contain 65% of the world's proven hydrocarbon reserves. Few basins form during the subsequent continent–continent collision, but after collision a basin may form on land, such as the Ganges basin to the south of the Himalayas. Basins on land may also occur in continental rifts: The Rhine and Rio Grande rifts have small fields, as do some in Africa, though none has yet been found in the East African Rift (Chapter 21). However, yields are low, for not only are the fields small but the sediments are not marine, which give the richest source rocks. The North Sea – the subject of two case studies later in this chapter – is intermediate between a rift and a passive margin basin, as will be explained in Section 22.5. A number of small and complex fields have been found both on- and offshore associated with transform faults in the eastern Pacific off the California coast.

There are also some important fields that appear to be unrelated to plate boundaries. There are large areas of North America which, by repeated invasion over millions of years by shallow seas resulting from periodic subsidence, have accumulated sediments deep enough for oil to form; traps are often stratigraphic (Fig. 7.26).

(a) seismic traverses

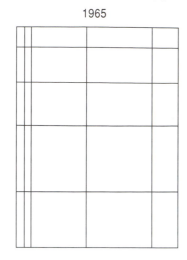

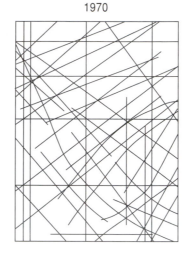

Figure 22.7 Comparison of seismic coverage and resulting structural maps, 1965 and 1970.

(b) base of Zechstein

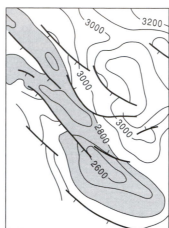

contours are metres depth

Production from West Sole started in 1967, and by 1970, 21 production wells had been drilled, the gas being brought ashore to Easington by pipeline. The reserves are estimated to have been about 60 bcm, and the field has been delivering about 2.3 bcm each year since 1972. Within a year of this discovery several other large gas fields had been discovered in the southern North Sea, including Leman (165 bcm), Hewett (120 bcm), Viking Group (100 bcm), Indefatigable (50 bcm), Barque (40 bcm), and a number of smaller ones with reserves of less than 10 bcm. Many other discoveries have been made since the 1960s, and new fields are still being found. The reserves of natural gas in the North Sea are now estimated to be about 4000 bcm, roughly 3% of the global total. By 1996 over 25% of this had been extracted from the Rotliegendes reservoir in the Southern North Sea province.

22.6 The Forties oil field of the northern North Sea: A case study

22.6.1 Discovery and initial development of the field

Within four years of the first discovery of gas in the southern North Sea in 1965, so many other gas fields had been discovered that there was little incentive to find more. The consequent fall in demand for exploration led to seismic survey ships being available cheaply, and some were employed on more speculative surveys in the more northerly parts of the North Sea, where the geological history was rather different.

During the Jurassic and Lower Cretaceous, thick marine shale sequences, notably the Kimmeridge Clay, the major source of oil in the North Sea, were deposited in the graben described in Section 22.5, with coarser sand sequences near their margins. Rifting then ceased and was followed by cooling of the thinned lithosphere, which led to subsidence (Section 17.3.2). The effects of rifting were much more noticeable than in the southern North Sea.

During the Upper Cretaceous the sea level was high, and as there was little land to provide clastic sediments, nonclastic sedimentation dominated and the chalk was deposited. During the Tertiary, when sea level fell again, clastic sedimentation resumed in the basin, providing the shale caprock. Oil and gas were discovered in a chalk reservoir in Denmark in 1966. In the following year a further discovery was made in a Palaeogene sandstone offshore of Norway, though geological complexities prevented its extraction. These discoveries gave reason for optimism about finding oil in the northern North Sea, and then, in 1969, the substantial Ekofisk field was discovered in Norwegian waters in a reservoir of chalk. This discovery led to a major shift of interest from the Southern to the Central and Northern North Sea Basins.

The first major discovery in U.K. waters, late in 1970, was the Forties field (Fig. 22.4). Its geological structure was initially mainly deduced from detailed seismic surveys. Figure 22.8 shows the variation of depth (shown as isochrons, lines of equal TWT) of a reflector overlying the reservoir rock. Well 21/10–1 was located near the crest of a low-amplitude structure and at 2098 m entered oil-bearing Palaeocene sands and encountered the oil–water contact at 2217 m. On testing, it produced low-sulphur oil at 4730 barrels/day. To test that oil could be extracted at commercially valuable rates from other parts of the field, appraisal wells 2, 3a, and 5 in block 21/10 and 1 in 22/06 were drilled and confirmed the discovery of a major oil field.

Fields in the northern North Sea have been exploitable only because drilling technology has also

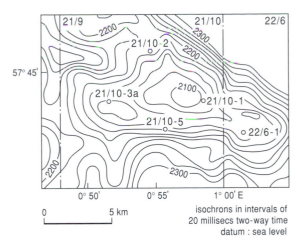

Figure 22.8 Seismic isochrons on a seismic reflector overlying reservoir rock, Forties field.

advanced greatly. Drilling for the Forties field was done using a semisubmersible rig, which could operate in waters over 130 m deep and where winds may be up to 130 mph and waves 30 m high. This ability to drill in difficult conditions has been paralleled by impressive advances in the precision with which drill holes can be directed: today, holes may be guided to within a metre or two of the chosen target, which may be several kilometres from the drilling platform, and holes may even snake subhorizontally through narrow undulating beds. This high precision is needed to extract the maximum amount of oil from complex fields. In turn, this has to be matched by the precision and accuracy with which targets can be identified, which requires very high-resolution seismic sections plus precise navigation.

Usually, to identify lithologies, SP and resistivity logs are used (Sections 18.5.2 and 18.5.3 and Fig. 18.2), but as most of the wells in the Northern North Sea were drilled with mud made with sea water, which has a high electrical conductivity, these logs were not very useful. Much more useful were the γ ray and sonic logs, shown in Figure 22.9. They reveal the fine structure of the geological sequence. As explained in Chapter 18, the natural γ ray log in sedimentary sequences gives the highest responses opposite shale beds, and here the value is about 75 API units. The lowest is opposite the limestones, which is about 15 API units, as at the bottom of the sequence. Near the top of the sequence, at a depth of 2030 m, the highest response is recorded opposite an ash band. The sands have intermediate values of about 30 units but with fluctuations due to the presence of variable amounts of shale, which reduces porosity and can be quantified by simple proportion. The sonic log shows the variation in transit time in the usual units of ms/ft, the inverse of velocity. This log shows two main features: One is the general decrease of transit time with depth due to increase in compaction; the other is the variation with lithology. The highest times are associated with shales, the shortest with the limestones, with the sandstones being intermediate. Thin limestone and shale bands are responsible for the spikes on the log.

From the borehole samples and geophysical logs (Fig. 22.9) the reservoir rock was identified as massive Palaeocene sandstones at depths between 2098 to 2380 m. Although these are interbedded with mudstones over most of the field, clean homogeneous sandstones up to 80 m thick are dominant in the upper part of the depth range. The cap rock is a shale rich in montmorillonite (a clay mineral) in the interval 2066–2098 m. Immediately above this, in the interval 2037–2066 m, is a mudstone containing volcanic ash; this has a characteristic sonic log pattern, with its velocity increasing with depth. It gives a prominent reflection that is useful for correlation and is the reflector of Figure 22.8.

The velocities found from the sonic log were used to convert two-way times to accurate and precise depths, from which the structural map of the top of the reservoir and section shown in Figure 22.10 were deduced. It shows a broad dome with an area of about 90 km², roughly 16 km by 8 km, and elongated in the E–W direction. The oil–water contact at 2217 m is also shown. The porosities are between 25 and 30%, with permeabilities up to 3900 md (millidarcies – see Section 18.4). The recoverable reserves were estimated to be 334 Mt, making it one of the largest fields in the North Sea, though only of moderate size compared to the giant fields of the Middle East and some other areas. Production started in 1975 and built to a peak of about 25 Mt/year in 1979, but as the field became depleted, declined to about 5 Mt/year in 1996. By this time the Forties field had produced over 300 Mt, about 6% of the total oil produced from the U.K. sector of the North Sea so far.

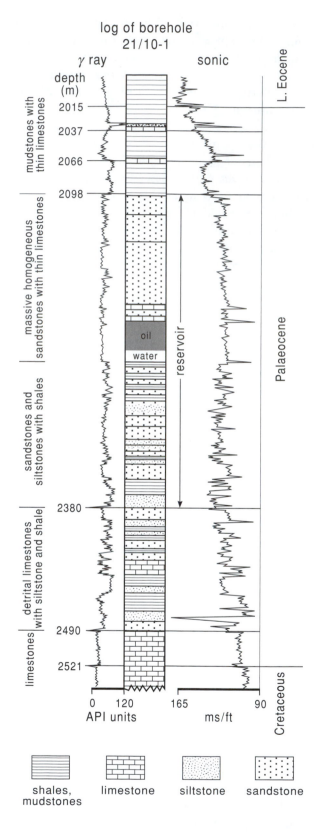

Figure 22.9 Geological, sonic, and γ ray logs through the Forties reservoir rock.

(a) structural contours

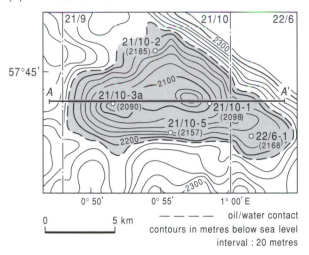

0   5 km   ― ― ― ―   oil/water contact
                              contours in metres below sea level
                                          interval : 20 metres

(b) section

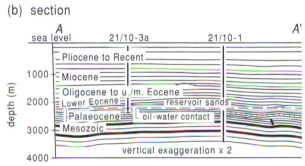

Figure 22.10 Structural map of the top of the reservoir, Forties field.

22.6.2 Further development of the Forties field

Following the drilling of exploration wells to find reservoir rocks containing hydrocarbons, appraisal wells are drilled to test that they can be extracted at commercially useful rates. These are drilled in various parts of the field using mobile drill rigs, but once a viable field has been proved the production wells from which the oil is extracted are drilled from only a few platforms, to minimise the cost of drilling and the number of places from which the oil has to be collected, the wells being drilled obliquely down from the well head in a roughly radial arrangement, as shown in Figure 22.15. Figure 22.11 shows the five platforms, Alpha to Echo, of the Forties field, plus the exploration and appraisal wells.

The production wells are for bringing the oil to the surface, and this may be achieved in various ways. Often the underground pressure of water or

gas is sufficient to force the oil up to the surface. If the pressure is not great enough initially or falls with production, it can be raised artificially by injecting water or steam into the reservoir through other wells. Sometimes, mechanical devices (e.g., electronic submersible pumps) are used to help raise the fluid to the surface. The water-injection wells used in the Forties field to enhance the recovery are also shown in Figure 22.11.

During the late 1980s, by when the production rate had fallen considerably (Fig. 22.12), it became apparent that the wells drilled according to the original development plan of 1975 would not be able to extract oil from significant parts of the reservoir. Before the positions of new wells that would extract this oil could be chosen, a better understanding of the reservoir was needed, and a new exploration and development program was put together. A key element of it was the acquisition of 3D seismic data to obtain a better model of the depositional environment to identify where best to drill the new wells. The 3D data was acquired in three months in 1988, but interpretation could not begin for a year while the hardware necessary to handle the large quantity of data was assembled and the processing programs developed. It then took two years to interpret the data. The first new well was drilled in 1992, four years after the 3D survey had been made.

A typical section from the 3D survey (Fig. 22.13a) shows the form of the top and base of the reservoir lens. Sediments are compacted by the weight of later sedimentation, and since different lithologies compact by different amounts a section can be very different today from when it was formed. To understand the depositional history better, sections can be 'decompressed' according to the type of sediments and weight of overburden, and this has been done to give Figure 22.13b. From this the depositional model of Figure 22.14 has been deduced, giving a clearer understanding of a complex reservoir and so improving the planning of the extraction. Initially, it was planned to drill 16 infill wells between the existing wells to reach the unextracted oil, but as the first few of these were very successful the number was increased. By mid-1997, 17 infill wells had been drilled, plus 25 wells to replace old production wells that, after a long period of use, were no longer producing oil efficiently for mechanical reasons.

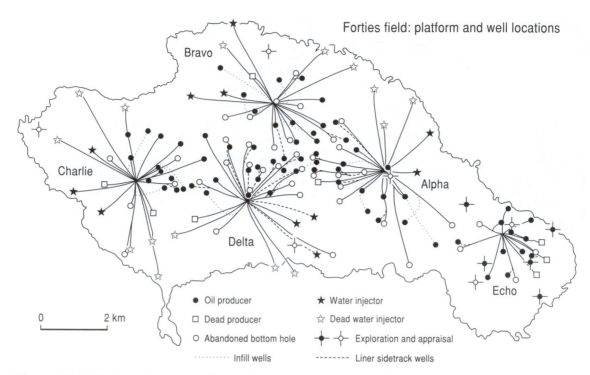

Figure 22.11 Exploration, production, and water-injection wells in the Forties field.

Together, these 42 new wells accounted for almost 90% of the production from the field in 1997, the infill wells accounting for 35%. These new wells will extend the life of the field by several years (Fig. 22.12).

Over 90% of the currently recoverable reserves of the Forties field have now been extracted, so the remaining targets for further infill wells are small and harder to exploit; if the past success is to continue, understanding of the reservoir needs further improving.

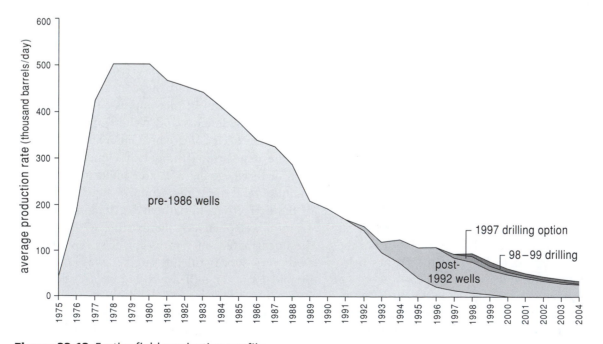

Figure 22.12 Forties field production profile.

(a) Forties reservoir seismic section

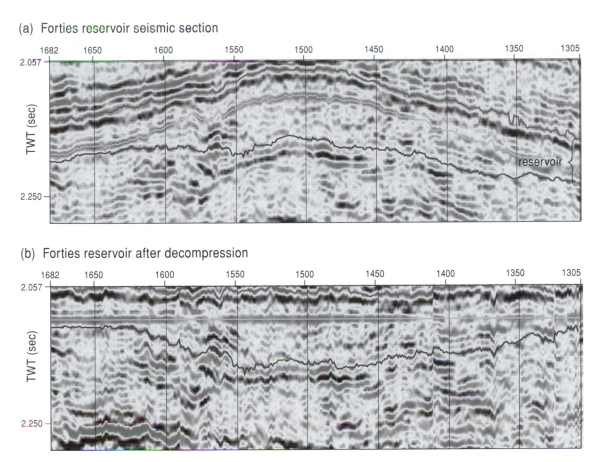

(b) Forties reservoir after decompression

Figure 22.13 Seismic section before and after decompression.

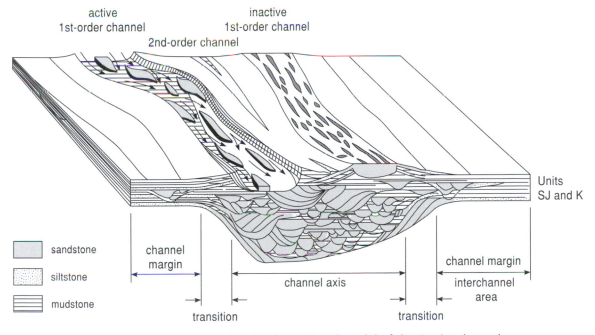

Figure 22.14 Schematic drawing showing the depositional model of the Forties channel.

To this end, the original 3D data was reprocessed using improved processing methodology, and a second survey was carried out in 1996. This second survey took only a month to acquire and was available for interpretation within six months, about one fifth of the time of the first survey. Already, it has given a 20% improvement in resolution, with more detail of the reservoir architecture – the arrangement of channels in which porous detrital sediments were deposited. In addition, by comparing seismic data acquired at successive times, the areas from which oil has been extracted can be recognised and so used to identify remaining targets. Plate 8 clearly shows the reduction in oil remaining, resulting from five years of extraction.

22.7 The future

In a world that is changing rapidly technologically, politically, and in other ways, predictions can be no more than tentative. For example, crises in the Middle East (which has 65% of known petroleum reserves) and the unexpected breakup of the former Soviet Union have affected prices and consumption in the past. However, the trend has been increasing consumption of petroleum products (Fig. 22.1). At present, per capita consumption by the peoples of North America and Europe is six times the world average and accounts for half the global consumption. Despite pressures to stabilise and even reduce consumption (to contain global warming resulting from increased concentration of carbon dioxide in the atmosphere), so far this has not been achieved, and consumption is still rising. As the less-developed parts of the world strive to improve their standards of living, they are also likely to consume more. Thus increasing consumption is probable, and a global increase in the next two decades of 70% has been predicted.

However, most of the world's major oil fields have probably been found, and of these many have had most of their oil and gas extracted. Currently, petroleum is being found less rapidly than it is being consumed. Though there still remain huge reserves in the Middle East, the cost of importing large amounts of oil, combined with the political uncertainties of the area, has encouraged countries to be more self-sufficient in hydrocarbons, which

may be achieved in various ways. One is improved recovery; for example, the standard recovery in the North Sea is 45%, but in the Forties and some other fields the use of water injection to drive oil through the reservoir has improved the figure to 60%. Another is to exploit fields hitherto too small to be economic; an example is the development of the Andrew and Cyrus oil fields, which were discovered in 1974, but were then uneconomic because of the thin oil column. However, the ability to drill horizontally has allowed a few wells guided by 3D seismic images to tap into Andrew field's recoverable reserve of 16 Mt, and the oil is delivered by a pipeline connected to the Forties field only 50 km to the southwest (Fig. 22.15).

The even smaller Cyrus field, with reserves of 2.2 Mt, has been connected to it, at much lower cost than building a new facility. There is also exploration for smaller and more elusive fields in increasing difficult locations, such as the deeper parts of continental shelves. To handle the more dif-

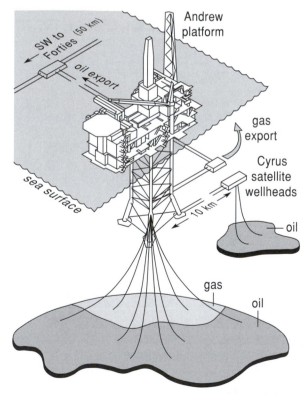

Figure 22.15 The Andrew–Cyrus development in the North Sea.

ficult conditions and to contain the huge costs of exploration and extraction will require advances in identifying the presence of hydrocarbons and of drilling and extraction technology, which in turn will require increasingly precise surveying of the subsurface. Presumably, geophysical advances will be mainly in seismic reflection imaging, though other techniques may be developed to identify remotely the presence of hydrocarbons. Meanwhile, the advances in methodology resulting from developing the North Sea and other fields are being applied elsewhere in the world, such as South America, the former Soviet Union, and the Gulf of Mexico.

Further reading

Stoneley (1995) provides an excellent introduction in nontechnical language to petroleum exploration, covering the origin, accumulation, and exploration of petroleum, and Tiratsoo (1973) gives a good, if somewhat old, overview of the world's oil fields. DTI (1997) is the primary source of information on the United Kingdom's oil and gas resources, and Walmsley (1975) provides an early account of the discovery of the Forties field. BP (1997) is an authoritative compilation, in the form of tables and graphs, of the world's reserves, production, and consumption of oil, gas, coal, nuclear energy, and hydroelectricity from 1965 to the end of 1996.

chapter 23

Exploration for Metalliferous Ores

Many of the minerals of economic importance are ores of various metals, particularly sulphides. Traditionally, they have been discovered from their surface outcrops, but more recently geochemical surveys have been used to detect above-average concentrations of relevant elements in near-surface samples. However, as shallow deposits are extracted there is a need to explore to greater depths. Geophysical surveying potentially can be used because many ores have sufficiently different properties from their surroundings – particularly electrical and magnetic ones – to be detectable.

This chapter gives an outline of the different ways in which orebodies originate and the role of geophysics in exploring for them. Much of the chapter describes the exploration and evaluation of the Elura orebody in Australia, a massive sulphide body upon which many geophysical methods have been employed.

23.1 Introduction: Metalliferous and other ore deposits

Geophysics has an important role to play in the exploration for ore deposits, a term that is being used here to refer principally to those containing metals that can be profitably extracted (diamonds are included though not a metal). Ore deposits are mined for the precious metals gold, silver, and platinum, and for many of the raw materials for manufacturing industries, including aluminium, cobalt, chrome, iron, lead, manganese, molybdenum, nickel, thorium, uranium, zinc, and zirconium, which are used variously for making iron, special steels, refractory materials, pigments, special glasses, and solder, plus numerous alloys and chemicals. In turn, these are incorporated into such manufactured items as aeroplanes, jet engines, saucepans, corrugated iron, furnace linings, cars, batteries, electrical and electronic devices of all kinds, and computers.

In short, our modern industrial economy would be impossible without them.

New ore deposits are often found in remote areas and can have an important bearing on the development of a region, for their extraction may require development of roads, water, power supplies, and communications. These activities trigger demands for services and for supplies of raw materials of a nonmetallic nature, such as sand and gravel for building, road metal, clays for bricks, limestone for cement, and so on. These industrial minerals are sometimes also referred to as ore deposits, but as they are usually extracted at the surface and geophysics is seldom required to locate them, they are excluded from the considerations that follow.

23.2 The formation of ores and their geophysical properties

Metals occur widely in the crust and mantle but usually only in trace quantities – parts per million, or even less – and so are uneconomic to extract unless they are concentrated into ores. Concentration occurs by a variety of mechanisms, most of which depend on fluids, and also depends upon both the chemical properties of the metal and the source from which it is to be concentrated.

As will be explained in the next section, an exploration programme may involve many methods, including some geophysical ones. Which – if any – geophysical methods are likely to be useful depends upon whether the ore contrasts physically with its surroundings, so physical properties will be mentioned briefly in the following descriptions of various types of ore deposits. The most widely used methods depend on an ore being electrically conducting, and if such an ore is at least partly disseminated, IP is likely to be more useful than resistivity or e-m methods. Also of potential use for discovery of ores are magnetic properties, their often high density, and radioactivity. Successful investigation also depends, of course, upon the depth, size, and shape of the orebody and upon the concentration of the ore. The main processes that form ore deposits follow.

(i) Crystallisation from basic magma. The source of many ores is magma, because during the crystallisation of basic and ultrabasic magmas, many metals form dense minerals, particularly compounds of sulphur and oxygen, which settle through the still-liquid magma to form layers. For instance, about half the world's chromium comes from bands in layered intrusions of Precambrian age; a notable example is the Bushveldt Complex of South Africa, very large deposits that contain up to 28 layers, each about a metre thick and containing about 50% chromite, Cr_2O_3, and extending laterally for tens of kilometres. In most of the remaining occurrences of chromium, it occurs as pods in Phanerozoic ophiolites with about 30% chromite, in deposits of 0.1 Mt or more. These podiform deposits were formed originally by settling in oceanic magmas but subsequently were tectonically emplaced. They are more difficult to locate and trace than layered deposits, but as they are dense, and often magnetic and conducting, gravity, magnetic, and electromagnetic surveys are often used.

Nickel sulphide is another mineral concentrated by crystallization in magmas, sometimes after extrusion, and the world's major deposits occur near the base of thick lavas, or intrusions in greenstone belts. Deposits are relatively small, 1 to 5 Mt with 2 to 3% Ni (plus valuable amounts of gold and platinum), though larger deposits with lower concentration occur. The ore is usually dense, conducting, and magnetic, but if it is disseminated IP is likely to be a better geophysical technique than gravity, resistivity, or e-m.

Diamonds are precipitated from kimberlite magma, which originates at depths of more that 150 km beneath stable Precambrian cratons, mainly in Africa but also in Australia, Canada, the former Soviet Union, and the United States. The kimberlites reach the surface in pipes that are typically 500 m across at the surface and taper downwards like a carrot. Although the diamonds themselves cannot be detected remotely, the pipes containing them can be found because they also contain magnetic and radioactive minerals; they are magnetic because they usually contain 5 to 10% of iron oxide, and often radioactive due to the presence of mica, which contains several percent of potassium. They have also been observed to be more electrically conducting than the surrounding rocks.

(ii) Hydrothermal residues of acidic magmas. Silica-rich igneous bodies may also give rise to mineralization. When these are emplaced high in the crust and crystallisation is nearly complete, residual magmatic fluids deposit metals as disseminations in porous zones in the igneous body or as hydrothermal veins extending into the country rocks. These give rise to porphyry deposits, which are the main sources of copper and molybdenum, and they also contain gold in economic quantities. Deposits are very large, typically about 1000 Mt, but as this ore is distributed within a vertical cylinder 1 to 2 km across, concentrations are less than 1% and disseminated, so IP surveys are commonly used in their exploration.

(iii) Oceanic and other hydrothermal processes. In another kind of hydrothermal process, massive sulphide deposits of copper, lead, and zinc, often with gold and silver, are formed in association with volcanics in marine environments. They are deposited from hydrothermal waters formed when sea water percolated downwards to high-temperature zones above submarine magma chambers. Several types have been identified. One is associated with basic volcanics formed at an oceanic spreading ridge, since made accessible by obduction as an ophiolite (Section 20.8), as in Cyprus. Another is associated with island-arc volcanism, which sometimes has an early stage of calc-alkaline volcanism – as at Besshi in Japan – and others with late-stage, more felsic volcanism – as at Kuroko in Japan. An additional, more primitive (i.e., less evolved), type has been proposed to account for the numerous zinc–copper deposits of the Canadian Archaean.

Worldwide, massive sulphide deposits are among the most numerous and most economically important ore deposits. They typically have grades, of all their metals combined, of about 6%, and over 80% of them are between 0.1 and 10 Mt in size. These sulphides, being massive rather than disseminated, are usually excellent conductors, and some of the electrical and electromagnetic methods described in Chapters 13 and 14 were developed primarily to prospect for them. They are also usually magnetic.

In another kind of process, the declining heat of an intrusion may be sufficient to 'power' hydrothermal circulation of groundwater through the intrusion, where it dissolves compounds of gold, silver, lead, cobalt, tungsten, zinc, and uranium, later to deposit them in veins a long way from the source. Globally, these are not as important as porphyries and the massive volcanogenic sulphides mentioned above, except for tungsten, much of which comes from this source. Locally, they may be important for gold, and in some countries for uranium, used for power generation. The minerals in veins are often disseminated, so IP is appropriate, and more radioactive than the host rocks due to small amounts of uranium and thorium in the mineralising fluids.

(iv) Precipitation from solution in groundwater. Some ore minerals may be dissolved in groundwater and then redeposited elsewhere, as in the important sedimentary ironstone and manganese ores. The principal ores of iron (sometimes banded) were formed in this way and contain about 30% iron, mainly in magnetite, haematite, and siderite (iron carbonate). They occur throughout the geological column but are most common in the early Proterozoic due to a favourable combination of structural, geochemical, and biological factors. Extensive deposits have been found in Africa, the former Soviet Union, North and South America, and Australia, ranging in size upwards from 1 Gt (10^9 tonnes) to 100 Tt (10^{14} tonnes). (Smaller amounts of iron are mined from ores formed from magmas or from the heat of contact metamorphism.) Because of the great size of the deposits, they have mostly been found by geological mapping. However, their magnetic properties offer geophysical detection and, indeed, early explorers supplemented visual observation with simple magnetic instruments, including the dip needle (Fig. 10.3). Today, more sophisticated magnetic instruments are used to map concealed extensions, on the ground and from the air.

Precipitation has also formed extensive deposits of copper, lead, and zinc. About one third of the world's copper reserves have been formed in this way in sandstones and calcareous shales, as in the Proterozoic Zambia–Zairean copperbelt of Africa and the Permian Kupferschiefer of Poland–Germany. The average grade is about 2.5%, and individual deposits range in size from 1 to 1000 Mt. In other well-known places, such as Mt. Isa in Australia and Tynagh in Ireland, lead and zinc are the dominant metals, being 1 to 10% and 5 to 20% respectively in bodies about 100 Mt in size. Most of the lead and zinc of the United States comes from carbonate-hosted deposits of 20 to 500 Mt in the Mississippi Valley with 1 to 15% zinc and 0.5 to 5% lead. The ore fluids seem to have originated in intracratonic basins, so magnetic and gravity surveys aimed at locating regional structures have been useful in the initial stages of prospecting. For the detailed location of specific bodies, IP in combination with gravity has been useful.

(v) Mechanical weathering and deposition. Rock debris produced by weathering may be transported, comminuted, and sorted mechanically by water according to density, and then deposited to form placer deposits. This is an important source of materials resistant to abrasion and chemical weathering, notably diamonds, platinum, gold, tin, ilmenite, monazite, and zircon. Placer deposits, being formed at the surface, are vulnerable to erosion, so most of those remaining are Tertiary or Quaternary in age, older ones mostly having been destroyed. They occur throughout the world and range in size from small pockets in hollows close to the source, through large ones in river deposits, to very large ones in beach sands in which heavy minerals have been concentrated at the high-water mark by surf action. The heavy minerals are sometimes magnetic or radioactive and show up as linear anomalies on ground and aerial surveys.

(vi) Chemical weathering residues. Chemical – rather than physical – weathering of rock exposed at the surface and the removal of the products to leave a residue can be important under the hot, wet conditions of the tropics. This is the main source of bauxite, which may contain alumina – aluminium oxide, the main source of aluminium – in concentrations over 50%. Over 80% of the world's bauxite reserves, estimated to be over 30,000 Mt, overlie stable continental platforms in West Australia, South America, India, and Jamaica. Though bauxite is readily identified on the surface, there are problems with estimating the size of a deposit due its irregular base, but this can be mapped by seismic and electrical surveys supplemented by drilling. Individual deposits are typically a few hundred megatonnes. Another consequence of weathering is the concentration of nickel in laterites (e.g, from about 80 ppm in the ultrabasic rocks to over 1% in the residual deposit in New Caledonia, where the total reserve is estimated to be over 1 Gt). About half the world's nickel output comes from sources of this type.

(vii) Chemical deposition. It was explained in Section 13.2.1 that shallow sulphide orebodies can generate self-potentials, probably by oxidation of parts above the water table. The chemical reactions move iron to the surface as goethite, sometimes with haematite, which together with silica, forms the brightly coloured gossans that commonly overlie sulphides containing copper, lead, and zinc, which are deposited in a zone of enrichment at the water table; SP, electrical, e-m, and magnetic methods are routinely used to explore such regions.

(viii) Metamorphic processes. Finally, there are some important ore deposits, notably of copper and tungsten, that resulted from metamorphism at low temperature and pressure in the vicinity of plutonic intrusions. Less fluid is usually involved than with porphyry deposits. An important class of these are associated with skarns, a Swedish term used to describe the calcium silicate gangue formed by metamorphism of limestones. Most of the world's tungsten comes from low-grade (less than 1%) bodies of a few megatonnes in Canada, the western United States and Tasmania. The skarn-type deposits formed by the Copper Mountain plug of Quebec contain about 1.5% copper and useful amounts of gold, silver, molybdenum, lead, and zinc. Magnetite is a common by-product of the metamorphism, so ground and airborne magnetic surveys have been useful aids to prospecting. The Marmora body, used as an example of a magnetic anomaly in Figure 11.10c, is believed to be an iron-enriched body of this type. Gravity surveys have also been useful in locating granite plutons, at the margins of which these deposits might be expected.

23.3 Where ores form

When searching for orebodies it is obviously useful to have a general idea of where they are likely to have occurred, which depends on the processes that formed them. In the outline that follows, the processes – as listed in the previous section – probably involved are given, though this is still often in debate because of the great variety of ore deposits.

Most of the processes of igneous origin described above are associated with plate boundaries, as this is where igneous activity is concentrated. Most magma is generated at ocean spreading ridges (Section 20.8),

leading to ore formation by process (iii), or (i) where there is crystallisation of basic magma. Hot springs produced by sea water entering hot rocks beneath ocean ridges have been observed depositing metallic sulphides in the sea floor as black smokers (Section 20.8), giving metal-rich sediments. These are often associated with lenses of pyrite overlying pipes with ore minerals. Some of the copper-containing sulphide deposits of Cyprus (after which copper is named) are believed to have been formed in this way.

The initiation of ocean ridges by the splitting of a continent (Section 20.7) can often be traced back to a continental rift, in turn resulting from the action of an underlying plume, as was described in Section 21.6, forming ores by process (i). The tin deposits of Nigeria and Brazil occur in granites believed to have been formed by the melting of continental crust by such a plume. The alkaline lavas that occur in this setting give rise to carbonatites and kimberlites, which sometimes contain diamonds, as in East and South Africa. As oceans develop, the rift sides become passive margins with a continental shelf on which sediments are deposited. These sediments are often hosts for the deposition of ores of the 'Mississipi Valley type' (process iv), which are the world's principal sources of lead and zinc. Extensive banded iron formations are also believed to have been formed in this environment, by process (ii).

At subduction zones, many kinds of orebodies may form. Cyprus-type bodies, originating at ridges, may be transported on the downgoing slab and be trapped at the junction with the continent. Later, as the descending slab heats up and begins to melt, it gives rise to plutonic as well as volcanic activity in the overlying continent; massive sulphide bodies with copper and zinc, of the Besshi type (process iii), and porphyry copper deposits with copper, gold, and molybdenum are formed. In the final stage of ocean closure, when continents collide (Section 20.4.3), the subducted plate melts and yields differentiated granites with, by process (ii), concentrations of tin, tungsten, and other metals in the overlying continent.

There is no clear association between mineral deposits and transform faults, for these have little igneous activity, but the intersection of major lineaments, as at Olympic Dam in Australia, have long been recognised as pathways for mineralising fluids, forming ores by process (iv). The last three processes described in the previous section do not depend on thermal processes and so are not particularly associated with plate margins.

23.4 Exploration for orebodies

Some orebodies may be recognised on the surface, either because they outcrop or – if they are sulphide orebodies at very shallow depths and contain iron – because they have oxidised into coloured gossans, characteristic of ores. Once located, orebodies have traditionally been investigated by drilling to determine the geometry, size, and grade of the mineralisation. This knowledge is essential for deciding whether to develop a mine in an orebody, for there must be enough of the minerals present to produce a reasonable annual return on investment over a few decades.

Though most of the economic deposits with obvious surface expression have been found, extensions of these and new bodies in their vicinity can be investigated by further drilling. However, deciding where to drill for new bodies with no visible surface expression, in areas not known to contain ore deposits, requires the application of some or all of the methods of geology, remote sensing, geochemistry, and geophysics in a systematic way.

How these methods are integrated to plan and execute a mineral exploration programme depends on who the explorer is as well as what the minerals of interest are. The range is wide: At one extreme is the solitary prospector or small company interested only in a single mineral such as gold within a small prospect area. In this case, digging pits or trenches, with the occasional shallow drill hole by a contractor, may be all that the exploration programme requires. At the other extreme, governments may be interested in carrying out a mineral inventory survey of the whole country for planning long-term development or to stimulate private investment. In this case, there is no limit to the size of the area or the minerals of interest, although there will be priorities dictated by the demand for specific minerals, both within the country and abroad. Between these extremes are the multinational mining companies looking for investment opportunities to develop mines in marketable commodities anywhere. As the time between exploration and production can be a decade or more, with a cost of over $10 million for a medium-sized mine, programs need to allow for

changes, political as well as economic, that may take place in that time. There is therefore a trend towards diversification of interests to allow for changes in demand, and towards exploration in developed countries with political and economic stability, even though these increasingly have restrictions being imposed by environmental considerations. In view of this diversity of scenarios, the importance of geophysics in mineral exploration can only be indicated.

Once the decision has been made to explore a large region or country, the area of search for orebodies has to be reduced in a series of stages culminating in the selection of sites for drilling (the ideal of drilling the whole region on a closely spaced regular grid would be hugely expensive).

The first stage is a planning one in which the limits of the area, the minerals of interest, the budget, and the duration of the exploration are specified. At this stage, all the existing geological, geochemical, geophysical, topographic, remote sensing, drill-hole, and mining data will be assembled and studied. In some countries or regions, these data may be sufficient to permit the selection of areas for follow-up work or even the selection of some drilling targets, but in other regions there may be little previous work, so a reconnaissance of the whole area may be necessary. At this stage, aerial surveys are particularly cost-effective, and these could include magnetic, radiometric, electromagnetic, and photographic surveys. However, ground surveys are necessary for other methods, which could include geological traverses, stream sediment geochemistry, gravity surveying, boulder tracing, and heavy-mineral analyses. The ground and airborne data are combined to produce outline maps showing the main geological and structural features. The photographs, obtained primarily for photogeology, may also be used for photogrammetry if adequate topographic maps do not already exist.

These regional data are used to select areas for follow-up work, using our knowledge of the geological environment in which the different kind of ore deposits formed, as outlined in Section 23.2. At this level there may still be large areas to be investigated, so further aerial magnetic, radiometric, and electromagnetic surveys, as well as photography, at higher resolution achieved by having closer line spacing and lower elevation, as illustrated in Figure 11.5, may be advantageous, using either fixed-wing aircraft or helicopters, depending on the nature of the terrain. Ground work might include geological mapping; geochemistry of streams, soils, water, rocks, and plants; and appropriate large-scale ground magnetic, electromagnetic, induced polarisation, self-potential, gravity, radiometric, and seismic surveys. There may also be some exploratory diamond drilling to investigate interesting anomalies at this stage. If ore deposits are found, they may be investigated further by more holes or small exploratory mines.

The anomalies revealed by these geochemical and geophysical surveys are used to select areas for more detailed ground surveys with yet closer line spacing and station interval, with the aim of deciding on the location, depth, and direction of further exploration holes. The physical properties of samples from boreholes, outcrops, and the exploratory mines may be measured at this stage and used to refine the interpretation of the geophysical surveys. Once boreholes become available, they can be logged or used for further surveys to extend the investigation laterally or downwards, and, in the light of what is found, to plan the location of other boreholes.

The above idealised account of successive stages from prospecting to evaluation is often only partly followed, usually for a variety of economic and political reasons. The following section describes the discovery of a large deposit using geophysical surveys. The area was also used as a test bed for geophysical prospecting methods and to compare the instrumental systems that were emerging at the time.

23.5 The Elura orebody, New South Wales, Australia: A case study

Geophysical methods employed: Mainly magnetic (aero- and ground); also gravity, electrical (VES and mise-à-la-masse), and e-m (TEM).

23.5.1 Background and reconnaissance surveys

The Cobar mineral field of New South Wales in Australia (Fig. 23.1) has been an important producer of gold and base metals since 1870. It is the biggest source of copper in the state. Zinc was also exploited at the CSA mine, while lead and zinc were also encountered at some of the mines further south.

In 1970, the Electrolytic Zinc Co. decided to explore for further base-metal resources to meet the growing demand. Mining experience in the Cobar region indicated that the mineralisation is stratabound (i.e., associated with particular strata), being hosted in graded siltstones and shales belonging to the Silurian–Lower Devonian CSA siltstone unit of the Cobar supergroup. It was therefore natural to pursue exploration northwards following the arcuate trend of the group. However, the geology is concealed beneath about 2 m of regolith, plus up to 10 m of unconsolidated alluvium in drainage depressions, so the geology is known mainly from drill holes and pits. A further complication is that the top ½ m of bedrock is normally iron stained or ferruginised, and sometimes includes gravels with the magnetic mineral maghaemite. Some structures suggested that the group has a northeast trend, but it was appreciated that this was uncertain and that it might continue north or NNW.

All the known bodies in the orefield had magnetic signatures, typically aeromagnetic anomalies of about 40 nT in amplitude, so airborne magnetometry was selected as the primary reconnaissance method. Geological mapping, with the aid of colour aerial photographs, was carried out at the same time for control purposes. The first magnetic survey was flown in May 1972 using a proton precession magnetometer, with a ground clearance height of 90 m and a line spacing of 300 m. Twenty-five anomalies were selected for ground follow-up and listed in order of priority, using aboriginal names alphabetically. The fifth, Elura – meaning 'pleasant place' – was the only one substantiated by the later ground magnetic surveys, the others being attributed to maghaemite on the surface, as explained below.

After removal of the total Earth's field of 51700 nT, a small 'bull's-eye' aeromagnetic anomaly was revealed, as shown on Figure 23.2, with the flight lines also shown. Figure 23.3 shows a profile for the flight line through the centre of the anomaly. The anomaly appears as a 55-nT positive anomaly superimposed on a regional gradient of 45 nT/km, which was removed to reveal the anomalous field. As a first approximation, the anomaly was modelled (Fig. 23.4) in the simplest possible way as a spherical body (see, e.g., Section 11.2.4) using a magnetic susceptibility of 0.11×10^{-3}, based on knowledge of other bodies in the region. The model gave its radius as 118 m, with 250 m as the depth to its centre (Blackburn, 1980).

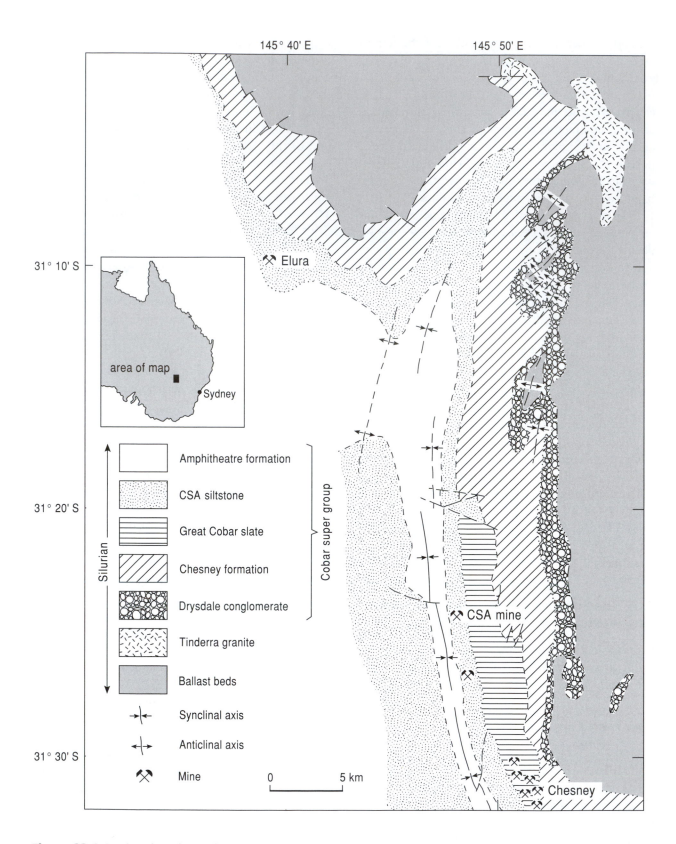

Figure 23.1 Regional geology of the Cobar area.

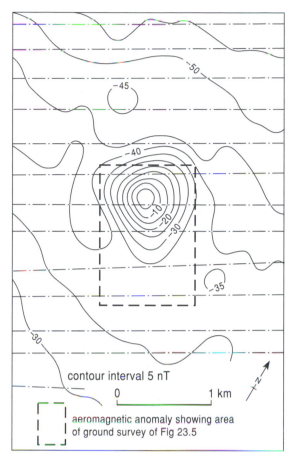

Figure 23.2 Aeromagnetic anomaly over the Elura orebody.

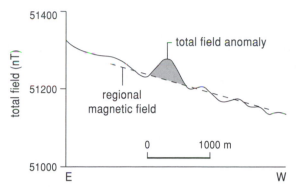

Figure 23.3 Aeromagnetic profile over the Elura orebody.

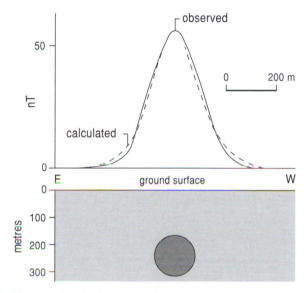

Figure 23.4 Calculated and observed magnetic anomaly over the Elura orebody.

23.5.2 Initial surveys of the Elura orebody

It was realised that if the magnetic anomaly was really due to a metalliferous orebody and had a higher density (c. 4 Mg/m^3) than the host rock (c. 2.4 Mg/m^3) it would produce a measurable gravity anomaly on the surface. So, in 1973–1974, a detailed gravity survey was carried out in which 374 measurements were made at 25 m intervals on a rectangular grid. A network of 57 base stations was set up, and the measuring stations were arranged to be in loops of 11, tieing in to a base station every hour, to allow instrumental drift to be measured (Blackburn, 1980). Drift averaged 0.05 mGal/hr, and the precision of the readings was estimated to be 0.04 mGal; the elevations of all stations were surveyed, and the elevation errors were equivalent to 0.02 mGal, giving an overall error of about 0.05 mGal. The data were reduced, using a density of 2.6 Mg/m^3, based on laboratory determinations, for the Bouguer correction. The

residual Bouguer anomaly map is shown in Figure 23.5 (together with the location of the orebody as proven later). The pattern of the gravity anomalies suggested that the sphere approximation could be used, and the depth to the centre was estimated from the half-width to be 209 m.

As few profiles of the aeromagnetic survey crossed the anomaly, a more detailed survey was carried out on the ground using the same 25-m grid as for the gravity survey, but the reading interval was reduced to only 5 m where the values changed rapidly. The regional field of 51700 nT was removed from the observations to leave the anomalous field, which is shown in Figure 23.6 as stacked profiles. Two kinds of variations are evident: There are short-wavelength variations, which appear as spikes, and these are superimposed on a long-wave-length magnetic high most clearly seen on profiles 20100–20300E. The spikes were found to be due to highly magnetic maghemite pebbles in the regolith and so were treated as noise, but the longer-wave-length feature was attributed to a deeper source.

The data were filtered (Section 3.2) to remove the noise and contoured to give the map shown in Figure 23.7. It has a clear maximum coinciding with the gravity maximum, plus a smaller minimum to the south, as would be expected from a sphere mag-netised by induction in the Earth's magnetic field at this latitude, as shown in Figure 11.10a. This was modelled by a sphere magnetised by induction and gave a depth to its centre of 194 m. As this is close to the estimate using gravity, it suggests that the gravity estimate and magnetic anomalies are due to the same body. Chemical analyses of auger samples from depths of 2 m or more were anomalously high in lead, supporting the idea of a mineral deposit. The decision to drill was based on these data.

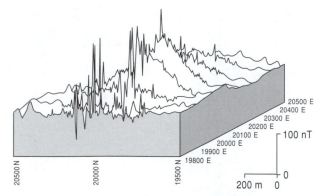

Figure 23.6 Stacked magnetic field profiles over the Elura orebody.

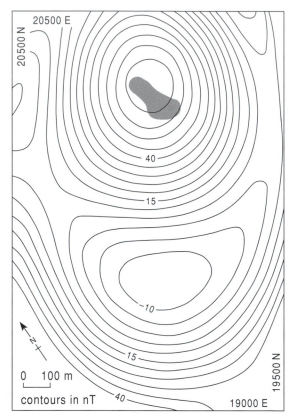

Figure 23.7 Smoothed ground magnetic anomaly contours over the Elura orebody.

23.5.3 Evaluation of the deposit

While exploration titles were being processed, fur-ther detailed geophysical surveys were carried out to plan the drilling programme. A total-count scintil-lometer survey on the ground produced only a very weak radiometric anomaly over the most intense

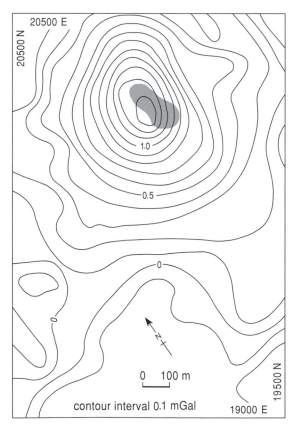

Figure 23.5 Residual Bouguer anomaly over the Elura orebody.

geochemical anomaly, so no more work was done with this method. Electrical and e-m surveys were then carried out, as the ore was likely (from knowledge of those found elsewhere in the region) to contain sulphides, which are usually conducting. Three methods were used initially. A e-m survey using a tilt-angle method (see Section 14.5 and Box 14.2) revealed only a weak anomaly near the magnetic maximum. An IP survey using the gradient array (Section 12.4.2) was carried out with the current electrode separated by 1000 m, the potential electrodes 50 m apart, and readings made every 25 m. This gave encouraging IP anomalies at the locations shown on Figure 23.8, which also shows the lead anomaly contours. (The results of a later IP survey using the dipole–dipole array are shown in Fig. 13.4.) Vertical electrical soundings using resistivity suggested a conductive weathered layer with a thickness of about 90 m.

Based on these results, the drilling programme was started with hole DDH1 in February, 1974 'to test the source of the magnetic, gravity and IP anomalies beneath a surface lead–arsenic anomaly'. The presence of mineralisation was confirmed

when it intersected gossan at 102 m on the seventh day, and then sulphide mineralisation (mainly pyrite with galena, sphalerite, and minor chalcopyrite) at 133.5 m on the ninth; it was stopped in weathered mineralisation at 152 m. Hole DDH2 intersected 80 m of mineralisation, while holes DDH3, DDH4, and subsequent holes parallel to DDH4 were drilled to evaluate the extent and grade of the deposit.

The density of the ore found by measuring drillhole samples was combined with the gravity anomaly to estimate the mass of the ore, as explained in Section 8.9. With a mean density of the ore of 4.2 Mg/m^3, compared to the 2.6 Mg/m^3 of the host rock, the mass was estimated to be about 27 Mt.

For planning the development of a mine, it is necessary to have as much information as possible about the form of the body. As explained in Section 12.4.2, one way of investigating this is to use the mise-à-la-masse resistivity technique, in which one current electrode is implanted in the orebody and the distribution of potentials is mapped using surface or borehole electrodes. At Elura, surveys were limited to the surface because when the conductive metal casing was removed from the holes to do the survey, most of the holes collapsed.

The result of a survey using hole DDH2 is shown in Figure 23.9. The contour pattern reflects the shape of the body and indicates a northward elongation. The combined application of geophysics, drilling, and chemical and physical analyses had resulted in the section through the body shown in Figure 23.10. It has the form of a vertical pipe elongated N–S with maximum dimensions 200×100 m, extending to a depth of over 500 m and with zones of different mineralogy. This shape is different from that of the early spherical models, but the volume is not too different. It demonstrates the usefulness of using even simple models to make initial estimates of ore quantities before proceeding with further exploration.

The body contains an estimated 27 Mt of massive sulphide ore valued at about £100/tonne, with 8.3% zinc, 5.6% lead, 140 g/tonne of silver, and less than 0.2% copper. The capitalisation cost of the mine in 1980 was about £100 million and it has been producing about 1 Mt/year.

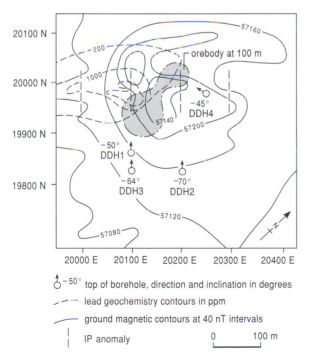

Figure 23.8 Location of magnetic, IP, and lead anomalies, and drill holes at Elura.

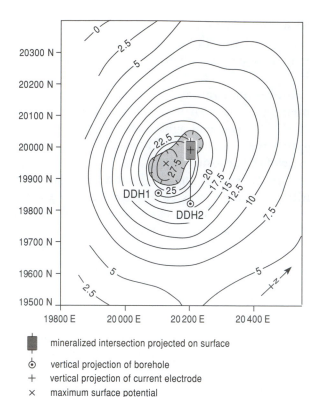

mineralized intersection projected on surface

⊙ vertical projection of borehole
+ vertical projection of current electrode
× maximum surface potential
⌢ approximate outline of Elura orebody at 100 m depth

contour interval 2.5 mV   earthing in DH2 at drill depth 315 m

Figure 23.9 Surface mise-à-la-masse results contoured, Elura orebody.

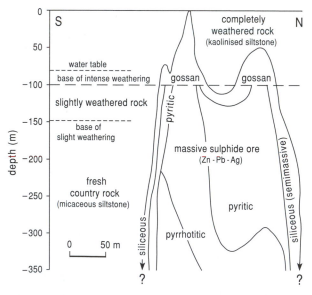

Figure 23.10 Schematic longitudinal section through the Elura orebody.

23.5.4 Assessment of geophysical surveying methods

The Elura discovery was based on an airborne magnetic survey guided by geological considerations and then followed up by ground magnetics, gravity, IP, geochemistry, and drilling. It was successful despite the depth of the mineralisation, deep weathering, and high magnetic noise. In this example, the most useful method proved to be magnetic surveying, initially as a reconnaissance airborne survey, but had the ore been less magnetic or the development of maghaemite more extensive it might have failed. The gravity survey was a useful complement to help confirm the existence of an orebody and later to help provide an estimate of its total mass, though the high cost of the precise surveys needed for such a small anomaly could limit their use in other cases. E-m surveys were less successful, largely because the depth of the body and the difficulty of penetrating conductors in the overburden. Though at Elura, IP surveys gave anomalies over ten times the background, it is generally unreliable for such deep bodies.

In the five years following the discovery of the Elura orebody, a wide range of new geophysical methods and techniques were employed at the location to assess how useful – and cost-effective – they could be for detecting other orebodies of a similar type, with concealed geology, high magnetic and telluric current noise, and inhomogeneities, which are often encountered in difficult arid terrains in many parts of the world. Methods included airborne e-m as well as more magnetics; ground surveying using magnetics, gravity, IP, TEM, and resistivity; and borehole measurements using magnetics, resistivity, mise-à-la-masse, IP, and TEM. In addition, the density, magnetic, and electrical properties of samples were studied in laboratories to aid quantitative interpretation. The results are presented in a special volume of the Australian Society of Exploration Geophysicists (Emerson, 1980), which has been useful in the planning of geophysical exploration surveys elsewhere.

Further reading

Evans (1993) is a well-known undergraduate text with clear accounts of the nature, origin, and economics of the main types of mineral deposits. It is not an exploration book and so does not contain geophysics. Dixon (1979) gives presentations in the form of maps, plans, and sections to illustrate the geology, discovery, and mines in 48 mineral deposits of different kinds throughout the world.

Edwards and Atkinson (1986), aimed at undergraduates, is on the geology of ore deposits and its influence on mineral exploration. It discusses the origin and distribution of ore deposits and the methods, including geophysics, that have been used to find them. Parasnis (1997) describes the principles and application of the geophysical methods used in mineral exploration.

Evans (1995) is an authoritative multiauthored text in two parts on the exploration of metallic and nonmetallic deposits. The first part deals with the economics, mineralogy, occurrence, and evaluation of the main types of deposits and the principles of the methods (including geophysics) used at various stages of exploration for them. The second part consists of case studies, mainly of well-known metalliferous orebodies.

chapter 24

Volcanoes

We are increasingly aware that we should not take our environment for granted. Part of the concern is about damage caused by human activity; part is the realisation that nature is not always benign and may produce catastrophes to which we are increasingly vulnerable as populations grow and modern civilisation with its industry and transport systems becomes more complex. Some of these catastrophes can be investigated by solid earth geophysics, notably earthquakes (discussed briefly in Section 5.10) and volcanoes.

Volcanoes can inflict great catastrophes on mankind. Historically, they have caused far fewer deaths than earthquakes (or storms or flooding), but this may not be true in the future, for the geological record reveals vastly greater eruptions in the past. For example, the 1980 eruption of Mt. St. Helens, cause of the greatest volcanic damage in the United States in the 20th century, erupted 1 km³ of material, but in 1783–1784 Laki, Iceland, poured out 15 km³ of basalts, and Krakatoa in 1883 blew out a similar volume, while in 1815 Tambora, Indonesia, erupted about 50 km³ of magma. But 700,000 years ago, the Long Valley Caldera of California (not yet extinct!) discharged 500 km³ of ash, and even larger eruptions have occurred. In contrast, it is unlikely that tectonic earthquakes much larger than those experienced this century can occur.

On the brighter side, eruptions rarely occur without being preceded by increased activity long enough in advance to allow time for evasive action, one reason for their relatively low death count. The aim must be to learn how to recognise when activity provides warning of an eruption, and to forecast its size, type, and duration, as well as when it will occur. In the longer term this requires a better understanding of the processes underlying volcanoes, but meanwhile there is a need to develop practical ways of predicting eruptions, based on the knowledge we have. Geophysics has a vital part to play in both of these aims, as the case studies that follow illustrate.

24.1 Introduction: Types of eruptions and damage

There are many types of volcanic eruptions, from sluggish lava flows that one can outwalk to incandescent clouds of ash and gas that travel many kilometres at more than 100 km/hr. The first step is to understand what produces this range. The most important factor is the viscosity of the magma, which in turn depends on its composition. Basaltic magma, with about 50% silica, is runny, but as the silica content increases, through andesitic (roughly 60%) to rhyolitic (70%), the viscosity increases by many times. Consequently, central volcanoes formed from basaltic lavas do not have the conical shape often regarded as typical of volcanoes, but are dome-shaped shield volcanoes, far wider than they are high; central volcanoes collectively are called strato-volcanoes, in contrast to the nearly flat outpouring of fissure eruptions, which usually are basaltic.

The type of eruption and its violence are mainly determined by viscosity, partly because the more viscous magmas tend to block the conduit and more particularly through its effect on the release from the magma of dissolved volatiles, mostly water and carbon dioxide, but including sulphur dioxide and other noxious gases. As magma nears the surface the decreasing pressure allows the volatiles to come out of solution, just as bubbles form within a fizzy drink when it is uncapped. If the viscosity is low the gases escape from the magma *comparatively* peacefully, but in high-viscosity magma they often form bubbles, leading to cinders (scoria) or pumice so light that it floats (after the Krakatoa eruption there were rafts 3 m thick on the sea). If the conduit is blocked the increasing pressure of released gases may lead to explosions, blasting out ash, cinders, and blocks (collectively called tephra) up to tens of kilometres into the air, from where they shower down as pyroclastic falls as far as hundreds of kilometres away. Tephra can also travel over the surface, supported by fluids, usually entrained air, forming pyroclastic flows. These range from avalanches, like those that occur on nonvolcanic mountains, to hot debris flows or incandescent nuées ardentes. In 1902, a single nuée ardente destroyed in minutes the city of St. Pierre in the Caribbean and all but a handful of its 30,000 inhabitants. These forms of pyroclastic flow travel very quickly (sometimes exceeding 100 km/hr) but usually stop when the slope becomes near to horizontal; however, ignimbrites are supported by their own volcanic gases and are able to continue on the level and even ascend slopes; fortunately, they are rare. Another type of flow is a lahar, or mud flow, where water is the supporting fluid. These can form when an eruption blasts up through a crater lake, or later when rain saturates ash that has been deposited on the slopes of a volcano. In 1985, a lahar swept down a canyon in the slopes of Nevado del Ruiz, Colombia, to emerge above the town of Armero as a 40-m-high wave that killed 21,000 persons.

Volcanoes can also generate large earthquakes. If a volcano is beside the sea, tsunamis (giant waves – see Section 5.10.1) can be generated by earthquakes, explosions, or landslides; tsunamis were the main cause of death in the Krakatoa eruption, killing 36,000 persons in coastal villages. Finally, the atmospheric pollution produced has a range of effects, from affecting breathing to climatic cooling; the severe winter that followed the 1783–1784 basaltic eruption of Laki caused a famine in northern Europe, in addition to the fifth of Iceland's population who were killed in various ways, while more recently the 1991 eruption of Pinatubo in the Philippines temporarily caused global cooling of ½°C. The climatic effects of some of the huge eruptions that occurred in the past must have been extremely severe, and they have been suggested as a cause of global extinctions (see Sections 25.1 and 25.4).

24.2 Methods for investigating volcanoes and monitoring activity

Geological mapping can, of course, reveal the types and extents of past eruptions, and so provide a guide to future activity, while study of dissected volcanoes shows details of the 'magma plumbing' that cannot be accessed in an active volcano. As with other types of geological mapping, geophysics can assist. Seismic reflection and tomography plus attenuation of seismic waves can reveal the extent of magma chambers and hot rocks (their use to investigate the structure below oceanic ridges was discussed in Section 20.8), while resistivity has been used to locate geothermal areas and active lava tubes by their low resistivities. Gravity surveys can

help reveal the deep structure, and magnetic surveys can be used to map intrusions and extrusions. Radiometric dating can help unravel the history of a volcano; for example, dating of mapped units showed that El Chicon volcano in Mexico erupted about every 600 years during the Holocene. Palaeomagnetism has been used to measure temperatures of intrusions or emplacement of pyroclastics (Section 10.3.4), and magnetic fabrics can reveal the flow directions of cooled lavas and dykes (Section 10.7).

Turning to monitoring, the most used technique is to record seismicity. Tremors can provide early warning of increasing activity, though not all eruptions give such warning, while pinpointing its location within the volcano can reveal the depth of activity and the movement of magma. As magma nears the surface, it often causes the volcano to swell, and this deformation can be detected by precise surveying or by measuring tilting or strain using tilt or strain meters. The addition of microgravity surveys shows if there are also density changes: for example, magma moving into preexisting fissures produces a significant increase in g but little inflation, while vesiculation of magma by gas coming out of solution adds no mass and so causes inflation with little change in g.

Magnetic and electrical readings vary during some eruptions, though why is poorly understood at present. Analysis of gases is becoming increasingly important, particularly their ratios; for example, carbon dioxide (CO_2) is less soluble in magma than sulphur dioxide (SO_2) and so is released earlier as magma rises, and thus their relative proportions change as the magma nears the surface.

At present, eruptions are so variable in their geophysical and other types of activities that it is desirable to use several types of measurements and over a period to establish the variation of 'background activity' so that the abnormal activity related to an eruption can be better recognised. Applications of some of these methods are described in the two case studies that follow.

24.3 The 1989–1990 eruption of Redoubt Volcano, Alaska: A case study

Geophysical methods employed: Mainly seismicity, but also K–Ar dating and magnetic susceptibility.

24.3.1 Background

This volcano is one of many stratovolcanoes along the Aleutian arc (Fig. 24.1a), which result from the subducting Pacific Plate, about 100 km below (Fig. 20.12). Its cone rises about 1500 m above granitic continental rocks, reaching 3108 m, with an ice-filled summit crater, which is breached to the north.

Geological mapping, combined with K–Ar dates, shows that the volcano began to form about 0.9 Ma ago, commencing with an explosive stage that gave way to cone building. Its composition has varied considerably but is predominantly of intermediate composition (around 60% silica), whose relatively high viscosity accounts for its steep sides. In post-glacial times, over 30 pyroclastic eruptions have occurred, many leading to debris flows and lahars, some of which extended over 30 km from the volcano; melting of the numerous glaciers at the top of the volcano would have helped produce these. Much of the geological record close to the volcano has been destroyed by erosion and eruptions, so a detailed history of the volcanism of the area in the past 500 years has been deduced from cores taken from a lake 50 km away: tephra layers (not apparent to the eye) have been conveniently recognised from magnetic susceptibility variations (see Section 10.6), and they have been related to their parent volcanoes by electron microprobe analysis. Ashfalls reached the lake every 50 to 100 years, with Redoubt Volcano one of the chief sources; nearer, in Cook Inlet, the interval is only 10 to 35 years, showing that additional smaller eruptions occurred. Although activity of Redoubt Volcano has been observed several times in the past 250 years, tephra falls are known only for 1902 and 1966.

24.3.2 The 1989–1990 eruption

The first obvious sign of volcanic activity occurred five days before the first eruption, as a white plume rising from the crater rim, but as there was no apparent increase in seismicity it was thought to have little significance (a network of 5 seismographs – later increased to 9, of which 8 are shown in Figure 24.1c – had become operational only two months before the eruption, too short a period to establish the range of background seismic activity in the area, so minor premonitory activity might have escaped notice). Unusually, there was no obvious increased seismic activity until only hours before the first eruption, but when an intense swarm of small, shallow earthquakes began on December 13, 1989, a public

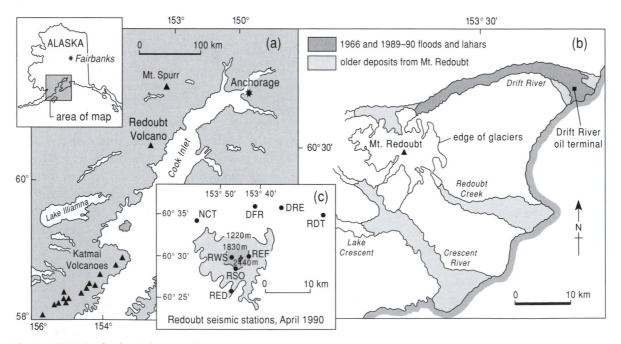

Figure 24.1 Redoubt Volcano, Alaska.

warning was given. Vent-clearing explosions began the following morning, with the largest occurring on the 15th, producing an ash column rising 12 km. Following vent clearing, a viscous andesitic dome slowly grew in the summit crater until it became unstable and collapsed, with accompanying explosions which ejected debris, producing a pyroclastic flow that reached 40 km away, and reexposed the vent (collapse of a dome was responsible for the major eruption of Mt. St. Helens in 1980). In the following six months 13 domes were formed successively and collapsed, at irregular intervals.

The 1989–1990 eruption was small, totalling only 0.1 to 0.2 km³ of ejected material, yet it produced 20 significant tephra falls and several pyroclastic flows; 0.08 km³ of ice and snow were melted, leading to floods a hundred times greater than any produced by normal rainfall; and over 700,000 tonnes of sulphur dioxide were released into the atmosphere. Five commercial jet airliners were damaged, including one that temporarily lost power in all four engines while approaching Anchorage to land, shortly after the largest explosion. The threat of floods and lahars curtailed oil production at a terminal on Cook Inlet (Fig. 24.1b). In all, the cost was about $160 million.

24.3.3 Monitoring of activity

Seismic monitoring did not consist simply of measuring the intensity of activity, for many tephra eruptions were successfully predicted only by recognising that there were two sorts of seismic events. Figure 24.2a shows the rapid oscillations of a short-period event, in contrast to the long-period oscillations of Figure 24.2b, which have a dominant period of ½ sec and also continue for longer. The short-period seismograms are typical of tectonically produced earthquakes and so were termed volcano-tectonic events. Their hypocentres occurred predominantly within a vertical column extending down to 10 km (Fig. 24.3b), before and during eruptions; as they were probably due to stresses produced by the movement of magma at these depths, they reveal the conduit followed by the magma in its ascent. Volcanotectonic events also occurred at shallow depths soon after eruptions, probably due to small collapses when the loss of magma and volatiles by eruption removed internal support of the vent sides.

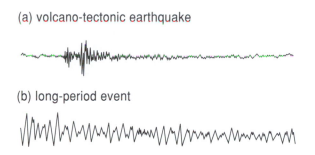

(a) volcano-tectonic earthquake

(b) long-period event

0   10 sec

Figure 24.2 Two sorts of seismogram, Redoubt Volcano.

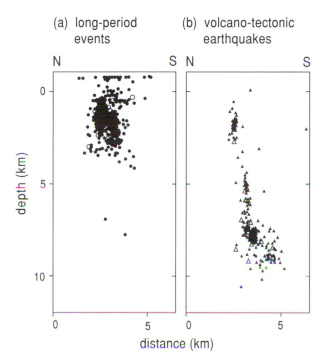

(a) long-period events   (b) volcano-tectonic earthquakes

Figure 24.3 Distribution of hypocentres, Redoubt Volcano.

The long-period events were no deeper than 3 km and close to the vent (Fig. 24.3a). They were probably due to oscillations in a column made semifluid by large amounts of volatiles, which led to subsequent explosions. Swarms of long-period events were the main predictor of eruptions; a total of 11 swarms were recognised, and each was followed by an eruption, within 2 hr to 2 days; however, a few eruptions were not preceded by swarms. Recognising the types of event, often against a background of tremors, was not usually as easy as the examples of Figure 24.2 suggest, and a method of harmonic analysis (Section 3.1) was successfully developed to identify them. Successful prediction also required seismometers close to the vent, and the destruction of the only one handicapped predictions until it was replaced.

Because of the possible damage to aircraft, it was essential to know when ash clouds were being produced, but frequently these could not be seen, for the eruption was in midwinter, with daylight sometimes lasting no more than 5 hr, or even less when weather was bad. Lightning is common in eruption columns but, of course, can also occur in ordinary storms. A lightning detector was installed, and when seismicity within the volcano and lightning above it occurred together, it was practically certain that an ash cloud was present, providing warning both for aircraft and of impending ash falls. Further details are available in Miller and Chouet (1994).

24.4 Etna lava eruption 1991–1993: A case study

Geophysical methods employed: Microgravity, deformation, and seismicity.

24.4.1 Background

Etna, on the east coast of Sicily, reaches about 3300 m and is the highest European volcano, with a long history of activity. It owes its existence to the complex tectonics of the Mediterranean, which result from the northwards collision of Africa with Europe (see Section 20.7), and it lies at the intersection of two major structural lineaments. Its magma is basaltic, so eruptions are not particularly violent, the major activity often being lava flows, which can engulf fields and villages on its slopes. These flows originate from both summit and flank eruptions. Because of its easy access and its threat to a densely populated region, and because it is one of the most active volcanoes in the world, it has been intensively studied.

There are several craters on the summit of Mt. Etna (Fig. 24.4). To their east is a great natural amphitheatre, the Valle del Bove, bounded by cliffs, which has probably been formed by slumping of the eastern part of the mountain towards the sea. After an eruption in 1986–1987, the volcano was quies-

cent for about two years, but late in 1988 there was seismic activity below 10 km depth, which continued the following year but then was mostly above this depth, indicating the rise of magma, an inference strengthened when the mountain inflated in the spring of 1989. In September 1989 a major eruption began – preceded and accompanied by tremors and earthquake swarms – with lava fountains up to 600 m high from the southeastern crater. Two fractures developed on the flanks at a late stage of this eruption, one to the northeast of the crater, producing a modest volume of lava (0.02 km^3), and a second to the SSE, which extended for 7 km roughly along the western wall of Valle de Bove (Fig. 24.4) but did not produce an eruption. The latter fracture provided conduit for the eruption of 1991–1993, as will be explained later.

Between June and November 1989, gravity at the summit increased by 75 μGal. This was computed to be a mass increase of 0.1 to 1 Mt (as explained in Section 8.9), and interpreted as intrusion of magma. This roughly matches the subsequent eruption in January 1990 of about 1 Mt of lava, after which gravity returned to near its former value. Mt. Etna then became quiescent – apart from continual degassing from the summit craters – until late in 1991, when major activity resumed. The key to understanding this second eruption depended on combining gravity and deformation measurements, which will be explained before the 1991–1993 eruption is described further.

24.4.2 Deformation and microgravity

The ascent of magma into the higher parts of a volcano often inflates the summit region as the magma forces a space for itself. However, things may not be so simple. If the magma intrudes into existing cavities or fractures there may be no inflation; conversely, vesiculation as dissolved gases form bubbles can expand the volume of the magma and produce inflation without influx of new material from below.

It is not obvious how the gravity reading at a station will change for all these possibilities because, though additional magma will increase a reading, inflation, by raising the gravimeter, will decrease it. Suppose a gravity station is on a wide flat area such as *A* in Figure 24.5, and the magma

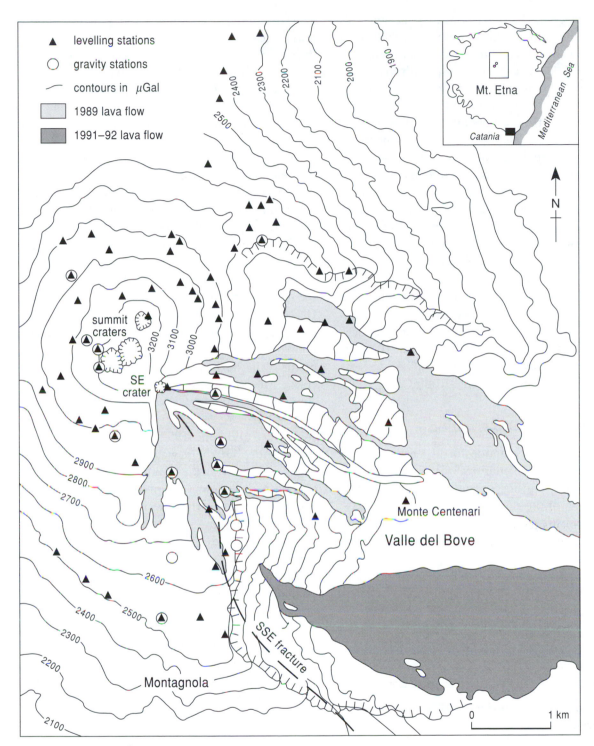

Figure 24.4 Map of Etna.

below expands, raising the gravimeter to C. We can consider this in two stages: Simply raising the gravimeter from A to B decreases the value of g by the free-air correction (Section 8.6.1). If next the magma expands below the gravimeter, there is no change in the mass and so g does not change further; the overall result is a decrease equal to the free-air correction. However, if the uplift is due to influx of magma, the mass does increase and this outweighs the decrease due to uplift, and the overall result is an increase equal to the Bouguer correction (Section 8.6.1). These various possibilities and resulting changes in volume and gravity are shown in Table 24.1. (The real situation, of course, may be a combination of these possibilities.) (The amount of change in g depends not only on the amount of magma or its expansion but also on the shape of the volcano. The value for the free-air correction given in Section 8.6.1 assumes the gravity meter is on an extensive plateau, but this is seldom the case for a volcano, and a revised value is found by actually measuring g at different heights. However, this correction does not affect Table 24.1.)

Over the period 1989–1994, changes in gravity due to uplift, or subsidence, were usually much less than those due to mass changes, so gravity readings are given without correction for elevation changes (elevation changes were generally less than 100 mm, which changes g by only about 20 μGal, whereas measured changes often exceeded 100 μGal).

24.4.3 The 1991–1993 eruption

This was the largest eruption for at least a century and continued for 15 months, yet it was not preceded by either unusual seismic activity or significant inflation of the summit region. However, gravity measurements revealed the movement of magma. Before the eruption, between June 1990 and June

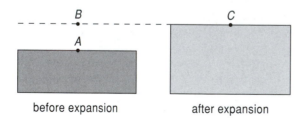

Figure 24.5 Change in gravity reading due to expansion of layer.

1991 (readings are compared at yearly intervals) gravity readings increased in the summit area (Fig. 24.6a) by a maximum of 400 μGal, with a lesser increase along the SSE fracture. As elevation changes were small, nearly all of the gravity increase must have been due to intrusion of magma into existing voids, probably the vertical magma conduit and the SSE fracture. From the change in gravity, its mass was calculated to be at least 10 Mt, ten times the amount in 1989. Modelling suggests the vertical conduit was 50 m in diameter and that the SSE fracture contained a 4-m-wide dyke (Fig. 24.7).

Between June 1991 and June 1992 there was little change in g at the summit, despite the cessation of the summit eruption when the SSE fissure opened down the flank of the volcano. Then, late in 1991, the upper part of the SSE fracture reopened, and on December 14 small flows issued from it. The fracture propagated downslope, and the next day lava poured from where the fracture intersected the wall of the Valle de Bove, about 600 m below the SE crater. After this there were no more lavas from the summit area, presumably because magma could now erupt more easily at the lower level. Between June 1991 and June 1992 a large volume of lava flowed out, but there was little change in either height or gravity at the summit (Fig. 24.6b), consistent with the magma supply route bypassing the summit. However, the lower part of the fracture widened, suggesting widening of the dyke feeding the eruption. In the following year, which saw the eruption end, gravity in the summit area decreased by up to 170 μGal (Fig. 24.6c) – again with only small elevation changes – interpreted as withdrawal of magma about 500 m down the vertical conduit. By the end of the eruption in 1993, 0.235 km³ of lava had issued, far larger than the amount that had inflated the volcano in 1990–1991, so most of the lava must have been fed directly from below the volcano, rather than from a near-summit magma chamber as had occurred in the 1989–1990 eruption.

Table 24.1 Magma, deformation, and gravity readings

Possible development	Mass of volcano	Deformation of volcano	Gravity reading
New magma forcefully intruding	increase	inflation	increase
New magma intruding pre-existing cavities	increase	no change	increase
Expansion of existing magma	no change	inflation	decrease

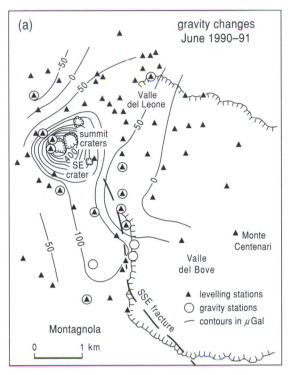

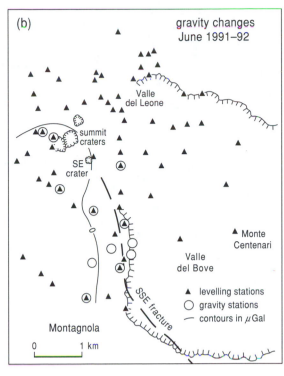

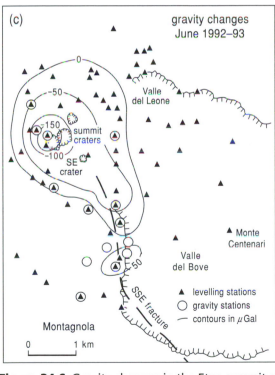

Figure 24.6 Gravity changes in the Etna summit area.

Further details can be found in Rymer et al. (1995). This case study is only one of the numerous investigations of Etna, but it shows the contribution of combined gravity and deformation studies to an understanding of internal processes. Seismic activity and tomography studies are described in Cardaci et al. (1993), Castellano et al. (1993), and Ferrucci and Patanè (1993).

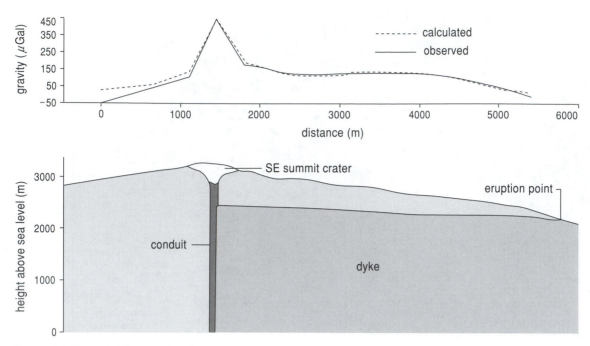

Figure 24.7 Model for gravity changes of Etna.

Further reading

McGuire et al. (1995) is an advanced book that discusses the techniques used for monitoring volcanoes, with numerous examples. Francis (1993) gives a general account of volcanoes, but with little geophysics; Scarth (1994) describes volcanoes from a geographic viewpoint, and includes a section on prediction, but no geophysics.

The Hawaiian volcanoes have been extensively studied and their internal plumbing deduced in detail. An introductory account is given by Dvorak et al. (1992) and a more advanced one by Ryan et al. (1981). An account of the 1783–1785 Laki eruptions is given in Thordarson and Self (1993).

chapter 25

The Chicxulub structure and the K/T mass extinction

At intervals in the Earth's history there have been mass extinctions, when the number of species of animals and plants was drastically reduced in a time that was geologically short. There has been little agreement about their cause, with suggestions ranging from catastrophes to the cumulative effect of changes in factors such as temperature. It is accepted that abrupt and violent processes, such as meteorite impacts and great volcanic eruptions, do occur from time to time, but it is also being appreciated that the environment, particularly the climate, is less stable than had been thought, so that the cumulative effect of comparatively small, steady changes may have large and abrupt consequences. Therefore, an appreciation of the environmental effects of impacts and volcanism will increase our understanding of the processes at work in the world we inhabit today, as well as possibly accounting for some extinctions in the past.

This chapter examines some of the evidence that the K/T (end-of-Cretaceous) extinction was due to the impact of an extraterrestrial body that produced the Chicxulub structure, and looks briefly at the competing theory that volcanism was responsible.

25.1 Introduction

Throughout the Earth's history the forms of living organisms have changed, and the ever-changing mix of organisms has permitted the Phanerozoic timescale to be constructed (Section 15.11). The changes have not been steady: At intervals there have been mass extinctions, when many species died out in a geologically short time. The extinctions are used to divide the timescale into its major though geologically rather arbitrary intervals. The second-largest extinction – the largest was at the end of the Permian – divides the Cretaceous from the Tertiary period, known as the K/T boundary. It saw the end of the dinosaurs, as is well known, but also of the ammonites and the flying pterosaurs, and saw drastic reductions in many other species, particularly marine microfossils. In all, 75% of species are estimated to have died out.

Many theories have been put forward to explain this extinction, but the ones currently most discussed are impacts and volcanic eruptions. An impact is due to a body from space, termed a bolide, hitting the Earth; the energy released can equal many megatonnes of explosives, and its release in seconds obviously has disastrous results, though whether sufficient to cause a mass extinction is debated. Somewhat less dramatic are huge eruptions of lavas or pyroclastics, as mentioned in Chapter 24. The volumes of the largest individual eruptions have exceeded 1000 km^3, and eruptions of a series of lavas can be much larger, with the accompanying dust and gases severely affecting the climate of the whole Earth and so perhaps causing mass extinctions. The problem is not to show that impacts and huge eruptions have occurred – we know they did – but to link them to extinctions as causes. This requires showing that they occurred close in time to the extinctions, and that their likely effects match the evidence in the geological record.

25.2 Impacts and craters

The sources of the bodies that hit the Earth are of two kinds, comets and asteroids. The Solar System (Fig. 25.1) probably formed from a disc of gas with a few percent of solid particles. Most of the mass was in the centre of the disc and evolved to become the Sun, but further out in the disc the solid particles progressively agglomerated into larger and larger bodies, called planetesimals, culminating in the planets. Huge numbers of bodies were left over from the intermediate stages. Many impacted the planets during the first few hundred million years of their existence, producing the numerous huge craters seen on the Moon and Mercury (and no doubt on the Earth as well, but they have been erased by erosion). Others were deflected into orbits beyond Jupiter, the largest planet, and periodically a few are deflected back, appearing as comets, which travel in very oblique orbits about the Sun. Comets are loose aggregates that tend to break up as they pass close to the Sun or Jupiter, increasing the chances of impacts. A spectacular example occurred in 1994 when 21 parts of Comet Shoemaker–Levy 9 crashed into Jupiter; lines of craters on earth probably had a similar cause.

The other source of bolides is asteroids, planetesimals that failed to aggregate into a full-sized planet. They are more solid than comets, with many probably having a core and mantle, the result of being molten in the past. Most orbit between Mars and Jupiter (Fig. 25.1), where many thousands are known; the largest are several hundred kilometres across, but most are much smaller, often just irregular lumps. Some may get deflected out of their orbits, or they may collide to produce fragments; either way, bodies may get into elongated orbits that approach closer than the Earth to the Sun, and occasionally some collide with the Earth. The smallest bodies are vaporised – perhaps after breaking up – when they enter the Earth's atmosphere, but larger ones reach the Earth's surface and are then called meteorites. Many thousands, up to a few tonnes in weight, have been found on the surface, but larger ones penetrate the surface and produce a crater.

The chance of a collision with a body a few tens of metres or more across is slight, for space is very large and empty, and a collision requires not only that the orbits of the Earth and the body intersect, but also that they arrive at the intersection at the same time. But impacts do occur from time to time, for geological time is very long. Bodies tens of metres or more across are far less common than the size of meteorites seen in museums, and we do not know how many there are, for most are too small to be seen unless they approach close to the Earth; new ones are being discovered all the time, and are sometimes reported in the media.

The most recent definite collision of a large bolide with the Earth occurred in 1908 and caused the Tunguska event. A huge fireball was seen over Russia, but the damage was in a part of Siberia so remote that it was not investigated until many years afterwards, when trees were found felled over an area about 30 km across. It is thought that the body – perhaps part of a comet – was too small to reach the Earth's surface and instead produced an airburst; even so, the energy is estimated as about that of 20,000 tonnes of explosive, the size of the Hiroshima nuclear explosion.

The best-known impact crater is the Meteorite (Barringer) Crater in Arizona. It is a simple crater about 1200 m across and 170 m deep, with its rim rising above the surrounding plains. It was formed about 50,000 years ago, and fragments of iron–nickel metal suggest the bolide was made of iron. Calculations give its energy as about that of 20 Mt of explosive, which corresponds to a bolide only 60 m across.

Over 100 terrestrial impact craters up to 200 km or more across and with ages back to nearly 2000 Ma ago are known. They can be detected in various ways. If not deeply buried they may show up as

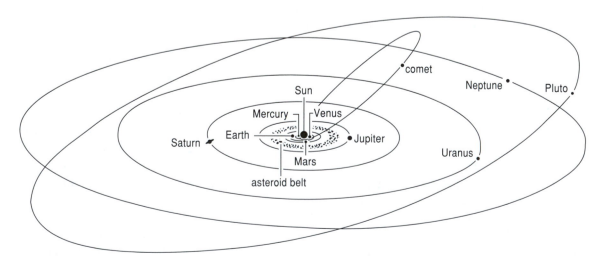

Figure 25.1 The Solar System.

rings in pictures taken from satellites, but more often they are not so obvious. They may still be recognised as circular features on geological maps or as circular magnetic or gravity anomalies. To distinguish an impact from the eroded remains of a volcano – which also often have a circular structure – further information is needed, such as the presence of characteristic shock textures in minerals or characteristic high-pressure minerals. The deep structure can be revealed by geophysical surveys, supplemented by drilling.

Another indication of an impact is the presence of ejecta. However, the source may not be obvious, for deposits can be found hundreds or thousands of kilometres away. Generally, deposits thicken and coarsen towards the source, which can help locate the source. When a possible impact site has been found, its rocks should match those of the ejecta in type and age. When tektites are found – molten droplets formed by the impact and blasted out of the crater – their radiometric age should equal the age of the impact, while ejecta that have not been reset should have the same age as the rocks of the crater.

25.3 The Chicxulub structure

Geophysical methods employed: Gravity, aeromagnetic and seismic reflection surveys, and radiometric dating by the Ar–Ar and U–Pb methods.

25.3.1 Background

The idea that the K/T boundary extinction was due to an impact was put forward in 1980 by Alvarez and others (Alvarez et al., 1980). Their principal piece of evidence was an 'iridium anomaly' found in sediments close to the K/T boundary. Iridium is a very rare element on Earth but is considerably more abundant in meteorites; their idea is that when the bolide hit the Earth much of it was pulverized or vaporized and spread over the Earth, to be deposited in sediments formed at that time. From the amount of iridium in the anomaly, the bolide was estimated to be 10 to 15 km in diameter.

This idea led to several lines of enquiry. One was whether the iridium anomaly is found worldwide at the K/T boundary (wherever suitable sediments

occur), and whether it could have other causes, such as extensive volcanic eruptions, the contending theory, or be concentrated from terrestrial sources into the sediments in some way. It is now the consensus that the anomaly is indeed found worldwide, and probably only a bolide could have provided sufficient iridium, though iridium is released from a few volcanoes and also is sometimes concentrated by seafloor processes.

No impact crater was recognised when the idea was first put forward, and it was later suggested that the site was in ocean floor that has since subducted. However, ejected particles found in sediments indicated, by variation in their size and abundance across the Americas, that the source was near Central America, and in 1990 the Chicxulub structure, on the tip of the Yucatan Peninsula, Mexico (Fig. 25.2, inset) was suggested as the source. This structure had already been recognised from its gravity and magnetic anomalies (Figs. 25.2 and 25.5). The gravity anomaly forms a circular structure except to the northwest, probably partly because there were fewer gravity stations at sea than on land, but also because of truncation by a ENE-striking lineation roughly 30 km offshore. The smaller aeromagnetic anomaly is not affected by either of these effects.

The structure had been interpreted as an impact crater in 1981 but not suggested as the source of the K/T iridium anomaly until 1990. To relate it firmly to the K/T boundary, it had to be shown to have the same age; it had also, of course, to be confirmed that it really was due to an impact and not to another cause, such as a volcano.

25.3.2 The structure of Chicxulub

The pre-Tertiary rocks of the Yucatan area in which the structure is found consist of several kilometres of carbonates and evaporites resting on a basement in which granitic gneiss predominates. The structure is covered by about a kilometre of Tertiary sediments, which is why it was not recognised until the geophysical surveys were made. Boreholes, put down for petroleum exploration, show layers of breccia above what was thought to be a volcanic rock but has since been identified as melt rock, whose andesitic composition matches that of the basement. Quartz crystals in clasts of the breccia show several sets of planar deformations, attributed to shock metamorphism.

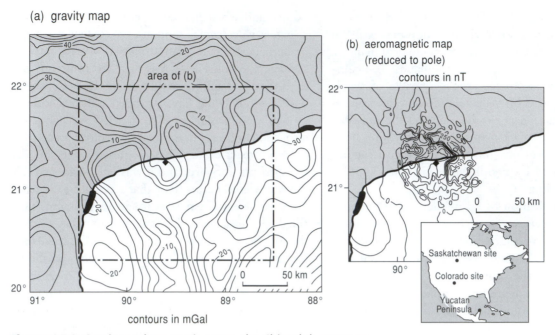

Figure 25.2 Gravity and magnetic maps, the Chicxulub structure.

Though the gravity anomaly is everywhere negative it shows concentric rings (Fig. 25.5), which are believed to reflect ring faults, expected if the impact is a large one. A comparatively small bolide, such as the one that produced Meteorite Crater, produces a simple saucer-shaped crater, but large impacts produce structures that are complex and not fully understood. Initially, the bolide compresses the area of impact, developing huge pressure and shock waves, which then blast out material, forming a transient crater (Fig. 25.3a). Melt rocks and brec-cias are formed. In large craters the transient crater is too deep to be stable and within about 10 min its floor rises, while its rim tends to subside. On a much longer timescale further adjustment takes place: The central rise may partly subside, forming a peak ring, while outside the impact crater material subsides, forming a series of concentric ring faults (Fig. 25.3b). Working out the size of the crater is important because it indicates the size and energy of the bolide and hence its environmental impact.

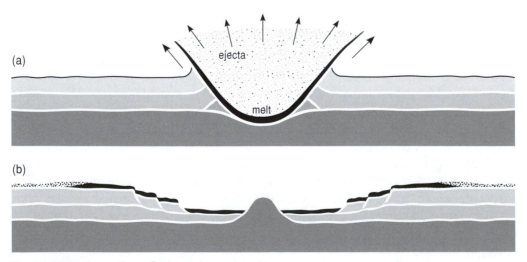

Figure 25.3 Formation of a large-impact crater.

(a) gravity profile

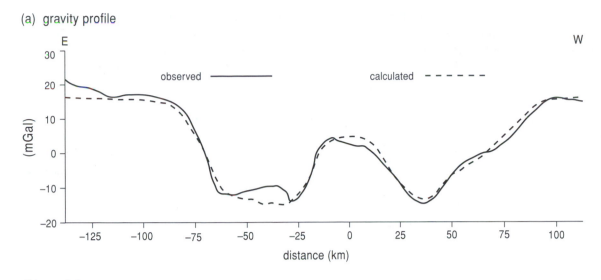

(b) model

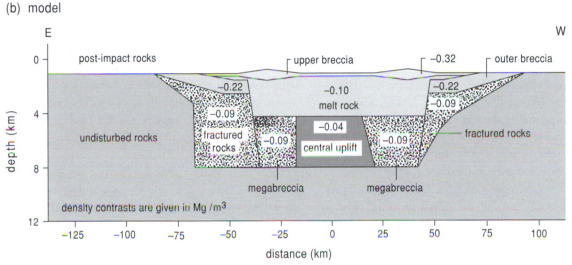

(c) seismic section

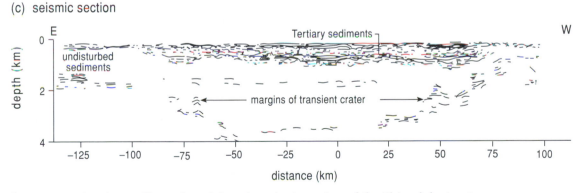

Figure 25.4 Gravity profile and model, and a seismic section of the Chicxulub structure.

Results of geophysical surveys. The gravity and magnetic maps have been shown (Fig. 25.2). Figure 25.4a shows a gravity profile roughly along the north coast, while Figure 25.4b shows a model that accounts for it, taking into account evidence from the boreholes and the structures of proven impact structures found elsewhere. The anomaly is negative because the density of the rocks has been lowered either by fracturing or, in the case of melt rock near the centre, by vesiculation; however, the central peak within the negative trough of the anomaly is attributed to denser uplifted rocks.

Figure 25.4c shows a marine seismic reflection section surveyed to the north of Figure 25.4a and 25.4b, roughly along the line of the lineation. It extends down only to about 4 km but shows the top of the impact structure, which is the base of the overlying Tertiary sediments. There are few reflectors below these sediments, but these outline the transient crater and show that most of the reflectors were disturbed by the impact, except those beyond the crater to the east.

The magnetic anomaly (Fig. 25.2b) is mainly due to remanent magnetisation, for core samples have a Königsberger ratio (Section 11.4) greatly exceeding one. The remanence is reversed with an inclination of $-43°$, consistent with magnetisation at the time of the K/T boundary (chron 29R; see Fig. 10.21) at the latitude Yucatan is known, from plate tectonic reconstructions, to have had at that time (Fig. 20.26). The magnetisation is probably thermal remanent magnetisation (TRM), possibly enhanced by hydrothermal alteration converting the minerals to more magnetic forms, acquired when the melt cooled, and so the magnetic anomaly shows the position of the melt rocks and provides a minimum size for the transient crater. This is why the magnetic anomaly is smaller than the gravity one, which is also due to slumping at ring fractures. Magnetic models are not shown because they are poorly constrained, for there are few core samples.

The size of the structure. Its size determines the amounts of carbon and sulphur dioxides released from the carbonates and evaporites, plus particles, and whose effect upon the atmosphere would have been like that produced by burning fossil fuels, but vastly greater. The impact structure has been variously estimated as 170 to 300 km in diameter, the larger size requiring an impact roughly ten times as energetic as the smaller.

The size of the structure would best be found using extensive seismic reflection surveys supplemented by boreholes, but, as coverage by these methods has been limited, most estimates have depended upon recognising ring faults using gravity maps. However, picking out all the 'steps' on the gravity map is difficult because of variations in the regional gravity anomaly due to preexisting underlying struc-

tures. One way to accentuate the steps is to take the horizontal gradient of the gravity, a form of data processing that emphasises where the greatest lateral changes in the density of the near-surface rocks occur (Section 3.2.5). Figure 25.5a shows the result, dark areas being where gravity is changing most rapidly, with a total of six ring faults. The innermost ring is thought to be due to uplift, the next to mark the transient crater, while the remaining ones are attributed to slumping at ring faults. In the field, there is a zone of cenotes – water-filled karst features – just inside the southwest part of prominent ring 5; these are thought to have been formed by solution by groundwater flowing along a ring fracture, and springs occur where the zone meets the coast. Ring 5 is taken to be the size of the crater. To find its position more precisely, five approximately radial traverses (shown on Fig. 25.5a) were surveyed with closely spaced stations to find the position of maximum slope more precisely, and profiles of two are shown in Figure 25.5b. Ring 5 has a diameter of about 180 km, and though considerably smaller than some estimates, still makes the Chicxulub structure one of the largest known terrestrial craters.

Subsequently, more detailed seismic reflection surveys have been carried out over the submerged part of the crater, showing considerably more detail, revealing a peak ring 80 km across and dipping reflectors that probably mark the edge of the transient crater (Dressler and Sharpton, 2000).

25.3.3 Ages of the Chicxulub structure and ejecta

Critical to the case that the structure caused the K/T extinctions is that its age is the same as that of the boundary. Breccia from above the melt rock has been claimed, from its fossils, to be pre-Tertiary in age, though doubt has been cast on the dating of the fossils; in any case, blocks may have been lifted from their true stratigraphic position by the violence of the impact. Radiometric dating is clearly needed to establish the date.

Ar–Ar step-heating dating (Section 15.4.2) on milligram-sized samples of melt rock gave near-ideal age spectra, giving a mean date of $64.98 ± 0.05$ Ma (Swisher et al., 1992); one age spectrum is shown in Figure 25.6a. These dates are not significantly different from those of tektites from Haiti (Fig. 25.6b) and also from Mimbral, Mexico.

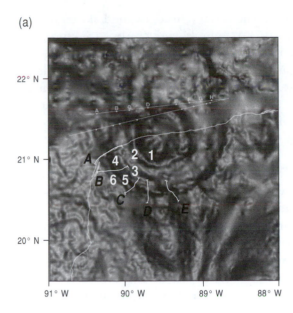

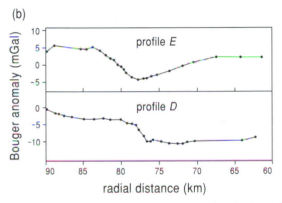

Figure 25.5 Gravity gradients and radial profiles, Chicxulub structure.

Shocked zircons from the breccia in the Chicxulub structure have been analysed using the U–Pb method. They do not give concordant (i.e., consistent) dates, but when the data are plotted on a concordia plot (Fig. 25.6c) they give a straight discordia line (Section 15.9.5), showing they have been reset in a brief event. The upper intercept of about 550 Ma matches the age of the basement rock originally bearing the zircons, while the age of the lower intercept, 59 ± 10 Ma – the time of partial resetting – is consistent with the impact age. Tiny shocked zircons have also been found in K/T boundary layer sediments remote from Chicxulub. Some of these from sites 2300 and 3500 km away – Colorado and Saskatchewan (Fig. 25.2, inset) – weighing only a few micrograms each, also give concordia plots with intercepts as above. The zircons that have been more reset (i.e., are closer to the lower intercept) are more shocked. These results are not consistent with an explosive volcanic eruption, which would be expected to produce unshocked zircons dating from the time of the eruption (Kamo and Krogh, 1995).

The radiometric results place the Chicxulub structure at the K/T boundary and also link it to widespread ejecta, which demonstrate a very large impact.

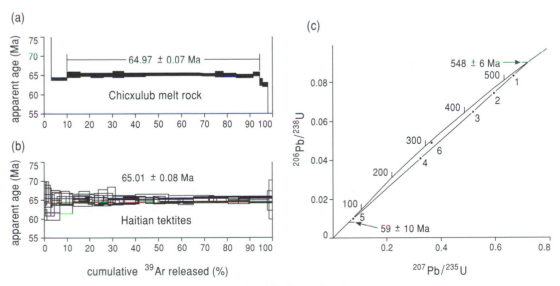

Figure 25.6 Argon–argon age spectra and U–Pb discordia plot.

25.3.4 The Manson Crater

At one time this crater, in Iowa, was thought to be related to the K/T extinctions. As it is only 35 km in diameter, the impact that formed it was considered too small by itself to have caused the extinctions, but perhaps it was one of several impacts resulting from the breakup of a comet, like the Shoemaker–Levy 9 impacts on Jupiter.

The main reason for linking it to the K/T extinctions was because its age, measured by the Ar–Ar method on a core sample, seemed to be indistinguishable from that of the K/T boundary. However, palaeomagnetic measurements showed it to have formed during a normal polarity interval, whereas the K/T boundary occurs in a reversed one. Nor do the upper intercept dates of shocked zircons, measured as reported above, match the Precambrian age of the rocks that underlie the Manson Crater. Further, Sr and Nd isotopic ratios (and the results of other analyses) of impact glasses in sediments from the K/T boundary in Haiti do not match those of melt rocks from the Manson Crater, though they do match those from Chicxulub. Subsequently, new Ar–Ar dates on less altered samples have given a well-defined age of 74 Ma, and shocked crystals stratigraphically of this age have been found in sediments in adjacent South Dakota. It now seems clear that the Manson Crater is unrelated to the K/T extinctions.

25.4 Giant eruptions

As mentioned in Chapter 24, the geological record shows there have been far larger eruptions than those experienced historically. Largest are continental flood basalt successions, which can have individual lavas exceeding 1000 km^3 in volume, the total sometimes topping 1 million km^3. The largest known are the Siberian basalts, which formed around the end of the Permian and exceeded 2 million km^3 in volume. All the largest outpourings seem to have been due to hot spots or plumes (Section 20.9.2): For example, the Columbia River Basalts are related to the Yellowstone hot spot.

There is evidence that the global effects of large lavas are more severe than those of the same volume of pyroclastics, because gases and aerosols are released throughout the eruption, which can last for

months. The severe effects of the Laki eruption of 1783–1784, which was 'only' 15 km^3, were described in Section 24.1. Gases include vast amounts of the greenhouse gas carbon dioxide and large amounts of sulphur dioxide, which converts into sulphuric acid, plus poisonous fluorine and chlorine compounds. Consequently, it is argued that huge outpourings of continental flood basalts would have had a devastating effect.

The flood basalts that are possibly linked to the K/T extinctions are the Deccan 'Traps', which, with an area of 800,000 km^2, cover about a third of peninsula India, and reach a thickness of 2400 m. Underlying sediments are of Maastrictian age – the last stage of the Cretaceous period – and dinosaur fossils have been found in sediments between the flows. Argon–argon step-heating analyses (few of which produced ideal age spectra) have given a range of ages, mostly in the range 64 to 68 Ma, which straddles the K/T boundary. Most lavas are reversely magnetised, and as there is no evidence of the lavas being formed during more than three polarity intervals, N–R–N, it is likely most of the lavas were poured out in a single R interval. They are consistent with a K/T age, for the magnetic polarity chron that straddles the K/T boundary is reversely magnetised (chron 29R, which has a duration of only 0.3 Ma). As globally there was much other volcanic activity at about this time, the Deccan volcanism may have so increased already existing environmental stress that it caused many extinctions.

Iridium, though unusual in volcanoes, has been found in gases from Hawaiian volcanoes and it has been suggested that it may have been released by the Deccan lavas and so account for the iridium anomaly. It has also been suggested that shocked grains and tektites could be produced by volcanic activity, but there is little evidence for this.

25.5 Conclusions to date

The evidence that the Chicxulub structure is a very large impact crater with age close to the K/T boundary is now accepted. Less clear is its relation to the extinctions, partly because there is a considerable amount of evidence that extinctions of many species do not seem to have occurred at the boundary but either preceded it or occasionally followed it, by up to millions of years. A further problem – for any theory – is to account for the pattern of extinctions (e.g., about 95% of microfossils were eliminated, yet deep sea benthic foraminifera and freshwater plants and animals were hardly affected).

Emphasis has shifted from proving the origin and age of the structure to understanding how it relates to the impact and what the environmental consequences were. And as the only known well-preserved multiring crater on Earth (others are known on other planets but obviously are difficult to study) it offers the chance to learn more about the formation of such craters.

There is the broader question of what caused other extinctions. Are any others linked to impacts, of which about 100 are known? One in Chesapeake Bay has been tentatively linked to the Eocene/Oligocene extinction, the largest since the K/T one (Kerr, 1995). The same question applies to volcanism. For instance, was the end-of-Permian extinction related to the approximately simultaneous vast outpourings of lava in Siberia, each the largest of their kind? Many other theories have been offered for extinctions, and only painstaking collection of evidence will settle the mat-

ter. Evidently, we have much to learn about the causes of extinctions, which are likely to be complex.

Further reading

The most comprehensive account of recent investigations of the Chicxulub crater is given in Dressler and Sharpton (2000), while Hildebrand et al. (1998) describes recent gravity and seismic reflection data. Other papers have been mentioned in the text. More generally, Benton and Little (1994) and Powell (1998) give historical accounts of the Chicxulub Crater.

Sharpton and Ward (1990) is a conference volume that covers impacts and volcanism, with many papers on the K/T extinction. Alvarez and Asaro (1990) and Courtillot (1990) give popular accounts of the competing impact and volcanic eruption theories as the cause of extinctions, while 'What really killed the dinosaurs?' (*New Scientist*, **155**, 23–27, 1997) considers other theories as well. The Manson impact crater is discussed in Koeberl and Anderson (1996).

chapter 26

Hydrogeology and Contaminated Land

One of the most important natural resources is fresh water, essential for growing crops, for many industries, and of course for drinking and other personal uses; it is also the basis of many leisure activities, from fishing to water sports. In many parts of the world demand now rivals the natural supply of water, leading to a need for better understanding of aquifers, as well as for building dams and for more recycling. A separate problem is pollution, which has many causes, ranging from influx of saline water due to excessive extraction of fresh water, to contamination by sewage, agricultural and industrial chemicals, or leachate from landfill sites. Hydrogeology is concerned with these problems, and geophysics is an increasingly valuable aid.

26.1 Introduction

The most obvious water source is surface water in rivers and lakes, but these derive much of their supply from groundwater, while much water is extracted directly from the ground by boreholes (over half the population of the United States gets its water this way). Therefore, hydrogeology is mainly concerned with the hidden resource of groundwater. The goals of hydrogeology are (i) locating new groundwater resources, (ii) developing schemes for the best utilization of known sources, (iii) proposing measures for protection against contamination and overextraction, and (iv) monitoring potential or known sources of contamination.

26.2 Aquifers

Groundwater moves through aquifers, which are often subhorizontal layers of permeable rock such as porous sands and sandstones, but including crystalline rocks with interconnected fractures and fissures. The top of the water-saturated volume is termed the water table. Aquifers are bounded by lower permeability rocks such as clays and shales. Important aquifers sometimes reach considerable depths in synclines; an example is the London Basin, where Cretaceous Chalk occurs beneath the impermeable London Clay at a depth of up to about 140 m, and water is extracted through boreholes.

An important task of hydrogeology is to map aquifers, their depth, lateral extent and features such as faulting and dykes that could affect water flow. In addition, their porosity, permeability, and degree of water saturation (terms explained in Section 18.4) and water quality need to be known.

Most aquifers are recharged where they are connected to the surface; for instance, the chalk syncline below London is recharged where its limbs are exposed, to the north of London in the Chiltern Hills and the North Downs to the south. (Some aquifers, such as huge ones beneath parts of the Sahara Desert, are no longer being recharged; their 'fossil' water was acquired when the climate was wetter.) Aquifers are vulnerable to surface contamination in the recharging regions; in addition, excessive extraction may reverse the direction of groundwater flow and cause polluted or saline water to enter from adjacent regions, such as the sea. It may take years before contamination reaches an extraction point and much longer thereafter to eliminate it, so monitoring for contamination is another important task.

26.3 Geophysical methods useful in hydrogeology

The most widely used method is electrical resistivity, either using a resistivity array or some form of e-m surveying; this is because the presence of water greatly increases the conductivity of most rocks. The presence of some pollutants – particularly saline water – may be detected because they change the conductivity.

However, other methods can be useful. Seismic refraction and reflection can reveal buried river channels or vertical offsets in aquifers caused by faulting. They can also measure the depth to the water table, for it may form a seismic interface. However, in fine-grained rocks the water table may be a gradational boundary up to some metres thick, due to capillary action; this will prevent reflection if it is thicker than a quarter of the wavelength of the seismic – or ground-penetrating radar – waves being used, Section 7.8.2. Gravity too can map buried channels, and more cheaply if less precisely than seismology, or reveal basement topography that may partly determine the thickness or course of an over-lying aquifer. Magnetic surveys may reveal the presence of faults or of dykes that sometimes act as barriers to groundwater flow. They may also detect the presence, in landfill sites, of drums of noxious chemicals, for these are often made of steel. Moving groundwater containing salts can produce small p.d.'s that can be measured using SP, while some pollutants that do not change the resistivity of the host rocks significantly can be detected using IP. Ground-penetrating radar (GPR) is useful for detecting some types of interfaces when they are no more than a few meters deep, such as the water table, which often forms a good reflector, as does the interface of hydrocarbon liquids floating on ground-water. In addition, saline contamination may some-times be detected by the high absorption of radar waves due to its higher conductivity.

26.4 GPR surveying of the water table, the Netherlands: An example

Geophysical method employed: GPR.

26.4.1 Background

This example utilises the property of the water table sometimes to be a strong reflector of radar waves. Provided GPR can penetrate to the depths of interest, it can provide more precise information than other geophysical methods. However, its depth of penetration is at most a few tens of metres, and it may be reduced to a fraction of a metre by the strong attenuation of clays, silts, or saline water, which are much more conductive than most aquifers. Lower frequencies give deeper penetration but at the expense of resolution (Section 14.8.2), so the aim is usually to choose the highest frequency that allows penetration to the required depth.

In the Netherlands, poor penetration makes GPR unsuitable in many areas, but it has been used suc-cessfully in sands, to as much as 40 m (using the low frequency of 25 MHz), and where other geophysical methods – electrical, seismic refraction, and reflec-tion – were inappropriate. It has been used to map offsets of the water table, to recognise perched water tables (a lens of water perched on a low-per-meability layer above the main aquifer), to find dip-ping impermeable layers that obstruct an aquifer, and to map sedimentary and tectonic structures, all of which are relevant to water extraction. This example is of an offset.

26.4.2 Offsets of the water table

In one area of the Netherlands, water is extracted from a push moraine (a glacial deposit emplaced by the advance of an ice sheet), and more details of its structure were needed for modelling water flows. As the water table was known to be only 5 to 10 m below the surface, the moderately high frequency of 100 MHz was chosen to increase resolution (a Pulse EKKO IV system was used). Traverses were made at intervals of 250 m, and one resulting section is shown in Figure 26.1; the depth scale is only approximate, as times were converted to depths using a single average velocity.

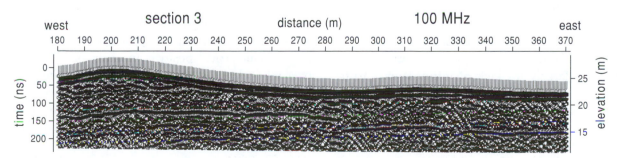

Figure 26.1 Ground-penetrating section through a push moraine, Netherlands.

The water table shows up as a strong reflector roughly halfway down the section, and has an obvious offset at about 285 m and a lesser one at about 330 m (note the diffraction hyperbolas where the reflectors have abrupt ends due to offsets). A possible cause of the offset could be a steep, relatively impermeable barrier impeding water flow from left to right, but no such barrier is visible. An alternative is that the offset is an artifact due to an abrupt lateral change in velocity in the layers above, causing the radar waves to take longer to reach the water table where it seems deeper (pull-down, Section 7.8.1), but this also was discounted, by a velocity analysis that showed no change above the offsets and by the presence of the diffraction hyperbolas, so the reality of the step – presumably due to a small fault – was accepted. The step was traced for about a kilometre along strike from parallel profiles. Further details of this and other surveys are given in van Overmeeren (1994).

26.5 Structural control of aquifers in East Anglia, England: A case study

Geophysical methods employed: Gravity, resistivity VES (Schlumberger array), and TEM.

26.5.1 Introduction

The two case studies that follow are both in East Anglia (Fig. 26.2, inset), a flat area with little land rising more than 50 m above sea level. The solid geology (Fig. 26.2a) is largely obscured by Quaternary glacial deposits but is known to have basins of Crag (Plio–Pleistocene marine deposits) up to 80 m thick resting on the Cretaceous Chalk, which underlies much of East Anglia; both the Crag and Chalk are aquifers. The Crag basins were thought to be at least partly fault-bounded, but it was not known if they are structurally controlled by the faults extending up from the underlying basement of Palaeozoic and possibly older rocks, whose surface was assumed – largely for lack of information – to be flat. However, the regional gravity map suggested some structure, for a positive anomaly trending NE–SW coincides approximately with the fault-bounded Sudbury–Bildeston Ridge of chalk (Fig. 26.2b). Detailed geophysical surveys were carried out to investigate the basement.

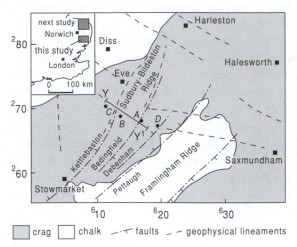

(a) geology

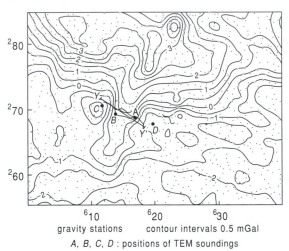

(b) gravity

Figure 26.2 Simplified geology and regional gravity maps of part of East Anglia.

26.5.2 Geophysical surveys

The regional gravity data had only about one station per square kilometre so a detailed gravity profile was measured across the Sudbury–Bildeston Ridge with stations only 50 m apart. This shows a nearly symmetrical positive anomaly of 2.7 mGal over the Ridge (Fig. 26.3a). It was thought unlikely that it could be due to the Crag here having a higher density, for this would require an implausibly large increase over densities measured elsewhere, so a deeper origin was likely. Because of the inherent non-uniqueness of gravity anomalies (Section 8.8.1), it could not be modelled without additional information, so resistivity surveys were carried out. (Seismic reflection would have provided more precise depths to interfaces but is more expensive than resistivity; resistivity also has the advantage that it provides more direct information about aquifers and particularly the presence of any saline water.)

Schlumberger depth soundings (VES) were made, with centres 600 m apart along the traverse and end-to-end lengths up to 500 m, giving penetration to a depth of about 50 m; this was sufficient to reach the Chalk but not penetrate through it into the basement. These surveys were useful for mapping the extent and thickness of the glacial till (Fig. 26.3b), for, due to its high clay content, its resistivity is significantly lower (15–28 ohm-m) than that of the underlying Crag (40–100 ohm-m). However, the interface between the Crag and the Chalk could not be mapped because their resistivities overlap, so it was mapped using such borehole data as were available; the depths to the Chalk are given by the numbers on Figure 26.3b.

To have penetrated deeper using the Schlumberger array would have required impractically long arrays, so TEM (Section 14.3) was used, for it has considerably deeper penetration than a resistivity array that can be fitted into the same space. Soundings were made at four places, A to D, along the gravity traverse (Fig. 26.2), where fields were large enough to lay out the square transmitter coil, whose side ranged from 150 to 400 m in length. Readings were repeated many times and stacked to improve the signal-to-noise ratio (Section 2.3). The results are shown in Figure 26.4 as apparent resistivity (not conductivity) versus delay time, with longer delay times corresponding to greater depths. The observed data are shown only for sounding D, with the extent of the boxes indicating the size of the errors; the curves are for models that match the data for each of the four soundings.

The curves have the same general shape, and the number of peaks and troughs show the presence of *at least* four layers, as explained in Section 12.3.5, though five layers were needed to match the data for soundings A and C. The most prominent feature is the low resistivity trough; this is deepest and widest for sites C and D, showing that the low resistivity layer (LRL, Fig. 26.3b) is thicker here. As the basement rocks, believed to lie directly beneath the Chalk, are expected to have a higher resistivity than the Chalk, the low-resistivity layer was likely to be due to saline water in the lower part of the Chalk; this would also explain its upper surface being nearly horizontal, unlike the other interfaces.

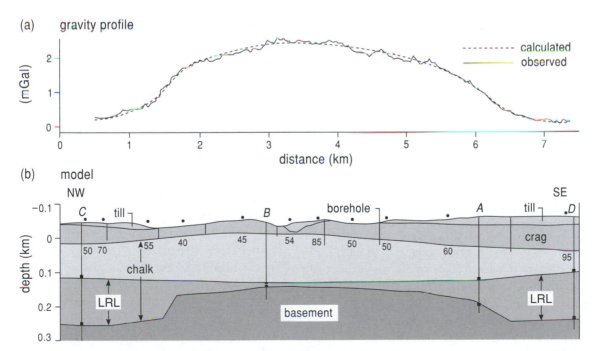

Figure 26.3 Gravity profiles and model, Sudbury–Bildeston Ridge.

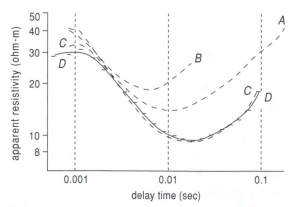

Figure 26.4 Four TEM curves; locations are shown in Figure 26.3

This interpretation was supported when similar resistivities were found in a TEM survey carried out in the Chalk about 70 km to the north (the Trunch Borehole site), where saline water – beneath fresh water – was proved by a detailed study to be present and of formational origin.

A model based on all the evidence and using densities of 2.7 Mg/m³ for the basement and 2.1 for the younger rocks fits the observed gravity values well (Fig. 26.3a), and provides support for the presence of a basement ridge. The overlying Sudbury–Bildeston Ridge is therefore probably structurally controlled by northeast-trending basement faults, which may have been reactivated to affect the Crag sequence. The discovery of saline water in this area may limit future extraction of water from the Chalk. Further details are given in Cornwell et al. (1996).

26.6 Saline contamination of the Crag aquifer, East Anglia: A case study

Geophysical methods employed: Seismic reflection, Slingram e-m, and VES resistivity (Offset Wenner).

26.6.1 Background

This investigation, like the previous one, was in East Anglia but further to the north (Fig. 26.5, inset). It is concerned with the Crag aquifer, rather than the underlying Cretaceous Chalk, from which it is separated in this area by the Paleogene London Clay, except perhaps along its western margin. As was mentioned in the previous case study, in some areas

the Chalk contains saline formational water, which may restrict future water extraction. Consequently, if demand increases there is likely to be a need to extract more water from the Crag aquifer, though its fine-grained sands, with horizons of clay and silt, have been difficult to develop.

As there had been no integrated investigation of the Crag, geophysical, hydrogeological, and hydrochemical measurements were made of part of it to increase knowledge of its hydrogeological behaviour. The River Thurne catchment area was chosen as containing a variety of environments, including gently undulating hills, drained marshes, and a coastal margin (Fig. 26.5). Part of this investigation was to find the source of saline contamination in the central, low-lying part of the area, and this forms the main concern of this case study.

The River Thurne is part of Broadland, which contains wetlands of international importance. At the centre of the catchment area are several 'broads', very shallow lakes occupying medieval peat workings. Parts of the surrounding area are intensively cultivated, with lush pastures or arable farming, but these are possible only because the water table has been lowered by pumping and is now below sea level. This has caused shrinkage of peat such that the river now flows above the fields, between embankments. The River Thurne drains to the southwest, though previously it drained northeast to the sea through Hundred Stream. The catchment area is bounded to the north, west, and south by low hills with a maximum height of 23 m above Ordnance Datum (which is close in height to mean sea level), while to the east sand dunes separate it from the sea.

The salinity of the broads has been slowly increasing for most of the 20th century, from about 2½% to 10% of the salinity of sea water. This has sometimes resulted in algal blooms that have killed fish and some other aquatic creatures. The obvious source of the saline contamination is the sea, but the concentration of salt in surface waters is highest in the centre of the area and decreases towards the sea. To investigate the structure of the Crag and the cause of the salinity in the marshes, a number of geophysical and other surveys were made. In this study, the Crag is considered to include the overlying glacial sands and gravels in addition to the Plio–Pleistocene marine deposits of sands with interbedded clays and silts.

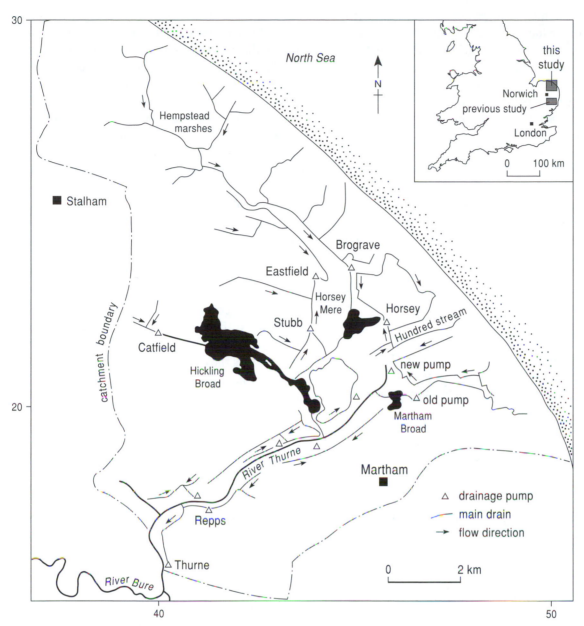

Figure 26.5 River Thurne catchment area.

26.6.2 Geophysical surveys

To provide information on the thickness and internal features of the Crag, particularly in the coastal marshes where there are few boreholes of sufficient depth, seismic refraction and reflection surveys were chosen initially. But the refraction surveys were soon discontinued because the seismic velocity of the London Clay is only slightly greater than that of the Crag, so to receive refracted arrivals from the base of the Crag would have required a very long line,

and then arrivals would have been difficult to observe because there is considerable attenuation. However, there was sufficient contrast in acoustic impedance (Section 7.8.1) to permit seismic reflection, and lines were shot using a hammer or buffalo gun (Section 7.12) as source, at 4 m spacing, and with geophones every 2 m (in clusters of 24 to improve the signal-to-noise ratio), giving six-fold coverage.

Four separate sections were shot, filtered, and

corrected for moveout, which gave the velocities needed for deducing depths; part of the Brograve section (located on Fig. 26.7) is shown in Figure 26.6. The reflector at 63 m was identified by comparison with borehole logs as the base of the Crag. The strong reflector at about 28 m is likely to be one of the clay horizons known to be present, and natural γ ray logs for two boreholes in the area show a higher activity in the layer, consistent with an increased clay content compared to adjacent depths (Section 18.5.4). As this layer is also seen in other sections, it probably continues below much of the area. Other reflectors are weak and short in length, suggesting thin, clayey layers of small extent.

Brograve marshes

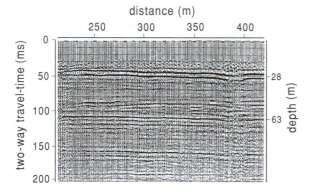

Figure 26.6 Part of Brograve seismic reflection section.

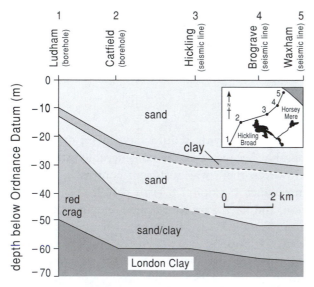

Figure 26.7 Cross-section through the Crag aquifer.

The results of the seismic, γ ray and available driller's log data have been combined in Figure 26.7, which indicates that all along the profile the Crag is underlain by the London Clay and never directly by the Chalk, and that both it and the major layers within it deepen towards the coast.

The groundwater salinity was investigated using electrical surveys, for conductivity is greatly increased (resistivity is decreased) by the presence of salt. First, to provide a rapid survey of the lowest-lying, saline-contaminated part of the area, a Slingram e-m survey (using a Geonics EM-34 instrument, shown in Fig. 14.4c) was made, for it is fairly quick to use as it needs no contact with the ground. To provide an indication of how conductivity changes with depth, coil separations of 10, 20, and 40 m were used; these separations have nominal depths of penetration of 7½, 15, and 30 m, though actual depths depend on the conductivities and are less when the subsurface conductivity is higher than average, as is the case here (the coils were used with their axes horizontal, for this reduces the effect of variable conductivity of the surface layers, though it reduces the depth of penetration). Figure 26.8 shows contours of the apparent resistivities measured by the 10- and 40-m separation surveys; as the larger separation gave a higher conductivity, conductivity increases with depth, which in this area is almost certainly due to increased salinity, rather than other possible factors such as increased clay content. For a given coil separation, apparent conductivities were highest where the ground was lowest, beneath the lowest marshes, due either to the saline layer being closer to the surface or to a higher salt content.

To provide more information about the vertical variation of conductivity, 114 resistivity depth soundings (VES) were measured in the catchment, using the Offset Wenner system (Section 12.3.6). Expansion of the array had to stop when the value of $\Delta V/I$ reached the lower limit of the resistivity meter, due to a combination of the geometrical decrease resulting from expansion (Section 12.3.3) and increasing penetration into a very low resistivity layer at depth. The results were used to construct layered sections which – in the absence of borehole information to constrain the models – were made to match between soundings; one section is shown in Figure 26.9.

(a) 10-m coil separation   (b) 40-m coil separation

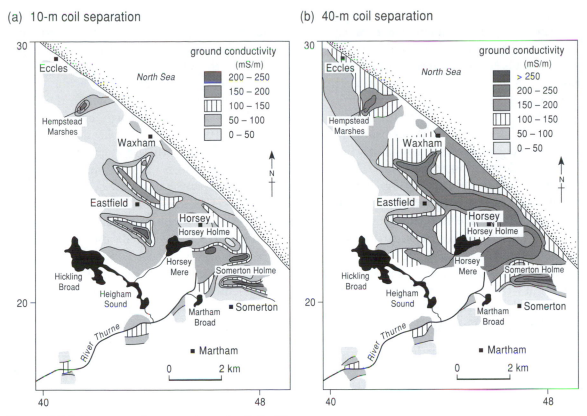

Figure 26.8 Contours of apparent conductivity, Slingram survey.

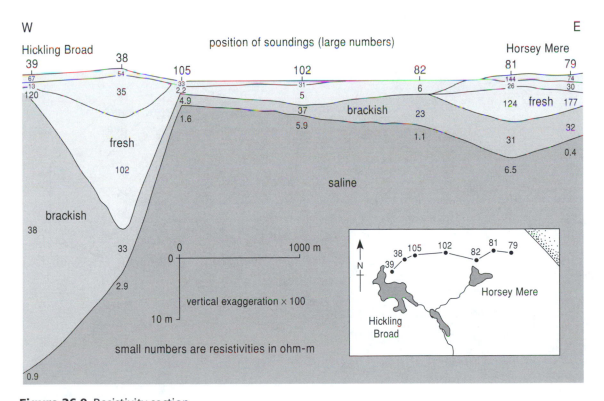

Figure 26.9 Resistivity section.

Layers are interpreted successively downwards as containing fresh, brackish, and saline water, the last layer having a resistivity of only 1 ohm-m or less. Though the models assume sharp boundaries between layers with uniform conductivity, these are probably approximations to gradational boundaries.

In the section, and elsewhere, the interface of fresh water with the denser saline water below it is not level, as one might expect, but highest where the land surface is lowest, and vice versa. This is because when fresh water flows through a thick aquifer it follows curves that penetrate far below where it enters and leaves the subsurface, as shown in Figure 26.10a (just as electric current flows in arcs below the electrodes, e.g., Fig. 12.5). This can depress the top of the saline layer below sea level by up to 40 times the height of the fresh-water table above sea level (this is known as the Ghyben–Herzberg relation). As the fresh-water

table is maintained by rainfall, it is highest where the land is highest, which is around the edges of the catchment area, including the coastal dunes, and so the depth to the saline layer is greatest here; conversely, the saline layer approaches the surface in the lowest part of the area, where the fresh-water table has been lowered by pumping.

Figure 26.10b summarises the flows of fresh and sea water beneath the area. The coastal marshes between the sea and the brackish drains remain fresh despite being below sea level because the saline layer is depressed below them by the flow of fresh water; in addition, in some parts of the area clay layers impede the upward flow of the saline water. These factors explain why the highest salinity is found in the lowest part of the catchment rather than nearer the coast. Further details are given in Holman and Hiscock (1998) Holman et al. (1999).

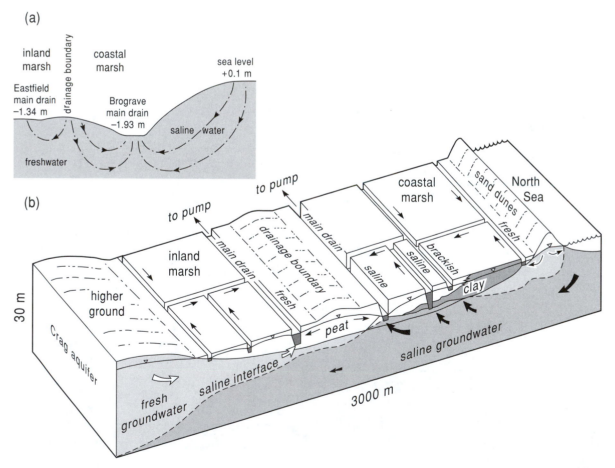

Figure 26.10 Water flow lines beneath area.

26.7 Landfill sites and contaminated ground

26.7.1 Introduction

Many waste products are disposed of in landfills, often utilising worked-out quarries or other excavations. Some of this waste is inert – used building materials, for example – but part is organic, such as waste food, which decomposes and thereby can cause two possible hazards: (i) Methane gas forms and may leak into adjacent areas, where it has been known to result in explosions in houses, while (ii) a liquid – known as leachate – can form and escape to contaminate groundwater. Worst of all, poisonous industrial chemicals may have been dumped; though these should be disposed of only in properly constructed and licensed sites, often they are not, or the fills may develop leaks. Most of the disposal sites in the United States are known or suspected to leak.

There are various other sources of contamination: agricultural fertilisers and animal wastes, industrial processes that use poisonous chemicals and contaminate the factory site, mining operations and smelting (for example, waste from copper mining often contains arsenic compounds), tailings ponds, and so on. The ill health caused in the 1970s to people living in the Love Canal area in New York State, by chemicals leaking from a waste burial site, necessitated 2500 houses being abandoned; it stimulated awareness of such hazards and led to environmental legislation.

Investigations of contaminated sites are increasingly needed, both because of the pressure to reuse land and because of increasingly stringent legislation to monitor contamination. Often little is known about what has been dumped in landfills. Traditionally, they have been investigated by drilling, but this has several drawbacks: Apart from the cost of the close-spaced drilling needed for such heterogeneous sites, it may aggravate the problem by bursting drums containing poisons or by piercing the lining, while obstructions may prevent individual holes from reaching their intended depth. Geophysics can give results quickly, and its noninvasive nature is a great advantage.

Geophysical methods used. A number of methods have been used, including magnetic surveys – gradiometer surveys are appropriate when fills are shallow – which can detect buried drums that may con-

tain poison, for example. Gravity has been used to map the limits of a fill of unknown size and, from the variations of g within these limits, help characterise the nature of the fill. Ground-penetrating radar is able to provide more detail than gravity or magnetics, but a layer of clay – often used to cap a fill – will prevent its use. (Gas that has leaked into adjoining areas is best detected using geochemical sampling.) The most commonly used methods are resistivity and e-m, for the water table is a common target, while the presence of leachate can be recognised by its high conductivity, at least ten times greater than normal groundwater.

26.7.2 Investigation of a landfill in northern England: A case study

Geophysical methods employed: Slingram e-m and VES resistivity (Offset Wenner).

The fill – in a former quarry whose boundaries but not depth were known – contained a variety of materials and was thought to overlie glacial till (but see below), which in turn was underlain by Triassic Sherwood Sandstone, an important aquifer. Surveys were made to find the thickness of the fill and of the underlying boulder clay, and also the distribution of materials within the fill.

First, the site was surveyed on a grid to provide a qualitative picture of conductivity variations. A Slingram e-m system (Geonics EM34) was chosen for its quickness of use, with a coil separation of 40 m to ensure penetration through the thickness of the fill, which could have been as much as 20 m. Readings were taken every 20 m along lines 40 m apart, and the resulting apparent conductivity values were contoured (Fig. 26.11a). Values show considerable variation over the site, which could be due to variation either in thickness or in type of fill; to find out which, resistivity depth soundings were made (BGS Offset Wenner system). The values of the offset pair of readings (D_1 and D_2; see Section 12.3.6) often differed considerably, as illustrated in Figure 26.12; this showed the nature of the fill varied laterally, but for modelling they were averaged, one of the advantages of this system. The soundings were combined to give the four-layered section shown in Figure 26.11b. Layers were initially interpreted as (i) a thin layer of unsaturated fill overlying (ii) saturated fill (17–20 ohm-m), then (iii) a low-resistivity layer (7–9 ohm-m) of boulder clay (its resistivity, however, was lower than typical values – c. 30 ohm-m – measured elsewhere in the area), and last (iv), sandstone saturated with clean water, accounting for its higher resistivity. However, subsequent drilling, while confirming the thickness of the fill, showed that there was no boulder clay below it; therefore the low-resistivity layer was reinterpreted as sandstone aquifer contaminated by leachate from the unsealed fill above. Further details are given in Barker (1990).

26.7.3 Landfill monitoring: A case study

Geophysical methods employed: Resistivity.

The previous case study shows that leachate may escape from a landfill and contaminate an aquifer. To prevent this, landfills are often constructed with an impermeable plastic liner. Checks for leaks in the liner just after its construction have often been made using resistivity equipment, but a recent development allows monitoring to be continued long after

(a) apparent conductivity contours (in mS/m)

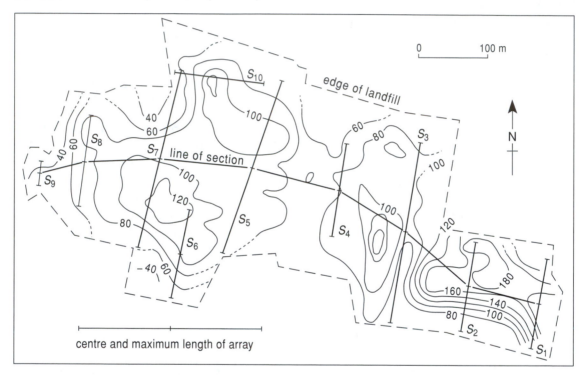

(b) resistivity section (values in ohm-m)

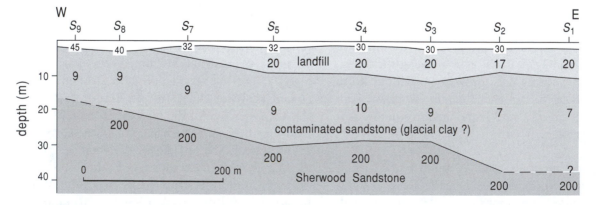

Figure 26.11 Apparent resistivity map and section, landfill site.

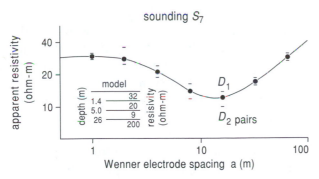

Figure 26.12 Resistivity sounding curve and model.

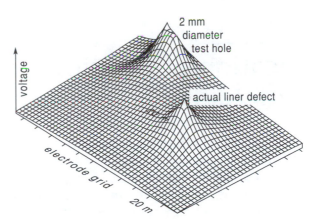

Figure 26.14 Voltage distribution below liner.

the site has been filled. Essentially, it is a multielectrode resistivity array, but with most of the electrodes buried below the liner.

It has been employed in a large landfill site constructed in the midlands of England; as this site overlies a sandstone aquifer, it was particularly important to prevent leaks. The liner is a plastic sheet 2½ mm thick which was laid on 30 cm of sand which had bentonite clay mixed with it to reduce its permeability, while on top of the liner there is 50 cm of sand to protect the liner from damage and to allow leachate to be drained (Fig. 26.13). Electrodes were placed below the liner on a 20-m grid and individually connected via multicore cables and switches to the monitoring equipment.

To test for leaks in the liner before filling, one current electrode was placed in the sand protection layer and the other outside the liner, while voltage was measured between another electrode outside the liner and each of the subliner electrodes in turn. To test the sensitivity of the method, a 2-mm hole was made; it was easily detected, and its position located to within a few metres, by the large potentials at the

electrodes surrounding it, as shown in the isometric projection of Figure 26.14. Figure 26.14 also shows a genuine defect, which proved to be a 1-cm cut.

The site has since been filled and capped, with one of the current electrodes placed within the fill. Should a leak occur it will probably be impossible to repair, but it is expected that the system will provide early warning and allow assessment to be carried out long before any contamination reaches the conventional monitoring boreholes placed around the site. Further details may be found in White et al. (1997).

Further reading

Many hydrogeological surveys are carried out for commercial purposes and remain confidential, so relatively few are described in the scientific literature. Some examples may be found in Reynolds (1997) or in Kelly and Mareŝ (1993). Van Overmeeren (1994) discusses the uses of GPR in hydrogeology.

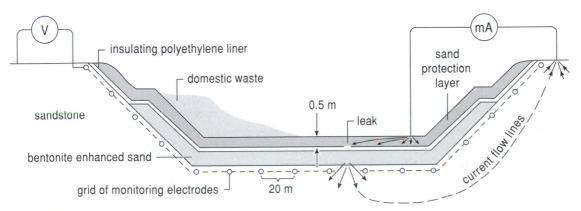

Figure 26.13 Landfill liner and monitoring.

chapter 27

Location of Cavities and Voids

Civil engineering often needs detailed information about the subsurface before starting construction of dams, bridges, roads, airports, buildings, tunnels, and so on. In the past, site investigation relied heavily on drilling, but though drilling can provide essential information it can miss important features between boreholes, and it does not give information about hazards such as earthquakes. A better strategy is first to carry out a combined geological and geophysical survey, and then concentrate drilling where the survey shows it will be most useful. For successful results, geophysicists and geologists need to be clear what information the engineer needs, while the engineer needs to understand what kinds of information the geophysicist and geologist can offer, and their limitations.

Applications where geophysics can be of use to civil engineers include mapping earthquake probability and severity; measuring the depth of unconsolidated cover or weathering, or the extent of infilling; detecting fracture zones; finding pipes and buried objects; locating voids, caves, and old mine workings; and investigating contaminated ground. Some of these have already been considered briefly, such as earthquakes (Chapter 5) and contaminated ground (Section 26.7). This chapter considers only the investigation of cavities and voids, which offers straightforward applications of geophysics.

27.1 Introduction

Unrecognised cavities beneath a site could lead to settling or collapse of the structure, or, in the case of dams and settling ponds, allow the contents to escape. Cavities may be natural or artificial. Natural ones are particularly common in limestone, for it is dissolved by acid groundwater, resulting in caves and swallow holes; swallow holes may subsequently have been filled by poorly consolidated sediments, which, though not a void, are likely to lack the strength of solid rock. ('Void' is used when the space is air or water filled; 'cavity' also includes a filling of weaker material and so is the more general term). Artificial cavities are left by mining or tunnelling, or are due to old wells and shafts. Shafts and wells may have only a weak capping or covering; some have been closed by jamming a tree in their upper part and then filling the space above, which becomes poorly supported when the tree subsequently rots away. The requirement of an engineer may be, first, to see if cavities are present; if they are, to find their depth and extent; and sometimes to check that voids have been completely filled. The engineering effect of a cavity may extend beyond its walls, for cracks in the surrounding rocks may have been formed or opened by the excavation.

27.2 Possible geophysical techniques for locating cavities

A cavity is likely to have a large contrast with the surrounding rock for many physical properties, so most geophysical techniques are *potentially* of use. However, the contrast will depend upon what fills the cavity, which may be air, water, or back-filling or poorly consolidated sediments, and upon the nature of the surrounding rock. The usefulness of a technique also depends upon the shape, size, and depth of the cavity (the depth to the top of a cavity usually exceeds its diameter, for otherwise it is likely to have collapsed already). Though it is obviously hard to detect a small, deep cavity, one that is small and deep may pose no problem of collapse, so it is important to estimate if a given technique can meet the engineer's requirements for a particular site.

27.2.1 Seismic methods

A cavity is likely to have a much lower seismic velocity than its surroundings, but this is not sufficient to ensure success by the basic seismic methods of refraction and reflection, for these are primarily for detecting subhorizontal interfaces; however, reflection may sometimes reveal a cavity by the interruption it produces in reflections from interfaces *below* the cavity. Tomography is a more generally applicable technique.

Seismic tomography. Fan shooting (Section 6.11) is a simple form of tomography that is likely to reveal a cavity provided the cavity interrupts the refracting interface (see Fig. 6.21a). Thus shafts and wells are suitable targets but caves and tunnels are not. A possible drawback is that the site may not be sufficiently large for the source–receiver distance to significantly exceed the crossover distance, roughly ten times the depth to the interrupted interface. More sophisticated forms of seismic tomography are probably too expensive to justify their use.

Borehole methods. Borehole-to-surface or crosshole surveys (Section 18.8) are able to locate cavities quite precisely if the boreholes have been suitably placed, but as the method is expensive it is only likely to be employed late in a survey, if at all.

In general, seismic methods are unlikely to be the first choice.

27.2.2 Electrical methods

An air-filled cavity will have a higher resistivity than the surrounding rock, but if the rock is dry its resistance may be so high that the contrast is too small to be exploitable. A water-filled cavity, in contrast, usually has a lower resistivity than its surroundings. Resistivity, e-m (electromagnetic), and GPR methods all have potential use. SP has also been employed.

Resistivity and e-m. Since a cavity forms a lateral variation in resistivity, a profiling technique is indicated. With resistivity arrays and Slingram instruments, the spacing must be large enough to penetrate to the depth of the cavity, but not so large that the cavity forms only a tiny part of the volume through which the current flows, while traverses need to be close enough not to risk missing cavities between them.

E-m has the advantage over resistivity arrays when the near-surface material has very high resistivity, such as desert sand, and conversely when it is very conductive. Among e-m systems, TEM surveying is more compact than most Slingram systems, an advantage in confined sites, but is less quick to employ. A more powerful technique is to produce a pseudo- or true imaging sections. As this is more costly, it may be better employed after a cavity has been found, to deduce its dimensions more precisely.

GPR. This technique resembles seismic reflection in detecting reflected waves, but responds to discontinuities in relative permittivity rather than in acoustic impedance. Its shallow penetration (a few tens of meters in limestone or sand, much less if clay or saline water is present) may not be a limitation for many engineering applications, while its high resolution – centimetres compared to metres for seismic reflection – is an advantage, and it is widely employed to detect small cavities below roads, runways, and so on, or graves, which are too small and near to the surface to be detectable using seismic reflection. Targets are generally revealed by diffracted waves in the case of small targets such as pipes, or the interruption to a deeper reflector for more extended targets (an example of the latter is given in Section 28.5.2), rather than by a near-horizontal reflection.

27.2.3 Magnetic methods

Air and water are both nonmagnetic, but the anomaly due to a void is likely to be inconspicuous even when the surrounding rock is relatively magnetic, and in nonmagnetic rocks such as in limestone, for instance, is likely to be undetectable. For this reason, a magnetic survey is not usually the first choice for detecting a natural cavity, unless it is likely to contain a relatively magnetic sediment surrounded by nonmagnetic rocks, as may occur with sinkholes. Magnetic surveys have their greatest success detecting artificially cavities, where abandoned ferrous objects, such as iron pipes and rails, can produce large signals. Ferrous rubbish in the infill of a shaft can be similarly detectable, while brick linings extending to near the surface, though producing a much smaller anomaly, can also be detected. The success of magnetic surveys therefore depends largely upon the presence of materials which are not an intrinsic part of the cavity, and they can be frustrated by magnetic noise due to near-surface pipes or shallowly buried ferrous objects; but as they can be carried out quickly they may be worth trying. A gradiometer survey (Section 11.7) would be preferred for targets only a metre or two deep as it would largely ignore more distant magnetic sources. An example follows.

Part of a former coal mining area in south Wales was known, from an old plan, to contain at least one shaft, but its position needed to be confirmed and the area checked for others (Fig. 27.1). Both magnetometer and gradiometer surveys were made, using proton precession sensors. Stations were at intervals of 1 m along traverses 2 m apart, over an area centred on the location indicated on the plan; readings were taken periodically at a base station to allow correction for diurnal variation. The total intensity results (Fig. 27.1a) reveal no anomaly at the supposed location – shown by the star – but there is clearly an anomaly just off the southeast edge. The total intensity survey was extended in this direction and revealed a pair of strong positive and negative anomalies due to a magnetic dipole (Fig. 27.1b); as its amplitude is so large, it was most likely due to iron objects in the shaft. The gradiometer survey (shown earlier as Fig. 2.7) confirmed this anomaly but also revealed a smaller anomaly to the northwest. Both anomalies occurred where there were depressions in the ground surface, so hidden shafts are likely to lie below.

27.2.4 Gravity methods

Both air- and water-filled cavities will have a density contrast larger than all other types of target except for very dense orebodies, and this favours detection by gravity surveys; a cavity filled with

rocks or soil will have a much reduced but possibly still significant contrast. Even so, anomalies will be small; for example, a spherical void 5 m in diameter centred 10 m below the surface, in rock with density 2.7 Mg/m^3, produces an anomaly of only about 12 μGal (in practice, it is likely to be larger, because of the 'halo effect' described in the next section). For this reason a microgravity survey is needed, as described in Section 8.10. This requires both a very sensitive gravimeter and great care in making the various corrections, particularly in measuring heights of gravity stations to a precision of a few millimetres. However, as only local variations in g are needed, results are reduced to a local datum, not sea level.

The gravity anomaly of a buried mass extends beyond the edges of the body (see, e.g., Fig. 8.9), as do the anomalies of some other methods; though this reduces the sharpness of anomalies, it means that the gravimeter need not be over a cavity to detect it, which can be an advantage (see example in Section 27.3.2). Gravity has a major advantage over other geophysical methods for detecting voids, in that a lower density is an intrinsic property of the target. It can also be used to give the mass of 'missing' material; in turn, this can be used to calculate the amount of material needed to fill the void or to check that it has been filled.

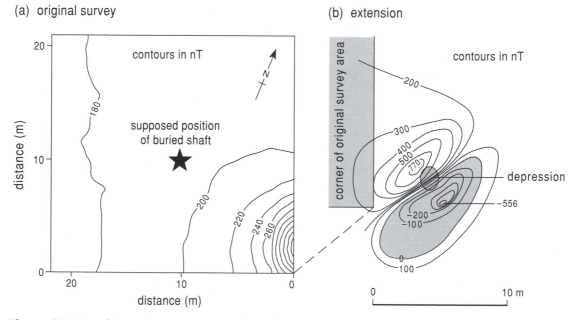

Figure 27.1 Total intensity magnetic maps, mined area.

27.2.5 Fracturing around cavities

Anomalies are sometimes larger than predicted by simple theory, and this is attributed to there often being a 'halo' of fractured rock around a cavity. These cracks lower the seismic velocity – especially V_S – but as their thickness is small compared to the seismic wavelength they will not produce reflections (Section 7.8.2). Dry cracks can reduce the conductivity of the rock around a dry cavity or increase it when filled with water, effectively enlarging the target. Cracks are unlikely to have much effect on the magnetic anomaly but can significantly enlarge the volume of reduced density to which gravity surveys respond; this can increase the gravity anomaly considerably, for some of the fractured volume will be nearer to the surface than the cavity proper, and gravitational attraction depends on the inverse square of the distance.

All of the above techniques have had their successes in detecting cavities, though sometimes failures too. Which technique to try partly depends upon the local circumstances: the likely size and depth of the cavity, the properties of the rocks, limitations of the site, and, of course, cost. However, the most widely applicable method is gravity because, as pointed out above, it responds directly to the absence of material and so the size of anomaly expected can be calculated. The two case studies described in the following sections both employed gravity surveys but illustrate different aspects of its use.

27.3 Collapses in buried karstic terrain, Kuwait: A case study

Geophysical method employed: Microgravity.

27.3.1 Background

The Al-Dahr residential area is one of the largest in Kuwait, with 2500 houses plus schools, and so on, in six sectors, subdivided into quadrants (Fig. 27.2). It was constructed on an area of clastic sands and gravels, and for years there was no apparent subsidence until, in 1989, a resident saw his car sink into the ground as a hole formed. During the next few

Figure 27.2 Part of Al-Dahr residential area, Kuwait.

days the void enlarged to a cylindrical hole 15 m in diameter and 31 m deep. In the subsequent 15 months three further holes appeared in Quadrant 3 (Fig. 27.2), and though none was as large as the first one, one house had to be demolished.

Though the holes were entirely in the sands and gravels, the area is underlain by Eocene limestones of the Damman Formation (Fig. 27.3a). This formation was exposed to erosion and karstic formation during the period from Late Eocene to the end of the Oligocene, and it is thought that voids formed in the limestone then. The sands and gravels that overlie the limestone unconformably are of Late Oligocene to Pleistocene age and are poorly cemented by carbonates and sulphates. Though the area is a local topographic high ('Al-Dahr' means 'high relief'), the sand and gravels are at their thinnest here, about 35 to 40 m thick, because the underlying limestone forms a structural high beneath the area.

It is thought that the disturbance caused by the activities of those living in the area, particularly the watering of gardens, caused the weak intergranular cement to dissolve, allowing the sands and gravels

to flow down, like sand in an hourglass, into voids in the underlying limestone. *A, B, C,* and *D* of Figure 27.3b illustrate the likely development of a surface hole. A broken water pipe found in the first hole may have been leaking for some time and dissolving the cement.

27.3.2 Gravity survey

Geophysical surveys were needed to find out if voids existed under other parts of the area, for these might similarly lead to subsidence. Most techniques were ruled out for a variety of reasons: Magnetic anomalies would be unlikely to exist and anyway would be masked by ferrous objects such as reinforcement in the buildings; similarly, electrical anomalies would be small and masked by the effects of roads, houses, gardens, wires, and pipes, and resistivity arrays would be difficult to lay out; seismic lines of sufficient length would also be difficult to lay out, and cultural noise would be high. Ground-penetrating radar could probably have penetrated deep enough, but the area was obstructed with buildings. However, gravity was feasible: As it can detect a void without being vertically above it, stations could be placed where convenient among the buildings, and though the masses of the buildings were sufficient to produce anomalies, these were not large and could be corrected for.

Readings taken in Quadrant 3 (Fig. 27.2), which was where the holes had occurred, were spaced as near as possible to a 7-m grid (half the separation of the houses), but reduced to 3 m near the holes; readings were taken in gardens and inside houses, using a Lacoste & Romberg D meter, which can measure down to 1 μGal. A number of secondary base stations were used, as shown in Figure 27.2, to reduce the time spent returning to base station; they were 'tied' to the primary base station by separate measurements. A Bouguer correction was made using a density of 2.0 Mg/m^3 for the sands and gravels, but terrain corrections were not needed as the relief of the area is very gentle (but see below for the effects of the buildings).

Figure 27.4a shows the Bouguer anomaly map; its main feature is a regional gradient with *g* decreasing from the southwest the northeast, but smaller anomalies clearly are present. The regional anomaly, approximated by a plane dipping to the

(a)

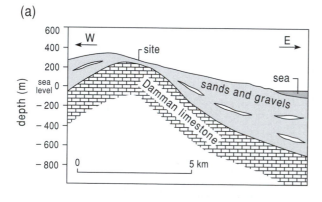

(b)

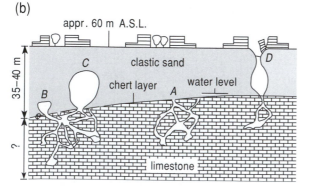

Figure 27.3 Cross-sections through region.

(a) Bouguer anomaly

(b) Residual anomaly

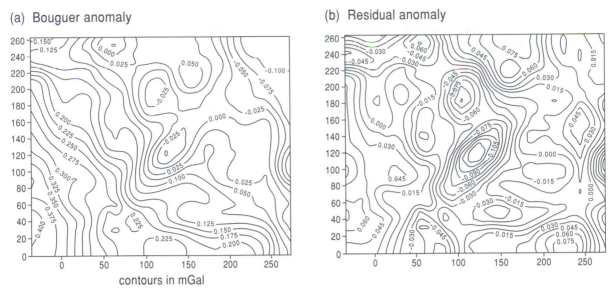

contours in mGal

Figure 27.4 Bouguer and residual anomaly maps, Al-Dahr area of Kuwait.

northeast, was subtracted, to give the residual anomaly map shown in Figure 27.4b. Now the principal feature is a central negative anomaly, but there are others more or less regularly spaced; these are due to the regularly spaced houses and appear in the residual maps of the other quadrants. Rather than apply terrain corrections for the houses

(which would have been possible but laborious), advantage was taken of the symmetry of the layout: Quadrant 2 is a mirror image of Quadrant 3. Therefore, the residual anomaly map of Quadrant 2 was first reflected and then subtracted from the residual anomaly map of Quadrant 3; their difference is shown in Figure 27.5. Now the central anomaly is

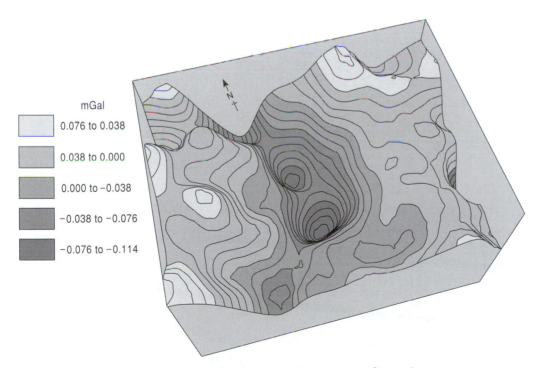

mGal

	0.076 to 0.038
	0.038 to 0.000
	0.000 to −0.038
	−0.038 to −0.076
	−0.076 to −0.114

Figure 27.5 Corrected residual gravity anomaly, Al-Dahr area of Kuwait.

dominant, larger than before (about −100 μGal) and aligned closer to N–S, but it is part of a low that curves across the figure. This central anomaly lay under a number of houses, including where one of the major collapses had occurred.

A number of holes were drilled, including three in Quadrant 3, numbered after the nearest house, as shown on Figure 27.2. Borehole 21, near the centre of the biggest anomaly, encountered a number of voids in the underlying limestone, probably part of an inter-connected system, as the drilling fluid rapidly lost pressure. Borehole 44, near the anomaly at the right-hand edge of Figure 27.5, also encountered voids, near the contact with the limestone. Borehole 7, drilled where there was no anomaly, found no voids. Drilling in other areas where there were negative anomalies with magnitude greater than 60 μGal also encountered voids, near the contact with limestone.

The total missing mass in Quadrant 3, estimated from the volume of the anomaly of Figure 27.5 (as explained in Section 8.9), was 9000 tonnes, even though several thousand tonnes of sand, gravel, and concrete had been poured into the first cavity before the gravity survey. Further details are given in Bishop et al. (1997).

27.4 Land reclamation, south Wales: A case study

Geophysical method employed: Microgravity.

27.4.1 Background

The site to be reclaimed was level 'made ground', consisting of colliery spoil that had recently been 'reworked', overlying glacial deposits. Beneath the glacial deposits are two coal seams, the upper and lower seams, with an ironstone band between them (Fig. 27.6). Both coal seams and possibly the ironstone had been worked, but to what extent was not clear. The reclamation proposal included a small lake, and there was concern that cavities beneath it might allow subsidence and cause rupture of its water-retaining lining. Possible workings in the upper seam were not a problem, for it was planned to remove the remaining coal of this seam and its overburden, but any workings in the ironstone or lower seam needed to be located. A gravity survey was carried out to detect cavities in the ironstone or lower seam but was made more difficult by the likely presence of cavities in the upper seam.

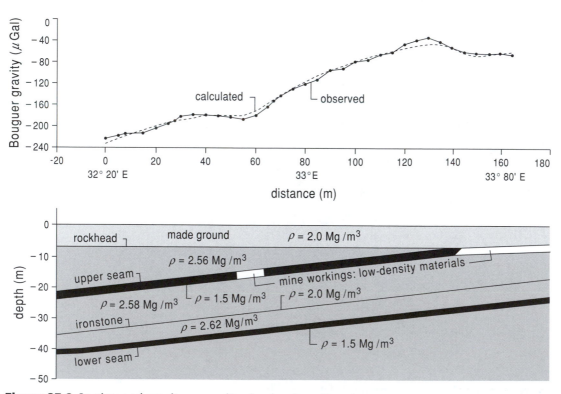

Figure 27.6 Section and gravity anomalies, land reclamation area.

(a) Bouguer anomaly

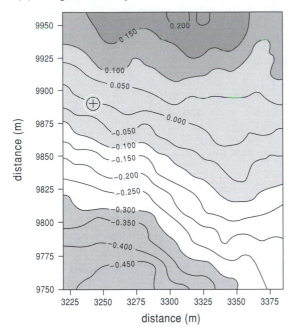

Figure 27.7 Bouguer and residual anomaly maps, land reclamation site.

(b) residual anomaly, unfiltered

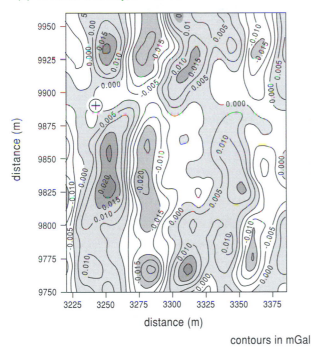

(c) residual anomaly, filtered

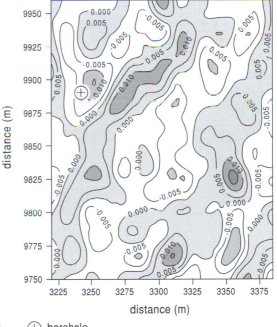

contours in mGal   ⊕ borehole

27.4.2 Gravity survey

Gravity was measured over the area on a grid with 5-m spacing, and the resulting Bouguer anomaly map is shown in Figure 27.7a. The main feature is the steady decrease in g from roughly south to north (reflecting the direction of dip of 186°), which is therefore the regional anomaly. This was approximated by a plane, dipping surface, and after its removal two negative anomalies became obvious (Fig. 27.7b), striking approximately N–S at about 3280 and 3365 m east, plus a probable third one just off the left of the figure. These linear anomalies are likely to be due to parallel tunnels in the coal, with a roughly equal width left unmined between to support the roof. Depth estimates based on E–W profiles across the anomalies give a value of about 14 m, which is the depth to the upper seam at this location.

The N–S negative anomalies are interrupted at roughly 9900 m north, and it is not clear if beyond it the more easterly of the two anomalies continues to the north or roughly northeastwards. The 'interruption' is unlikely to be a narrowing of the tunnels and so is probably at a different level – the lower seam or the ironstone being likely – and with a direction different from the tunnels in the upper seam. To emphasize these weaker anomalies the N–S anomalies due to tunnels in the upper seam were greatly reduced by using directional filtering (Section 3.2.5), revealing a set of small-amplitude anomalies running approximately northeast to southwest (Fig. 27.7c).

The borehole indicated by the crossed circle confirmed the presence of coal in the upper seam and a cavity in the lower seam at this position. Figure 27.6a compares a modelled E–W gravity profile, calculated using the data shown in the section below, with the measured one. They are in good agreement, giving support to the interpretation just given. Further details can be found in Bishop et al. (1997).

Further reading

Bishop et al. (1997) discuss detecting cavities using microgravity, and give several examples, including those described in Sections 27.3 and 27.4. Other surveys to find cavities (plus surveys to solve other types of problem) are described in McCann et al. (1997). Lowry and Shive (1990) discuss resistivity methods.

chapter 28

Archaeological Site Surveying

Archaeology is the study of how humans lived in the past, so no direct observation is possible. For this reason, information about how – and when – people lived has to be deduced from what they have left, both intentionally – such as monuments and written records – and, just as important, incidentally, in traces of buildings, fortifications, field systems, and so on, now often buried. Unravelling the sometimes long and complicated history of an archaeological site often requires the skills of a detective applied to meticulous and painstaking excavations. To aid the investigation, the archaeologist can call upon various techniques (i) to help find or map a site, (ii) to help date the site and its artefacts, and (iii) to help characterise artefacts, such as analysis of their materials to help find their source. Geophysics has a large contribution to make to the first two of these, and a lesser one to the third.

Dating is obviously important to understanding how cultures develop or relate to one another. For example, it was once thought that the builders of Stonehenge derived their culture from the ancient cultures of the eastern Mediterranean, such as ancient Greece, until carbon dating showed Stonehenge to pre-date them. The dating methods most used in archaeology are those suitable for younger materials (described in Section 15.12), but the potassium–argon method is the main method for dating early hominids, mainly in Africa where they extend back over 4 Ma. Magnetic stratigraphy (Section 10.5) has also been used to establish the age of hominid remains and buried artefacts.

Characterisation uses a wide variety of methods, particularly chemical trace element analysis, but magnetic properties, such as susceptibility, have been used to characterise the material of artefacts, while palaeomagnetism has been employed to show whether artefacts were heated by fire, deduced from the effect it had on the remanent magnetisation (Section 10.3.4).

Both dating and characterisation applications of geophysics are straightforward, commonly using only a single technique, and will not be considered further. This chapter is about site investigation where it is often advantageous to use more than one technique, either for reconnaissance followed by detailed mapping or to detect different kinds of feature.

28.1 Site surveying

Figure 28.3b and c shows results of two geophysical surveys of an archaeological site. There are obviously features present, but what are they likely to be? Though some features are common to both figures, others are not: Why is this and have others been missed? This is equivalent to asking, What – if any – geophysical signals will various types of archaeological features produce, and how are these signals best measured and displayed? This chapter begins by answering these questions and then goes on to show results for actual sites.

The targets of most archaeological sites are at shallow depth, usually within a metre of the surface, and so usually within the overburden, which, for many geological investigations, merely obscures the features of interest below. Therefore, surveying methods have to be adapted to look into the overburden, rather than through it, and so usually have to be capable of detecting small variations at very shallow depths. The great potential value of geophysical surveying to the archaeologist is to show the likely presence or extent of an archaeological site, perhaps producing a map of it, and so allowing the painstaking process of excavation to be concentrated where it is most likely to produce valuable results, if it is needed at all. This role of geophysical surveying is particularly important for 'rescue archaeology', where there is limited time before an area is to be quarried or built upon.

Surveying of archaeological sites usually has to be intensive, so as not to miss small or narrow features such as pits and walls. Readings are therefore often taken at spacings of a metre or less, so that many thousands of readings are needed to cover even a small site. It is obviously desirable that readings be taken rapidly, and – rather than being written down – be recorded in a form that can be fed into a computer. Developments of both microelectronics and the design of instruments has advanced to the stage where, for some techniques, thousands of readings can be taken each day, and often displayed before leaving the site, at least in a preliminary form. This ease of use is important when choosing which method to use but, obviously, the overriding consideration is whether a given technique can detect likely features of interest. The following section will describe some common archaeological features and how they might be detected geophysically.

Anyone contemplating a survey should be aware that many sites are under state protection (certainly in Britain) and geophysical surveys may not be carried out without permission, even though they do not disturb the subsurface.

28.2 Archaeological features and their geophysical expression

28.2.1 Ditches, pits, and postholes

These are features cut into the ground. Ditches were used variously as part of defences, to mark field and other boundaries, and for drainage. Pits were dug for storage and other purposes, including burials, while postholes once anchored posts that perhaps supported a building but have long since rotted away. The main difference between ditches and pits and postholes, for geophysical surveying, is their extent: The long continuous anomaly of a ditch is much more easily found than the pointlike one of a pit or posthole, so more closely spaced readings are needed for the last two.

Surveying has to utilise the physical difference between the properties of the hole and the surrounding undisturbed soil or rock, due either to the initial digging or to infill accumulated over the passage of time. Often infill is less consolidated and richer in organic and some mineral constituents, including ferromagnetic minerals, which do not dissolve easily and so tend to be concentrated when rocks weather. Nearly always, it will have a lower resistance than its surroundings, so that resistivity or e-m surveying are often successful, though the electrical contrast will vary with the seasons.

The ferromagnetic minerals in an infill give it a higher susceptibility than its surroundings, so that it produces positive magnetic and susceptibility anomalies. For example, this has allowed what is proba-

bly a large wood henge, consisting of posts in concentric circles, to be found by magnetic surveying (Aveling, 1997).

Ground-penetrating radar (GPR) is often able to detect disturbed ground, either because the infill or buried objects reflect the radar waves, or because original, reflecting boundaries within the ground are interrupted on the record. When GPR is applicable, its high resolution and continuous profiling are advantages.

28.2.2 Foundations

If the lower part of a wall survives below ground, it is more consolidated than its surroundings and therefore contains less moisture, giving rise to a positive resistivity anomaly, though it will be less prominent in dry conditions or if the surroundings have a high resistivity, such as gravel. However, anomalies may be made more difficult to recognise by the 'noise' of any buried blocks from the vanished upper part of the wall, plus any clay used to plug gaps between blocks.

Magnetically, a wall is likely to produce a weak negative anomaly, for most rocks have lower susceptibility and remanence than soils. Blocks of basic volcanic rock, such as basalts, have a considerable remanent magnetisation, so a detectable field will exist close to individual blocks; however, the wall as an entity will be much less magnetic, because the blocks are likely to have their remanences in different directions, producing fields that partially cancel at a distance. Similarly, bricks have individual magnetisations (acquired on cooling; see next section) which may be detectable close to. Susceptibility contrasts may be detected down to about half a metre, while magnetometers may detect differences in magnetisation to over a metre, in favourable situations.

28.2.3 Furnaces, fireplaces, and kilns

The geophysical significance of these types of targets is that their magnetisation usually will have been increased by the heating, partly by oxidising nonmagnetic minerals to magnetic ones, but mainly by giving the material a TRM (thermal remanent magnetisation) when it cools in the Earth's field. As the material is likely to have remained in situ – unlike bricks removed to make walls – it can produce a considerable magnetic anomaly, probably the largest on an archaeological site (apart from those of ferrous objects, most of which will be recent rubbish). These targets are likely to give a positive resistivity anomaly.

This brief account describes only the main types of targets and merely touches on the role of the surrounding material in producing the physical contrast that results in an anomaly; for example, the very different resistivities of wet gravel and clay will affect the size and even sign of resistivity anomalies.

28.3 Geophysical methods useful for archaeological surveying

In practice, for the reasons outlined in the previous sections, most archaeological geophysical surveys have exploited either magnetic or resistivity (or conductivity) contrasts between a target and its surroundings, though other physical properties are sometimes utilised (Sections 28.3.3 and 28.3.4). For many surveys, it has been sufficient merely to detect the presence of subsurface features, but interest is turning to deducing their depths as well, which requires more sophisticated survey techniques.

28.3.1 Magnetic and susceptibility surveys

The most commonly used method is magnetic, partly because of its speed, and this is described first, together with susceptibility surveying. Since the targets are very shallow compared to those in most other geophysical surveys and the anomalies are small – often only a few nT – a magnetic gradiometer has the advantage over a single magnetometer sensor, as explained in Section 11.7. Instruments are usually fluxgate gradiometers, and readings are often taken over a grid, for many features – foundations, holes, and so on – would often be missed by widely spaced traverses.

The small magnetic anomalies of archaeological sites can be due to variation in either or both remanent magnetisation and susceptibility, but more commonly the latter. Variations in susceptibility are due to variations in the concentrations and types of magnetic minerals. This can be because magnetic minerals are more concentrated in the infilling of holes or because fire enhances susceptibility, as explained above; it can also be due, as mentioned in Section 10.3.6, to bacterial action (which may have been stimulated by organic residues accumulated in

archaeological sites) forming magnetite or maghaemite. Remarkably, susceptibility anomalies in the top few centimetres can persist for hundreds or even thousands of years in grassland, especially on level ground. These can also be measured using a susceptibility meter, which works by producing an alternating field in the ground; as the coil used to produce the field is typically only about 20 cm across, only the susceptibility of the topmost 10 cm is measured.

28.3.2 Resistivity surveys

The resistivity array most commonly used is the twin array, which is rapid to use and produces simple anomalies. It has a pair of moving electrodes and a pair of fixed ones (Fig. 28.1); it differs from the dipole–dipole array in having a current and potential electrode in each pair. The fixed electrodes are pushed separately into the ground and are connected by a cable to the two moving electrodes, which have a constant separation (usually 0.5 or 1 m). The moving electrodes form the feet of a frame that also carries the resistivity meter and a data logger to record the results (Fig. 28.1a). Though the moving probes are held rigidly, they still make contact with uneven ground because there are only two of them, though sometimes the frame needs to be pressed down to make good electrical contact. The circuitry is designed to take a reading and record it automatically (and alert the operator that it has been done) once good contact has been made, so that surveying, which needs only one person, is very simple and quick; for example, the 400 readings of a 20-m-square grid with 1-m separation can be taken in less than 30 min.

The fixed electrodes are placed at a distance of at least 30 times the spacing of the moving electrodes from the nearest station; as readings cannot be more than the cable length (commonly 50 m) from the fixed electrodes, this array is better for measuring over a grid than along a profile. If the fixed electrodes are placed as shown in Figure 28.1b, three squares with 20-m sides can be reached without moving the fixed electrodes. To extend the survey beyond three squares, the moving electrodes are left at their last station while the fixed electrodes are moved from position 1 to position 2 – as at the right of Figure 28.1b – and then their separation is adjusted to match the reading before they were moved.

(a) power supply and data logger

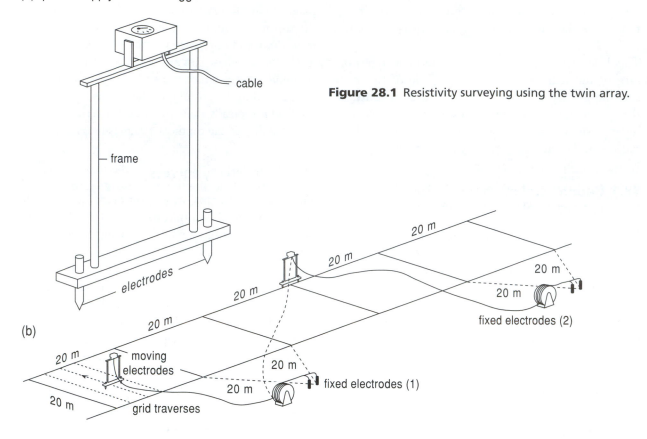

Figure 28.1 Resistivity surveying using the twin array.

It produces simple anomalies, in contrast, for example, to the Wenner array (Fig. 12.22), and can be very effective is revealing linear features (e.g., Fig. 28.3a). It gives little depth information, though some indication can be gained by repeating with a different separation (e.g., half or twice). As the values are not strictly apparent resistivities, they cannot be used to identify the nature of the subsurface.

For more detailed depth information, electrical imaging, which measures both vertical and lateral variations, is better. This is conveniently carried out using a line of equally spaced electrodes (1 or 2 m apart) along a profile, individually connected to switches and thence to a resistivity meter, as described in Section 12.5. Sets of 4 electrodes are selected by switching, to produce a profile first with one electrode spacing, and then repeated with progressively larger spacings. The results can be displayed as a pseudosection, but results are clearer if they are computer processed to give a true resistivity section. This method is relatively quick, but as it provides a section rather than cov-

ers an area, it is best employed across a linear feature of interest, perhaps one found using the twin array.

28.3.3 Ground-penetrating radar (GPR)

The limited depth penetration of GPR is less often a limitation for archaeological surveys, where the shallow depth of most targets permits the highest frequencies (shortest wavelengths) to be used, with resolutions of a few tens of centimetres. This allows suitable targets to be 'seen' in considerable detail. Results are clearest when closely spaced parallel profiles are used to build up a 3D picture, rather like 3D seismic-reflection surveys. Ground-penetrating radar responds to variations in relative permittivity, but how this relates to human constructions and activities is not fully understood. The technique has had successes in locating, sometimes in detail, disturbed ground such as graves and kilns, and is being increasingly used when the extra information it provides, compared to other surveying methods, justifies its higher cost.

28.3.4 Other techniques

Resistivity contrasts are occasionally measured using e-m (electromagnetic) instruments of the Slingram type, with coils mounted in a rigid boom up to a few metres apart (Fig. 14.4a shows an example). As they are self-contained and have no electrodes, they are quick to use, a considerable advantage for reconnaissance profiling, though less decisive when grid surveying, for laying out the grid takes a considerable fraction of the total time. As the resolution is not as good as for resistivity, it is not usually the first choice, except in very dry or perhaps frozen ground where it is difficult to make electrical contact. Pulsed e-m (TEM) has been used on occasion. Induced polarisation (IP) – which uses apparatus similar to that of conventional resistivity, rather than the specialist twin array described in Section 28.3.2 – may have advantages over resistivity in a few situations.

Because archaeological targets are usually small and shallow, gravity is not useful except to detect occasional deeper targets, such as caves and tunnels or perhaps an infilled lake. Similarly, seismic reflection is of limited use on land, but Chirp, a method which, like Vibroseis, uses swept frequencies, has been used for shallow marine targets such as buried sunken ships; by using high frequencies (1–12 kHz), resolution as small as 15 cm and penetration of up to 50 m – depending on type of sediment – has been achieved. Seismic refraction has been used to measure the depth of ditches or vallums (defensive ditches), particularly when these are cut into the bedrock, which gives a good velocity contrast, but as it is slow and provides only a profile, it is best limited to occasional locations where depths are valuable and the velocity contrast is likely to be large.

28.3.5 Display of data

Resistivity or magnetic *profiles,* and GPR or seismic *sections,* are displayed as described in Part I of this book, but results for resistivity or gradiometer *grids* – which is how these methods are most often used – are usually shown as dot density or grey-scale plots. In a dot density display (Fig. 28.2a) the map of the surveyed area is divided into small squares and into each square is put a number of dots in proportion to the strength of the anomaly there, so the squares look darkest where the anomaly is highest; the dots are distributed randomly within the square, so that they do not line up to produce spurious lines. Just how many dots should correspond to what size of anomaly is partly a matter of trial and error to produce the clearest figure, but the method can allow the eye to recognise patterns within quite a small range of variation.

Grey-scale plotting (Fig. 28.2b) is an extension of dot density plotting, made feasible by advances in laser printers: Instead of having the *number* of dots in a square proportional to the size of the anomaly, a single dot is used with its *size* proportional to the size of the anomaly. This is the way that black-and-white pictures are printed in newspapers; if the dots are closely spaced they merge to give a range of greys (halftone reproduction). As with dot density, some experimentation is needed to produce the clearest plots. Dot density and grey-scale plots of gridded data are often superior to contoured plots (even with colour) or other presentations for revealing archaeological features (Fig. 28.2).

(a) dot density   (b) grey scale   (c) contours   (d) stacked profiles

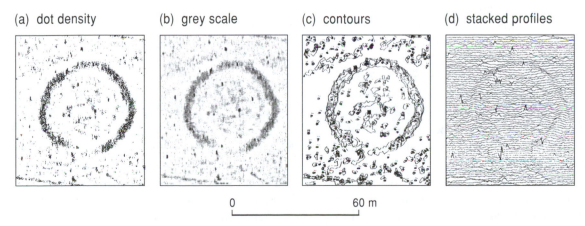

0   60 m

Figure 28.2 Various ways of displaying gridded data.

28.4 A possible Roman villa: A case study

Geophysical methods employed: Magnetic gradiometry, magnetic susceptibility, and resistivity.

28.4.1 Background

The Charlton Villa site lies in a field on the northern edge of Salisbury Plain, an area well known for its prehistoric remains, including, of course, Stonehenge. However, the interest of the site in question was the possibility of its concealing a Roman villa; a number of settlements of Romano-British date have been found in the area, so associated villas – which ranged in function from a country house to a working estate – are expected to have existed, but few are known. The site was chosen for investigation because fragments of Romano-British pottery and building tiles had been found on the surface. There was no other indication of a villa, but aerial photography had showed several circular features (ring ditches), some with double ditches, in a corner of the field; these are probably of Bronze Age date, suggesting a long history of occupation of the site. The surface drift is gravel, overlying chalk, both of which have low magnetism.

28.4.2 Geophysical surveys

The first geophysical survey was a reconnaissance one, using magnetic susceptibility, because susceptibility is often increased by human occupation of an area. Measurements were taken with the coil placed on the bare soil, at 20-m intervals, over an area of 6 ha; the results are shown in Figure 28.3b. The highest values were mostly found towards the southwestern corner of the field, so further surveys were limited to there. As the selected area was *without* known ring ditches it suggested the occupation was not a Bronze Age one.

Magnetic gradiometry and twin array resistivity were both chosen for more detailed surveying because they respond to different kinds of archaeological feature (Section 28.2). The dot density plot of the gradiometer results (Fig. 28.3c) shows a number of obvious lines, some areas with many small features whose arrangement is hard to discern (in the positions of 1 and 3 on Fig. 28.3d), plus one or more circles in the northeastern corner additional to those seen from the air.

The resistivity results, shown in the grey-scale plot (Fig. 28.3a), reveal that areas 1, 2, and 3 are buildings, some quite complex. The dark area at the southeastern end of area 1 may be due to a dense mass of masonry, such as a bathhouse with underfloor heating (hypocaust); this is one of the areas where tiles were found, which supports the interpretation. If so, the foundations show up by resistivity but not magnetometry because they have high resistivity but are not magnetic; the featureless magnetic anomalies within the buildings are probably due to fallen bricks and tiles or thermoremanence (TRM) produced by fires. In contrast, the linear feature 4 can easily be seen on the gradiometer plot but not on the resistivity one, probably because all the ground there was wet. It is likely to be an infilled watercourse, which may have supplied water to the building or drained it. The dark areas on the western part of the resistivity survey are thought to be due to a former river meander.

In conclusion, it is probable that all the buildings are of the same age and, taking into account the presence of tiles and other nongeophysical evidence, probably formed part of a Roman villa, though this could only be confirmed by excavation, which has not been carried out. However, the survey helps build up a picture of the area in Romano-British times. Further details may be found in Corney et al. (1994).

28.5 Hudson's Bay Company fur trade post: A case study

Geophysical methods employed: GPR, magnetic gradiometry, and e-m (Slingram).

28.5.1 Background

At the end of the 18th century the competing North West and Hudson's Bay Companies each established trading posts on the North Saskatchewan River in western Canada, to trade with the Indians living in the Rocky Mountains to the west. Because of trade rivalry, hostility from the Indians, and shifting trade routes, the trading posts were repeatedly abandoned and reopened, but in 1835 the now-amalgamated companies built a new, wooden fort, called Rocky Mountain House (Fig. 28.4). This was almost continuously occupied until 1861, when the Blackfoot Indians forced its evacuation and later burned it down. Yet another fort was built and occupied from 1865 to 1875.

(a) resistivity survey

(b) map and susceptibility survey

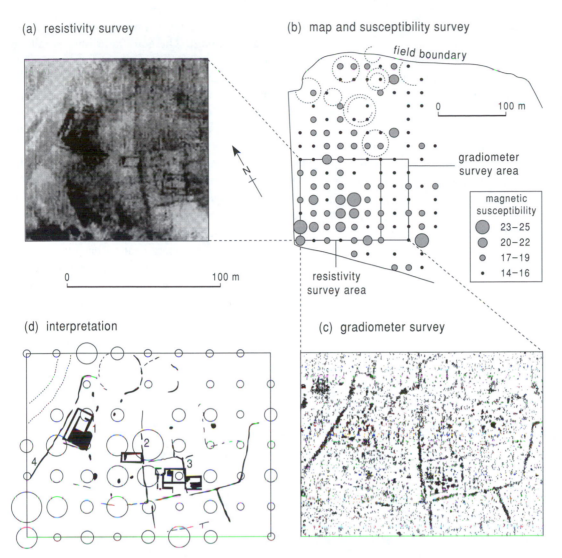

field boundary

0   100 m

gradiometer
survey area

magnetic
susceptibility

23–25
20–22
17–19
14–16

resistivity
survey area

0   100 m

(d) interpretation

(c) gradiometer survey

Figure 28.3 Charlton villa: Map and survey results.

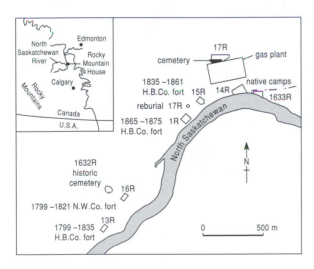

Figure 28.4 Map of the forts and other features.

Some idea of how the forts were laid out is known from descriptions and pictures, and the positions of the palisades are revealed by aerial photographs taken when the sun was low. A small part of the 1835–1861 fort was excavated, principally parts of the palisade walls and gateways; after excavation, the structures revealed were covered and the land returned to cultivation. Later, it was decided to carry out geophysical surveys, particularly to provide information about the positions of the rooms and palisades of the fort, and of various possible burial sites at some distance from the fort, plus Indian campsites; the objectives were to reveal more details of the area and to test how useful geophysical surveys could be for sites of this type.

28.5.2 Geophysical surveys

Ground-penetrating radar was chosen as having the highest resolution, required to reveal details of palisades and rooms, and for detecting the disturbed ground of graves. But as it was to be used only where targets probably existed – because of its cost – it needed to be preceded by surveys that could pick out likely target areas within the larger area of the site. Resistivity was an obvious choice for detecting excavations, such as postholes, cellars, and graves. Less obvious for detecting the remains of wooden constructions was magnetometry, but it is sensitive to ferrous objects, fireplaces, and – as was discovered – burnt areas.

Figure 28.5 shows the results of a magnetic gradiometer survey over the area of the 1835–1861 fort (carried out using proton precession sensors, with enhanced sensitivity). There are faint lines parallel to the top and base of the figure. As this was the direction of the traverses, these are likely to be artefacts of plotting, but the more prominent positive anomalies (i.e., where the magnetic field is above average), most of them on a 3-m grid, are real, and some of the stronger anomalies form lines. Some of these lines are known to be the positions of the palisade walls, which were often supported on king posts spaced 3 m apart. The positions of the posts probably show up because the burning of the posts produced a thermal remanence in the surrounding soil. Other anomalies are due to fireplaces where heating has given a thermal remanence to the sandstone rock and to the clay mortar used in their construction. Yet other anomalies may be due to buried ferrous objects.

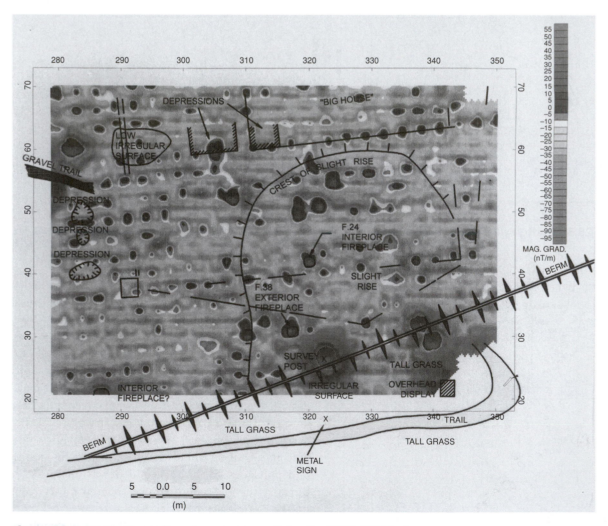

Figure 28.5 Magnetic gradiometer survey result for the 1835–1861 fort.

(a)

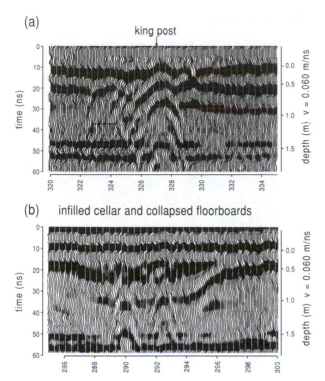

(b) infilled cellar and collapsed floorboards

Figure 28.6 Ground-penetrating radar sections within the 1835–1861 fort.

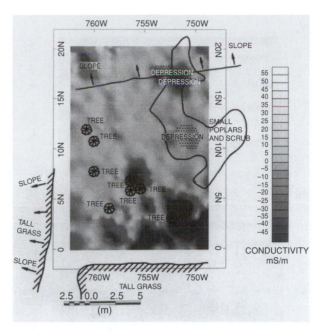

Figure 28.7 Electromagnetic conductivity survey of site 1632R.

GPR surveys were carried out within the fort area along parallel lines usually spaced 2 m apart. Hyperbolic reflections – produced by small, reflective objects (Section 7.5) – are common (Fig. 28.6a). Many are due to postholes, for they coincide with the positions of regular features found in the magnetic gradient survey. Others targets found were due to fireplaces or cellars; Figure 28.6b shows a partially filled cellar with the overlying floor half-collapsed into it (the hyperbolic reflections interrupting the sagging reflectors are due to small objects in the cellar).

A conductivity survey of burial site 1632R is shown in Figure 28.7; it was carried out using an e-m instrument of the Slingram type, designed for archaeological surveys with a coil separation of only 1 m (Geonics EM38 instrument), giving a penetration of about 1.5 m. A number of anomalies can be seen, chiefly at the southern end. Nine potential targets were identified; the negative conductivity anomalies shown are not physically real, but are an instrumental response to very high conductivities, which here suggests the presence of metal in the graves, later confirmed when excavation found iron and copper objects.

To find how well GPR works for burial sites it was used over part of the cemetery (site 17R). Graves were taken to be where horizontal reflections deeper than ½ m show gaps, due to the excavation. Approximately 50 possible burials were identified. Figure 28.8b shows two examples, which may be compared with Figure 28.8a, which was measured outside the cemetery. However, reflections from boulders or due to sideswipe from objects out of the plane of the section sometimes made it difficult to recognise which anomalies were due to graves, and to distinguish one grave anomaly from another. Later excavation of the site showed that some of the GPR anomalies were caused by large rocks, and of six possible burial sites that had been identified, only two were confirmed by excavation.

The survey showed the value of employing several geophysical methods. Magnetic and conductivity surveys were not only used to select areas for GPR surveying, they sometimes gave clearer results. Overall, GPR produced many images of archaeological targets, though not all were as clear or as easily interpreted as those shown, and some reflections were not due to targets of archaeological interest. Conversely, at some sites GPR failed to find anticipated targets. Further details may be found in Bauman et al. (1994).

(a)

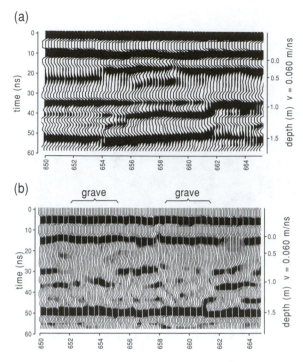

(b)

Further reading

Scollar et al. (1990) gives a comprehensive account of site surveying, of which over half is about geophysical methods, while Clark (1990) gives a shorter and less formal account. Leute (1987) describes the range of physical and chemical measurements available to assist the archaeologist, including dating and characterisation, with a relatively small section on site surveying. Advances may be followed in the *Journal of Archaeological Science* and the *Journal of Field Archaeology*. Parent and O'Brien (1993) describe the Chirp system.

Figure 28.8 Ground-penetrating radar sections at burial site.

Appendix A

List of Symbols and Abbreviations

(Chemical elements are not included)

Latin characters

A	area
a	amplitude of waves
a	spacing of electrodes of Wenner resistivity array
a.f.	alternating magnetic field
APW	apparent polar wander (in palaeomagnetism)
bcm	billion cubic metres (of gas)
Bq	Becquerel (unit of radioactivity)
CDP	common depth point (in seismic reflection)
Ci	Curie (unit of radioactivity)
CMP	common mid-point (in seismic reflection)
CRM	chemical remanent magnetisation
c	specific thermal capacity
c	velocity of light
D	displacement of a fault rupture
D	number of daughter atoms (in radioactivity)
DRM	detrital or depositional chemical remanent magnetisation
d	depth, separation
E	electric field
e-m	electromagnetic
emf	electromotive force
F	force
F	formation resistivity factor
FE	frequency effect (in IP surveys)
f	frequency of waves and oscillations
G	'big Gee', universal gravitational constant
GPR	ground-penetrating radar
g	'little gee', acceleration due to gravity
g.u.	gravity unit
H	magnetic field
Hz	Hertz, unit of frequency, measured in cycles/sec
h	depth, depth to earthquake, thickness
I	angle of inclination of Earth's magnetic field
I	electric current
IGRF	International Geomagnetic Reference Field
IP	induced polarisation
IRM	isothermal remanent magnetisation
i	angle (particularly of refraction or reflection)
i	electric current density
i_c	critical angle (in seismology and GPR)
K	thermal conductivity
K	temperature coefficient affecting SSP
k	permeability of rock
k	resistivity constant
KRISP	Kenya Rift International Seismic Project
K/T	Cretaceous/Tertiary (geological boundary)
L	inductance
L	length, length of fault rupture
M	magnetisation
M	mass
M	chargeability (in IP surveying)

M_E	mass of the Earth
M_o	seismic moment of earthquake
M_L	local magnitude of earthquake
M_S	surface wave magnitude of earthquake
M_W	moment magnitude of earthquake
MF	metal factor (in IP surveying)
MT	magnetotelluric
MTOE	million tonnes oil equivalent
m	(small) mass
m_b	body-wave magnitude of earthquake
mGal	milliGal, unit of gravity
md	millidarcy, unit of permeability
NMO	normal moveout (in seismic reflection)
NRM	natural remanent magnetisation
nT	nanoTesla, unit of magnetic field
P	number of parent atoms (in radioactivity)
P	pressure
P_g	P-wave through upper crust
$P_{i_1}P$	P-wave reflected from upper crust
$P_{i_2}P$	P-wave reflected from middle crust
$P_{i_3}P$	P-wave reflected from lower crust
P_mP	P-wave reflected from Moho
P_n	P-wave head wave from Moho
PFE	percentage frequency effect (in IP surveying)
p	ray parameter in global seismology
p.d.	potential difference (in electricity)
Q	Königsberger ratio (in magnetism)
Q	quantity of heat
q	heat flux
q_r	reduced heat flux
R	electrical resistance
R	reflection coefficient (in seismology and GPR)
R_E	radius of the Earth
S_h	hydrocarbon saturation
S_w	water saturation
S/m	Siemens/metre, unit of electrical conductivity
SP	self-, or spontaneous, potential
SSP	static SP (in well logging)
T	temperature

T	time interval
T	transmission coefficient (in seismology and GPR)
T_b	blocking temperature (in palaeomagnetism)
T_c	Curie temperature (in palaeomagnetism)
T_c	closure temperature (in radiometric dating)
TEM	Transient electromagnetic method
TRM	thermal remanent magnetisation
TWT	two-way time (in seismic reflection and GPR)
t	time (particularly to reach seismic or GPR receiver)
t	thickness
t_r	reduced seismic travel-time
t_o	time for vertical reflection
V	volume
V	voltage, voltage difference
VES	vertical electrical sounding (in resistivity surveying)
VRM	viscous remanent magnetisation
v	seismic velocity
v_P	velocity of P-waves
v_{rms}	root mean square velocity (in seismic reflection)
v_r	reduced seismic velocity
v_S	velocity of S-waves
W	down-dip length of fault rupture
WWSSN	World Wide Standard Seismological Network
x	distance (particularly to seismic or GPR receiver)

Greek characters

Δ	epicentral angle
Δm	mass difference
ΔT	temperature difference
Δt	transit time (in sonic well logging)
ΔV	value of potential difference
α	angle (of dip, of azimuth)
α_{95}	angle of 95% confidence (in palaeomagnetism)

α-particle	positively charged particle emitted by radioactive elements	ρ_{ac}	apparent electrical resistivity measured using alternating current
β-ray	electron emitted by radioactive elements	ρ_{dc}	apparent electrical resistivity measured using direct current
γ ray	e-m radiation emitted by radioactive elements	ρ_b	bulk density (of porous rock)
δ	delay time (in seismic refraction)	ρ_g	resistivity of gas (in well logging)
δg	difference in g	ρ_i	resistivity of invaded zone (in well logging)
$\delta\rho$	difference in density	ρ_m	resistivity of mud (in well logging)
ϕ	magnetic flux	ρ_{mc}	resistivity of mudcake (in well logging)
ϕ	porosity of a rock	ρ_{mf}	resistivity of mud-filtrate(in well logging)
ϕ_N	neutron porosity (obtained with neutron logging tool)	ρ_o	resistivity of formation fully saturated with water (in well logging)
ε	electromotive force	ρ_s	resistivity of adjacent beds (in well logging)
ε_r	relative permittivity (in connection with GPR)	ρ_t	formation resistivity (i.e., resistivity of uninvaded zone) (in well logging)
θ	an angle		
κ	compressibility, or bulk, modulus	ρ_W	resistivity of pore water (in well logging)
λ	angle of latitude	ρ_{xo}	resistivity of flushed zone (in well logging)
λ	decay constant (in radioactivity)		
λ	wavelength	σ	electrical conductivity
μ	rigidity modulus	χ	magnetic susceptibility
ρ	density	τ	period of (seismic) oscillation
ρ	electrical resistivity	Ω	ohm (unit of resistance)
ρ_a	apparent electrical resistivity	Ω-m	ohm-m (unit of resistivity)

Appendix B

Answers to Problems, Chapters 2–20

Note: Answers to part or all of some questions have not been provided if a graph or diagram is needed, if a short answer cannot be given, or if the answer is given explicitly in the text.

Chapter 2

1: A positive one is above the surrounding average, a negative one below it. **2:** Any of them. **3:** i, ii (the improved signal-to-noise ratio makes the result clearer). **4:** i. **5:** i or v would be best, though ii, iv, vi, and viii would also work. **6:** Show negative values (or ones below the average) in a different colour; shade area with negative values; use much wider contour intervals for positive values. **7:** iii. **8:** Any, but iv would be clearest.

Chapter 3

1: Period of 1 day, which equals a frequency of 0.0000115 Hz. **2:** (a) Aliasing. (b) Slow forward rotation, slow backward rotation. (c) 45°, 90°, 135°, ... **3:** 25Hz, low-pass. **4:** The deeper one would have a proportionately wider anomaly. **5:** ii, iii. **6:** 10 m, 10 m, 90 m, infinity, 110 m. **7:** Maxima and minima are, respectively, +4, –4; 2.67, –3; 2.2, –2.6. The most obvious wavelengths are 2 m and about 15 m, but the amplitude of the 2-m wavelength is progressively reduced by the 3- and 5-point filters.

Chapter 4

1: ii. **2:** vi. **3:** ii (total internal reflection). **4:** i. **5:** i. **6:** 50°: 9; 90°: 13; 98°: 14; 142°: 19.5, 180°: 20, 183°:

20. **7:** iv, vii. **8:** i. **9:** iii. **10:** Reflections and refractions at interfaces generate them by wave conversion. **11:** (a) Equation 2 of Box 4.1 shows that increasing density lowers the seismic velocity. (b) To compensate for this, the elastic moduli of granite must be much larger than those of water (obviously μ is). **12:** iii. **13:** ii. **14:** iv. **15:** iv.

Chapter 5

1: iv. **2:** It is the point of initial rupture. **3:** Aseismic slip allows the strain to be relieved. **4:** The Richter scale measures magnitude. **5:** Waves generated by different parts of the fault have to travel different distances to the receiver; reflections produce 'echoes'; defocusing of ray paths may exist between the hypocentre and the receiver. **6:** 24° (measuring ray *F* on Fig. 4.20c). **7:** Any of them are possible, though Love surface waves (v) usually generate the largest horizontal amplitudes of motion. **8:** N or S. **9:** iv. **10:** The two planes are, approximately; strike 44, dip 80, and strike 136, dip 80. It is approximately strike-slip and dextral. **11:** The two planes, are approximately, strike 92, dip 32, and strike 120, dip 60. It is approximately normal. **12:** The two planes are, approximately, strike 52°, dip 44° roughly NW with slip angle of 38°, and strike 110°, dip 62° roughly south with slip angle of 48°. **13:** A magnitude increase of 1 equals an energy increase of about 30 times.

Chapter 6

1: Refer to Section 6.2. **2:** Close to the source the direct ray is first and the reflected second, and there is no refracted arrival. At the critical distance the direct ray is first, and the refracted and reflected rays arrive later at the same time. At the crossover distance the direct and refracted rays arrive together, with the reflected ray later. At a large distance the refracted ray is first, then the direct ray, with the reflected ray only a little after it. **3:** The time spent in them can be significant. **4:** Refer to Section 6.4. **5:** The crossover distance is about 15 m. There should be several geophones either side of it to delineate the direct and refracted lines, e.g., 3, 6, 9, 12, 15, 18, 21, 25, 30, 35, 40 m. **6:** With hammer seismics it is easy and cheap to repeat the 'shot' for different distances of the receiver, but much less so when using an explosive. **7:** (i) Yes. (ii) No (sand likely to have slower velocity than the clay). (iii) Yes. (iv) Yes, provided there is a considerable velocity difference between the two rocks. (v) No (the strata are unlikely to contain sufficiently large velocity contrasts, and the dip may be too steep for rays to return to the surface). (vi) Yes. (vii) Yes, using fan shooting. (vii) Yes, using the plus–minus method. **8:** For a line along strike the rays recorded have travelled in a plane perpendicular to the interface, whereas for the other two cases they travel in a vertical plane. However, for the along-strike line the forward and reversed travel-time plots are the same, like those of a horizontal interface. To detect that the interface is tilted, a second reversed line should be shot perpendicular to the first.

Chapter 7

1: (a) Liquids do not transmit S-waves. (b) Hydrophones are normally within the critical distance. **2:** Correction is made for rays reflected at an angle to the vertical from a tilted interface. They would be the same if the interfaces are horizontal. **3:** Moveout for primary is less than for the multiple, so it stacks at a higher velocity. **4:** No. **5:** Less. For rest of answer refer to Section 7.4. **6:** The source is not impulsive but continues with changing frequency. It is quick to deploy, has good noise rejection, and by releasing energy at a relatively low rate causes little damage. **7:** Improved signal-to-noise ratio; moveout allows velocity analysis to be carried out. **8:** The layer below the interface has a lower acoustic impedance than the layer above. The polarity of the reflected wave is inverted. **9:** An anticline is broad-

ened; a syncline will be narrowed if the centre of curvature is above the surface, but appears as an anticline if the centre is below the surface. **10:** The reflection and transmission coefficients are, respectively, 0.1065 and 0.8935. The ray reflected from the top interface has amplitude reduced by one reflection only, but those from deeper interfaces also are reduced by transmission through all interfaces they pass through, both going down and returning to the surface. Amplitudes, assuming an initial value of 1.00, are 0.107, –0.085, 0.068, –0.055. **11:** (a) Reflections perpendicular to the slope when plotted appear as a shallower dip – see Fig. 7.9. (b) Synclines may appear narrower or as bow ties, while anticlines are broadened – see Fig. 7.10. (c) It may be an interference effect – see Fig. 7.22. (d) Sections are usually given in TWTs; if layers have different velocities their TWTs will not be in true proportion. **12:** Shorter-wavelength (higher-frequency) pulses or waves give higher resolution. Shorter wavelengths penetrate less far because of greater attenuation, so a compromise may be needed. **13:** May be formed by interference from several closely spaced interfaces. **14:** The different acoustic impedances of gas and water give a large reflection coefficient, and the surface of a liquid is horizontal. However, travel times to the interface may vary because of variation in the layers above (pull-up or pull-down). **15:** Same V_{rms}. **16:** A seismic reflector may be due to interference from more than one interface, while the acoustic impedances on either side of a lithological boundary may differ too little to give a noticeable reflection. For examples, see Section 7.8.1. **17:** Higher resolution is often needed, requiring high-frequency sources and receivers; ground roll is proportionately larger. **18:** About 14 m using an average frequency of about 50 Hz. **19:** (a) There is a velocity inversion (low-velocity layer); the layer below the interface is very thin (hidden layer); the interface separates layers with the same velocities but different acoustic impedances. (b) The interface is gradational and exceeds $\lambda/2$.

Chapter 8

1: iv. **2:** It is undetectable regardless of its depth as it produces no anomaly, only a uniform increase in g. **3:** See Section 8.4. **4:** (a) 0. (b) Increase of 0.812 mGal. (c) Decrease of 0.812 mGal. (a) 0 m. (b) Lowering of

2.63 m. (c) Rise of 2.63 m. **5:** Weight is due to the pull of the Earth. (a) Measures weight. (b) Measures mass. **6:** iv. **7:** iii. **8:** v. **9:** ii, increase. **10:** Correcting for the depression would decrease the Bouguer anomaly values and so might eliminate the dip in the profile. **11:** iii. **12:** v. **13:** (a) Approximating the chamber by a hollow sphere of 2 m radius with its centre 5 m down gives a negative anomaly of about 18 μGal. (b) As the anomaly is very small, spacing should be less than the half-width of the anomaly, which roughly equals the depth of the 'body', so choose (c), 3 m. **14:** Refer to Section 8.10 and the examples in it; see particularly Section 27.4. **15:** 5.5; about twice. **16:** A spring balance measures weight, which depends upon g, and this decreases with altitude; but a pair of scales measures (compares) masses and is independent of g. **17:** v. **18:** A fifth.

Chapter 9

1: S-waves can transmit only through a solid, whereas rebound requires a yielding material. This is possible because on a bulk scale the asthenosphere is rigid to brief stresses but yields to long-sustained ones. In turn, this is because solid-state creep permits steady deformation while maintaining a largely intact crystal structure. **2:** It would have no isostatic gravity anomaly. **3:** i. **4:** If peaks are eroded less rapidly than uplift due to rebound following removal of lower parts of mountains. **5:** iii. **6:** (a) Higher, because of higher sea level. (b) Lower, because of isostatic rebound. **7:** i, iii. **8:** (a) 25 cm. (b) 22 cm. (c) 22 cm. **9:** Rises 0.45 km. **10:** 0.625 km. **11:** Rises by 0.105 m. **12:** 5.5 km, greater. **13:** Be less. **14:** It is a gradational change, not a seismic discontinuity. **15:** Nearly 1 km. **16:** ii, iii. **17:** i.

Chapter 10

1: Rotation of the rock, secular variation, anisotropy. **2:** See Section 10.1.2. **3:** See Section 10.4.2. **4:** iii. **5:** Demagnetisation first removed an approximately reversed component. **6:** Earth's interior is too hot; geomagnetic field varies with time. **7:** ii. **8:** Curie temperature. **9:** It also depends on the magnetic mineral the iron is in, grain size of the mineral, strength of the field when it was magnetised, and magnetising mechanism; additionally, the remanence of one rock could have been changed by partial remagnetisation. **10:** Relative rotation of the rocks about a vertical axis. **11:**

Compasses are balanced for a limited range of latitudes. **12:** The field is an axial and geocentric dipole. This is closely true provided directions are averaged over periods of at least 10,000 years (its application also assumes that the remanence has remained unchanged since formation). **13:** 72°. **14:** The plot shows an approximately horizontal line, then one with an essentially constant slope; the lines intersect at about 180 Ma ago. Interpreted as continental movement, the continent remained at the same latitude (about 35° S) until 180 Ma ago; then it moved steadily northwards until the present. **15:** This shows two lines with somewhat different slopes that join at about 40 Ma ago; the direction for 140 Ma ago is anomalous but lies on the line if its direction is reversed. Interpretation as continental drift: The anomalous direction for 140 Ma is because the rock was magnetised during a time of reversed magnetisation. The continents moved northwards from about 53° S but slowed somewhat about 40 Ma ago. There was rotation after about 40 Ma ago, of 10° clockwise. It might have separated from a larger area about 40 Ma ago. **16:** (a) Dyke magnetisation primary (baked contact test); no check on sandstone's. (b) Neither, because dyke and sandstone have probably been remagnetised. **17:** The dyke was intruded during a time of reversed magnetisation; after allowing for this, the dyke and lava have essentially the same direction. Since the Jurassic, both bodies have probably moved from a latitude of about 11° N and rotated counterclockwise by about 20°. **18:** (a) The lavas were probably erupted during three successive polarity intervals, N, R, N. If so, their duration was at least as long as the R interval but no more than the duration of N–R–N. From Fig. 10.21, in the Palaeocene, R intervals lasted as little as c. 0.25 Ma, while a N–R–N succession could have taken no more than c. 3 Ma; i.e., the lavas took 0.25 to 3 Ma to form. Eruption could have been at almost any time during the Palaeocene. (b) In the Campanian there is only one R interval, so the same argument gives a minimum duration of c. 4 Ma and a maximum of the whole of the Campanian (c. 9 Ma). The R rocks lasted c. 83 to 79 Ma ago so the eruption must have included this interval.

Chapter 11

1: See Figure 11.9. **2:** Take readings along S–N line, extending at least 100 m either side of peak and/or trough, with stations as little as 1 m apart where readings vary rapidly, decreasing to 5–10 m elsewhere. **3:** Rapid coverage of a large area, and cheaper than ground surveying if area is large and/or ground access difficult, less affected by ferrous objects; anomalies are smaller and less 'sharp'. **4:** Symmetrical, negative. **5:** It standardises the anomaly of a body, regardless of its latitude and strike; this makes it easier to visualise and model. **6:** (a) Important because anomaly would probably not be large and the Earth's field would vary considerably during the time needed to carry out the large survey. (b) Not important because the anomaly probably will be large and measured quickly. (c) Probably important because the anomaly is likely to be very small and also, if a gridded survey is used, the survey time could allow significant variation. **7:** Proton sensor does not need precise orienting and does not drift, but it can give only the total field without direction; readings take a few seconds and the instrument is heavier. **8:** iv. **9:** vi. (The field of a dipole decreases as the inverse cube, as reported in Section 11.6.3.) **10:** The basement rocks are likely to produce anomalies that are shorter in wavelength and larger in amplitude than those over the basin. **12:** (a) No. (b) Yes. (c) Yes. (d) No. **13:** This would produce a large variation in the magnitude of the anomaly; if the remanent direction were roughly antiparallel to the induced component, the values of the ratio over about 1 would result in the total field anomaly changing sign. **14:** The strength but not the direction of the field has been measured.

Chapter 12

1: VES when subsurface likely to have near-horizontal layers; profiling when there is lateral variation; imaging when there is both lateral and vertical variation. **2:** Layers close to horizontal (dips no more than 10°), each with a uniform resistivity, and therefore with abrupt interfaces. **3:** (a) 3, (b) 4, (c) 3, (d) 3, (e) 5. Only obvious layers can be recognised without a modelling program. **4:** See Section 12.3. **5:** A profile along a traverse oriented N–S along the east side of the area; 'a' spacing c. 5 m – close enough not to miss the vein and wide enough to penetrate the overburden. Drawbacks: takes 4–5 persons, anomaly complex; use Schlumberger array. **6:** (a) An electrical uniform subsurface with resistivity of 100 ohm-m. (b) The subsurface is uniform with this resistivity down to at least 30 m. **7:** Greater: dissem-

inated pyrites, clean gravel, dry sand, dry sand, sphalerite ore, granite; less: clay, massive galena ore, coastal sand, shale. **9:** The depth scale is arbitrary, and the value plotted at a given point reflect resistivities of the surrounding area as well (even if the structure is truly 2D). **10:** (i) As Fig. 12.9a. (ii) As Fig. 12.9b. **11:** To avoid the effect of contact resistance of the electrodes. **12:** See Sections 12.3.3, 12.3.6, and 12.4.2.

Chapter 13

1: The disseminated particles can each store and release charge, but as the conducting particles are separated the bulk conductivity of the rock remains low. Massive sulphides are often surrounded by disseminated ore. **2:** ii. **3:** i. **4:** The current is alternating, so any steady self-potential will average to zero. **5:** See Section 13.2.2. **6:** The same array can be used and voltage generators and current measuring instruments are available, so little adaptation and extra time are needed; the resistivity results can be important when interpreting the IP results. **7:** Membrane and electrode polarization – see Section 13.1.1. Electrode polarization. **8:** (a) i. (b) i. (c) iii. **9:** (a) i. (b) iii. (c) ii. **10:** In resistivity surveying the current is reversed several times a second, and the potential meter is reversed at the same time, so self-potentials, being constant, are averaged out. **11:** iii (but not all sulphide ores give an SP). **12:** IP and resistivity both need arrays with four electrodes, with two used to inject current – which is periodically reversed – into the ground, but IP needs a more powerful current supply and the ability to measure the decay of current after it is switched off. SP uses two non-polarising electrodes and does not inject current. **13:** The value of FE but not MF would increase.

Chapter 14

1: Assuming coil axes are vertical, the anomaly is zero when either the transmitter or the receiver is over the sheet, with its largest value – a trough – when the coils are symmetrical about the sheet. **2:** E-m but not resistivity operates when the surface layer is very resistive. Conversely, it is has less penetration if the surface layer is very conductive; resistivity gives more precise location of bodies; e-m is usually easier to use as electrodes do not have to be moved.

3: See Section 14.5. 4: 180°, 180°. 5: vi. 6: Subhorizontal interfaces between layers with contrasting velocities/relative permittivities, not more than a few metres deep, depending on the conductivity of the layer(s), which preferably should be low. Water table, sand/clay interfaces. 7: Measures resistivities down to tens or hundreds of kilometres; uses low-frequency natural variations of the magnetic field. 8: (a) VLF or EM16. (b) Expanding Slingram system. (c) Profiling using Slingram system with separation of, say, 10 m. VLF with transmitter roughly aligned along strike of fault. (d) TEM sounding. (e) Profiling using Slingram system with separation of, say, 4 m. (f) Expanding Slingram system up to 40 m or more separation; TEM. 9: Slingram, TEM (Turam and VLF cannot be used in the form described in the book but aerial modifications exist). 10: Geometrical spreading; attenuation in conductive rocks. 11: Equally good, but should keep to one direction, because sign of signal would invert. 12: As readings are taken after the transmitter is switched off, there is no need to cancel accurately the transmitter signal at the receiver, and sensitivity can be higher; these are particularly valuable advantages for aerial surveys. 13: Need high frequency for high resolution, but higher frequencies are more rapidly attenuated by conductive overburden, so may need to decrease frequency to get required penetration.

Chapter 15

1: vi. 2: iii. 3: iv (actually, 0.000976 . . .). 4: iv. 5: iv. 6: iv. 7: iii. 8: iv (nearly 55 half-lives have elapsed, so fraction is 2^{-55}). 9: iv. 10: iii. 11: iv. 12: ii. 13: viii. 14: iii. 15: (a) i, ii, iii. (b) i, ii. 16: iii. 17: Both need a suite of samples and, having long half-lives, are unsuitable for young rocks, but whereas Rb–Sr is applied to acid rocks, Sm–Nd is applied to basic and ultrabasic rocks. 18: vi. 19: (a) C-14: constant rate of production of C-14 (as it isn't a correction is made, if possible). (b) No detrital Th. 20: 2 half-lives or about 11,460 years. 21: (a) It gives no indication what fraction of the gas is released in each step. (b) ^{40}Ar may have leaked out, giving an under-indication of fraction of gas released. 22: iii. 23: 35%; e.g., whether a lava is pre-Pleistocene. 24: ii, iii. 25: i, iii, vi. Primeval isotopes must have a half-life comparable with the age of the Earth and so can be used to date old rocks, but often not young ones.

Chapter 16

1: iv. 2: (a) To avoid the need for complicated geometrical factors. (b) Weathering can remove radioactive elements and change their ratios. 3: (a) ×2, (b) ×1.5, (c) c. 1, (d) ×0.25. 4: It has to be long enough that the counting error (equal to the square root of the count) is sufficiently low for the purpose of the survey. 5: See Section 16.2.3. 6: Hydrothermal alteration, surface weathering. 7: The sandstone is rich in potassium in the mica, giving it its high count; however, there is far more thorium, on detrital minerals, in shales than in sandstones. 8: Airborne surveys cover much larger areas, and quickly, but only if activity is fairly high; resolution is poorer. 9: As it is a gas and also inert, it is free to leak through the ground or be carried by water to where it can be breathed in.

Chapter 17

1: In conduction, heat travels *through* the material, but in convection it travels *with* moving material. In thermal convection the material (a fluid) moves because temperature differences in it cause some parts to be less dense (generally the hotter parts) and these rise, and conversely for colder parts; in forced convection the material is made to move by an external force; example of conduction: the increase of temperature with depth in nonvolcanic areas; example of convection: hot springs, heat transport in the mantle. 2: The value is so small that if one tried to measure it directly by the rate at which it heats up an object in contact with the ground the inevitable heat losses would invalidate the measurements. The formula used to deduce the heat flow assumes only conduction. 3: ii and iii, which are the same. 4: See Fig. 17.4. They differ because the amount of heat generated within oceanic lithosphere is very small but is significant in continental lithosphere. 5: See Figs. 17.9 and 17.10. 6: Only a small proportion is concentrated into small and accessible areas from which it can be extracted. 7: Extraction removes heat faster than it is generated within the rocks or can be conducted up from below, so the rocks cool. 8: v (see e.g., Figure 17.6; a heat flux value of 40 mW/m^2 gives 50,000 km^2). 9: Unlike the temperature outside, their temperature remains nearly constant because of the insulation of the surrounding rock. 10: See Section 17.5. 11: Because of the large masses of rock involved and their low thermal conductivity. **12:** ii.

Chapter 18

1: Boreholes only sample locally; core samples may be altered by the drilling, e.g., loss of original fluids; some formations may not be sampled at all due to wash out. **2:** In a cased hole or a dry hole, or a hole with oil-based mud. **3:** When the hole does not contain a conducting fluid; it gives deeper penetration. **4:** Porosity, water saturation, water resistivity. **5:** (a) Locating shale–sand contacts, and measuring shale contamination of sand, needed to evaluate the hydrocarbon saturation. (b) To locate most massive sulphide ores (not sphalerite) and graphite. **6:** It responds to natural radioactivity, due to K, U, and Th in total. It helps identify lithology. The spectral γ ray log measures the relative amounts of these elements and so is often able to distinguish shale from other radioactive rocks, or to recognise shales with low radioactivity. **7:** Use a sonic log/neutron log cross plot: expect values to fall between limestone and dolomite lines; deduce composition in proportion to position between the two lines along a line of equal porosity. **8:** There is very little signal in dry sand, a strong one in wet sand, a weak one in gas-filled sand. Neutrons slow mainly by collision with H atoms in water or hydrocarbons; once slowed they are absorbed by other elements and lead to emission of a γ ray which may reach the detector. Dry sand has very little hydrogen and so neutrons will penetrate too far for any γ rays to reach the detector. Gas, having a low density has only a few H atoms. **9:** It is adjacent to the formation whose velocity is being measured, and it measures the velocity directly (using seismic refraction methods) rather than indirectly, through moveout; because it does not need to penetrate far, much higher frequencies can be used, giving higher resolution. **10:** (a) 6.9%. (b) There is fluid in fractures as well as between grains in the sandstone. **11:** SP (low value between higher ones), natural γ ray (high reading), calliper (wide diameter). **12:** 12.5%. **13:** $S_w = (1 - S_{hc})$. **14:** See Fig. 18.2 and Sections 18.5.2 and 18.5.3.

Chapter 19

1: i, iv. **2:** ii only. **4:** The method used needs to be quick and cheap as well as able to detect the sink-

holes. A Slingram e-m system with coils separated by several metres would penetrate sufficiently deeply to detect the higher conductivity of the clay; a magnetometer might detect the induced magnetisation of the clay but would be affected by any ferrous objects along the route. **5:** Saline water could be detected by its high conductivity/low resistivity: Successive surveys over a grid using a Slingram system with different spacings could detect whether saline water is present and whether at a shallow or deeper level, but more detailed vertical resolution would best be done using an expanding resistivity array. To detect a fault with a vertical component of offset (which could allow saline water into the aquifer) seismic refraction would be cheap and effective; shallow seismic reflection is another possibility but more expensive. **6:** The igneous dykes would probably give an obvious magnetic anomaly, so a magnetometer survey would be used. The saline water could be detected by conductivity methods, perhaps e-m Slingram methods to detect if high conductivity (i.e., saline water) is present, followed by expanding resistivity surveys to determine its depth. **7:** g (this would probably detect the infilled trench and perhaps even the pipe); f or h might detect a difference between the granite and infilled trench and if so would be cheaper than g. **8:** f (or e if a gradiometer is not available). **9:** Bodies producing gravity and magnetic anomalies can vary in shape, depth, size, density, and magnetic susceptibility. A unique interpretation cannot be made using the geophysical data alone as there are an infinite number of combinations of these variables which can give the same anomaly. However, interpretation becomes possible if some of the variables can be constrained by geological considerations, measurements of physical property on samples, and borehole and other geophysical data. **10:** Magnetics, e-m, γ ray, and sometimes gravity. They provide quick coverage of an area, and will be cheaper than ground surveying if the area is large or access if difficult, but resolution is poorer.

Chapter 20

1: See Sections 20.1 and 20.4.3. **2:** iv. **3:** iii (see Fig. 20.25). **4:** ii. **5:** iv. **6:** iii, iv, vii (see Fig. 20.25). **7:** iii. **8:** i, ii, v. **9:** i. **10:** ii. **11:** For forces, see Section 20.9. Drive forces: slab pull at subduction zones and ridge push at spreading ridges; for evidence, see Section

20.9. Retarding forces: Where plates slide past one another at transform faults; where they converge at subduction zones; resistance by the mantle to subduction, especially if a plate reaches the 660-km discontinuity, and basal drag beneath plates. **12:** ii. **13:** See Section 20.5.1. **14:** i. **15:** The north pole. **16:** v. **17:** See Section 20.8. **18:** See Section 20.2.1. **19:** ii. **20:** iv. **21:** See Section 20.3. **22:** Ridges are higher because of expansion and buoyancy of the material below, due to its higher temperature. **23:** A moves 8.5 cm/year northeastwards with respect to C and so must be subducting obliquely. **24:** 5.2 cm northwards, so it is a dextral strike-slip conservative boundary. **25:** See Section 20.9.1. **26:** See Section 20.9.4.

Bibliography

Abrams, M., Buongiorno, F., Realmuto, V., and Pieri, D. 1994. Mt. Etna lava flows analyzed with remote sensing. *Eos,* **75,** 545, 551.

Adams, R. L., and Schmidt, B. L. 1980. Geology of the Elura Zn–Pb–Ag deposit. *Bull. Aust. Soc. Explor. Geophys.,* **11,** 143–6.

Al-Rifaiy, I. A. 1990. Land subsidence in the Al-Dahr residential district in Kuwait: A case history study. *Q. J. Engl. Geol.,* **23,** 337–46.

Alvarez, L. W., Alvarez, W., Azaro, R., and Michel, H. V. 1980. Extraterrestrial cause for the Cretaceous–Tertiary extinction. *Science,* **208,** 1095–1108.

Alvarez, W., and Asaro, F. 1990. An extraterrestrial impact. *Sci. Am.,* **263,** no. 4, 44–52.

Arab, N. 1972. Seismic reflection and gravity investigations of the 'Widmerpool Gulf' in the East Midlands, with a study of linear seismic sources and data processing techniques involving computer graphics. Ph.D. thesis, Univ. of Leicester, UK.

Armstrong, R. L., Monger, J. W. H., and Irving, E. 1985. Age of magnetization of the Axelgold Gabbro, north-central British Columbia. *Can. J. Earth Sci.,* **22,** 1217–22.

Aveling, E. 1997. Magnetic trace of a giant henge. *Nature,* **390,** 232–3.

Baker, B. H., and Wohlenberg, J. 1971. Structure and evolution of the Kenya Rift Valley. *Nature,* **229,** 538–42.

Balling, N. 1995. Heat flow and thermal structure of the lithosphere across the Baltic Shield and northern Tornquist Zone. *Tectonophysics,* **244,** 13–50.

Baltosser, R. W., and Lawrence, H. W. 1970. Application of well-logging techniques in metallic mineral mining. *Geophysics,* **35,** 143–52.

Barker, R. D. 1990. Improving the quality of resistivity sounding data in landfill studies. *In:* S. H. Ward, ed., *Geotechnical and environmental geophysics, Volume 2: Environmental and groundwater.* Society of Exploration Geophysics, 245–52.

Barker, R. D. 1995. Apparatus and methods for simple resistivity experiments. *Teaching Earth Sciences,* **20,** 88–96.

Batchelor, A. S. 1987. Hot dry rock exploitation. *In:* M. Economides and P. Ungemach, eds., *Applied geothermics,* Wiley, Chichester, UK., 221–34.

Bauman, P. D., Heitzmann, R. J., Porter, J. E., Sallomy, J. T., Brewster, M. L., and Thompson, C. J. 1994. The integration of radar, magnetic and terrain conductivity data in an archaeogeophysical investigation of a Hudson's Bay Company fur trade post. *GPR 94, Proceedings of the fifth international conference on ground penetrating radar,* vol. 2, June, Kitchener, ON, Canada, 531–46.

Beer, K. E., and Fenning, P. J. 1976. *Geophysical anomalies and mineralisation at Sourton Tors, Okehampton, Devon.* Report of the Institute of Geological Sciences, No. 76/1, HMSO.

Begét, J. E., and Nye, C. J. 1994. Postglacial eruption history of Redoubt Volcano, Alaska. *J. Volcan. Geotherm. Res.,* **62,** 31–54

Bellerby, T. J., Noel, M., and Branigan, K. 1990. A thermal method for archaeological prospection: Preliminary investigations. *Archaeometry,* **32,** 191–203.

Beltrami, H., and Chapman, D. 1994. Drilling for a past climate. *New Sci.,* **142,** 36–40.

Benton, M. J., and Little, C. T. S. 1994. Impact in the Caribbean and death of the dinosaurs. *Geol. Today,* **10,** Nov–Dec, 222–7.

Bertin, J., and Loeb, J. 1976. *Experimental and theoretical aspects of induced polarisation,* vol 1. Geoexploration Monograph Series 1, no. 7, Gebruder Borntrager, Berlin.

Birt, C. S. 1996. Geophysical investigation of active continental rifting in southern Kenya. Ph.D. thesis, Univ. of Leicester, UK.

Birt, C. S., Maguire, P. K. H., Khan, M. A., Thybo, H., Keller, G. R., and Patel, J. 1997. The influence of pre-existing structures on the evolution of the southern Kenya Rift Valley – Evidence from seismic and gravity studies. *In:* K. Fuchs, R. Altherr, B. Måller, and C. Prodehl, eds., Structure and dynamic processes in the lithosphere of the Afro-Arabian rift system. *Tectonophysics,* **278,** 211–42.

Bishop, I., Styles, P., Emsley, J., and Ferguson, N. S. 1997. The detection of cavities using the microgravity technique: Case histories from mining and karstic environments. *In:* D. M. McCann, P. J. Fenning, and G. M. Reeves, eds., *Modern geophysics in engineering geology.* Geological Society Engineering Geology Special Publication No. 12, 155–68.

Blackburn, G. 1980. Gravity and magnetic surveys – Elura Orebody. *Bull. Aust. Soc. Explor. Geophys.,* **11,** 159–66.

Bolt, B. A. 1982. *Inside the Earth.* Freeman, San Francisco.

Bolt, B. A. 1993a. *Earthquakes and geological discovery.* Scientific American Library, Freeman, New York.

Bolt, B. A. 1993b. *Earthquakes: A primer,* 3rd ed., Freeman, New York.

Bolt, B. A. 1999. *Earthquakes,* 4th ed. Freeman, New York.

BP statistical review of world energy. 1997. Group Media and Publications, BP Co., plc.

Brabham, P. J., and McDonald, R. J. 1992. Imaging a buried river channel in an intertidal area of South Wales using high-resolution seismic techniques. *Q. J. Engl. Geol.,* **25,** 227–38.

Brown, G. C., and Mussett, A. E. 1993. *The inaccessible Earth,* 2nd ed. Chapman and Hall, London.

Brown, G., Hawksworth, C., and Wilson, C. 1992. *Understanding the Earth,* 2nd ed. Cambridge Univ. Press, Cambridge, UK.

Butler, R. F. 1992. *Paleomagnetism: Magnetic domains to geologic terranes.* Blackwell, Oxford, UK.

Cardaci, C., Coviello, M., Lombardo, G., Patanè, G., and Scarpa, R. 1993. Seismic tomography of Etna volcano. *J. Volcan. Geotherm. Res.,* **56,** 357–68.

Cassidy, J. 1979. Gamma-ray spectrometric surveys of Caledonian granites: Method and interpretation. Ph.D. thesis, Univ. of Liverpool, UK.

Castellano, M., Ferrucci, F., Godano, C., Imposa, S., and Milano, G. 1993. Upwards migration of seismic focii: A forerunner of the 1989 eruption of Mt Etna (Italy). *Bull. Volc.,* **55,** 357–61.

Chapellier, D. 1992. *Well logging in hydrogeology.* A. A. Balkema. Rotterdam, The Netherlands.

Chiu, J.-M., Isacks, B. L., and Cardwell, R. K. 1991. 3-D configuration of subducted lithosphere in the western Pacific. *Geophys. J. Int.,* **106,** 99–111.

Chouet, B. A., Page, R. A., Stephens, C. D., Lahr, J. C., and Power, J. A. 1994. Precursory swarms of long-period events at Redoubt Volcano (1989–1990),

Alaska: Their origin and use as a forecasting tool. *J. Volcan. Geotherm. Res.,* **62,** 95–135.

Clark, A. J. 1990. *Seeing beneath the soil: prospecting methods in archaeology.* Batsford, London.

Cliff, R. A. 1985. Isotopic dating in metamorphic belts. *J. Geol. Soc. Lond.,* **142,** 97–110.

Collier, J., and Sinha, M. 1990. Seismic images of a magma chamber beneath the Lau basin back-arc spreading centre. *Nature,* **346,** 646–8.

Condie, K. C. 1989. *Plate tectonics and crustal evolution,* 3rd ed. Pergamon, Oxford, UK.

Coontz, R. 1998. Like a bolt from the blue. *New Sci.,* **160,** 36–40.

Corney, M., Gaffney, C. F., and Gater, J. A. 1994. Geophysical investigations at the Charlton Villa, Wiltshire (England). *Arch. Prospection,* **1,** 121–8.

Cornwell, J. C., Kimbell, G. S., and Ogilvy, R. D. 1996. Geophysical evidence for basement structure in Suffolk, East Anglia. *J. Geol. Soc. Lond.,* **153,** 207–11.

Courtillot, V. E. 1990. A volcanic eruption. *Sci. Am.,* **263,** no. 4, 53–60.

Cox, A. 1973. *Plate tectonics and geomagnetic reversals.* Freeman, San Francisco.

Cox, A., and Hart, R. B. 1986. *Plate tectonics – How it works.* Blackwell, Palo Alto, CA.

Craig, L. E., Smith, A. G., and Armstrong, R. L. 1989. Calibration of the geologic time scale: Cenozoic and Late Cretaceous glauconite and nonglauconite dates compared. *Geology,* **17,** 830–2.

Dalrymple, G. B. 1991. *The age of the Earth.* Stanford Univ. Press, Palo Alto, CA.

Daly, M. C., Bell, M. S., and Smith, P. J. 1996. The remaining resource of the UK North Sea and its future development. *In:* K. Glennie and A. Hurst, eds., *NW Europe's hydrocarbon industry,* 187–93. Geological Society of London Special Publication.

Davis, L. W. 1980. The Discovery of Elura and a brief summary of subsequent geophysical tests at the deposit. *Bull. Aust. Soc. Explor. Geophys.,* **11,** 147–51.

Decker, R., and Decker, B. 1991. *Mountains of fire: The nature of volcanoes.* Cambridge Univ. Press, Cambridge, UK.

De Mets, C., Gordon, R. G., Argus, D. F., and Stein, S. 1990. Current plate motions. *Geophys. J. Int.,* **101,** 425–78.

Dickin, A. P. 1995. *Radiogenic isotope geology.* Cambridge Univ. Press., Cambridge, UK.

Dickinson, W. R. 1977. Tectono-stratigraphic evolution of subduction-controlled sedimentary assemblages. *In:* M. Talwani and W. C. Pitman, III, eds., *Island arcs, deep sea trenches and back-arc basins.* American Geophysical Union, Washington, DC, 33–9.

Dixon, C. J. 1979. *Atlas of economic deposits*. Chapman and Hall, London.

Doyle, H. 1995. *Seismology*. Wiley, Chichester, UK.

Dressler, B. O., and Sharpton, V. L., eds., 2000. *Large meteorite impacts and planetary evolution II*. Geological Society of America, Special Paper 339.

Duffield, W. A. 1972. A naturally occurring model of global plate tectonics. *J. Geophys. Res.*, 77, 2543–55.

Durrance, E. R. 1986. *Radioactivity in geology: Principles and applications*. Ellis Horwood, Chichester, and Halstead Press, New York.

Dvorak, J. J., Johnson, C., and Tilling, R. I. 1992. Dynamics of Kilauea Volcano. *Sci. Am.*, 267, no. 2, 18–25.

Eaton, J. P., O'Neill, M. E., and Murdoch, J. N. 1970. Aftershocks of the 1966 Parkfield–Cholame, California, earthquake. *Bull. Seis. Soc. Am.*, 60, 1151–97.

Economides, M., and Ungemach, P. 1987. *Applied geothermics*. Wiley, Chichester, UK.

Edwards, R., and Atkinson, K. 1986. Ore deposit geology and its influences on mineral exploration. Chapman and Hall, London.

Emerson, D. W. 1980. *Proceedings of the Elura symposium, Sydney, 1980*. The Elura Compendium. Australian Society Exploration of Geophysicists.

Emery, D., and Myers, K., eds. 1996. *Sequence stratigraphy*. Blackwell, Oxford, UK.

The energy report: Oil and gas resources of the United Kingdom, vol. 2. 1997. The Stationery Office, Norwich, UK.

Ernst, R. E., and Baragar, W. R. A. 1992. Evidence from magnetic fabric for the flow pattern of magma in the Mackenzie giant radiating dyke swarm. *Nature*, 356, 511–3.

Evans, A. M. 1993. *Ore geology and industrial minerals – An introduction*, 3rd ed. Blackwell Scientific, Oxford, UK.

Evans, A. M. 1995. An introduction to mineral exploration. Blackwell Scientific, Oxford, UK.

Farquharson, C. G., and Thompson, R. 1992. The Blairgowrie magnetic anomaly and its interpretation using simplex optimisation. *Trans. R. Soc. Edin. Earth Sci.*, 83, 509–18.

Faure, G. 1986. *Principles of isotope geology*, 2nd ed. Wiley, New York.

Ferguson, N.S. 1992. The detection and delineation of subterranean cavities by the microgravity method. M. Phil. thesis, Liverpool Univ, UK.

Ferrucci, F., and Patanè, D. 1993. Seismic activity accompanying the outbreak of the 1991–1993 eruption of Mt. Etna (Italy). *J. Volcan. Geotherm. Res.*, 57, 125–35.

Fitch, T. J., and Scholz, C. H. 1971. Mechanism of underthrusting in southwest Japan: A model of convergent plate interactions. *J. Geophys. Res.*, 76, 7260–92.

Floyd, J. D., and Kimbell, G. S. 1995. Magnetic and tectonostratigraphic correlation at a terrane boundary: The Tappins Group of the Southern Uplands. *Geol. Mag.*, 132, 515–21.

Forsyth, D., and Uyeda, S. 1975. On the relative importance of the driving forces of plate motion. *Geophys. J. R. Astr. Soc.*, 43, 163–200.

Fowler, C. M. R. 1990. *The solid Earth: An introduction to global geophysics*. Cambridge Univ. Press., Cambridge, UK.

Francis, P. 1993. *Volcanoes: A planetary perspective*. Clarendon Press, Oxford, UK.

Fuchs, K., Altherr, R., Måller, B., and Prodehl, C., eds. 1997. *Structure and dynamic processes in the lithosphere of the Afro-Arabian rift system. Tectonophysics*, 278.

Gaffney, C., and Gater, J. 1993. Development of remote sensing, Part 2: Practice and method in the application of geophysical techniques in archaeology. *In:* J. Hunter and I. Ralston, eds., *Archaeological resource management in the UK: An introduction*. Institute of Field Archaeologists. 205–14.

Gates, A. E., and Gunderson, L. C. S. 1992. *Geologic controls on radon*. Geological Society of America Special Paper no. 271.

Geyh, M. A., and Schleicher, H. 1990. *Absolute age determination*. Springer-Verlag, Berlin.

Giardini, D., and Woodhouse, J. H. 1984. Deep seismicity and modes of deformation in Tonga subduction zone. *Nature*, 307, 505–9.

Gidley P. R., and Stuart D. C. 1980. Magnetic property studies and magnetic surveys of the Elura prospect, Cobar, NSW. *Bull. Aust. Soc. Explor. Geophys.*, 11, 167–72.

Gillot, P-Y, and Keller J. 1993. Radiochronological dating of Stromboli. *Acta Vulcan.*, 3, 69–77.

Glazner, A. F., and Bartley, J. M. 1985. Evolution of lithospheric strength after thrusting. *Geology*, 13, 42–5.

Goldstein, S. J., Perfit, M. R., Batiza, R., Fornari, D. J., and Murrel, M. T. 1994. Off-axis volcanism at the East Pacific Rise detected by uranium-series dating of basalts. *Nature*, 367, 157–9.

Goguel, J. 1976. *Geothermics*. McGraw-Hill, New York.

Gough, D. I., and Majorowicz, J. A. 1992. Magnetotelluric soundings, structure, and fluids in the southern Canadian Cordillera. *Can. J. Earth Sci.*, 29, 609–20.

Gradstein, F. M., and Ogg, J. 1996. A Phanerozoic time scale. *Episodes*, 19, 3–5.

Grasmueck, M. 1996. 3D ground-penetrating radar applied to fracture imaging in gneiss. *Geophysics*, 61, 1050–64.

Grauch, V. J. S. 1987. A new variable-magnetization terrain correction method for aeromagnetic data. *Geophysics*, 52, 94–107.

Grieve, R. A. F. 1990. Impact cratering on the Earth. *Sci. Am., 262,* no. 4, 44–51.

Griffiths, D. H., and King, R. F. 1981. *Applied geophysics for geologists and engineers,* 2nd ed. Pergamon, Oxford, UK.

Gripp, A. E., and Gordon, R. G. 1990. Current plate velocities relative to the hotspots incorporating the NUVEL-1 global plate motion model. *Geophys. Res. Lett., 17,* 1109–12.

Gubbins, D. 1990. *Seismology and plate tectonics.* Cambridge Univ. Press, Cambridge, UK.

Hailwood, E. A. 1989. *Magnetostratigraphy.* Blackwell Scientific, Oxford, UK.

Haren, R. J. 1981. Large loop Turam electromagnetic survey of the Woodlawn orebody. *In:* R. J. Whiteley, ed., *Geophysical case study of the Woodlawn orebody New South Wales, Australia.* Pergamon, Oxford, UK, 219–32.

Harland, W. B., Armstrong, R. L., Cox, A. V., Craig, L. E., Smith, A. G., and Smith, D. B. 1990. *A geologic timescale 1989.* Cambridge Univ. Press, Cambridge, UK.

Harris, C., Williams, G., Brabham, P., Eaton, G., and McCarroll, D. 1997. Glaciotectonized Quaternary sediments at Dinas Dinlle, Gwynedd, north Wales, and their bearing on the style of deglaciation in the eastern Irish Sea. *Q. Sci. Rev., 16,* 109–27.

Harrison, T. M., and McDougall, I. 1980. Investigations of an intrusive contact, northwest Nelson, New Zealand – I. Thermal, chronological and isotopic constraints. *Geochim. Cosmochim. Acta., 44,* 1985–2003.

Heezen B. C., and Tharpe, M. 1968. The Atlantic Ocean. *National Geographic.*

Heirtzler, J. R., Le Pichon, X., and Baron, J. G. 1966. Magnetic anomalies over the Reykjanes Ridge. *Deep Sea Res. 13,* 427–43.

Heirtzler, J. R., Dickson, G. O., Herron, E. M., Pitman, W. C. III, and Le Pichon, X. 1968. Marine magnetic anomalies, geomagnetic field reversals and motions of the ocean floor and continents. *J. Geophys. Res. 73,* 2119–36.

Hildebrand, A. R., Pilkington, M., Connors, M., Ortiz-Aleman, C., and Chavez, R. E. 1995. Size and structure of the Chicxulub crater revealed by horizontal gravity gradients and cenotes. *Nature, 376,* 415–7.

Hildebrand, A. R., Pilkington, M., Ortiz-Aleman, C., Chavez, R. E., Urrutia-Fucuguachi, J., Connors, M., Graniel-Castro, E., Camara-Zi, A., Halfpenny, J. F., and Niehaus, D. 1998. *In:* M. M. Grady, R. Hutchinson, and G. J. H. McCall, eds., *Meteorites: Flux with time and impact effects.* Geological Society of London, Special Publication 140, 155–76.

Hill, D. P., Eaton, J. P., and Jones, L. M. 1990. Seismicity. *In:* R. E. Wallace, ed., *The San Andreas fault sys-* *tem, California.* U.S. Geological Survey Professional Paper 1515, 115–51.

Hill, I. A. 1992. Better than drilling? Some shallow seismic reflection case histories. *Q. J. Eng. Geol., 25,* 239–48.

Hodgson, B. D., Dagley, P., and Mussett, A. E. 1990. Magnetostratigraphy of the Tertiary igneous rocks of Arran. *Scott. J. Geol., 26,* 99–118.

Holman, I. P. 1994. Controls on saline intrusion in the Crag Aquifer of North-east Norfolk. Unpub. Ph.D. thesis, University of East Anglia, UK.

Holman, I. P., and Hiscock, K. M. 1998. Land drainage and saline intrusion in the coastal marshes of north-east Norfolk. *Q. J. Eng. Geol., 31,* 47–62.

Holman, I. P., Hiscock, R. M., and Chroston, P. N. 1999. Crag aquifer characteristics and water balance for the Thurne catchment, northeast Norfolk. *Q. J. Eng. Geol., 32,* 365–80.

Hornabrook, J. T. 1975. Seismic interpretation of the West Sole gas field. *Norges Geol. Unders., 316,* 121–35.

Irving, E., and Tarling, D. H. 1961. The palaeomagnetism of the Aden volcanics. *J. Geophys. Res., 66,* 549–56.

Isacks, B., Oliver, J., and Sykes, L. R. 1969. Seismology and the new global tectonics. *J. Geophys. Res., 73,* 5855–99.

Ivanovich, M., and Harmon, R. S., eds. 1992. *Uranium-series disequilibrium,* 2nd ed. Clarendon Press, Oxford, UK.

Jenson, H. 1961. The airborne magnetometer. *Sci. Amer. 204,* no. 6, 151–62.

Kamo, S. L., Czamanske, G. K., and Krogh, T. E. 1996. A minimum U–Pb age for Siberian flood-basalt volcanism. *Geochim. Cosmochim. Acta., 60,* 3505–11.

Kamo, S. L., and Krogh, T. E. 1995. Chicxulub crater source for shocked zircon crystals from the Cretaceous–Tertiary boundary layer, Saskatchewan: Evidence from new U–Pb data. *Geology, 23,* 281–4.

Kanamori, H., Hauksson, E., and Heaton, T. 1997. Real-time seismology and earthquake hazard mitigation. *Nature, 390,* 461–4.

Kasahara, K. 1981. *Earthquake mechanics.* Cambridge Univ. Press, Cambridge, UK.

Kearey, P., and Brooks, M. 1991. *An introduction to geophysical exploration,* 2nd ed. Blackwell Scientific, Oxford, UK.

Kearey, P., and Vine, F. J. 1996. *Global tectonics,* 2nd ed. Blackwell Science Ltd., Oxford, UK.

Kelly, W. E., and Mareš, S., eds. 1993. *Applied geophysics in hydrogeological and engineering practice.* Elsevier, Amsterdam.

Kennett, B. L. N., and Engdahl, E. R. 1991. Traveltimes for global earthquake location and phase identification. *Geophys. J. Int., 105,* 429–65.

Kent, P. E., and Walmsley, P. J. 1970. North Sea Progress. *Amer. Assoc. Petrol. Geol. Bull., 54,* 168–81.

Kerr, R. A. 1995. Chesapeake Bay impact crater confirmed. *Science, 269,* 1672.

Khan, M. A., Mechie, J., Birt, C., Byrne, G, Gaciri. S., Jacob, B., Keller, G. R., Maguire, P. K. H., Novak, O., Nyambok, I. O., Patel, J. P., Prodehl, C., Riaroh. D., Simiyu, S., and Thybo, H. 1998. *The lithospheric structure of the Kenya Rift as revealed by wide-angle seismic measurements.* Geological Society of London Special Publication, **164,** 257–69.

Koeberl, C., and Anderson, R. R. 1996. *The Manson impact structure, Iowa: anatomy of an impact crater.* Geological Society of America, Special Paper 302.

Kulhánek, O. 1990. *Anatomy of seismograms.* Elsevier, Amsterdam.

Lachenbruch, A. H. 1970. Crustal temperature and heat production: Implications of the linear heat-flow relation. *J. Geophys. Res., 75,* 3291–300.

Laughton, A. S., Whitmarsh, R. B., and Jones, M. T. 1970. The evolution of the Gulf of Aden. *Phil. Trans. R. Soc.,* **A267,** 227–66.

Lay, T. 1995. Slab burial grounds. *Nature, 374,* 115.

Lee, M. K., Smith, I. F., Edwards, J. W. F., Royles, C. P., and Readman, P. W. 1995. *Tectonic Framework of the British/Irish continental shelf from new gravity and aeromagnetic compilations.* European Association for Exploration and Geophysics, 57th Conference, Glasgow. Extended abstracts vol. 1, D27.

Leute, U. 1987. *Archaeometry: An introduction to physical methods in archaeology and the history of art.* Weinheim, New York.

Lewis, T.J., and Wang, K. 1992. Influence of terrain on bedrock temperatures. *Global Planet. Change, 6,* 87–100.

Lin, J., Purdy, G. M., Schouten, H., Sempere, J.-C., and Zervas. C. 1990. Evidence from gravity data focused magmatic accretion along the Mid-Atlantic Ridge. *Nature, 344,* 627–32.

Lister, C. R. B., Sclater, J. G., Davis, E. E., Villinger, H., and Nagihara, S. 1990. Heat flow maintained in ocean basins of great age: Investigations in the north-equatorial West Pacific. *Geophys J. Int., 102,* 603–30.

Loke, M. H., and Barker, R. D. 1995. Improvements in the Zohdy method for the inversion of resistivity sounding and pseudosection data. *Comput. Geosci., 21,* 321–32.

Loke, M. H., and Barker, R. D. 1996. Rapid least squares inversion of apparent resistivity pseudosections by a quasi-Newton method. *Geophys. Prospecting, 44,* 131–52.

Long, L. E. 1964. Rb–Sr chronology of the Carn Chuinneag intrusion Ross-shire, Scotland. *J. Geophys. Res., 69,* 1589–97.

Lowry, T., and Shive, P. N. 1990. An evaluation of Bristow's method for the detection of subsurface cavities. *Geophysics, 55,* 514–20.

McCann, D. M., Eddlestone, M., Fenning, P. J., and Reeves, G. M., eds. 1997. Modern geophysics in engineering geology. Geological Society, Engineering Geology Special Publication No. 12.

McClelland Brown, E. 1981. Paleomagnetic estimates of temperatures reached in contact metamorphism. *Geology, 9,* 112–6.

Macdonald, K. C., and Fox, P. J. 1990. The mid-ocean ridge. *Sci. Am., 262,* no. 6, 42–9.

McCulloch, M. T., and Bennett, V. C. 1998. Early differentiation of the earth: An isotopic perspective. *In:* I. Jackson, ed., *The Earth's mantle,* Cambridge Univ. Press, Cambridge, UK.

McDougall, I., and Harrison, T. M. 1988. *Geochronology and thermochronology by the $^{40}Ar/^{39}Ar$ method.* Oxford Univ. Press, New York, and Clarendon Press, Oxford, UK.

McElhinny, M. W. 1973. *Palaeomagnetism and plate tectonics.* Cambridge Univ. Press, Cambridge, UK.

McEvilly, T. V., Bakun, W. H., and Casaday, K. B. 1967. The Parkfield, California, earthquakes of 1966. *Bull. Seis. Soc. Am., 57,* 1221–44.

McFadden, P. L. 1977. A palaeomagnetic determination of the emplacement temperature of some South African kimberlites. *Geophys. J. R. Astr. Soc., 50,* 587–604.

McGuire, W. J., Kilburn, C., and Murray, J., eds. 1995. *Monitoring active volcanoes: Strategies, procedures and techniques.* UCL Press, London.

McKenzie, D. 1969, Speculations on the consequences and causes of plate motions. *Geophys. J. R. Astr. Soc., 18,* 1–18.

McQuillin, R., Bacon, M., and Barclay, W. 1984. *An introduction to seismic interpretation,* 2nd ed. Graham & Trotman, London.

The MELT Seismic Team. 1998. Imaging the deep seismic structure beneath a mid-ocean ridge: The MELT experiment. *Science, 280,* 1215–35.

Merrill, R. T., McElhinny, M. W., and McFadden, P. L. 1996. *The magnetic field of the Earth,* 2nd ed. Academic Press, San Diego, Ca.

Miller, T. P., and Chouet, B. A., eds. 1994. The 1989–1994 eruptions of Redoubt Volcano, Alaska. *J. Volcan. Geotherm. Res., 62,* 1–10.

Milsom, J. 1989. *Field geophysics.* Open University Press, Milton Keynes; and Halsted Press (Wiley), New York.

Milsom, J. 1996. *Field geophysics,* 2nd ed. Wiley, New York.

Minissale, A. 1991. The Larderello geothermal field: A review. *Earth-Science Rev., 31,* 133–51.

Minster, J.-F., Ricard, L.-P., and Allègre, C.J. 1982. ^{87}Rb-^{87}Sr chronology of enstatite meteorites. *Earth Planet. Sci. Lett., 44,* 420–40.

Morgan, J. 1972. Deep mantle convection plumes and plate motions. *Bull. Am. Assoc. Petrol. Geol., 56,* 203–13.

Müller, R. D., Roest, W. R., Royer, J-Y., Gahagan, L. M., and Sclater, J. G. 1997. Digital isochrons of the world's ocean floors. *J. Geophys. Res.,* **102,** 3211–4.

Mussett, A. E., Dagley, P., and Skelhorn, R. R. 1989. Further evidence for a single polarity and a common source for the quartz–porphyry intrusions of the Arran area. *Scott. J. Geol.,* **25,** 353–9.

Mutter, J. C., Carbotte, S. M., Su, W., Xu, L., Buhl, P., Detrick, R. S., Kent, G. M., Orcutt, J. A., and Harding, A. J. 1995. Seismic images of active magma systems beneath the East Pacific Rise between 17°05′ and 17° 35′ S. *Science,* **268,** 391–5.

Mwenifumbo, C. J. 1989. Optimisation of logging parameters in continuous, time domain induced polarisation measurements. *Proceedings of the 3rd international symposium of borehole geophysics for minerals, geotechnical and groundwater applications,* Las Vegas, NV.

Mwenifumbo, C. J., Killeen, P. G., and Eliott, B. E. 1993. Classic examples from the Geological Survey of Canada data files illustrating the utility of borehole geophysics. *Proceedings of the 5th international symposium of the minerals and geotechnical logging society,* Tulsa, OK.

Nafe, J. E., and Drake, C. L. 1957. Variation with depth in shallow and deep water marine sediments of porosity, density and the velocities of compressional and shear waves. *Geophysics,* **22,** 523–52.

Nettleton, L. L. 1976. *Gravity and magnetics in oil prospecting.* McGraw-Hill, New York.

Odin, G. S. 1982. *Numerical dating in stratigraphy,* 2 vol. Wiley, Chichester, UK.

Olsen, K. H. 1995. *Continental rifts: Structure, evolution and tectonics.* Elsevier, Amsterdam.

Opdyke, N. D., and Channell, J. E. T. 1996. *Magnetic stratigraphy.* Academic Press, San Diego, CA.

Oversby, V. M. 1971. Lead (82). *In:* B. Mason, ed., *Elemental abundances in meteorites.* Gordon and Breach, London, 499–510.

Parasnis, D. S. 1973. *Mining geophysics.* Elsevier, Amsterdam.

Parasnis, D. S. 1986. *Principles of applied geophysics,* 4th ed. Chapman and Hall, London.

Parasnis, D. S. 1997. *Principles of applied geophysics,* 5th ed. Chapman and Hall, London.

Parent, M. D., and O'Brien, T. F. 1993. Linear-swept FM (Chirp) sonar seafloor imaging system. *Sea Tech.,* **34,** 49–55.

Park, R. G. 1988. *Geological structures and moving plates.* Blackie, Glasgow.

Pietilä, R. 1991. The application of drillhole magnetometry and mise-à-la-masse in the exploration for nickel sulphides, Finland – discovery of the Telkkälä orebody. *Expl. Geophys.,* **22,** 299–304.

Pilkington, M., and Hildebrand, A. R. 1994. Gravity and magnetic field modelling and structure of the Chicxulub Crater, Mexico. *J. Geophys. Res.,* **99,** 13147–62.

Piper, J. D. A. 1987. *Palaeomagnetism and the continental crust.* Open Univ. Press, Milton Keynes, UK.

Pollack, H. N., and Chapman, D. S. 1993. Underground records of changing climate. *Sci. Am.,* **268,** no. 6, 16–22.

Powell, J. L. 1998. *Night comes to the Cretaceous.* Freeman, New York.

Power, J. A., Lahr, J. C., Page, R. A., Chouet, B. A., Stephens, D. H., Murray, T. L., and Davies, J. N. 1994. Seismic evolution of the 1989–1990 eruption sequence of Redoubt Volcano, Alaska. *J. Volcan. Geotherm. Res.,* **62,** 69–94.

Pozzi, J-P., and Feinberg, H. 1991. Paleomagnetism in the Tajikistan: Continental shortening of European margin in the Pamirs during Indian Eurasian collision. *Earth Planet. Sci. Lett.,* **103,** 365–78.

Prévot, M., and McWilliams, M. 1989, Paleomagnetic correlation of Newark Supergroup volcanics. *Geology,* **17,** 1007–10.

Prodehl, C., Keller, G. R., and Khan, M. A., eds. 1994. Crustal and upper mantle structure of the Kenya rift. *Tectonophysics,* **236.**

Prodehl, C., Mechie, J., Achauer, U., Keller, G. R., Khan, M. A., Mooney, W. D., Gaciri, S. J., and Obel, J. D. 1994. Crustal structure on the northeastern flank of the Kenya rift. *In:* C. Prodehl, G. R. Keller, and M. A. Khan, eds., Crustal and upper mantle structure of the Kenya rift. *Tectonophysics,* **236,** 33–60.

Prodehl, C., Ritter, J. R. R., Mechie, J., Keller, G. R., Khan, M. A., Fuchs, K., Nyambok, I. O., Obel, J. D., and Riaroh, D. 1997. The KRISP 94 lithospheric investigations in Southern Kenya – The experiments and their main results. *In:* K. Fuchs, R. Althen, B. Muller and C. Prodehl, eds., Stress and stress release in the lithosphere. *Tectonophysics,* **278,** 121–47.

Reed, L. E. 1981. The airborne electromagnetic discovery of the Detour zinc–copper–silver deposit, northwestern Québec. *Geophysics,* **46,** 1278–90.

Renkin, M. L., and Sclater, J. G. 1988. Depth and age in the North Pacific. *J. Geophys. Res.,* **93,** 2919–35.

Reynolds, J. M. 1997. *An introduction to applied and environmental geophysics.* Wiley, New York.

Riddihough, R. 1984. Recent movements of the Juan de Fuca plate system. *J. Geophys. Res.,* **89,** 6980–94.

Rider, M. H. 1986. *The geological interpretation of well-logs.* Wiley, Blackie, Glasgow.

Rider, M. H. 1996. *The geological interpretation of well-logs,* 2nd ed. Whittles Publishing, Caithness, Scotland.

Robinson, E. S., and Coruh, C. 1988. Basic exploration geophysics. Wiley, New York.

Roy, R. F., Blackwell, D. D., and Birch, F. 1968. Heat generation of plutonic rocks and continental heat flow provinces. *Earth Planet. Sci. Lett.,* **5,** 1–12.

Ryan, M. P., Koyanagi, R. Y., and Fiske, R. S. 1981. Modeling the three-dimensional structure of macroscopic magma transport systems: Application to Kilauea volcano, Hawaii. *J. Geophys. Res.,* **86,** 7111–29.

Rymer, H., Cassidy, J., Locke, C. A., and Murray, J. B. 1995. Magma movements in Etna volcano associated with the major 1991–1993 lava eruption: Evidence from gravity and deformation. *Bull. Volcan.,* **57,** 451–61.

Sandwell, D. T., and Smith, W. H. F. 1997. Marine gravity anomaly from Geosat and ERS – satellite altimetry. *J. Geophys. Res.,* **102,** 10039–45.

Sayre, W. O., and Hailwood, E. A. 1985. The magnetic fabric of early Tertiary sediments from the Rockall Plateau, northeast Atlantic Ocean. *Earth Planet. Sci. Lett.,* **75,** 289–96.

Scarth, A. 1994. *Volcanoes.* UCL Press, London.

Schlumberger 1958. *Well log handbook.* Schlumberger Limited, 277 Park Ave., New York, NY, 10017.

Schlumberger Limited. 1972. *Log Interpretation, Volume 1: Principles.* New York.

Scholz, C. H. 1990. *The mechanics of earthquakes and faulting.* Cambridge Univ. Press, Cambridge, UK.

Scholz, C. H., Aviles, C. A., and Wesnousky, S. G. 1986. Scaling differences between large interplate and intraplate earthquakes. *Bull. Seis. Soc. Amer.,* **76,** 65–70.

Sclater, J. G., Parsons, B., and Jaupart, C. 1981. Oceans and continents: similarities and differences in the mechanisms of heat loss. *J. Geophys. Res.,* **86,** 11535–52.

Scollar, I., Tabbagh, A., Hesse, A., and Herzog, I. 1990. *Archaeological prospecting and remote sensing.* Cambridge Univ. Press, Cambridge, UK.

Searle, R. C. 1970. Evidence from gravity anomalies for thinning of the lithosphere beneath the rift valley in Kenya. *Geophys. J. R. Astr. Soc.,* **21,** 13–31.

Sharma, P. V. 1984. The Fennoscandian uplift and glacial isostasy. *Tectonophysics,* **105,** 249–62.

Sharpton, V. L., and Ward, P. D. 1990. Global catastrophes in Earth history: An interdisciplinary conference on impacts, volcanism and mass mortality. Geological Society of America Special Paper 247.

Sharpton, V. L., Burke, K., Camargo-Zanoguera, A., Hall, S. A., Lee, D. S., Marin, L. E., Súarez-Reynoso, G., Quezada-Muñeton, J. M., Spudis, P. D., and Urrotin-Fucugauchi, J. 1993. Chicxulub multiring impact basin: Size and other characteristics derived from gravity analysis. *Science,* **261,** 1564–7.

Sheriff, R. E., and Geldart, L. P. 1995. *Exploration seismology,* 2nd ed. Cambridge Univ. Press, Cambridge, UK.

Shimazaki, K. 1986. Small and large earthquakes: The effects of the thickness of the seismogenic layer and the free surface. *In:* S. Das, J. Boatwright, and C. H.

Scholz, eds., *Earthquake sources and mechanics.* American Geophysical Union, Washington, DC.

Shive, P. N., Steiner, M. B., and Huycke, D. T. 1984. Magnetostratigraphy, paleomagnetism, and remanence acquisition in the Triassic Chugwater Formation of Wyoming. *J. Geophys. Res.,* **89,** 1801–15.

Sinha, A. K., and Hayle, J. G. 1988. Experiences with a vertical loop VLF transmitter for geological studies in the Canadian Nuclear Fuel and Waste Management Programme. *Geoexploration,* **25,** 37–60.

Sinton, J. M., and Detrick, R. S. 1992. Mid-ocean ridge magma chambers. *J. Geophys. Res.,* **97,** 197–216.

Slack, P. D., and Davis, P. M. 1994. Attenuation and velocity of P-waves in the mantle beneath the East African Rift, Kenya. *In:* C. Prodehl, G. R. Keller, and M. A. Khan, eds., Crustal and upper mantle structure of the Kenya Rift. *Tectonophysics,* **236,** 331–58.

Smith, D. K., and Cann, J. R. 1993. Building the crust at the Mid-Atlantic Ridge. *Nature,* **365,** 707–15.

Smith, R. J. 1985. Geophysics in Australian mineral exploration. *Geophysics,* **50,** 2637–65.

Sowerbutts, W. T. C. 1987. Magnetic mapping of the Butterton Dyke: An example of detailed geophysical surveying. *J. Geol. Soc. Lond.,* **144,** 29–33.

Stauder, W. 1968. Mechanism of the Rat Island earthquake sequence of February 4, 1965, with relation to island arcs and sea-floor spreading. *J. Geophys. Res.,* **73,** 3847–58.

Stein, C. A., and Stein, S. 1994. Constraints on hydrothermal heat flux through the oceanic lithosphere from global heat flow. *J. Geophys. Res.,* **99,** 3081–95.

Stoneley, R. 1995. *An introduction to petroleum exploration for non-geologists.* Oxford Univ. Press, Oxford, UK.

Stuiver, M., and Pearson, G. W. 1993. High-precision bidecadal calibration of the radiocarbon time scale, A.D. 1950–500 B.C. and 2500–6000 B.C. *Radiocarbon,* **35,** 1–23.

Sullivan, W. 1991. *Continents in motion,* 2nd ed. American Institute of Physics, Woodbury, NY.

Swisher, C. C. III, Grajales-Nishimura, J. M., Montanari, A., Margolis, S. V., Claeys, P., Alvarez, W., Renne, P., Cedillo-Pardo, E., Maurrasse, F. J-M. R., Curtiss, G. H., Smit, J., and McWilliams, M. O. 1992. Coeval $^{40}Ar/^{39}Ar$ ages of 65.0 million years ago from Chicxulub Crater melt rocks and Cretaceous–Tertiary boundary tektites. *Science,* **257,** 954–8.

Sykes, L. R. 1967. Mechanism of earthquakes and nature of faulting on the mid-ocean ridges. *J. Geophys. Res.,* **72,** 2131–53.

Talwani, M., Sutton, G. H., and Worzel, J. L. 1959. A crustal section across the Puerto Rico Trench. *J. Geophys. Res.,* **64,** 1545–55.

Talwani, M., Le Pichon, X., and Ewing, M. 1965. Crustal

structure of mid-coean ridges. 2: Computed model from gravity and seismic refraction data. *J. Geophys. Res.,* **70,** 341–52.

Tarling, D. H. 1983. *Palaeomagnetism.* Chapman and Hall, London.

Tarponnier, P., Peltzer, G., Le Dain, A. Y., Armijo, R., and Cobbold, P. 1982. Propagating extrusion tectonics in Asia: New insights from simple experiments with plasticine. *Geology,* **10,** 611–6.

Taylor, B., Goodliffe, A., Martinez, F., and Hey, R. 1995. Continental rifting and initial sea-floor spreading in the Woodlark basin. *Nature,* **374,** 534–7.

Telford, W. M., Geldart, L. P., and Sheriff, R. E. 1990. *Applied geophysics,* 2nd ed. Cambridge Univ. Press, Cambridge, UK.

Thompson, R., and Oldfield, F. 1986. *Environmental magnetism.* Allen and Unwin, London.

Thordarson, Th., and Self, S. 1993. The Laki (Skaftá Fires) and Grímsvötn eruptions in 1783–1785. *Bull. Volcan.,* **55,** 233–63.

Tilton, G. R. 1988. Age of the solar system. *In:* J. R. Kerridge and M. S. Matthews, eds., *Meteorites and the early solar system.* Univ. of Arizona Press, Tucson.

Tiratsoo, E. N. 1973. *Oilfields of the world.* Scientific Press, Beaconsfield, UK.

Tongue, J. A., Maguire, P. K. H., and Young, P. A. V. 1992. Seismicity distribution from temporary earthquake recording in Kenya. *Tectonophysics,* **236,** 151–164.

Tsuboi, C. 1983. *Gravity.* Allen and Unwin, London.

Turcotte, D. L., and Schubert, G. 1982. *Geodynamics.* Wiley, New York.

Tyne, E. D. 1980. A review of mise-à-la-masse surveys at Elura. *Bull. Aust. Soc. Explor. Geophys.,* **11,** 186–7.

Van der Hilst, R. 1995. Complex morphology of subducted lithosphere in the mantle beneath the Tonga trench. *Nature,* **374,** 154–7.

Van der Hilst, R. D., Widiyantoro, S., and Engdahl, E. R. 1997. Evidence for deep mantle circulation from global tomography. *Nature,* **386,** 578–84.

Van Overmeeren, R. A. 1994. Georadar for hydrogeology. *First Break,* **12,** 401–8.

Verhoogen, J. (1980). *Energetics of the Earth.* National Academy of Science, Washington, DC.

Vine, F. J. 1966. Spreading of the ocean floor: New evidence. *Science,* **154,** 1405–15.

Walmsley P. J. 1975. The Forties Field. *In* Petroleum and the continental shelf of NW Europe, vol. 1. A. W. Woodland, ed., Applied Science Publishers, London, 477–86.

Wang, K., Lewis, T. J., and Jessop, A. M. 1992. Climatic changes in central and eastern Canada inferred from deep borehole temperature data. *Global Planet. Change,* **6,** 129–41.

Watts, A. B., and Daly, S. F. 1981. Long wavelength gravity and topography anomalies. *Annu. Rev. Earth Planet. Sci.,* **9,** 415–48.

Webster, S. S. 1980. The implications of a spectral IP survey at Elura. *Bull. Aust. Soc. Explor. Geophys.,* **11,** 201–7.

Wheildon, J., Morgan, P., Williamson, K. H., Evans, T. R., and Swanberg, C. A. 1994. Heat flow in the Kenya Rift Zone. *In:* C. Prodehl, G. R. Keller, and M. A. Khan, eds., Crustal and upper mantle structure of the Kenya Rift. *Tectonophysics,* **236,** 131–49.

White, C. C., Barker, R. D., and Taylor, S. 1997. Electrical leak detection systems for landfill liners: A case history. *Ground Water Monit. Remediation,* **17,** 153–9.

White, R. S., McKenzie, D., and O'Nions, R. K. 1992. Oceanic crustal thickness from seismic measurements and rare earth element inversions. *J. Geophys. Res.,* **97,** 19683–715.

Wilson, C. 1992. Sequence stratigraphy: an introduction. *In:* G. Brown, C. Hawksworth, and C. Wilson, eds., *Understanding the Earth,* 2nd ed. Cambridge Univ. Press, Cambridge, UK.

Wilson, R. L. 1970. Palaeomagnetic stratigraphy of Tertiary lavas from Northern Ireland. *Geophys. J. R. Astr. Soc.,* **20,** 1–9.

Woods, A. W., and Kienle, J. 1994. The dynamics and thermodynamics of volcanic clouds: Theory and observations from the April 15 and April 21, 1990 eruptions of Redoubt Volcano, Alaska. *J. Volcan. Geotherm. Res.,* **62,** 273–300.

Zalasiewicz, J. A., Mathers, S. J., and Cornwell, J. D. 1985, The application of ground conductivity measurements to geological mapping. *Q. J. Engl. Geol. Lond.,* **18,** 139–48.

Figure Sources

Plates

Plate 1: Modified after van der Hilst (1995), Fig. 3.

Plate 2: By permission of National Geophysical Data Centre, U.S. Dept of Commerce.

Plate 3a: By permission of Sandwell and Smith (1997).

Plate 3b: Modified after Müller et al. (1997), Plate 1(a).

Plate 4: From British Geological Survey (1997). Colour Shaded Relief Gravity Anomaly Map of Britain, Ireland, and adjacent areas. Smith, I. F., and Edwards, J. W. F. (compilers) 1:1,500,000 scale. British Geological Survey, Keyworth, Nottingham, United Kingdom. Lee et al. (1995).

Plate 5: From British Geological Survey (1998). Colour Shaded Relief Magnetic Anomaly Map of Britain, Ireland, and adjacent areas. Royles, C. P. and Smith, I. F. (compilers), 1:1,500,000 scale. British Geological Survey, Keyworth, Nottingham, United Kingdom. Lee et al. (1995).

Plates 6, 7, and 8 are published with the permission of BP Amoco Exploration, operator of the Forties Field, and their partners, Shell Exploration and Production and Esso Exploration and Production UK Limited.

Figures

Figure 2.5: Modified after Arab (1972), Fig. B.20.4.

Figure 2.7: Modified after unpublished report by Styles, Figs. 5 and 6.

Figure 3.13: Modified after Grauch (1987), Fig. 6 (part).

Figure 4.7: Modified after Robinson and Coruh (1988), Fig. 2–24.

Figure 4.14: Modified after Brown and Mussett (1993), Fig. 2.5.

Figure 4.15a: Modified after Kennett and Engdahl (1991), Fig. 1.

Figure 4.16: Modified after Bolt (1982), Fig. 2.1.

Figure 4.17: Modified after Kulhánek (1990), Plate 1.

Figure 4.20c: Modified after Gubbins (1990), Fig. 1.3.

Figure 5.4 Modified after an unknown source on the Web.

Figure 5.15: Modified after Eaton et al. (1970), Figs. 8 and 9.

Figure 5.16: Modified after Fitch & Scholz (1971), Fig. 10.

Figure 5.17: Modified after McEvilly (1967), Fig. 1.

Figure 5.19b: Modified after Shimazaki (1986), Fig. 1.

Figure 5.19c: Modified after Scholz et al. (1986), Fig. 4.13.

Figure 5.20: Modified after Hill et al. (1990), Figs. 5.4 and 5.7.

Figure 5.21: Modified after Bolt (1982), Fig. 2.4.

Figure 5.22: Modified after A. Nyblade, personal communication.

Figure 5.23: Modified after Bolt (1993b), Box 7.1.

Figure 5.24: Drawn from data in Kasahara (1981).

Figure 5.25: Partly drawn from data in Bolt (1993b), App. A, and partly from McEvilly et al. (1967), Fig. 7, modified.

Figure 7.2: Gardline Surveys.

Figure 7.9: Modified after McQuillin et al. (1984), Fig. 4/30.

Figure 7.10: Myanma Oil & Gas Enterprise and Atlantic Richfield Corporation.

Figure 7.12 : Modified after McQuillin et al. (1984), Fig. 3/1.

Figure 7.18: Courtesy Schlumberger Geo-Prakla.

Figure 7.22: Modified after McQuillin et al. (1984), Fig. 9/1.

Figure 7.25 Modified after McQuillin et al. (1984), Fig. 9/3.

Figure 7.26: Modified after McQuillin et al. (1984), Fig. 9/4.

Figure 7.27: Modified after McQuillin et al. (1984), Fig. 9/5.

Figure 7.28c, d: Modified after Wilson (1992), Fig. 20.7.

Figure 7.29: Modified after Wilson (1992), Fig. 20.5.

Figure 8.17a: Modified after Griffiths and King (1981), Fig. 7.4.

Figure 8.17b: Modified after Arab (1972), Fig. B.20.3.

Box 8.1, Figure 1a: Modified after Kearey and Brooks (1991), Fig. 6.2.

Figure 9.12: Modified after Watts and Daly (1981), Fig. 4.

Figure 9.14: Modified after Sharma (1984), Figs. 1 and 3.

Figure 10.5: Modified after McElhinny (1973), Fig. 4.

Figure 10.9: Modified after Butler (1992), Fig. 10.9.

Figure 10.10: Modified after McElhinny (1973), Fig. 130.

Figure 10.18: Modified after McClelland Brown (1981), Fig. 1.

Figure 10.19b: Modified after Pozzi & Feinberg (1991), Fig. 11.

Figure 10.21: Drawn from data in Harland et al. (1990).

Figure 10.22: Modified after Ernst and Barager (1992), Fig. 1.

Figure 11.5: Modified after Floyd & Kimbell (1995), Figs. 1 and 2.

Figure 11.10: Modified after Jenson (1961), p. 162.

Figure 11.18: Modified after Sowerbutts (1987), Figs. 2(b) and 3.

Figure 11.19: Modified after Farquharson and Thompson (1992), Figs. 1 and 2.

Figure 11.20: Modified after Farquharson and Thompson (1992), Figs. 3 and 6.

Figure 11.21: Modified after Farquharson and Thompson (1992), Fig. 7.

Box 11.1: Figure 1a: Modified after Robinson and Coruh (1988), Fig. 10–13.

Figure 12.13: Modified after Barker (1995), Appendix 1.

Figure 12.22: Modified after Telford et al. (1990), Figs. 8.34 and 8.36.

Figure 12.23: Modified after Pietilä (1991), Fig. 5.

Figure 12.24b, c: Bertin and Loeb (1976), Vol. 1, Fig. 103.

Figure 12.25: Modified after Loke and Barker (1996), Fig. 4.

Figure 13.3: Modified after Telford et al. (1980), Fig. 9.4.

Figure 13.4: Modified after Webster (1980), Fig. 3.

Figure 13.8: Modified after Beer and Fenning (1976), Fig. 4.

Figure 14.5a, b, c, d: Modified after Milsom (1989), Fig. 8.5.

Figure 14.5e: Modified after Parasnis (1986), Fig. 6.21.

Figure 14.6: Modified after Zalasiewicz et al. (1985), Figs. 5 and 6.

Figure 14.8: Modified after Haren (1981), Fig. 1.

Figure 14.9: Modified after Telford et al. (1990), Fig. 7.28.

Figure 14.10: Modified after Reed (1981), Fig. 5.

Figure 14.16c: Modified after Milsom (1989), Fig. 9.4.

Figure 14.17: Modified after Sinha & Hayles (1988), Fig. 7.

Figure 14.23: Modified after Gough & Majorowicz (1992), Figs. 2 and 8.

Figure 14.25: Modified after van Overmeeren (1994), Fig. 1.

Figure 15.9: Drawn from data in Long (1964).

Figure 15.11a: Modified after Tilton (1988), Fig. 5.2.2.

Figure 15.11b: Modified after Oversby (1971) and other sources.

Figure 15.12b: J. A. Miller, personal communication.

Figure 15.16a: Modified after Cliff (1985), Fig. 5.

Figure 15.16b: Modified after Harrison and McDougall (1985), Fig. 3.

Figure 15.19: Modified after Gradstein and Ogg (1996).

Figure 15.22: Drawn from data in Stuiver and Pearson (1993).

Figure 16.5: Modified after Smith (1985), Fig. 15.

Figure 16.6: Modified after Cassidy (1979), Figs. 5.1, 5.5, 5.6, 5.7, 5.8.

Figure 16.7: Modified after Cassidy (1979), Figs. 6.1, 6.4, 6.5, 6.6, 6.7.

Figure 17.5: Modified after Stein and Stein (1994), Fig. 6.

Figure 17.6a: Modified after Roy et al. (1968), Fig. 2.

Figure 17.6b: Modified after Balling (1995), Fig. 8.

Figure 17.10: Modified after Glazner & Bartley (1985), Fig. 2.

Figure 17.11: Modified after Sclater et al. (1981), Fig. 7.

Figure 17.12: Modified after Batchelor, A. S. (1987), Fig. 1.

Figure 17.14a, b: Modified after Lewis & Wang (1992), Fig. 11.

Figure 17.14c: Modified after Wang et al. (1992), Fig. 2(b).

Box 17.1: Figure 1: Modified after Goguel (1976), Fig. 1–1.

Box 17.1: Figure 2: Modified after Lister et al. (1990), Fig. 2.

Figure 18.3: Modified after Schlumberger (1972).

Figure 18.5: Modified after Schlumberger (1972), Fig. 2.1.

Figure 18.6: Modified after Schlumberger (1972), Fig. 2.3.

Figure 18.7: Modified after Schlumberger (1972), Fig. 3.3.

Figure 18.8: Modified after Schlumberger (1972), Fig. 4.2.

Figure 18.9: Modified after Schlumberger (1972), Fig. 5.1.

Figure 18.10: Modified after Schlumberger (1958).

Figure 18.11: Modified after Schlumberger (1972), Fig. 8.1.

Figure 18.12: Modified after Rider (1986), Fig. 9.2.

Figure 18.13: Modified after Schlumberger (1972), Fig. 7.1.

Figure 18.14: Modified after Schlumberger (1972), Fig. 12.1.

Figure 18.15: Modified after Killeen et al. (1993), Fig. 12.

Figure 18.16: Modified after Killeen et al. (1993), Fig. 11.

Figure 18.17: Modified after Killeen et al. (1993), Fig. 1.

Figure 18.18: Modified after Parasnis (1973), Fig. 23A.

Figure 18.19: Modified after Mwenifumbo et al. (1993), Fig. 14.

Figure 18.20: Modified after Baltosser (1971), Fig. 2.

Figure 18.21: Modified after Mwenifumbo (1989), Fig. 13.

Figure 20.1: Modified after Heezen and Tharp (1968).
Figure 20.2a, b: Modified after De Mets et al. (1990), Fig. 38.
Figure 20.2c: Modified after Heirtzler et al. (1966).
Figure 20.3: Modified after Talwani et al. (1965), Fig. 2.
Figure 20.5: Modified after Renkin and Sclater (1988), Fig. 18.
Figure 20.7: Modified after Sykes (1967), Figs. 4, 8, 9, and 10.
Figure 20.9: Modified after Giardini and Woodhouse (1984), Fig. 2.
Figure 20.10: Modified after Chiu et al. (1991), Fig. 4D.
Figure 20.11: Modified after Dickinson (1977).
Figure 20.12a: Modified after Stauder (1968), Fig. 4a.
Figure 20.12b: Modified after Isacks et al. (1969), Fig. 12.
Figure 20.12 inset: Modified after Heirtzler et al. (1968), Fig. 11.
Figure 20.13: Modified after Talwani et al. (1959), Fig. 3.
Figure 20.14: M. Atherton, personal communication.
Figure 20.15: Modified after Tarponnier et al. (1980), Fig. 1.
Figure 20.20: Modified after McKenzie (1969), Fig. 4.
Figure 20.23: Modified after Riddihough (1984), Fig. 1.
Figure 20.25: Modified after Morgan (1972), Fig. 3.
Figure 20.26: A. G. Smith, personal communication.
Figure 20.27: Modified after White et al. (1992), Fig. 1.
Figure 20.29a: Modified after Collier and Sinha (1990), Fig. 2.
Figure 20.29b: Modified after Sinton & Detrick (1992), Fig. 2.
Figure 20.30: Modified after Sinton & Detrick (1992), Figs. 9 and 10.
Figure 20.32: Modified after Forsyth and Uyeda (1975), Fig. 1.
Figure 20.33: Modified after Park (1988), Fig. 3.10.
Figure 20.34: Modified after Gripp and Gordon (1990), Fig. 1.
Figure 20.35: Modified after Forsyth & Uyeda (1975), Figs. 5, 7, and 8.
Figure 20.36: Modified after Lay (1995).
Figure 21.1: Modified after Laughton et al. (1970), Fig. 24.
Figure 21.2: Modified after Irving and Tarling (1961), Fig. 6.
Figure 21.3: Modified after Birt (1996), Fig. 1.4.
Figure 21.4: Modified after Birt (1996), Fig. 1.6.
Figure 21.5a: Modified after Searle (1970), Fig. 9.
Figure 21.5b: Modified after Baker and Wohlenberg (1971), Fig. 7.
Figure 21.6: Modified after Slack et al. (1994), Figs. 1, 4, and 8.
Figure 21.7: Modified after Prodehl et al. (1997), Fig. 1.
Figure 21.9: Modified after Birt et al. (1997), Fig. 3.

Figure 21.10: Modified after Birt (1996), Figs. 3a and 3.14.
Figure 21.11: Modified after Birt (1996), Fig. 7.1.
Figure 21.12: Modified after Birt (1996), Fig. 7.5.
Figure 21.13: Modified after Birt (1996), Fig. 5.1.
Figure 21.14: Modified after Birt (1996), Fig. 5.3.
Figure 21.15: Modified after Wheildon et al. (1994), Fig. 1.
Figure 21.16: Modified after V. Sakkas, personal communication.
Figure 22.1: Modified after BP (1997), Fig.1.
Figure 22.3: Modified after Nettleton (1976), Figs. 8.19 and 8.20.
Figure 22.4: Modified after Stoneley (1995), Fig. 5.2.
Figure 22.5: Modified after Kent and Walmsley (1970), Fig. 8.
Figure 22.6a: Modified after Hornabrook (1975), Plate 3.
Figure 22.6b: Modified after Kent and Walmsley (1970), Fig. 7.
Figure 22.7: Modified after Hornabrook (1975), Plate 2.
Figure 22.8: Modified after Walmsley (1975), Fig. 5.
Figure 22.9: Modified after Walmsley (1975), Fig. 4.
Figure 22.10: Modified after Walmsley (1975), Figs. 6 and 7.
Figures 22.11–14: Published with the permission of BP Amoco Exploration, operator of the Forties Field, and their partners Shell Exploration and Production and Esso Exploration and Production UK Limited.
Figure 22.15: Modified after Daly et al. (1996), Fig. 11.
Figure 23.1: Modified after Adams and Schmidt (1980), Fig. 2.
Figure 23.2: Modified after Gidley and Stuart (1980), Fig. 2.
Figure 23.3: Modified after Blackburn (1980), Fig. 1.
Figure 23.4: Modified after Blackburn (1980), Fig. 2.
Figure 23.5: Modified after Blackburn (1980), Fig. 7.
Figure 23.6: Modified after Blackburn (1980), Fig. 8.
Figure 23.7: Modified after Blackburn (1980), Fig. 9.
Figure 23.8: Modified after Davis (1980), Fig. 3.
Figure 23.9: Modified after Tyne (1980), Fig. 1.
Figure 23.10: Modified after Emerson (1980).
Figure 24.1a: Modified after Woods and Kienle (1994), Fig. 1.
Figure 24.1b: Modified after Begét and Nye (1994), Fig. 1.
Figure 24.1c: Modified after Woods and Kienle (1994), Fig. 1.
Figure 24.1 inset: Modified after Woods and Kienle (1994), Fig. 1.
Figure 24.2: Modified after Power et al. (1994), Fig. 5.
Figure 24.3: Modified after Chouet et al. (1994), Fig. 10.
Figure 24.4: Modified after Rymer et al. (1995), Fig. 1, and Abrams et al. (1994), Fig. 2.
Figure 24.6: Modified after Rymer et al. (1995), Fig. 2.

Figure 24.7: Modified after Rymer et al. (1995), Fig. 4.

Figure 25.2a, b: Modified after Pilkington et al. (1994), Figs. 2 and 5.

Figure 25.2 inset: Modified after Kamo & Krogh (1995), Fig. 1.

Figure 25.3: Modified after Grieve (1990), p. 48.

Figure 25.4a, b: Modified after Pilkington et al. (1994), Fig. 7.

Figure 25.4c: Modified after Hildebrand, personal communication (with permission of the Geological Survey of Canada).

Figure 25.5: Modified after Hildebrand et al. (1995), Figs. 1 and 2.

Figure 25.6a: Modified after: Swisher (1992), Fig. 2.

Figure 25.6c: Modified after Kamo & Krogh (1995), Fig. 2.

Figure 26.1: Modified after Van Overmeeren (1994), Fig. 5.

Figure 26.2: Modified after Cornwell et al. (1996), Figs. 1 and 2.

Figure 26.3: Modified after Cornwell et al. (1996), Fig. 4.

Figure 26.4: Modified after Cornwell et al. (1996), Fig. 5.

Figure 26.5: Modified after Holman & Hiscock (1998), Fig. 1.

Figure 26.6: Modified after Holman et al. (1999), Fig. 4.

Figure 26.7: Modified after Holman et al. (1999), Fig. 7.

Figure 26.8: Modified after Holman and Hiscock (1998), Fig. 5.

Figure 26.9: Modified after Holman and Hiscock (1998), Fig. 7.

Figure 26.9: Modified after Holman (1994), Fig. 4.25.

Figure 26.10a: Modified after Holman (1994), Fig. 5.25.

Figure 26.10b: Modified after Holman & Hiscock (1998), Fig. 10.

Figure 26.11: Modified after Barker (1990), Figs. 3 and 5.

Figure 26.12: Modified after Barker (1990), Fig. 4.

Figure 26.13: Modified after White and Barker (1997), Fig. 1.

Figure 26.14: Modified after White and Barker (1997), Fig. 7.

Figure 27.1: Modified after Styles, personal communication.

Figure 27.2: Modified after Bishop et al. (1997), Fig. 3.

Figure 27.3: Modified after Al-Rifaiy (1990), Figs. 8 and 11.

Figure 27.4: Modified after Bishop et al. (1997), Fig. 4.

Figure 27.5: Modified after Ferguson (1992), Fig. 6.10.

Figure 27.6: Modified after Bishop et al. (1997), Fig. 10.

Figure 27.7: Modified after Bishop et al. (1997), Figs. 7 and 9.

Figure 28.2: Modified after Gaffney and Gater (1993), Fig. 18.5.

Figure 28.3: Modified after Corney et al. (1994), Figs. 3, 4, 5, and 6.

Figure 28.4: Modified after Bauman et al. (1994), Fig. 1.

Figure 28.5: Modified after Bauman et al. (1994), Fig. 3.

Figure 28.6: Modified after Bauman et al. (1994), Figs. 4 and 7.

Figure 28.7: Modified after Bauman et al. (1994), Fig. 14.

Figure 28.8: Modified after Bauman et al. (1994), Figs. 9 and 12.

Index

Page numbers in **bold** indicate where a term is defined or explained. Page numbers followed by f refer to a figure, those followed by a t to a table.

slow, **335**, 336f
stacked profiles, **11**, 175f, 176, 215, 385, 386f
stacking, **9**, 74, 87–8, 92–3, 310
static correction, **93**
static SP, SSP, **290**
station, 7, 10, 59, 118, 176, 310; *see also* base station
steady-state, thermal, **277**
step age, *see* argon-argon dating
stereoplot, **143**
streamer (seismic), **91**
streaming potential, **206**
subduction, **321**, 328, 331, 336
subduction zone, 135, 256, **321–5**, 328, 329–30, 331, 337, 340, 381
 Cascadia, 330
sulphide, 196, 197f, 202, 245, 301, 333, 379, 381
 disseminated, **202**, 203, 204
 massive, 197, **202**, 207, 208f, 215, 216f, 300, 379, 387, 388
 volcanogenic, 300
S-wave, *see* wave, transverse

target, **8**
tektite, **401**, 404, 405f, 406
teleseismic distance, **39**
teleseismic ray, **39**
teleseismic survey, 350, 351f
telluric current, 206, **225**, 226f, 388
temperature, **269**
 emplacement, 150–1
 reheating, 149
temperature gradient, **270**; *see also* temperature profile
 adiabatic, *see* adiabat
temperature profile, **271**; *see also* temperature gradient
 after overthrusting, 279
 after underthrusting, 279
 in the Earth, *see* geotherm
terrain correction, *see* gravity corrections
Tethys, 332
^{230}Th/^{232}Th dating, **254–5**
thermal capacity, **277**
 specific, **277**
thermal conductivity, K, **272**, 273, 274
thermal convection, **270**, 270–1, 279, 340
thermochronometry, **249–50**
thorium, chemistry, 254–5, 264
thoron, 266
time-distance (t-x) diagram, 67, 72, 351
time domain (in IP measurements), **203**
time-lapse modelling, **99–100**
tomography, electrical, 198, 303
tomography, seismic, 39, 81, 310, 331, 334, 335, 340, 341f, 350, 391, 397, 421, Plate 1
 cross-hole, **303**
Tonga-Kermadec subduction zone, 321, 322f, 328, 340, 341f, Plate 1
tool, **287**; *see also* sonde
toplap, 102

total field anomaly, *see* magnetic anomaly, total intensity
total intensity anomaly, *see* magnetic anomaly, total intensity
total internal reflection, **65**
transient e-m (TEM) system, **216**, 215–17, 411–12, 421
transient state, thermal, **277**
transition zone, in borehole, **286**
transition zone, in mantle, **37**
transit time, Δt, in sonic logging, **297**, 372
transmission coefficient, **95**
transmitter, e-m surveying, 210, 411
trap, *see* hydrocarbon trap
travel-time, **31**
 reduced, **174**
traverse (geophysical), 7, 176, 214, 227, 310, 369, 404, 409, 411, 422, 436
trench, oceanic, 135, **322**, 323f, 324, 340, 341f, Plates 2, 3a
triple junction, **328–31**, 332, 345, 346, 367
troilite, 245
Turam e-m system, **215**, 216f
twin array, *see* resistivity array, twin
two-way time, TWT, **84**, 227, 372

^{234}U/^{238}U dating, **254**, 255
uninvaded zone, **286**
U-Pb dating, *see* uranium-lead dating
upward continuation, **174–5**
uranium, chemistry, 254, 264, 266
uranium-lead dating, **235–6**, 256–7
 discordia, **250**, 405
uranium-series dating, **253–4**, 255, 336

variable area display (re. seismic record), **93**
vector component diagram, **150**, 177, 178f
velocity, *see* seismic velocity, radar velocity
velocity-depth structure in Earth, 31–3; *see also* Earth, structure
vertical electrical sounding, VES, **186–94**, 310, 387, 411, 414, 417–18
vertical field magnetic anomaly, *see* magnetic anomaly
Vibroseis system, **94–5**
Viking Graben, *see* North Sea
VLF, very low frequency (in e-m surveying), **219–21**, 223–4, 224–5
void, 396, 420–8
volcano, 238, 330, 338, 346f, 347, 348f, 390–8
volt, V, **182**

water saturation, S_w, 184, **288**, 294, 408
water table, 74, 199, 229, 230, 409–10, 416
wave conversion, **34**
wave, electromagnetic, 217–19
wave front, **25**
wavelength, λ, **13**
wave, radar, *see* ground-penetrating radar
wave, seismic, 24–25, 26–8
 body, 56
 conversion, 34